T0228985

MANAGEMENT AND CONTROL OF PRODUCTION AND LOGISTICS 2004
(MCPL 2004)

A Proceedings volume from the IFAC/IEEE/ACCA Conference,
Santiago, Chile, 3 – 5 November 2004

Edited by

GASTON LEFRANC
Pontificia Universidad Catolica de Valparaiso,
Valparaiso, Chile,

Published for the

INTERNATIONAL FEDERATION OF AUTOMATIC CONTROL

by

ELSEVIER LIMITED

ELSEVIER Ltd
The Boulevard, Langford Lane
Kidlington, Oxford OX5 1GB, UK

Elsevier Internet Homepage
http://www.elsevier.com

Consult the Elsevier Homepage for full catalogue information on all books, journals and electronic products and services.

IFAC Publications Internet Homepage
http://www.elsevier.com/locate/ifac

Consult the IFAC Publications Homepage for full details on the preparation of IFAC meeting papers, published/forthcoming IFAC books, and information about the IFAC Journals and affiliated journals.

First edition 2005

Library of Congress Cataloging in Publication Data

A catalogue record for this book is available from the Library of Congress

British Library Cataloguing in Publication Data

A catalogue record for this book is available from the British Library

The
publisher's
policy is to use
paper manufactured
from sustainable forests

ISBN 0-08-044484 9
ISSN 1474-6670

Transferred to digital print, 2007
Printed and bound by CPI Antony Rowe, Eastbourne

To Contact the Publisher

Elsevier welcomes enquiries concerning publishing proposals: books, journal special issues, conference proceedings, etc. All formats and media can be considered. Should you have a publishing proposal you wish to discuss, please contact, without obligation, the publisher responsible for Elsevier's industrial and control engineering publishing programme:

Christopher Greenwell
Senior Publishing Editor
Elsevier Ltd
The Boulevard, Langford Lane
Kidlington, Oxford
OX5 1GB, UK

Phone: +44 1865 843230
Fax: +44 1865 843920
E.mail: c.greenwell@elsevier.com

General enquiries, including placing orders, should be directed to Elsevier's Regional Sales Offices – please access the Elsevier homepage for full contact details (homepage details at the top of this page).

IFAC/IEEE/ACCA CONFERENCE ON MANAGEMENT AND CONTROL OF PRODUCTION AND LOGISTICS 2004

Sponsored by:
International Federation of Automatic Control (IFAC)
IFAC Technical Committee on Large Scale Complex Systems
and
Seccion Chile del IEEE
Instituto de Ingenieros de Chile
SOFOFA Chile
APEC Chile
Embajada Françia
Gobierno de Chile

National Sponsors:
Colegio de Ingenieros de Chile
Universidad de Las Américas

Co-sponsored by
IFAC Technical Committees on:
- Manufacturing Plant Control
- Manufacturing Model, Management and Control.
- Components and Instruments
- Robotics
- Cost Oriented Automation
and
ACCA: Asociación Chilena de Control Automático
IEEE Chile Section
IEEE Chilean Chapter on Control, Robotics and Cybernetics

Supported by:
Duran San Martin
Huawei
La Nación
Logistec
Mc Graw Hill
Pearson Education

Pepsi
Siemens
Sodexo
Yew
Yokogawa

Organized by
Universidad de Las Américas – Chile
Engineering Sciences Faculty

International Programme Committee (IPC)
Gastón Lefranc (Chile), Chairman
Zdenek Binder (France), Co-Chairman

Héctor Kaschel
José Ceroni

Bernard Decotes-Genon
Shimon Nof

PREFACE

It is an honour for me to introduce you the Proceedings of the Conference on Management and Control of Production and Logistics (MCPL). This Version was requested in Grenoble with my friend Dr. Nestor González, actual Academic Vice Rector of Universidad de Las Américas. I want to say thanks to IFAC (International Federation of Automatic Control), to the seven IFAC Technical Committees: Large Scale Complex Systems, Manufacturing Plant Control, Manufacturing Model, Management and Control, Components and Instruments, Robotics, Cost Oriented Automation, IEEE Chile and ACCA, for trusting us to organize MCPL2004.

This third version of MCPL is organized by ACCA, IEEE Chilean Chapter on Control, Robotics and Cybernetics, IFAC and its Technical Committees, and the Universidad de Las Américas. MCPL is held in parallel with XVI Congress of Chilean Association of Automatic Control ACCA (Asociación Chilena de Control Automático), in its 30 years of existence.

Since 1974, ACCA organize Congress (every 2 years), Tutorial Courses for continuing education, Seminars, Workshops, Exposition of Systems and Equipments and our Magazine Automatica and Innovation. All of these activities have results: important influence in all ambits, consolidating to ACCA as a point of meeting of people from industries, from private and public institutions, from the academics world, and from suppliers of systems and equipments. In 1976, ACCA had its first Program Committee to select papers, and its first Proceedings.

The 62 accepted papers selected by the International Program Committee, have been presented in Technical Sessions where new ideas, critical comments, and the beginning of cooperation leading to future projects will take place. The papers come from Argentina, Austria, Brazil, Belarus, Chile, Cameroon, England, France, Germany, Poland, Romania, Slovakia, and Spain. The papers have been organized in 16 Sessions called FMS & Plant Control, Application/Case Study, Supply-Chain Management, Diagnosis & Reliability ,Management techniques based on agents and evolution, SME Network & Information Support System, Factory Automation & Robotics, Modelling and Control Tools, Planning & Scheduling, Discrete event & Petri Nets. IPC proposed 26 papers to be published in some of the IFAC Magazines.

MCPL have Keynotes Speakers for Plenary: Dr. François Vernadat from France, Dr. Florin Filip from Romania, Dr. Shimon Nof from USA and Dr. Philippe Dupont, from France. Dr. Zdenek Binder organises a Round Table to discuss about productivity, interested in Chile and Latin American professionals. Thanks to all of them: Special thanks for the people who support the Conference and the IPC Committee.

Gastón Lefranc Hernández
Chairman of International Program Committee
IFAC ACCA IEEE MCPL Conference

CONTENTS

FACTORY AUTOMATION AND ROBOTICS

FMS AND PLANT CONTROL

INVITED PAPERS
MANAGEMENT TECHNIQUES BASED ON AGENTS AND EVOLUTION

MODELLING AND CONTROL TOOLS

PLANNING AND SCHEDULING

SME NETWORKD AND INFORMATION SUPPORT SYSTEM

SUPPLY-CHAIN MANAGEMENT

PUBLICATIONS
www.elsevier.com/locate/ifac

ADVANCED DECISION SUPPORT SYSTEMS FOR PLANNING AND CONTROL OF MANUFACTURING PROCESSES

F. G. Filip [1,2]

1.The National Institute for R&D in Informatics-ICI,
M. Averescu Ave. 8-10, 71316 Bucharest, Romania
2. The Romanian Academy, Calea Victoriei 125, Bucharest
URL:Http://www.ici.ro/ici/filipf/homepage.html

Abstract. A Decision Support System (DSS) is an anthropocentric, adaptive and evolving information system meant to implement several of the functions of a "Human support system " (or team) that would otherwise be needed to help the decision-maker to overcome his/her limits and constraints he/she may face when approaching decision problems that count in the organisation. These systems are very suitable information means one can utilize when approaching various planning and control problems that are complex and complicated. In manufacturing systems DSSs are frequently utilized to facilitate solving problems under time pressure when "crisis situations" show up. This paper aims at presenting several aspects concerning the technology of DSSs with particular emphasis on the mixed-knowledge solutions that combine numerical methods with AI-based tools.

Keywords. Artificial intelligence, decision, human factors, manufacturing, models, real-time.

1. INTRODUCTION

The role and place of the human operator in industrial automation systems started to be seriously considered by engineers and equally by psychologists towards the middle of the 7th decade. Since then, this aspect has been constantly and growingly taken into consideration in view of famous accidents of highly automated systems and of incomplete fulfilment of hopes put in CIM systems [Johannsen, 1994; Martenson, 1996; Orasanu, Fisher, Davison, 2002;].

The evaluation of the place of man in the system has known a realistic evolvement, triggered not only by practical engineer experience but also by the debates from academia circles. A long cherished dream of control engineers that of developing "completely automated systems where man would be only a consumer" (Bibby, Margulis, Rijndorp, 1976) or "unmanned factories", tends to fade away – not only due to *ethical* or *social motivations,* but more important because the technical realization of this dream proved to be *impossible.* Many years ago Bainbridge (1983) defined the *irony of automation* as "a combination of circumstances, the result of which is the direct opposite of what we expected". Starting from the engineers, desire to replace the human operator (who is unreliable and inefficient) by an automated device, Bainbridge identified two ironies of automation. The first one is the result of the fact that the designer of the device is also an imperfect person and, as a consequence, his/her design errors can become a source of new operating problems. The second irony consists in the fact that the designer (that aims at eliminating the human operator from the system) still leaves somebody to perform the tasks he/she is unable (or even does not think) to automate by utilizing traditional analytic methods or numerical models and corresponding existing solvers.

A possible modern solution seems to be the use of *artificial intelligence* (AI) methods (such as knowledge based systems) in the control of industrial systems, since these methods imitate the thinking process in the left hemisphere of the human brain. Artificial neural networks, functioning similar with the right hemisphere of the human brain, became since 1990 also increasingly attractive, especially for problems that cannot be efficiently formalized with present human knowledge. Even so, "on field", due to strange combinations of external influences and circumstances, rare or new situations may appear that were not taken into consideration at design time. Already in 1990 Martin and colleagues showed that "although AI and expert systems were successful in solving problems that resisted to classical numerical methods, their role remains confined to support functions, whereas the belief that evaluation by man of the computerized solutions may become superfluous is a very dangerous error". Based on this observation, Martin and colleagues (1991) recommend "appropriate automation", integrating technical, human, organizational, economical and cultural factors.

Decision Support Systems (DSS) represent a class of information systems that can be utilized in supervisory control and planning applications in the milieu of manufacturing systems. This paper aims at surveying, from an anthropocentric perspective, several DSS concepts and technologies with particular emphasis on the combination of traditional numerical methods and AI based tools.

In the sequel the paper is organised as it follows. First a review of the typical decision problems that can be met in manufacturing systems is done. The limits and constraints of the human-decision maker are then described. Several issues concerning the modern trends to build anthropocentric systems are reviewed next. Then the paper surveys several widely accepted concepts in the field of decision support systems. Several artificial intelligence techniques and their applicability to decision problems are reviewed next. The possible combination of artificial intelligence technologies with traditional numerical models within advanced decision support systems is eventually presented.

2. DECISIONS AND DECISION-MAKERS

One can speak about a *decision problems* when: a) a situation that requires action shows up, b) there are several action alternatives, and c) one or several persons are empowered to chose an alternative.

2.1. Typical Decision Problems in Manufacturing Systems

The decision-making problems in the manufacturing milieu are complex and complicated. Moreover, sometimes, one may face "crisis situations", when the problems must be solved under time pressure.

McMahon and Browne (1995) identify several classes of decision problems that are typical for manufacturing systems.

At the *strategic level*, business planning decision problems concern manufacturing strategy. This class includes problems such as a) assigning products, clients or markets to production facilities, b) "making or buying", c) capacity evolution (when/where/how to add a new production capacity or give up an existing one), d) defining processes (specifying the level of product customisation, volume of production and setting the requirements for the qualification of the personnel), e) up-grading the infrastructure, f) building up human resources.

At the *tactical level*, the main decisions to be made lead to Master Production Schedule (MPS) and Material/Manufacturing Requirements Planning (MRP).

At the *Shop floor Control level*, the typical decision problems that can be identified are in connection with *Factory Coordination* (production environment design, factory scheduling, dispatching and monitoring) and *Production Activity Control* (operative scheduling, dispatching, material movement coordination, process control and supervision).

2.2. Decision Making Models

Decision- making (DM) process is a specific form of information processing that aims at setting- up an action plan under specific circumstances. Several models of a DM are reviewed in the sequel.

Nobel Prize winner H. Simon (1960) identifies three steps of the DM process, namely: a) "intelligence", consisting of activities such as setting the objectives, data collection and analysis in order to recognize a decision problem, problem statement ;b) "design", including activities such as identification (or production) and evaluation of various potential solutions to the problem, model building ; c) "choice", or selection of a feasible alternative to be released for implementation. Subsequently, Simon introduced a fourth step, implementation of the solution chosen and review of the results.

If a decision problem cannot be entirely clarified and all possible decision alternatives cannot be fully explored and evaluated before a choice is made, then the problem is said to be "unstructured " or "semi-structured". If the problem were completely structured, an automatic device could have solved the problem without any human intervention. On the other hand, if the problem has no structure at all, nothing but hazard can help. If the problem is semi-structured a computer-aided decision can be envisaged.

The *"econological" model* of the DM assumes that the decision-maker is fully informed and aims at extremizing one or several performance indicators in a rational manner. In this case the DM process consists in a series of steps such as: problem statement, definition of the criterion (criteria) for the evaluation of decision alternatives, listing and evaluation of all available alternatives, selection of the "best" alternative and its execution.

It is likely that other DM models are also applicable such as: a) the *"bounded rationality"* model, that assumes that decision-making considers more alternatives in a sequential rather than in a synoptic way, use heuristic rules to identify promising alternatives and make then a choice based on a "satisfycing" criterion instead of an optimisation one; b) the *"implicit favourite"* model, that assumes that the decision-maker chooses an action plan by using in his/her judgment and expects the system to confirm his choice.

While the DSSs based on the "econological" model are strongly normative, those systems that consider the other two models are said to be "passive".

In many problems, decisions are made by a group of persons instead of an individual. Because the *group decision* is either a combination of individual decisions or a result of the selection of one individual decision, this may not be "rational" in H. Simon's acceptance. The group decision is not necessarily the best choice or combination of individual decisions, even though those might be optimal, because various individuals involved might have various perspectives, goals, information bases and criteria of choice. Therefore, group decisions show a high "social" nature including possible conflicts of interest, different visions, influences and relations (De Michelis, 1996). Consequently, a group DSS needs an important communication facility.

2.3. Decision-Makers

The term "decision-maker" may denote one individual or a "decision unit" composed of several participants (Holsapple, Whinston, 1996; Marakas, 2003; Filip 2004). Several classes of decision-makers are presented in Table 1.

Each person that *takes* decisions can be supported in performing his/her activities by a team of assistants or external consultants that know the problem domain and or master decision methods and associated IT tools. They together form a *hierarchical team* or a *Human Support System (HSS)*. The typical functions of a HSS are (Holsapple, Whinston, 1996):

. *Receiving and accepting decision maker's requests* for information (such as: problem data, results of an analysis, clarification of a previous response, helps to formulate a question...) or commands to acquire new information from various sources.

. *Issuing outputs*, that can represent feedbacks to decision-maker's requests or unsolicited (proactive) messages when decision situations or undesirable behaviours of the decision-maker are perceived.

. *Maintaing and processing* its own knowledge base

2.4. Limits and Constrains of Human Decision Makers

The work of the decision-maker is influenced by several limits and constraints. Though these depend on the characteristic feature of each individual decision-maker and his/her decision context, several classes of limits and constraints can be identified (Holsapple, Whinston, 1996; Filip, 2004) as it follows.

. *Cognitive* limits concerning decision problem data available and decision procedures, methods and techniques mastered by the human decision-maker.

. *Costs* of assistants or external consultants that support the work of the decision-maker. Also the *increased dependence* of the solutions chosen on the assistants' services may be viewed as a limit.

. *Temporal* constraints that must be observed when urgent decisions are to be made or several decision problems are to be solved at the same time.

. *Communication/collaboration* limits and constraints that show up when several persons are involved in making a decision or/and implementing a chosen solution.

. *Low trust* of human decision-makers in the solutions recommended by computerized decision methods and costs of the IT products.

3. ANTHROPOCENTRIC SYSTEMS

3.1. Anthropocentric Manufacturing Systems

Anthropocentric manufacturing systems (AMS) emerged from convergent ideas with roots in the social sciences of the '50s. Kovacs and Munoz (1995) present a comparison between the *anthropocentric approach* (A) *and the technology-centered approach* (T) along several directions: *a) role of new technologies*: complement of human ability, regarding the increase of production flexibility, of product quality and of professional life quality (A), versus decrease of worker number and role (T); *b) activity content at operative level*: autonomy and creativity in accomplishing complex tasks at individual or group level (A), versus passive execution of simple tasks (T); *c) integration content and methods*: integration of enterprise components through training, development of social life, of communication and co-operation, increased accesses to information and participation in decision taking (A), versus integration of enterprise units by means of computer-aided centralization of information, decision and control; *d) work practice*: flexible, based on decentralization principles, work multivalence, horizontal and vertical task integration and on participation and co-operation (A), versus rigid, based on centralization, strict task separation at horizontal and vertical level associated with competence specialization.

Table 1. Classes of decision-makers

Attribute Decision-maker type	Number of persons	Equal powers?	Stable composition?	Co-operation ?	Support team?
Individual	1	NA	NA	NA	No
Unilateral	1	NA	NA	NA	Yes
Group of peers	*n*	Yes	Yes	Yes	No/Yes
Collectivity	*n*	Yes/No	No	No/Yes	Yes/No
Organizational hierarchical Group	*n*	No	Yes	Yes	Yes/No

3.2. Human-Centered Information Systems

Johannsen (1994) shows that " failure and delay encountered in the implementation of CIM concepts must be sought in organizational and personnel qualification problems. It seems that not only CIM must be considered but also HIM (human integrated manufacturing)". In a man-centered approach integration of man at all control levels must be considered starting with the early stages of a project.

In Filip (1995), three *key questions* are put from the perspective of the "man in system" and regarding the man – information tool interaction: a) does the information system help man to better perform his tasks? b) what is the impact of man- machine system on the performance of the controlled object? c) how is the quality of professional life affected by the information system?

Most of the early information systems were not used at the extent of promises and allocated budget because they were *unreliable, intolerant* (necessitating a thread of absolutely correct instructions in order to fulfill their functions), *impersonal* (the dialogue and offered functions were little personalized on the individual user) and *insufficient* (often an IT specialist was needed to solve situations). It is true that most of this problems have been solved by IT progresses and by intense training, but nevertheless the problem of personalized systems according to the individual features of each user (such as temperament, training level, experience, emotional state) remains an open problem especially in process control applications.

The second question requires an analysis of *effectiveness* (supply of necessary information) and *efficiency* (supply of information within a clear definition of *user classes* (or roles) and real performance evaluation for *individuals* (or actors) who interact with the information tool along the dynamic evolution of the controlled object). In the case of industrial information systems, the safety of the controlled object may be more important than productivity, effectiveness or efficiency. As Johannsen (1994) pointed out, "in a technology-oriented approach the trend to let the information

system take over some of the operator tasks may lead to disqualification and even to boredom under normal conditions and to catastrophic decisions in crisis situations". This last observation is also a part of the answer to the last question, which answer holds an ethical and social aspect besides the technical one. Many years ago, Briefs (1981) stated, rather dramatically, that" automation is a major threat to human creativity and to the conscious development of the individual". This remark was motivated by the trend to polarize people into two categories. The first one groups IT specialists, who capitalize and develop their knowledge and creativity by making more and more sophisticated tools. The second one represents the broad mass of users, who can accomplish their current needs/jobs quickly and easy, without feeling tempted to develop an own in-depth perception of the new and easy means of production.

As Filip (1995) shows, "it is necessary to elaborate information systems that are not only precise, easy to use and attractive, all at a reasonable cost, but also stimulating to achieve new skills and knowledge and eventually to adopt new work techniques that allow a full capitalization of individual creativity and intellectual skills". The aim to develop anthropocentric information systems applies today as well, but the designer finds little use in generally formulated objectives with no methods to rely on. It is possible to formulate derived objectives representing values for the attributes of information systems: a) *broad service scale* (not "Procustean") – for the attribute "use / physiology"; b) *transparency* of system structure in regard to its capability to supply explanations – for the attribute "structure / morphology", and c) growing development and learning capabilities – for the attribute "construction / ontogeny"

4. DSS - BASIC CONCEPTS

The DSS appeared as a term in the early '70ies, together with managerial decision support systems. The same as with any new term, the significance of DSS was in the beginning a rather vague and controversed. While some people viewed it as a new redundant term used to describe a subset of MISs, some other argued it was "a new label" abusively used by some vendors to take advantage of a term in fashion. Since then many research and development activities and applications have witnessed that the DSS concept definitely meets a real need and there is a market for it.

4.1. Definition and Characteristic Features

A plethora of definitions have been proposed so far for Decision Support Systems (DSS). In (Filip, 2004) a DSS is viewed as an anthropocentric, adaptive and evolving information system which is meant to implement the functions of a HSS that would otherwise be necessary to help the decision-maker to overcome his/her limits and constraints he/she may encounter when trying to solve complex and complicated decision problems that count.

The main characteristic features of a DSS are synthetically presented in table 2.

Table 2. DSS Characteristics

Mission	To relax the limits & constraints of the human decision-maker in making and taking a decision
Attributes: Clientele"	Decision-makers, assistants, other knowledge workers
Qualities	Anthropocentric, adaptive, evolving
Stored Knowledge	Descriptive, procedural, reasoning, communication acquired from internal/external sources or internally produced
Functions	Computerized version of functions of the HSS

4.2. DSS technology

A distinction should be made between a *specific* (application-oriented) *DSS*s (SDSS) and *DSS tools*. The former are used by particular decision-makers ("final users") to perform their specific tasks. Consequently, the systems must possess *application-specific* knowledge. The latter are used by "system builders" to construct the application systems. A special class of tools is composed of DSS *generators* (DSSG), which are prefabricated systems oriented towards various application domains and functions and can be personalized for particular applications within the domain provided

they are properly customized for the application characteristics and for the user's specific in needs. The DSS *basic construction tools* can be general-purpose or specialized information technology tools. The first category covers hardware facilities such as PCs, workstations, or software components such as operating systems, compilers, editors, database management systems, optimisation libraries, expert system shells, and so on. Specialized technologies are hardware and software tools such as sensors, specialized simulators, spreadsheets, computerized decision trees and multiattribute models, report generators, etc, that have been created for building new application DSSs or for improving the performances of the existing systems. An application DSS can be developed from either a system generator, to save time, or directly from the basic construction tools to optimise its performances.

The generic framework of a DSS, proposed by Bonczek, Holsapple, and Whinston (1980), the so-called *BHW model,* is quite general and can accommodate the most recent technologies and architectural solutions. It is based on three essential components. The first one is the *Communication (Language* and *Presentation) Subsystem* (LS). This is used for: a) directing data retrieval, allowing the user to invoke one out of a number of report generators; b) directing numerical or symbolic computation, enabling the user either to invoke the models by names or construct model and perform some computation at his/her free will; c) maintaining knowledge and information in the system; d) allowing communication among people in case of a group DM, and e) personalizing the user interface. The *Knowledge Subsystem* (KS) normally contains: a) "descriptive knowledge " about the state of the application environment in which DSS operates and the results of various computations; b) "procedural knowledge", including basic modelling blocks and computerized simulation and optimisation algorithms to use for deriving new knowledge from the existing knowledge; d) "meta-knowledge" (knowledge about knowledge) supporting model building and experimentation and result evaluation; e) "linguistic knowledge" allowing the adaptation of system vocabulary to a specific application , and f) "presentation knowledge" to allow for a most appropriate information presentation to the user.

The third essential component of a DSS is the *Problem Processing Subsystem* (PPS), which enables combinations of abilities and functions such as information acquisition, model formulation, analysis, evaluation, etc.

The BHW model can accommodate, as a particular case, the well-known D/IDM (Dialogue/interface, Data, Models) *paradigm* of Sprague (1980) (Fig. 1).

Table 3. Task assignment in DSS (adapted from Dutta, 1996)

Decision steps and activities	H	NM	ES	ANN	CBR
Intelligence					
Perception of DM situation	I/E		P		
Problem recognition	I/P				I
Design					
• *Model selection*	M/I		I		I
• *Model building*	M		I	P	
• *Model validation*	M				
Choice					
• *Model solving*		E		P	
• *Model experimentation*	I/M		M/I		
• *Solution adoption*	E		P		
Release for implementation	E				

Legend. NM - numerical model, ES - rule based expert system, ANN - artificial neural network, CBR - case based reasoning, H - human decision-maker, P - possible, M - moderate, I - intensive, E - essential

De Michelis, G. (1996). Co-ordination with cooperative processes. In: *Implementing Systems for Support Management Decisions.* (P. Humphrey, L. Bannon, A. Mc. Cosh, P. Migliarese, J. Ch. Pomerol Eds)., 108-123, Chapman & Hall, London.

Dutta, A (1996). Integrated AI and optimization for decision support: a survey. *Decision Support Systems,* **18**, pg. 213-226.

Filip, F. G. (1995). Towards more humanized real-time decision support systems. In: *Balanced Automation Systems; Architectures and Design Methods.* (L. M. Camarinha – Matos and H. Afsarmanesh, Eds.),230-240, Chapman & Hall, London.

Filip, F.G. (2004). *Decision Support Systems.* Ed. TEHNICA, Bucharest (in Romanian

Filip, F. G., B. Barbat (1999). *Industrial Informatics.* Technical Publishers, Bucharest (In Romanian).

Filip, F.G., D.D. Donciulescu, Cr.I.Filip (2002). Forwards intelligent real-time decision support systems. *Studies in Informatics and Control – SIC 11, 4,* 303-312 (also: http://www.ici.ro/ici/revista/sic.html)

Holsapple, C.W., A.B. Whinston (1996). *Decision Support Systems: A. Knowledge – Based Approach.* West Publishing Co., Mineapolis.

Johannsen, G. (1994). Integrated systems engineering: the challenging cross discipline. In: *Preprints IFAC Conf. On Integrated Syst. Engng., 1-10,* Pergamon Press, Oxford.

Kovacs, J., A. B. Munoz (1995). Issues in anthropocentric production systems. In: *Balanced Automation Systems; Architectures and Methods* (L. Camarinha – Matos, H. Afsarmanesh, Eds.), 131-140, Chapman & Hall, London.

Martenson, L. (1996). Are operators in control of complex systems? In: *Preprints, IFAC 13th World Congress* (J. Gertler, J.B. Cruz; M.Peshkin, Eds.), vol B, 259 – 270.

Marakas, G.M. (2003). *Decision Support Systems and Megaputer.* Prentice Hall, Upper Saddle River, New Jersey.

Martin, T., J. Kivinen, J.E. Rinjdorp, M.G. Rodd, W.B. Rouse (1980). Appropriate automation integrating human, organisation and culture factors. In: *Preprints IFAC 11th World Congress vol 1 , 47 - 65.*

McMahon, C., J. Browne (1995). *CAD/CAM: From Principles to Practice.* Addison Wesley, Working ham.

Monostory, L., D. Barschendorff (1992). Artificial neural networks in intelligent manufacturing. *Robotics & Computer – Integrated Manufacturing, 9* (6), 421-437.

Orasanu,J. Fisher, J. Davison (2002). Risk perception: a critical element of aviation safety. *Preprints, IFAC 15th World Congress* (E.F, Camacho, L. Basanez, J.A. de la Puente, Eds.). *Vol.I: Plenary Papers, Survey Papers and Milestones.*

Power, D.J. (2002). *Decision Support Systems: Concepts and Resources for managers.* Quorum Pres, Westport, Connecticut.

Sheridan, T. (1992). Telerobotics, Automation and Human Supervizory Control.MIT Press, Cambridge.

Simon, H. (1960). *The New Science of Management Decisions.* Harper Row, New York.

Sprague, Jr., R.H. (1980). A Framework for the development of decision support systems. *MIS Quarterly, 4 (4).*

Sprague, R.H. (1987). DSS in context. *Decision Support Systems, 3,* 197-202, p29-40.

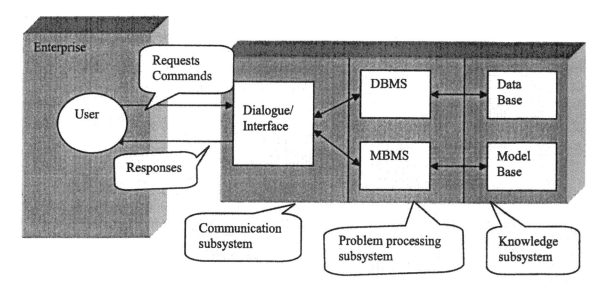

Figure 1. BHW model and D/IDM paradigm

Recently, Power (2002) expanded Alter's (1980) DSS taxonomy and proposed a more complete and up-to-date framework to categorise various DSS in accordance with one main dimension (the dominant component such as data, documents, models, communications, [reasoning] knowledge) and three secondary dimensions (the target user, the degree of generality, and the enabling technology)

4.2 Real time DSS for manufacturing

Most of the developments in the DSS domain have addressed business applications not involving any real time control. In the sequel the real time decisions in industrial milieu will be considered. Bosman (1987) stated that control problems could be looked upon as a "natural extension" and as a "distinct element" of planning *decision making processes* (DMP) and Sprague (1987) stated that a DSS should support communication, supervisory, monitoring and alarming functions beside the traditional phases of the problem solving process.

Real time decision making processes (RT DMPs) for control applications in manufacturing are characterized by several particular aspects such as: a) they involve continuous monitoring of a dynamic environment; b) they are short time horizon oriented and are carried out on a repetitive basis; c) they normally occur under time pressure; d) long-term effect are difficult to predict (Charturverdi and colleagues, 1993). It is quite unlikely that an "econological"' approach, involving optimisation, be technically possible for genuine RT DMPs. Satisfycing approaches, that reduce the search space at the expense of the decision quality, or fully automated DM systems (corresponding to the 10th degree of automation in Sheridan's (1992) classification), if taken separately, cannot be accepted either, but for some exceptions.

At the same time, one can notice that genuine RT DMP can come across in "crisis" situations only. For example, if a process unit must be shut down, due to an unexpected event, the production schedule of the entire plant might turn obsolete. The right decision will be to take the most appropriate compensation measures to "manage the crisis" over the time period needed to recompute a new schedule or update the current one. In this case, a "satisfycing decision" may be appropriate. If the crisis situation has been previously met with and successfully surpassed, an almost automated solution based on past decisions stored in the information system (IS) can be accepted an validated by the human operator.

On the other hand, the minimisation of the probability of occurrences of crisis situations should be considered as one of the inputs (expressed as a set of constraints or/and objectives) in the scheduling problem. For example in a pulp and paper mill, a *unit plant* (UP) stop may cause drain the downstream *tanks* (T) and overflow the upstream tank and so, shut/slow down the unit plants that are fed or feed those tanks respectively.

Subsequent UPs starting up normally implies dynamic regimes that determine variations of product quality. To prevent such situations, the schedule (the sequence of UP production rates) should be set so that stock levels in Ts compensate to as large extent as possible for UP stops or significant slowing down (Filip, 1995).

To sum up those ideas, one can add other specific desirable features to the particular subclass of information systems used in manufacturing control. An effective *real time DSS for manufacturing* (RT DSSfM) should support decisions on the preparation of "good" and "cautious" schedules as well as 'ad hoc", pure RT decisions to solve crisis situations (Filip, 1995).

5. AI BASED DECISION- MAKING

As discussed in the previous section, practical experience has shown that, in many cases, the problems are either too complex for a rigorous mathematical formulation, or too costly to be solved by using optimisation and simulation techniques. Moreover, an optimisation-based approach assumes an "econological" model of the DM process, but in real life, other models of DM, such as "bounded rationality" or "implicit favourite" are frequently met. To overcome these difficulties several alternatives based on artificial intelligence are used (Dhar and Stein, 1997, Filip, 2002). The term *Artificial Intelligence* (AI) currently indicates a branch of computer science aiming at making a computer reason in a manner similar to human reasoning.

5.1. Expert Systems

The *Expert System* (ES) is defined by E. Feigenbaum (the man who introduced the concept of "knowledge engineering") as "intelligent computer programs that use knowledge and inference procedures to solve problems that are difficult enough to require significant human expertise for their solution". As in the case of the DSS, one can identify several categories of software products in connection with ES: a) application ES or *Knowledge Based Systems* (KBS), that are systems containing adequate domain knowledge which the end user resorts to for solving a specific type of problem; b) system "shells", that are prefabricated systems, valid for one or more problem types to support a straightforward knowledge acquisition and storage; c) basic tools such as the specialised programming languages LISP, PROLOG or object-oriented programming languages.

One can easily notice the similarity of the ES and DSS as presented in Section 4.2. Also several problem types such as prediction, simulation, planning and control are reported to be solved by using both ESs and traditional DSSs. At the same time, one can notice that while there are some voices from the DSS side uttering that ESs are only tools to incorporate into DSSs, the ES fans claim that DSS is only a term denoting applications of ESs. Even though those claims can be easily explained by the different backgrounds of tool constructors and system builders, there is indeed a fuzzy border between the two concepts. However a deeper analysis (Filip and Barbat, 1999) can identify some real differences between typical ESs and typical DSSs such as: a) the application domain is well-focused in the case of ES and it is rather vague, variable, and, sometimes, unpredictable in the case of the ESs; b) the information technology used is mainly based on

symbolic computation in the ESs case and is heavily dependent on numerical models and databases, in traditional DSS cases; c) the user's initiative and attitude towards the system are more creative and free in the DSs case in contrast with ESs case, when the solution may be simply accepted or rejected.

5.2. Case-Based Reasoning

The basic idea of *Case-Based Reasoning* (CBR) consists in using solutions already found for previous similar problems to solve current decision problems. CBR assumes the existence of a stored collection of previously solved problems together with their solutions that have been proved feasible and acceptable. In contrast with the standard expert systems, which are based on deduction, CBR is based on induction.

The operation of CBR systems basically includes the first or all the three phases: a) selection from a knowledge base of one or several cases (decision situations) similar to the current one by using an adequate similarity measure criterion; b) adaptation of the selected cases to accommodate specific details of the problem to solve (this operation is performed by an expert system which is specialised in adaptation applications; differential" rules are used by the CBR system to perform the reasoning on differences between the problems); c) storing and automatically indexing of the just processed case for further learning and later use.

5.3. Artificial Neural Networks

Artificial Neural Networks (ANN) or connectionist systems are apparently a last solution to resort to when all other methods fail because of a pronounced lack of the structure of a decision problem. The operation of ANN is based on two fundamental concepts: a) the parallel operation of several independent information processing units, and the learning, and b) law enabling processors adaptation to current information environment

Expert systems and ANNs agree on the idea of using the knowledge, but differ mainly on how to store the knowledge. This is a rather explicit (mainly rules or frames) and understandable manner in the case of expert systems and implicit (weights, thresholds) manner, incomprehensible by the human in case of connectionist systems. Therefore while knowledge acquisition is more complex in case of ES and is simpler in case of ANN, the knowledge modification is relatively straightforward in case of ES but might require training from the very beginning in case a new element is added to ANN. If normal operation performance is aimed at, ANNs are faster, robuster and less sensitive to noise but lack "explanation facilities". (Monostory and Barschendorf,1992)

6. KNOWLEDGE BASED DSS

It has been noticed that some DSS are "oriented" towards the left hemisphere of the human brain and some others are "oriented" towards the right hemisphere. While in the first case, the quantitative and computational aspects are important, in the second, pattern recognition and the reasoning based on analogy prevail. In this context, there is a significant trend towards combining the numerical models and the models that emulate the human reasoning to build advanced DSS.

Over the last three decades traditional numerical models have, along with databases, been the essential ingredients of DSS. From an information technology perspective, their main advantages (Dutta, 1996) are: compactness, computational efficiency (if the model is correctly formulated) and the market availability of software products. On the other hand, they present several disadvantages. Because they are the result of intellectual processes of abstraction and idealisation, they can be applied to problems that possess a certain structure, which is hardly the case in many real-life problems. In addition, the use of numerical models requires that the user possesses certain skills to formulate and experiment the model.

As it was shown in the previous section, the AI-based methods supporting decision-making are already promising alternatives and possible complements to numerical models. New terms such as "tandem systems", or "expert DSS-XDSS" were proposed to name the systems that combine numerical models with AI based techniques. A possible task assignment is given in Table 1. Even though the DSS generic framework (mentioned in Section 4.2) allows for a conceptual integration of AI based methods, for the time being, the results reported mainly refer specific applications and not general ones, due to technical difficulties arising from the different ways of storing data or of communicating parameters problems, and from system control issues (Dutta, 1996).

7. CONCLUSIONS

Several important issues on the design of anthropocentric modern information systems were reviewed. DSS, as a particular kind of human centred information system, was described. A particular emphasis was put on the embedding the AI-based methods within DSS with the view to evolve DSS from simple job aids to sophisticated computerised decision assistants.

Several further developments can be foresighted such as:

- Incorporation and combination of newly developed numeric models and symbolic/sub-symbolic (connectionist) or agent-based techniques in advanced, user-friendly DSS will continue in an effort to reach the "unification" of man, numerical models, expert systems and artificial neural networks and other AI-based tools;

- Largely distributed group (or "multiparticipant") decision support systems that intensively use new, high-performance computer networks will be created so that an ever larger number of people from various sectors and geographical locations are able to communicate and make "co-decisions" in real time in the milieu of networked extended / virtual enterprises;

- Mobile communications and web technology will be ever more considered in DSS, thereby people will make co-decisions in "virtual teams", no matter where they are temporarily located;

- Other advanced information technologies such as virtual reality techniques (for simulating the work in highly hostile environments) or "speech computers" are likely to be utilised.

8. REFERENCES

Alter, S. (1980). *Decision Support Systems: Current Practices and Continuing Challenges*. Addision-Wesley. Reading, Massachusetts.

Bainbridge, L. (1983). Ironies of automation. *IFAC J Automatica, 19* (6), 775-779.

Bibby, K., F. Margulis, J.Rinjdorp (1976). Man's role in control systems. *Proceedings, IFAC 6th World Congress*, Pergamon Press, Oxford.

Bonczek, R.H., C.W. Holsapple, A.B. Whinston (1980). *Foundations of Decision Support Systems*. Academic Press, New York.

Bosman, A. (1987) Relations between specific *DSS, Decision Support Systems, 3,* 213-224.

Briefs, V. (1981). Re-thinking industrial work: computer effects on white collar workers. *Computers in Industry, 2, 76-89.*

Charturverdi, A.R., G.K. Hutchinson, D.L. Nazareth (1993). Supporting real-time decision-making through machine learning. *Decision Support Systems, 10*, 213-233.

Dhar,V, R. Stein (1997). *Intelligent Decision Support Methods: The Science of Knowledge Work*. Prentice Hall, Upper Saddle River, New Jersey

PUBLICATIONS
www.elsevier.com/locate/ifac

INTEROPERABLE ENTERPRISE SYSTEMS: PRINCIPLES, ARCHITECTURES, STANDARDS AND METRICS

F.B. Vernadat

*LGIPM: Laboratoire de Génie Industriel et Production Mécanique
ENIM/Université de Metz, Ile du Saulcy, F-57045 Metz Cedex France
vernadat@enim.fr*

*European Commission
Unit for Interoperability, Architecture and Methods
DG-DIGIT JMO C2/112, L-2920 Luxemburg.
Francois.Vernadat@cec.eu.int*

Abstract: M odern organizations, be they considered from the intra- or the inter-organizational point of view, need to be made interoperable both in terms of their business processes, their information systems and even their human resources. This concerns three fundamental types of flows in any organization: material or physical flows, information flows, and decision/control flows. This is one of the key aspects of Enterprise Integration. Interoperable enterprise systems (be they supply chains, extended enterprises or any forms of virtual organization) must be designed, controlled, and appraised from a holistic systemic point of view. Especially, Enterprise Integration recommends that the IT architecture and infrastructure be aligned with business process organization and control, themselves designed according to strategic decisions defined from the enterprise vision, mission and values as expressed in an Enterprise Architecture. The paper successively addresses different facets of the problem, including the challenges of interoperable systems in large organizations, the principles for enterprise integration and enterprise architectures, industrial standards for systems interoperability, and metrics for performance evaluation.

Keywords: Enterprise Integration, Enterprise Architectures, Interoperability, Industrial standards, Performance indicators

1. INTRODUCTION

With the advent of globalization, companies must develop tighter relationships than ever with other firms to be part of competitive enterprise networks or to build strong partnerships. This is true for industrial companies producing goods or services and focused on customer satisfaction. It also becomes a reality for administrations and government agencies at national and supra-national levels to better serve public and private sectors.

Until recently, only loose relationships between companies were necessary. They were materialized by input/output physical flows (paper flows, raw materials, components or semi-finished products, end products), essential information exchanges (such as purchase orders, delivery slips and invoices), and monetary flows. Each company was self-centric considering other companies in its business sector as competitors and only paying attention to potential suppliers, sub-contractors, and customers. Nowadays, companies have to closely cooperate with various partners (e.g., suppliers, service providers, engineering bureaus, sub-contractors, customers or retailers, and even former competitors) within tightly linked clusters of companies, also called networked organizations. Indeed, to face competition in a global economy it is necessary to put in common skills, competencies, capabilities, and knowledge coming from different business entities either in terms of product or service design and engineering, research and development, process design and planning, production planning, production, product or service distribution, and maintenance or after-sale support.

Networked organizations can take various forms. To name a few we can mention:
- The traditional supply chain: A supply chain (SC) is a networked organization made of nodes in which physical entities and information entities flow from suppliers to customers. Nodes of the SC can be raw material or component vendors, manufacturing units, warehouses and storage areas, final assembly plants, transporters, end product distributors or retailers, and customers.
- The extended enterprise: The extended enterprise (EE) has usually at the heart of its organization a large final assembly plant or a service company procured by its suppliers (1st tier suppliers, 2nd tier suppliers) and serviced by its engineering units, sales units, banks, etc. The idea is to provide the central node with all materials, skills, competencies, knowledge, and capabilities it requires at the right time. Material flows are usually optimized in just-in-time (JIT) mode. This is the case of the car industry, aerospace industry, naval industry, semi-conductor industry, etc.
- The virtual enterprise: The virtual enterprise (VE) has a more dynamic and less stable nature than the extended enterprise. Once more the idea is to put together competencies coming from several enterprises but no node in the network plays a central role. This is a cluster or temporary association of existing or newly created business entities taken from several companies to form a new viable business entity to satisfy a timely market need. An example has been the company that built the Channel Tunnel in Europe (now dismantled).

Networked organizations are characterized by distributed control, inter-organizational business processes (i.e., business processes that cross enterprise boundaries and therefore do not belong to one enterprise), various producer-consumer supply chains, and shared information and knowledge. The main challenge is the operations optimization via co-decision, co-ordination, and even negotiation mechanisms. The main advantage of the networked organization is its flexible structure (new nodes can be added or removed to the network) to be reactive to face changing economic conditions or to implement new business strategies.

Reactivity requires interoperable enterprise systems. In this context, interoperability means the capability to use functionality/service of another system or to be able to share information/knowledge with another system at the semantic level. Interoperability can thus happen at the system level, at the information system level, and at the people's level.

Interoperability is only one of the many facets of Enterprise Integration (EI). EI is the study of an organization, its mission, vision and values, its business processes, and resources, understanding how they are related to each other, and determining the enterprise

structure so as to efficiently and effectively execute enterprise goals as defined by the top management (AMICE, 1993; Petrie, 1992; Vernadat, 1996).

The aim of the paper is to propose a holistic approach based on systems engineering for Enterprise Integration and systems interoperability. The last 20 years of experience in systems and enterprise integration tell us that the fundamental claim to be advocated is that there must be a clear, communicated, statement of the business mission, vision, and values of the enterprise (be it a single or a networked enterprise) and there must be a sound Enterprise Architecture that will provide a structured framework for all management decisions and technical choices made in the company. More specifically, business process logic first and second the IT architecture must be aligned with the business mission and the Enterprise Architecture. Not vice-versa.

2. ENTERPRISE INTEGRATION AND INTEROPERABILITY

Enterprise Integration occurs when there is a need in improving the task-level interactions among people, systems, departments, services, and companies. Integration of (inter- or intra-) enterprise activities has long been, and often still is, confused with information systems integration under the influence of the computer science community. While it is true that at the end of the day the prime challenge of EI is to provide the right information at the right place at the right time, business integration must drive information integration. In other words, integration needs must be defined by business users, not by IT people or computer specialists! EI has therefore a strong organizational dimension relying on a management dimension and a technological dimension.

For Vernadat (1996), Enterprise Integration is concerned with facilitating information, control, and material flows across organization boundaries by connecting all the necessary functions and heterogeneous functional entities (information systems, devices, applications, and people) in order to improve communication, cooperation, and coordination within this enterprise so that the enterprise behaves as an integrated whole, therefore enhancing its overall productivity, flexibility, and capacity for management of change (i.e., reactivity). Li and Williams (2004) provide a broader definition of EI stating that Enterprise Integration is the coordination of all elements including business, processes, people, and technology of the enterprise(s) working together in order to achieve the optimal fulfillment of the business mission of that enterprise(s) as defined by enterprise management.

A holistic approach to integration is therefore necessary that takes into account the business strategy as defined from the enterprise vision, the business process definition and enactment, and the design and operation of interoperable enterprise systems as supported by a relevant and efficient IT infrastructure. In the talk, we will use the framework proposed by Giachetti (2004) for the assessment of EI approaches and technologies.

Enterprise Interoperability provides two or more business entities (of the same organization or from different organizations and irrespective of their location) with the facility to exchange or share information (wherever it is and at any time) and to use functionalities of one another in a distributed and heterogeneous environment. With the advent of A2A (application-to-application) and X2X technologies in business (B2B: business-to-business, B2C: business-to-customer, B2E: business-to-employee...) as well as in governments (G2B: government-to-business, G2C: government-to-citizen, G2G: government-to-government, G2N: government-to-non government organizations), there is a need for sound and efficient methods and tools to design and operate interoperable enterprise systems made of autonomous units.

Broadly speaking, interoperability is a measure of the ability of performing interoperation between two or more different entities (be they pieces of software, processes, systems, business units...). Thus, Enterprise Interoperability is concerned with interoperability between organizational units or business processes either within a large (distributed) enterprise or within an enterprise network. The challenge relies in facilitating communication, cooperation, and coordination of these processes and units.

Major principles that have prevailed in building interoperable enterprise systems so far have been:
- Data integration: At the heart of integration were common, possibly distributed, interconnected databases (CAD, CAPP, PDM, MRP, CAM...) and data exchange formats (IGES, STEP, EDI...). This was the technology used in CIM (Computer Integrated Manufacturing).
- Object-oriented approaches: The universal description of any kind of entities as object classes, encapsulating static object description as properties and dynamic object behavior as methods together with object class specialization/aggregation with property inheritance, has been the second major principle. This opened the door of open systems and reusability. The Object Management Group (OMG) played a central role with the definition of CORBA (Common Object Request Broker Architecture) and its associated IDL language in the 90's.
- Business process-oriented approaches: When the real needs of enterprise integration were better understood as being much more than just systems integration, it became obvious that business processes (i.e., the sequence of steps that govern enterprise operations to achieve business objectives) had to be modeled, re-engineered, and controlled. Thus, new approaches centered on business processes or workflow emerged, such as CIMOSA (AMICE, 1993).

- Service-oriented approaches: Service-Oriented Architectures (SOA) are emerging for building the new wave of interoperable, on-demand applications. It is about designing and building systems using heterogeneous network addressable software components (preferably over Internet). These components (Weston, 1999) provide services or functions used by the application. The components can be implemented as Web Services and, therefore, located anywhere on the Web (Kreger, 2001; Khalaf et al., 2004).

3. ENTERPRISE ARCHITECTURE PRINCIPLES

Architecture is foundational for managing modern enterprises and planning Enterprise Integration. An Enterprise Architecture (EA) is an organized collection of ingredients (tools, methodologies, modeling languages, models, etc.) necessary to architect or re-architect whole or part of an enterprise. For a given enterprise, the architecture describes the relationships among the mission assigned to the enterprise, the work the enterprise does, the information the enterprise uses, and the physical means, human labor, and information technology that the enterprise needs.

Having an Enterprise Architecture in place can provide the following benefits:
- provides a clear picture of the mission, strategies, business process map, and supporting technologies across the enterprise
- establishes a change control process over business, IT, or other types of projects
- enables reuse, reduces duplication of effort, leverages economy of scale
- promotes information and knowledge sharing
- communicates standards and guidance
The prime advantage of an Enterprise Architecture is to provide a common view (in the form of models) of what is going on in the enterprise to every actor of the enterprise. Thus, Enterprise Modeling (EM) is central to Enterprise Engineering and Enterprise Integration (Vernadat, 1996; Kosanke and Nell, 1997).

Nota Bene: an Enterprise Architecture is not an IT or computer architecture.

The International IFAC-IFIP Task Force on Architectures for Enterprise Integration has defined two types of architectures: Type 1 and Type 2 (Williams et al., 1996).
- Type 1 architectures are those which describe a particular architecture or physical structure of some component or part of the integrated system such as the computer system or the communications system.
- Type 2 architectures are those which present a reference architecture or structure of the project which develops the integration program, i.e., those that

illustrate the life cycle of the project developing the integrated enterprise.

Well-known Type 2 architectures include CIMOSA, PERA, GRAI, ARIS, or the Zachman Framework. GERAM, as proposed by the IFAC-IFIP Task Force is a generalization of those, covering the complete enterprise engineering life cycle.

In the talk, the Purdue Enterprise Reference Architecture (PERA) is first presented (Williams, 1992) because it shows the appearance of functions from mission, vision, and values and then requirements to become components of the physical architecture, the human architecture and the IT architecture. PERA also provides a fairly complete description of the enterprise engineering life cycle. Next, the Zachman Framework (Zachman, 1996) will be addressed as it defines a strong and logical connection between business processes, organization strategies, and enterprise architectures. In the Zachman Framework, the objective of Enterprise Modeling is to define the six perspectives of the *what, how, where, who, when,* and *why* of the *Enterprise Model, System Model, Technology Model* and *Component* level of an enterprise. The what defines entities and relationships of the business entity, the how defines the functions and processes performed, the where defines the network of locations and links of entities and agents, the who defines agents and their roles, the when defines time aspects and the schedule of events, and the why defines the strategy of the enterprise. CIMOSA (AMICE, 1993) is also of major interest because it identifies four essential views for Enterprise Modeling (Function, Information, Resource, and Organization) and because of the modeling constructs of its process-based event-driven approach.

4. INDUSTRIAL STANDARDS

When dealing with Enterprise Architecture, Enterprise Integration, and Interoperability, a certain number of standards or standardization efforts need to be considered. They have recently been surveyed by Chen and Vernadat (2004) and Giachetti (2004).

Regarding Enterprise Architecture, ISO15704:2000 will first be used as a guide to identify the right mix of Methodology, Models, Tools, and other components that an enterprise needs to embark on an Enterprise Integration program. Second, attention must be paid to the following two European standards, ENV 40003 (Conceptual Framework for Enterprise Integration) and ENV 12204 (Constructs for Enterprise Modeling) published by CEN (Comité Européen de Normalisation), and now respectively superseded by prEN/FDIS 19439 (aligned with ISO 15704) and prEN/DIS 19440 (metamodel defined as a UML package).

On the Object Management Group (OMG) side, the Model-Driven Architecture (MDA) is the latest initiative (OMG, 2001). It has been developed in order to support enterprises and organizations to integrate new applications into existing systems. MDA is a middleware framework and it acts as a high-level abstract architecture based on UML methodology and existing profiles. At the heart of MDA are the already defined Meta-Object Facility (MOF), CORBA, XMI/XML and Common Warehouse Metamodel (CWM). CWM defines a generic model (as a set of UML classes) that enables data exchange and sharing across databases and even data warehouses across enterprises (OMG, 2000). It is a new open industry standard recently adopted by some major companies such as Oracle, SAS, and others. CWM is also a common metamodel which is independent of any specific data warehouse implementation, but which becomes domain specific in association with precise and detailed domain specifications.

Concerning Integration Infrastructures (IIS), since the early 90's, EI has drastically evolved from specialized communication protocols (e.g. MAP, TOP, field-buses), diverse dedicated standard data exchange formats (e.g., IGES, STEP, EDI/EDIFACT, HTML…) and complex monolithic integration infrastructures for distributed computing environments (e.g., OSF/DCE in the Unix world, OLE, and DCOM in the MS Windows world, and OMG/CORBA in the OO world) proposed at that time. Regarding Enterprise Application Integration (EAI), the state of the art is now to use Message-Oriented Middleware (MOM) (either in stateless or state-full mode as well as in synchronous or asynchronous mode) on top of computer networks compatible with TCP/IP. The middleware must provide sufficient scalability, security, integrity, and reliability capabilities. Messages are more frequently in the form of XML documents. The current trend is to switch to Java programming (JSP, EJB) and apply the J2EE (Java to Enterprise Edition and Execution) principles, or the .NET approach in Windows environments to build integrated collaborative systems.

On top of these, business applications are implemented according to the 3-tier architecture (presentation – business logic – data storage) using a web architecture and a standard protocol (HTTP). A web user can access the application on his/her PC via HTTP using a standard HTML browser. The request is sent to a web server, which concentrates all requests and passes the request to the application server (AS). The AS processes each request using its local database server(s).

A new trend for the development of interoperable applications is to build them as a set of remote services accessible via the web, called *web services*. The client application does not need to know where the services are located on the web but it can request their use at any time (Service-Oriented Approach). Services need to be declared via WSDL (Web Service Description

Language) and registered in a common web repository, called UDDI.

Concerning message exchange, the trend is to make wide use of XML (eXtensible Mark-up Language) in order to neutralize data because of the ability of XML and XSL to separate the logic and structure of documents and data formatting from the data itself. This means that well-known data exchange formats used in industry (e.g., EDI, STEP, etc.) are going to be reworked in the light of XML (e.g., cXML, ebXML…).

Finally, concerning transport of messages, new protocols are being proposed including SOAP (Simple Object Access Protocol) (http://www.w3.org/TR/SOAP), RosettaNet (www.rosettanet.org), Bolero.net (www.bolero.net), Biztalk (www.biztalk.org), among others.

5. ENTERPRISE INTEGRATION METRICS

Enterprise Integration, and its supporting Enterprise Architecture, promises to the organization the achievement of strategic objectives on the basis of alignment between business strategy and enterprise operations. Therefore, there should be some mechanisms to measure and assess such alignment and objective achievement. This is the role of performance indicators and tableaux de bord.

For a long time, performance expressions have been purely financial ones (e.g., productivity ratios, stock levels, labor costs, yearly turnover…). The goal was targeted at maximization of profits according to the Taylorian organization prevailing at that time. With the new economic environment and the new forms of organizations, industrial performance can be defined in terms of numerous criteria to be synthesized for global control purpose, many of them being of complex nature, i.e., not related to one single physical measure (e.g., quality, cost, and delays). In addition to the mutli-criteria dimension of performance, the physical measures can be different in nature (e.g., numeric values, symbolic values, fuzzy values). Finally, new types of indicators need to be defined for the new forms of organizations, such as reactivity or responsiveness, flexibility, or system robustness.

In this context, determining a global performance expression raises the issue of information aggregation in defining and computing the performance indicator,. To address such an aggregation issue, so-called Performance Measurement Systems (PMSs) need to be implemented. The analysis of the corresponding literature leads to the conclusion that the majority of the proposed approaches either do not provide explicit aggregation mechanisms, or propose too simple methods to really cope with the complexity of the situations at hand. Only a few consider explicit adequate aggregation methods. Therefore,

performance expression aggregation can be approached based on four stages, namely extraction, representation, combination and interpretation, as proposed by the information aggregation community. It is proposed that the definition of a performance expression combination be based on two kinds of performance expressions commonly encountered: physical measures and performance evaluations, i.e., objective satisfaction degrees. Methodological guidelines for performance expression aggregation will be proposed in the talk as well as associated mathematical tools, especially the fuzzy Choquet integral to take criteria interactions into account as described in (Berrah *et al.*, 2004). An industrial case study will be used as an illustration, and some concluding remarks and emerging problems to be considered in the future, such as temporal aggregation, will be pointed out.

6. CONCLUSION

The new economic order imposed by globalization is forcing companies to change their strategic vision, to join forces by creating highly skilled, competitive and reactive enterprise networks, and constantly to adjust their objectives with fluctuating market conditions. Enterprise Integration, in its broad sense, is one way to achieve this goal and sustain management and control of these networked interoperable enterprise systems. The paper claims that to properly design and then control such systems, an Enterprise Architecture, defining the mission, vision, and values and ensuring alignment between business strategy and enterprise operations, is first needed. Then, application and information systems interoperability must be achieved. These solutions must take advantage of existing and emerging IT standards. Finally, performance metrics must be designed and made operational in the form of accurate and relevant performance indicator systems that will be part of the enterprise control loop

.

REFERENCES

AMICE (1993), *CIMOSA: CIM Open System Architecture*, 2nd edition, Springer-Verlag, Berlin.

Berrah L., Mauris G., Vernadat F. (2004), Information aggregation in industrial performance measurement : rationales, issues and definitions, *International Journal of Production Research*, **42**.

Chen D. and Vernadat F. (2004), Standards on Enterprise Integration and Engineering – A state of the art, *International Journal of Computer Integrated Manufacturing*, **17** (3), 235-253.

Giachetti R.E. (2004), A framework to review the information integration of the enterprise, *International Journal of Production Research*, **42** (6), 1147-1166.

Khalaf R., Curbera F., Nagy W., Mukhi N., Tai S., Duftler M. (2004), Understanding Web Services. In

Practical Handbook of Internet Computing (M. Singh, Ed.), CRC Press LLC, Boca Raton, FL.

Kosanke K. and Nell J.G. (Eds.) (1997), *Enterprise Engineering and Integration: Building International Consensus*, Springer-Verlag, Berlin.

Kreger H. (2001), Web Services Conceptual Architecture (WSCA 1.0), IBM Software Group, IBM, Somers, NY, USA.

Li H. and Williams T.J. (2004), A vision of Enterprise Integration considerations: A holistic perspective as shown by the Purdue Enterprise Reference Architecture, Proc. 4th Int. Conf. on Enterprise Integration and Modeling Technology (ICEIMT'04), Toronto, Canada, 9-11 October.

OMG (2000), Common Warehouse Metamodel (CWM) Specification. Object Management Group.

OMG (2001), Model-Driven Architecture (MDA). Object Management Group, Document 2001-07-01.

Petrie C.J., Ed. (1992), *Enterprise Integration Modeling*, The MIT press, Cambridge, MA.

Vernadat F.B. (1996), *Enterprise Integration and Modeling: Principles and Applications*, Chapman & Hall, London.

Weston R.H. (1999), Reconfigurable, component-based systems and the role of enterprise engineering concepts, *Computers in Industry*, **40**, 321-343.

Williams T.J. (1992), *The Purdue Enterprise Reference Architecture*, Instrument Society of America.

Williams T.J., Bernus P. and Nemes L. (1996), The concept of Enterprise Integration. In *Architectures for Enterprise Integration* (Bernus P., Nemes L. and Williams, Eds.), Chapman & Hall, London, pp. 9-20.

Zachman J. (1996), The Framework for Enterprise Architecture: Background, description and utility, Zachman Institute For Advancement, http://www.zifa.com.

www.elsevier.com/locate/ifac

Collaboration Principles for Networked Manufacturing & Logistics:
State of the art and challenges

Shimon Y. Nof

PRISM Center

Purdue University, West Lafayette IN 47907, USA

nof@purdue.edu

Abstract

The rapid development in information technology has transformed not only the way people work and interact with each other, but also the focus of management and research, needing more decision-support to handle complex automation and communication systems over distributed networks of facilities and enterprises, and over communication networks. Electronic media, such as e-mail, LAN, Intranet and Internet, enable collaboration support tools that help workers to coordinate, cooperate, and collaborate on their work and plans within their own organization and with other independent enterprises, including suppliers, customers, and other support organizations. The work activities and decisions that are related to all these electronic media are defined as e-Work, and its collaborative management is called "e-Work Collaboration". Its influence on manufacturing and logistics is significant.

This presentation addresses the emerging and future problems of collaborative e-Work in manufacturing and logistics systems, particularly, knowledge based decision-making models and activities, in both the development and implementation perspectives. The principles and characteristics of the emerging e-Work decision and collaboration support systems are presented based on conclusions from several Purdue University's PRISM Center research projects in planning distributed information and production/logistics systems. The opportunities and challenges for researchers and managers of e-Work of production, service, and logistics organizations are described, and a plan for a joint research project by international investigators on the design of next generation collaborative e-Work and robotics systems is also described.

Keywords: Agents, Collaborative Control, Collaboration Principles, e-Logistics, e- Manufacturing, e-Work, Integration, Sensor Fusion, Robot Systems, Protocols, Communication Protocols, Communication Networks, Fault-Tolerance.

Collaboration Principles for Networked Mfg. & Logistics:
State of the art and challenges

Shimon. Y. Nof

IFAC CC 5: Manufacturing & Logistics Systems
PRISM Center
Production, Robotics, and Integration Software for Mfg. & Management

Purdue University, W. Lafayette, IN, USA

MCPL, Santiago de Chile, November, 2004

1. e-Work and machine collaboration ("Smart Teams")
2. Networked e-activities
3. The explosive cost of conflict resolution/error recovery
4. Fault tolerance in sensor (robotic) arrays -- FTTP[pp]
5. Implications to mfg. and logistics

1

Mfg. and Logistics -- Why Collaborate?

◆ Need to exchange scientific, technological, business knowledge and management expertise

◆ Companies must look at the value chain through the customers' eyes to detect opportunities downstream

◆ Agility, quality, traceability, and service-ability all depend on enterprise networks

◆ Responding to unpredictable change

◆ Forming tactical and virtual partnerships

◆ Let's face it: We cannot do well alone

2

Downstream models depending on e-collaboration [Ceroni & Nof, 2001]

Downstream model	Characteristics	Example
Embedded Services	Embedding downstream services into the product, freeing customers from the need to perform them	Honeywell and its Airplane Information Management System (AIMS)
Comprehensive Services	Coverage of downstream services not possible to embed into the product, for example, financial services	General Electric in the locomotive market
Integrated Solutions	Combination of products and services for addressing customer needs	Nokia, array of products for mobile telephony
Distribution Control	Moving forward over the value chain to gain control over distribution activities	Coca-Cola and its worldwide distribution network

PRISM Lab/Purdue

3

Ceroni, J.A., and Nof, S.Y., 2001. The Parallel Computing Approach in Production Systems Integration Modeling, Proc. of MCPL 2, Grenoble, France, pp. 1257-1262 (Elsevier Science).

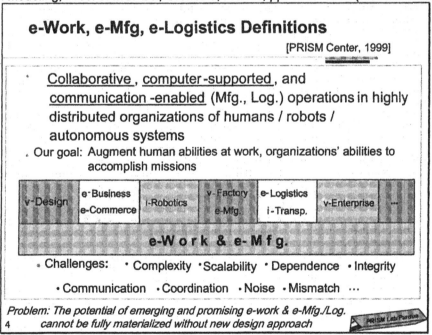

e-Work, e-Mfg, e-Logistics Definitions

[PRISM Center, 1999]

Collaborative, computer-supported, and communication-enabled (Mfg., Log.) operations in highly distributed organizations of humans / robots / autonomous systems

Our goal: Augment human abilities at work, organizations' abilities to accomplish missions

| v-Design | e-Business e-Commerce | i-Robotics | v-Factory e-Mfg. | e-Logistics i-Transp. | v-Enterprise | ... |

e-W o r k & e- M f g.

Challenges: • Complexity • Scalability • Dependence • Integrity
• Communication • Coordination • Noise • Mismatch ...

Problem: The potential of emerging and promising e-work & e-Mfg./Log. cannot be fully materialized without new design approach

PRISM Lab/Purdue

4

"As power fields, such as magnetic fields and gravitation, influence bodies to organize and stabilize, so does the sphere of computing and information technologies. It envelops us and influences us to organize our work systems in a different way, and purposefully, to stabilize work while effectively producing the desired outcomes."

[Nof, S.Y., 2003. Design of Effective e-Work: Review of Models, Tools, and Emerging Challenges, *Production Planning & Control* Special Issue on e-Work: Models and Cases, 14(8), pp. 681-703.]
Download the full article from gilbreth.ecn.purdue.edu/~prism

Fundamental Changes in the Evolution of Work Systems

Period	Technology Advancement	Human Work Augmented by	Fundamental Work System Principle	Engineering & Management Concerns	Work System Goal[1]
Before the Industrial Revolution	Hand-tools	Tooling	Manual and animal power	Work methods	Enable work functions
Before Computers	Engines; Machines	Power; Motions; Moves	Human-machine systems	Work flow	Reduce muscle load
Computer Age	Computers; Communication; Robots	Control; Data processing; Automation	Computer-aided and computer-integrated systems	Human-computer interaction; Human-robot interaction	Reduce information overload
Information Age	Telecom; Internet; Mobility	Cognitive skills; Collaboration	e-Work, e-Mfg.	e-Work design; Collaboration; Parallelism	Reduce information and task[2] overload

5 1 Goals beyond productivity, quality, safety... 2 Both cognitive and non-cognitive tasks   PRISM Lab/Purdue

Collaboration processes are fundamental to effective e-Work [Nof, 2003]

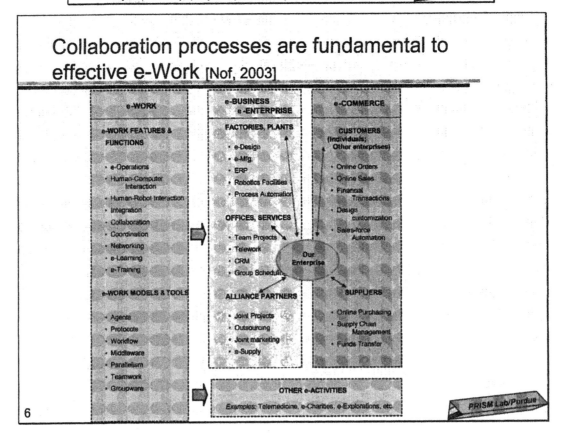

6

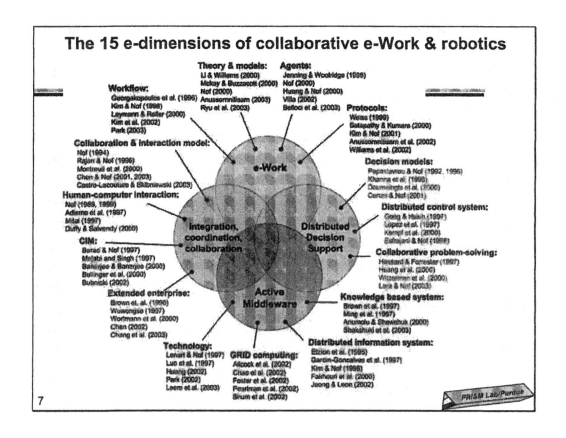

The 15 e-dimensions of collaborative e-Work & robotics

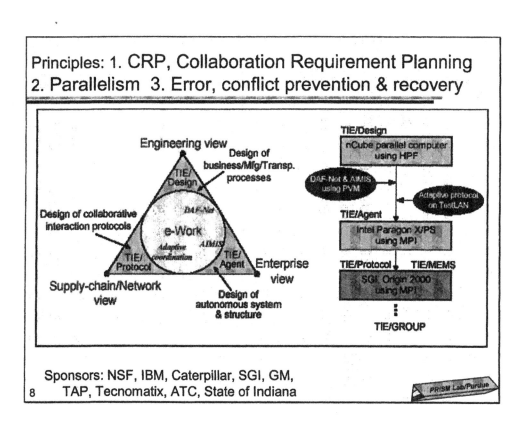

Principles: 1. CRP, Collaboration Requirement Planning 2. Parallelism 3. Error, conflict prevention & recovery

Sponsors: NSF, IBM, Caterpillar, SGI, GM, TAP, Tecnomatix, ATC, State of Indiana

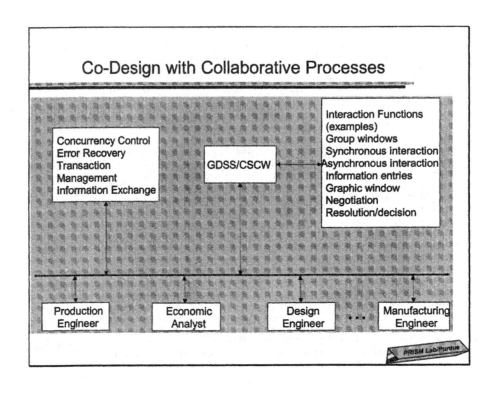

Co-Design with Collaborative Processes

Concurrency Control
Error Recovery
Transaction
Management
Information Exchange

GDSS/CSCW

Interaction Functions
(examples)
Group windows
Synchronous interaction
Asynchronous interaction
Information entries
Graphic window
Negotiation
Resolution/decision

Production
Engineer

Economic
Analyst

Design
Engineer

...

Manufacturing
Engineer

PRISM Lab/Purdue

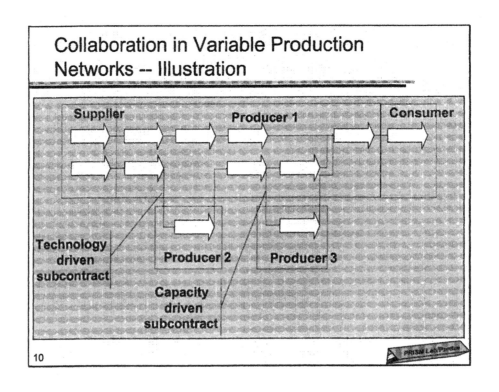

Collaboration in Variable Production Networks -- Illustration

Supplier

Producer 1

Consumer

Technology
driven
subcontract

Producer 2

Producer 3

Capacity
driven
subcontract

10

PRISM Lab/Purdue

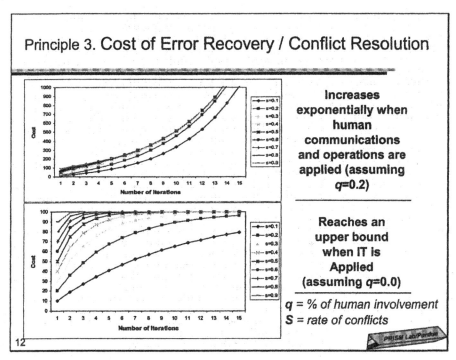

DPIEM: Integrated Optimization of collaborative design, mfg., logistics – Illustration   [Ceroni, 1999]

Summary: Local and Integrated Scenarios

	Φ	Π	T	Number of Sub-tasks
Local Scenario	1.9846	1.4239	0.5607	52
Enterprise A	1.3908	1.0350	0.3558	16
Enterprise B	0.5938	0.3889	0.2049	36
Integrated Scenario	1.8071	1.1666	0.6404	26

11   Optimize the DOP, Degree of Parallelism   PRISM Lab/Purdue

Ceroni, J.A., 1999. Models of integration with parallelism of distributed organizations, Ph.D. Dis., Purdue University, W. Lafayette IN, USA.

Ceroni, J.A., and Nof, S.Y., 2003. A Workflow Model based on parallelism for distributed organizations, *J. Int. Mfg.*, 13(6), pp. 439-462.

Principle 3. Cost of Error Recovery / Conflict Resolution

Increases exponentially when human communications and operations are applied (assuming $q=0.2$)

Reaches an upper bound when IT is Applied (assuming $q=0.0$)

q = % of human involvement
S = rate of conflicts

12   PRISM Lab/Purdue

Huang, C-Y., Ceroni, J.A., and Nof, S.Y., 2000. Agility of networked enterprises: Parallelism, error recovery and conflict resolution, Computers in Industry, 42(2-3), pp. 257-287.

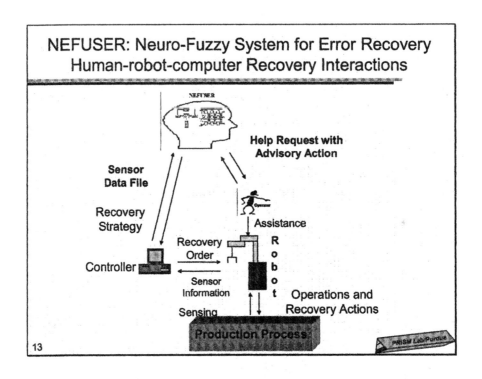

NEFUSER: Neuro-Fuzzy System for Error Recovery
Human-robot-computer Recovery Interactions

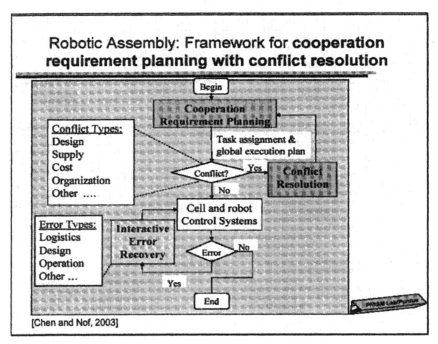

Robotic Assembly: Framework for **cooperation**
requirement planning with conflict resolution

[Chen and Nof, 2003]

Nof, S.Y., and Chen, J., 2003. Assembly and disassembly: An overview and framework for cooperation requirement planning with conflict resolution, *J. Int. & Robotics Sys.*, 37(3), pp. 307-320.

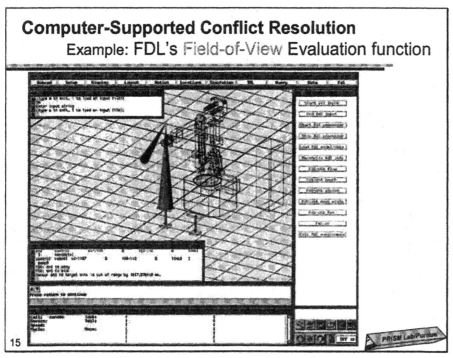

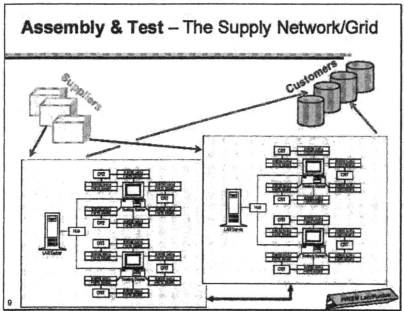

Case example: Extended enterprise research is a major outcome of advances in information technology. Powerful communication of information enables the integration of enterprise business processes beyond a single enterprise, including functions and software applications from end-to-end, not only internal but also for entire supply chains/networks. Example: From e-Work/e-Mfg. inside assembly facilities with TestLAN, to e-Work/e-Mfg. across supply networks and emerging supply grids (Nof, 2003)

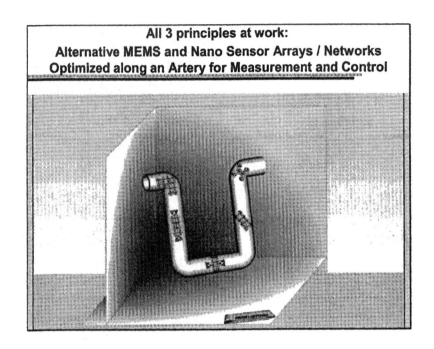

All 3 principles at work:
Alternative MEMS and Nano Sensor Arrays / Networks
Optimized along an Artery for Measurement and Control

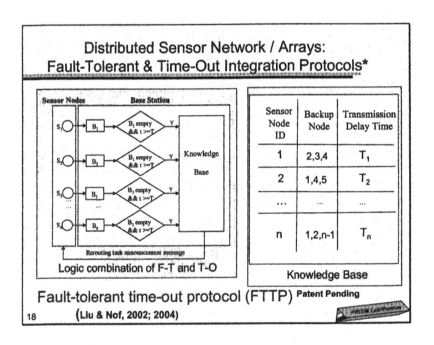

Distributed Sensor Network / Arrays:
Fault-Tolerant & Time-Out Integration Protocols*

Sensor Node ID	Backup Node	Transmission Delay Time
1	2,3,4	T_1
2	1,4,5	T_2
...	...	...
n	1,2,n-1	T_n

Logic combination of F-T and T-O

Knowledge Base

Fault-tolerant time-out protocol (FTTP) Patent Pending

18 (Liu & Nof, 2002; 2004)

Liu, Y., and Nof, S.Y., 2004. Distributed micro flow-sensor arrays and networks (DFMSN/A): Design of architectures and communication protocols, *IJPR*, 42(5), 3101-3115.

Collaborative Robotics and e-Work -- Challenges

Overall Trend: Smart robotic/agent teams (normal, micro, nano robots) will be able to interact as good as, or better than human teams; Collaborative control will enable collaboration support

Trend 1. Collaborative Coordination Control Theory

Why? Optimized coordination of e-Work interactions is key to effectiveness & competitiveness

Trend 2. Control Methods to Manage errors, conflicts, and interactions

Why? Complexity among distributed teams increases

Trend 3. Control Protocols for fault-tolerant, time-out integration

Why? Future collaborative robotics/agents will depend on cheaper, redundant arrays/networks (e.g., FTTP)

19

Smart human/machines/robotics/agents teams:
Challenges to mfg. & logistics

1. Optimized coordination of interactions is key to effectiveness and competitiveness

2. Manage errors, conflicts, and interactions' complexity among team of humans/machines/robots/agents; **CSS** (Collaboration Support Systems) for virtual and for human teams

3. Next generation collaborative systems will depend on fault-tolerant, time-out integration, e.g., FTTP[PP]

4. Smart machines/robot/agent teams will be able to interact better than human teams, and better *with* human teams

5. What are the priority issues? Let's collaborate . . .

20

Some priority issues . . . (incomplete, please add)

- Networked, decentralized decision support?
- Will workers and managers be willing to trust the results obtained and delivered by agents' work?
  By autonomous systems?
- Collaborative control theory? New modeling tools?
- New performance measures, e.g., Viability?
- Relations: Mechanism — Group behavior?
- Impacts on re-sequencing workflow cycles?
- Impact on education? On training?
- etc.?

21

PUBLICATIONS
www.elsevier.com/locate/ifac

PROBLEM SOLVING FOR NEW HIGH TECHNOLOGY PRODUCTS: THE CONTRIBUTION OF CRIMINAL AND INCIDENT INVESTIGATIONS METHODOLOGIES

AVRILLON Laetitia[(1,2)]   PILLET Maurice[(1)]   COMMERE Bruno[(2)]

[(1)]*LISTIC/University of Savoy*
BP 806
74016 Annecy Cedex France
Telephone: (+33) 4 50 09 65 80 Fax: (+33) 4 50 09 65 90
laetitia.avrillon@univ-savoie.fr
maurice.pillet@univ-savoie.fr

[(2)] *TRIXELL*
Z. I Centr' Alp 460, rue de Pommarin
38430 Moirans France
Telephone: (+33) 4 76 57 41 00 Fax: (+33) 4 76 57 40 48
laetitia.avrillon@trixell-thalesgroup.com
bruno.commere@trixell-thalesgroup.com

Summary: Product sophistication in terms of technology, coupled with the reduction of cycle design forces impose companies to constantly improve their ability to react faced with the appearance of a quality problem. Many problem-solving methods exist, but they are often limited in the context of new high technology products. After characterizing this particular context, this article suggests showing the limits of problem-solving methods when no or few experiments are allowed and a causes search is complex.
A parallel is proposed between this industrial context and two fields which have common points: criminal investigation and incident investigation. From this comparative study, points of improvement are proposed to approach problem-solving in an industrial context. An example of the implementation of the recommended method in the company Trixell is described.

Key words: Criminal investigation, High technology products, Incident investigation, New products, Problem-solving process.

1 – Introduction

To face the growing evolution of technical innovation and products sophistication, companies have to develop a true « intellectual productivity » with an exploratory field which is constantly increasing. Thus they need systematic methods to allow them to explore simultaneously and efficiently several fields, either to develop new products or to improve the quality of existing products. That is why the use of conception methods like TRIZ or problem-solving methods like Six Sigma is soaring today.
Many problem-solving methods for industry are available on the market and in literature. Each one has a particularity, either in its structure or in its toolbox. Two major orientations are however found (they can be combined):
- statistical analysis (e.g.: Six Sigma, Shainin) [De Mast 2003]. This orientation is the most widespread.
- combination of facts analysis, knowledge and creativity (e.g.: Kepner Tregoe, Japanese methods) [Hosotani 1997] [Kepner and al 1980].

These traditional tools and methods are however difficult to be apply in some particular industrial contexts. We propose to study particularities of these atypical contexts in order to establish a parallel between those and others fields and thus to transpose the link methods.
We chose to take for example the case of new high technology products (e.g.: detectors for medical imaging). The first part of this article presents the constraints of this context to direct us towards other fields to be studied, that is to say in this case criminal and incident investigations. Then, we will see how these methodologies are already connected with traditional problem-solving processes and how they differ. These differences will enable us to propose the contribution that they can make to the world of industry. Lastly, we will illustrate our proposal by an industrial application.
Hence, this article does not pretend to propose an nth still best method but presents the new light that non-industrial methodologies can cast on.

2 – Particular context and traditional problem-solving methods

First, let us study the specificities of new high technology products and see why traditional industrial tools and methods are difficult to use.

2.1 – New high technology products

The specificity of this context generates a huge complexity in terms of product design, manufacture and functioning. None or few products are strictly equivalent even at world level, comparisons are thus difficult to establish.

Indeed, the design of new high technology products requires developing technological blocks which are only useful for this company.

The singularity of the products also affects manufacturing processes, it is not possible to use standard processes. Moreover, the development of high technology products often requires several different trades: mechanics, electronics, chemistry, signal processing, computing...

This interaction between trades and new technologies also brings complexity for the product functioning: indeed, new properties are emergent.

Finally, the high technicality and the singularity of these products generates high cost for the design and manufacture of the finished products as well as for the purchased subsets (e.g.: electronic boards).

This specificity causes several constraints for problem solving:

- The complexity at all the levels (design, manufacturing, functioning) involves an important combination of possible causes. It is thus extremely difficult to draw an exhaustive list of potential causes.
- The participation of several high technicality trades in the product and the manufacturing processes development inevitably induces knowledge parcelling. Few people have global and a complete vision of the manufacturing process, the product and the application.
- The high cost of the products does not allow one to wait for sufficient statistics in order to start a problem-solving process. It is urgent to find the causes and to suggest solutions from the first occurrences of the defect.
- Because of the cost of the products, the step before the experimentation in problem-solving process becomes major. The number of experiments allowed is limited, a priori knowledge has to go as far as possible, the ideal aims is "no experiment".
- To have a continuous measure is even more important in this context in order to obtain statistical evidence.

Lastly, like any industrial context, it will be necessary to take into account constraints such as time, cost, available resources and priorities.

2.2 – Traditional problem-solving methods

Generally, problem-solving methods [Harry and al 2000] [Shainin 1995] [Kepner and al 1980] [Hosotani 1997] are known through a traditional sequence of successive stages:
- Definition of the problem
- (Definition of the measurement)
- Causes analysis
- Solutions analysis
- Solution implementation
- (Standardization).

Each one of these stages is exploited with adapted tools. There is a great quantity and a great diversity of quality tools that can be classified into three categories [Jayram and al 1997]:
- The first contains the tools which are based on data analysis (statistical tools such as comparison tests).
- The second uses graphic tools on the basis of diagrams (like concentration diagram).
- The third includes those that utilize people's opinion or knowledge (intuitive tools such as brainstorming).

Let us bring closer the constraints of our context with the first two categories of tools. It seems difficult to initially use these tools in a problem-solving process for two principal reasons:
- either relevant information is not recorded because it had not been identified like such (e.g.: new defect),
- either relevant information was traced but it is difficult to decide whether the observations are significant or not because of the small number of occurrences of defects and the limited number of possible experiments.

Concerning the last category of tools, the two major orientations in methods quoted previously are found. On the one hand, some methods [Hosotani 1997] encourage their use, especially to list the possible causes and the solutions to be considered. On the other hand, other methods completely refute their use like Shainin's one which assures that only the parts must talk (rather than engineers) [Shainin 1995]. In our context, we saw that it would also be difficult to start a problem-solving process with tools utilizing people's opinion given the complexity products and knowledge parcelling.

In fact, these three categories of traditional tools could be used later in the problem-solving process, i.e. when the quantity of data is more important and the exploratory field will be more targeted. It is thus interesting to study other fields which could bring us a methodological help for this first part of the problem-solving.

3 – Others fields methodologies and problem-solving process

Are there other fields that present similar constraints to the particular studied industrial context? By pushing to the extreme these constraints, we have to find contexts for which:
- the problem appeared only once,
- all the relevant data are not necessarily recorded,
- the search for causes does not allow any experiment,
- nobody has the complete knowledge of the defected object.

We could thus transpose these atypical contexts analysis to the industrial field. Two fields meet these criteria: incident investigation and criminal investigation (considered as "the causes of crime science").

3.1 – Incident investigation
[Livingston and al 2001]

The incident investigation process includes four main phases:
- On-site investigation, event sequence and circumstances description,
- Root causes determination,
- Recommendation development and report writing,
- Implementation, follow-up and monitoring.

The first phase involves obtaining a full objective description. The tools used are diagrams of sequences. They describe events, conditions, factors, actors... Different drawing codes are possible according to the presence of evidence or not. The construction of this diagram will only be a skeleton of the final product but it will ensure that valuable information available is not forgotten or lost during the investigation. It will progressively be modified and be updated as more information is gathered. These diagrams have the following advantages: they allow one to identify where gaps in the understanding of events chains lie, to emphasize obvious clues, to identify the causal factors.

The second phase is split up into two stages:
- Identification of critical events and actions (and thus of the direct causes). The tools used act as a filter to reduce the number of possible direct causes.
- Root causes identification (underlying direct causes). The main types of tools used are tree techniques and checklist methods.

A parallel process is carried out whose objectives are: investigation process review, points of lessons to remember and process improvement.

3.2 – Criminal investigation
[Prévost 1988] [Chartrand 2004]

Criminal investigation is a particular process searching for facts that starts from a certain number of clues and assumptions relative to a noted crime and that leads to the identification of a suspect and the gathering of evidence to sentence him. It is thus located at the origin of the penal justice system, because it precedes the legal phase then the magistrate's court phase.

The investigation is a rational and rigorous process that can be split up into the two following stages:
- Preliminary observations on the scene of the crime, commission of a crime declaration, identification of the victim, witnesses and suspects
- Gathering of the first statements, material traces gathering, processing and analysis, questioning, confrontations, reconstruction of the crime, construction of assumptions, investigation and checks, layout of the charge with the substitute of the Attorney General and testimony at the court.

One of techniques used for searching evidence is the criminal "modus operandi" (way of operating). Each criminal, by habit, by physical or psychological needs, indeed develops his own way of operating which ends up characterizing himself.

Greenwood and his collaborators works [Greenwood and al 1975] clearly showed that the majority of crimes solved by the police services are elucidated during the moments that just follow the perpetration of the crime. Other previous research had already highlighted that crime solving was more indebted to the information collected at the preliminary investigation than to the efforts made thereafter. The Stanford Research Institute (SRI) proposed consequently in 1973 [Greenberg and al 1973] a system to filter burglary investigations, based on mathematical weighting for various factors which can help solve this category of crime (filtering form of events). Thus, the investigation will continue only if it is a major crime or if the elements of evidence, in the case of a minor crime, are enough to believe in a possible solution.

The distribution of roles is perfectly established in the criminal investigation process.

Generally, an ordinary witness can only testify on facts of which he personally had knowledge.

Specialists in the technical and scientific police force have greater latitude especially for opinion testimony. They can, in certain conditions, draw conclusions and deliver their opinion. They do not investigate; they just bring a particular light on some aspects of the investigation.

The role of the expert is to draw the conclusions, to express his opinion (in front of the judge). The expert can only testify on a question which is a matter of his speciality and after being recognized expert by the judge.

3.3 – Parallel between these methodologies and the traditional problem-solving process

The following table presents the parallel between two methodologies (incident investigation and criminal investigation):

Problem-solving process	Incident investigation	Criminal investigation
	On-site investigation	Inspection
Definition of the problem	Event sequence and circumstances description	Preliminary inquire. Commission of a crime declaration.
Definition of the measurement		Filtering questioning of events
Causes analysis	Root causes determination	Clues collection and analysis, material proofs, interrogations, construction of assumptions...
Solutions analysis	Recommendation development and report writing	Legal phase
Solution implementation	Implementation, follow-up and monitoring	Magistrate's court phase
Standardisation		

Table 1: Parallel between problem-solving process, incident investigation and criminal investigation

The splitting up into phases between the three methodologies is quite similar; some tools used are common between the methods. It therefore appears quite relevant to use these two methods to complete traditional problem-solving.

4 - Contribution of other methodologies for problem-solving

We comment here on the parallel between methodologies seen in the previous part to highlight the possible contributions.

The analysis of table 1 shows differences on the one hand in terms of methodology and on the other hand in term of tools (little or not used in industry).

4.1 – Contribution in term of methodology

From a methodological point of view, firstly, an initial stage is additional in the two other processes: the investigation in the place where the problem was discovered. This first investigation particularly completes the problem definition to obtain a real description by collecting a maximum amount of information as soon as possible (material and testimonies). It thus enables one to overcome possible non-recording of relevant information and to visualize obvious clues on the site.

Secondly, the problem is seen with sequential vision in incident investigation as opposed to a "black box" vision, which is particularly interesting for the causes analysis. Indeed, in a traditional problem-solving process, the system is generally considered as a "black box", the aim is to find the input factors x_i which act on the output y [Pillet 2003]. Therefore, we have to find the function $y = f(x_i)$. The vision of incident investigation adds a notion of time: the sequencing of the x_i also has an influence on the output y. Figures 1 and 2 illustrate this difference in vision.

Finally, the distribution of roles in criminal investigation allows one to obtain a happy medium between the two "schools"(as mentioned in part 2.2): to primarily base solving on people's opinion (Japanese methods) and to only make the product talk (Shainin method). Indeed, witnesses and experts are perfectly distinguished concerning opinion testimony. Furthermore, the judge checks that proof of guilt is acceptable. The parallel with problem-solving process is easy. The testimony of the people who were present when the incident occurred (operators for example in the case of a production problem) is essential in order to collect all the facts relating to the appearance of the defect. The experienced operators or the specialized technicians in this field (experts) can give their opinions on these facts but do not necessarily take part in the entire problem-solving process. A third party person who is not implicated in the problem-solving could then play the part of the judge, i.e. validate or invalidate the detected cause.

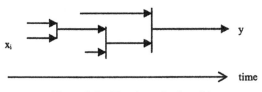

Figure 1: Traditional problem-solving vision

Figure 2: Incident investigation vision

4.2 – Contribution in term of tools

Others interesting contributions are to be noted in terms of tools.

The first technique that can be used is the criminal investigation one: the "modus operandi" (way of operating). Indeed, in industry, the study of the defect signature contributes to learning a lot about the nature of the real cause or even directly on the cause. For example, a single isolated defective pixel in a digital image is more likely to be due to defective photodiode than a particle above the diode that would generally affects the neighbouring pixels.

Then, the use of the filtering form of events mentioned in part 3.2 (criminal investigation) seems very useful to determine priorities for problem processing. Indeed, in the context of new high technology products, new problems are very frequent and it is impossible to solve them all due to the limited resources but also because of the difficulty in finding relevant information. Some problem-solving methods such as Kepner Tregoe already propose tools to determine priorities for problem processing according to gravity, urgency and trend of the defect [Kepner and al 1980]. However none takes into account this notion of relevant information availability. We propose (see table 2) an adaptation of the criteria used for criminal investigation in order to use it for problem-solving in our particular context.

Criteria for criminal investigation	Criteria for problem-solving	Mark
Crime with violence - in all the cases	Finished- product scrap - in all the cases	10
Past approximate period - 0 to 1 hour - 1 to 12 hours - 12 to 12 hours	Past approximate period - 0 to 6 hours - 6 to 12 hours - 24 to 48 hours	4 3 1
Identification of the suspects - clearly identified - known - already seen - complete description - description of the face - description of clothing	Identification of the causes - clearly identified - known - defect already seen - description of type of causes - description of great family of causes	10 7 5 7 3 3
Mode of moving - registration number - color - mark/model - particular clue	Mode of appearance - identified precisely - type of identified mode - - particular clue	10 2 2 2
Witness name or report - witness at the scene of the crime	Witness name or report - witness at the scene of the event	10
Technical proof - fingerprint - print of step - print of tyre	Technical proof - modus operandi known - modus operandi identified	7 5 5

Table 2: Parallel between the criteria and the notation of the filtering form of events

The third tool selected is usually used for incident investigation: diagram of sequence. It already enabled one to have an objective description of the problem and circumstances with a sequential vision for the causes search. It is also an excellent aid of centralisation and organisation for all information collected during the solving.

Lastly, identification techniques of critical events and tree techniques for incident investigation can provide a very useful framework for causes search. Indeed, the highlighting of the critical events allows one to identify the direct causes, and then to go back to the root causes. To draw a parallel with the industrial world, we thus propose to proceed in two steps: identify the critical process stages for the problem concerned in order to identify the direct causes (e.g.: product AAA missing) then go back to the root causes (e.g.: sensor of the product AAA level).

4.3 - Summary

The following table recapitulates contributions previously mentioned placing them in the DMAIC approach of Six Sigma problem-solving: Define, Measure, Analyze, Improve, Control [Harry and al 2000].

The filtering form of events thus allows one to determine if it is relevant to continue or not with the problem-solving. We will show an industrial application of these principles and tools real in the following part.

Stages	Contribution in terms of methodology and tools
Define	Preliminary inquire on the site (collection of data, testimonies) Outline of a diagram of sequence
Measure	Filtering form of events
Analyze	Diagram of sequence completed Techniques of identification of the critical events, tree techniques Intervention of the "judge" for final validation
Improve Control	

Table 3: Contribution in terms of methodology and tools in the DMAIC approach

5 - Industrial application

Let us illustrate our proposal by problem-solving carried out at the Trixell company[1].

---

[1] Trixell is a joint-venture company formed by Thales Electron Devices, Philips Medical Systems and Siemens Medical Solutions committed to the development and production of a complete family of X-ray flat panel digital detectors for the entire radiological imaging industry.

Foot-note: for reasons of confidentiality, we will mask some information by generic names.

Define

At the end of the year n, three consequent captors of the same type XYZ in the production cycle were rejected for the same signature that had never been seen until now. Then, the defect seemed to have disappeared.

Considering the low number of occurrences of the defect, it was difficult to consider the slightest use of statistical tools. However many people had an (different...) opinion on the question. We thus chose to employ the distribution of roles used for criminal investigation in order to distinguish facts, lame opinions and expert opinions. The interrogation of the witnesses (operators) made it possible to notice the abnormal absence of product AAA for a manufacturing step corresponding apparently to this period. The level sensor was out of order. This information is not traced. The product AAA had been replaced afterwards. But, according to experts, this did not explain the creation of this defect in particular. Moreover, the "on-site investigation" took place one week after, so one could not be sure connecting this absence of product to this particular defect.

Measure

With the aids of a questionnaire similar to the form used for the preliminary police investigation (see table 4), it was agreed (according to the total) that the information collected was not sufficient to solve this problem.

Criteria for problem-solving	Note
Finished product scrap - yes	10
Past approximate period - more than 48 hours	0
Identification of the causes - not identified - defect never seen	0
Mode of appearance - no particular clue	0
Witness name or reports - witness at the scene of the event	10
Technical proof - modus operandi unknown and not identified	0
Total	20

Table 4: First application of the filtering form of events

Define

At the beginning of the year n+1, four new scraps for the same defect appeared. The "on-site investigation" (in workshop) allowed them early on to note that once again the product AAA was missing...

Measure

A new iteration of the questionnaire was thus outlined in order to determine if this time the chances of success for solving were greater. Considering the result of the new notation (see table 5), "the investigator" decided to start a problem-solving process for this defect.

Criteria for problem-solving	Note
Finished product scrap - yes	10
Past approximate period - 0 to 6 hours	4
Identification of the causes - defect already seen	5
Mode of appearance - particular clue	2
Witness name or report - witness of the scene of the event	10
Technical proof - modus operandi unknown and not identified	0
Total	31

Table 5: Second application of the events filtering form

Analyze

A diagram of sequence of the operations was thus established to help experts explain this defect (see figure 3).

One operation of stage 1 weakened in fact a part of captor XYZ and made it sensitive to the absence of product AAA.

On the other hand, if it is possible to permute the two stages, the defect does not appear. We note here the great importance of the sequential vision of the x_i (compared to the "black box" traditional vision).

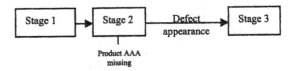

Figure 3: Part of manufacturing process

The search for causes being targeted and validated by the experts, it is possible then to verify it by traditional statistical tools (design of experiments). The "judge" thus recognised the guilt of the operation of stage 1 ...

5 – Conclusion

The problem-solving process in the context of new high technology products actually looks like investigators work (for crimes or incidents) a lot, it is thus natural to use their methods and tools.

We saw that the parallel between the various approaches (problem-solving, criminal investigation, incident investigation) reveals strong similarities, but also differences which are useful to exploit.

This work must still be completed, especially for defining the various roles (investigator, judge, experts, witnesses) better and how to adapt them to industrial people.

Finally, this work of transposition of methodologies from one field to another could be generalised in the other particular industrial contexts. We could adapt for example medicine methods of diagnosis in the context of unit-manufacture such as aerospace.

References:

[Chartrand 2004] Chartrand R., plan de cours SIP 2030 : Organisation de l'enquête, Ecole de criminologie, Université de Montréal, 2004.

[De Mast 2003] De Mast J., A methodological comparison of three strategies for quality improvement, International Journal of Quality and Reliability Management, Vol. 21, No. 2, p. 198-213, 2004.

[Greenberg and al 1973] Greenberg B., Enhancement of the investigative function: analysis conclusions, Menlo Park (California), Stanford Research Institute, 1973.

[Greenwood 1975] Greenwood P.W., Chaiken J., Petersilia J., Prusoff L., The criminal investigation process, Vol. III: Observations and analysis, Santa Monica (California), RAND, 1975.

[Harry and al 2000] Harry N., Schroeder R., Six Sigma: the breakthrough management strategy revolutionizing the world's top corporations, Editions Currency Doubleday, New York, 2000.

[Hosotani 1997] Hosotani K., Le guide qualité de résolution de problème : Le secret de l'efficacité japonaise, Editions Dunod, 1997.

[Jayram and al 1997] Jayram J., Handfield R., Ghosh S., The Application Of Quality Tools In Achieving Quality Attributes And Strategies, Quality Management Journal, No. 1, p. 75-100, 1997.

[Kepner and al 1980] Kepner C.H., Tregoe B.B., Le manager rationnel : méthodes d'analyse de problèmes et de prise de décision, Editions d'Organisation, 1980.

[Prévost 1988] Prévost L., Enquête criminelle, Editions Modulo Editeur, 1988.

[Livingston and al 2001] Livingston A.D., Jackson G., Priestley K, Root causes analysis: Literature review, WS Atkins Consultants Ltd for the Health and Safety Executive, Contract research report 325/2001, 2001.

[Pillet 2003] Pillet M, Six Sigma : comment l'appliquer, Chapitre 3 : étape 1 Définir, Editions d'Organisation, 2003.

[Shainin 1995] Shainin R.D., A common sense approach to Quality Management, 49th Annual Quality Congress Transactions, May 1995, Cincinnati OH, Vol. 0, No. 0, pp 1163-1169, May 1995.

ELSEVIER

IFAC
PUBLICATIONS
www.elsevier.com/locate/ifac

CALCULATION OF THE DEPENDANT DEMAND: THE CASE OF BATCH PROCESS INDUSTRIES

Frédéric David, Nicolas Gayot, Christophe Caux and Henri Pierreval

Equipe de recherche en systèmes de production de l'IFMA
LIMOS UMR CNRS 6158
IFMA – Campus des Cézeaux – BP 265
F-63175 Aubière – France

Abstract: In batch process industries it is often possible to manufacture several different kinds of finished products from a single upstream product. The consequence is that the relevance of classical dependant demand calculation (e. g., use of MRP concepts) may appear questionable. However, the calculation of the requirements in such a case has not (to the authors' knowledge) been addressed in the related literature. The article aims at proposing alternatives to compute the requirements in the case of a batch process industry, this in order to minimize the demand for upstream products. We identify specific features of batch process industries that are relevant regarding the problem of the dependant demand calculation. We formalise the problem. We propose heuristics methods to compute the dependant demand. We conclude on the benefits that can be expected from such approaches. Research perspectives are proposed.

Keywords: manufacturing systems, planning, process industry, optimization problem, heuristics.

1. INTRODUCTION

Classical requirements calculation strategies, such as those used by MRP systems, have been designed for discrete assembly industries. Typically, in discrete assembly processes, the dependant demand calculation is done using the concept of Bills Of Materials (BOM). BOMs enumerate the subcomponents that are required to manufacture each component. BOMs are described through a Directed Acyclic Graph. Each node of this graph represents either the final product, components or raw materials. It has a tree structure that converges toward the ending node, which is associated with the final product. The valuation of an edge between two nodes specifies the quantity of components that is required to manufacture the final product, and the time needed to produce them.

The production processes of batch process industries are very different from those of discrete assemble industries. Several times during the transformation process, the products can be split into different parts. As a result, the material flow appears to be divergent rather than convergent. It means that, from a particular product that has to be transformed, it is possible to manufacture simultaneously several

different finished products required by different work orders. For example, in metal industries, a slab can be cut into two parts, which will be rolled separately to produce several different kinds of sheets. This divergence of the material flow is very different from the classical description of discrete assembly processes. Figure 1 compares the divergent nature of the production process in a batch process industry to classical assembly processes. This important difference causes problems if the production management strategies rely on the concept of components and cannot take into account the divergence of the material flow. Indeed, David et al. (2003) have shown that the dependant demand calculation performed by an MRP based system can lead to such consequences as overproduction or increase of scrap and inventories in the case of a batch process industry.

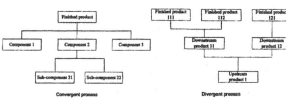

Fig. 1. Classical convergent process and divergent process of batch process industries

Classical requirements calculation strategies are not suited to compute the requirements in the case of batch process industries. Alternatives must be considered. As highlighted by the literature review of Potts and Kovalyov (2000), numerous research articles deal with the problems of scheduling in the case of batch processing. However, to the best of our knowledge, the problem of the dependant demand calculation in the case of batch process industries has not been addressed in the relevant literature. Therefore, the aim of this article is to propose alternatives to compute the requirements in the case of a batch process industry, this in order to minimize the demand for upstream products.

The article is organised as follows. First, we recall the characteristics of batch process industries that are relevant regarding the calculation of the dependant demand. Second, we address the computation of the requirements as an optimization problem, aiming at minimising the consumption of upstream products, under the constraints of meeting the combination rules that exist among products. Third, we propose heuristic approaches to solve this problem. Fourth, we present the application of those approaches to the case of an Aluminium Conversion factory. Fifth, we conclude on the benefits that can be expected from those approaches and outline research directions.

2. DIVERGENT PRODUCTION PROCESS AND ITS IMPACT ON THE REQUIREMENTS CALCULATION

The divergence of the material flow allows several downstream products to be manufactured simultaneously from one upstream product. The consequence is that there exist several possible choices regarding which downstream products will be produced from one upstream product. For example, one upstream product X can be used in full to obtain one kind of downstream product A, or used partially to obtain this downstream product A and partially to obtain another one B. This induces flexibility on the choice of the products that will be derived from one upstream product.

However, there are constraints that limit this flexibility. Indeed, it is not possible to manufacture any kind of downstream product from a particular upstream product. In fact, the downstream products must comply with metallurgical and process rules. Such rules, called combination rules, are based on the downstream products characteristics (such as alloy, thickness, width, quality or norms). They are used to identify the products that can be processed together in the earliest stages of the production process and differentiated later. Let us consider a simplified example. In a rolling facility, two sheets of different thickness cannot be produced from a single plate, as they cannot be rolled simultaneously, even if they require the use of the same type of plate as upstream material. Those combination rules result in a set of alternatives, which is illustrated in Figure

2. Ventola (1991) describes such combination rules in the case of Aluminium Conversion.

The flexibility that exists on the choice of the products that will be derived from one upstream product is commonly used. Indeed, many batch process industries produce according to a make-to-order (MTO) strategy. In a MTO strategy, the customer's requirements may not match the largest quantity of finished product that can be generated from the upstream product required. Therefore some remaining material may be generated. This remaining material can be used for producing more products than what is required by the customer. This overproduction can be stocked, and will hopefully be sold later. However, in many factories, there is a high product diversity. The consequence is that most of the products are not demanded repetitively. Therefore, stocking products may not be relevant. The remaining material can also be scraped, what is costly. An efficient way of using this remaining material is to use it to fulfil the requirements of other work orders. Such a solution is possible thanks to the flexibility that exist on the choice of the products that will be derived from one upstream product.

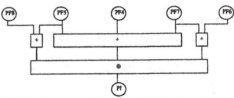

Fig. 2. Choices regarding the finished products (PF) obtained from a single upstream product (PI)

3. DESCRIPTION OF THE PROBLEM

The problem of the requirement calculation in the case of a batch process industry aims at minimizing the demand for upstream products (which is equivalent to minimising the quantity of remaining material) by using the flexibility that exists regarding the products that can be derived from an upstream product. This has to be done according to the requirements expressed in the following.

First, the products that would be obtained simultaneously from a single upstream product have to comply with the combination rules introduced in section 2.

Second, the whole range of planned orders has to be taken into account during the calculation. Indeed, some planned orders may not generate any remaining material. However, in batch process industries, there often exists orders and shipping tolerances. This means that it is allowed to deliver to the customer a quantity of products that can be different from the quantity ordered, this within a specified tolerance. The aim of such flexibility is not to optimize the demand of upstream products. It is allowed due to the fact that there exist uncertainty in the production process (i.e. it prevents re-launching a whole batch for a few quantity of products missing). However, this flexibility can be partially used to minimize the demand on upstream products: by producing less than expected for an order, it is possible to generate some remaining material that can be used to produce another product (which will no longer require additional upstream material). Practically, in most cases, it is not possible to decrease the planned quantities to be produced from more than 1~2%.

Third, we consider that once a planned order has been turned into a production order and launched on the shop floor, the requirements of this order cannot been changed. This is necessary both for organisational reasons (e.g. orders tracking, production of upstream material launched, etc.) and for technical reasons (the combination rules would have to be modified once the production has been started).

Fourth, we consider that the quantity of upstream products that is required to produce simultaneously various products cannot exceed a fixed quantity. This is necessary for organisational reasons, such as orders tracking, and management of planned orders.

4. PROBLEM FORMULATION

Let OF_1, OF_2, ..., OF_n be a set of N planned orders for finished products. Let PF_i $(1 \leq i \leq N)$ be the finished products to be manufactured from those orders. Let QOF_i $(1 \leq i \leq N)$ be the amount of products required for each of those products, and ΔQOF_i $(1 \leq i \leq N)$ the tolerance that exists regarding the quantity to be delivered. Let PA_i $(1 \leq i \leq N)$ be the upstream product required for the production of the finished product PF_i. Let QPA_i $(1 \leq i \leq N)$ be the quantity of finished products PF_i that can be manufactured from the transformation of the upstream product PA_i.

The requirement calculation problem consists in partitioning $\{OF_i\}$ into M $(1 \leq M \leq N)$ subsets G_j $(1 \leq j \leq M)$. Let G_j $(1 \leq j \leq M)$ be a partition of $\{OF_i\}$. We assume that the requirements for upstream products of G_j $(1 \leq j \leq M)$ are characterised by the upstream product that has to be used $PA_j (1 \leq j \leq M)$, and the quantity required, BG_j. *The objective is to determine the partition of* $\{OF_i\}$ *that minimize the requirements of upstream products,* $\sum_{j=1}^{M} BG_j$.

This has to be done according to the constraint that all the products within a subset can be produced together, in accordance with the combination rules. We assume that those combination rules are known. Let PF_a and PF_b $(1 \leq a \leq N$ et $1 \leq b \leq N$ et $a \neq b)$ be two finished products, let Y_{ab} be a binary variable so that:

Y_{ab} = 1 if PF_a and PF_b can be produced simultaneously from a single upstream product
       = 0 otherwise

As all the products within a subset have to comply with the combination rules, a partition G_j $(1 \leq j \leq M)$ of $\{OF_i\}$ verify:

$$\forall OF_a \in G_j, \ \forall OF_b \in G_j, \ Y_{ab} = 1 \qquad (1)$$

Another constraints is that a subset contains at least one order for which the requirements do not match the largest quantity of finished product that can be generated from the upstream product required, which can be expressed as follows:

$$\forall G_j / \dim(G_j) > 1, \exists OF_i \in G_j / \left(\frac{QOF_i - \Delta QOF_i}{QPA_i} \right) \succ \left(E(\frac{QOF_i - \Delta QOF_i}{QPA_i}) \right) \qquad (2)$$

where $E(x)$ represent the integer value of x.

In addition, the quantity of upstream products that is required by a subset cannot exceed a fixed quantity. Let QM be this quantity. This constraint can then be expressed as follows:

$$\forall G_j (1 \leq j \leq M) \ BG_j \leq QM \times QPA_j \qquad (3)$$

The requirements for upstream products of a subset G_j $(1 \leq j \leq M)$, BG_j, can then be calculated as follows:

$$BG_j = QP_j \times \left(ASUP(\frac{\sum_{k/X_{kj}=1}(QOF_i - \Delta QOF_i)}{QPA_j}) \right) \qquad (4)$$

where $ASUP(x/y)$ is used to return the next highest integer value of x/y.

Let X_{ij} be a decision variable:

X_{ij} = 1 if the planned order OF_i belongs to the subset G_j
       = 0 otherwise

Then, from (1):

$$\forall a(1 \leq a \leq N) \ \forall b(1 \leq b \leq N) \ \forall j(1 \leq j \leq M)$$
$$\left((X_{aj} + X_{bj}) \times \overline{y_{ab}} \right) \in \{0; 1\} \qquad (5)$$

In addition, from (4):

$$\forall j(1 \leq j \leq M)$$
$$\sum_{a=1}^{N} X_{aj} \times \left(\frac{QOF_a - \Delta QOF_a}{QPA_a} \right) \succ \sum_{a=1}^{N} X_{aj} \times \left(E(\frac{QOF_a - \Delta QOF_a}{QPA_a}) \right) \qquad (6)$$

Last but not least, it is necessary that all the customers' requirements are fulfilled, which necessitates that no plan orders are suppressed:

$$\forall OF_i \ (1 \leq i \leq N) \ \sum_{j=1}^{M} X_{ij} = 1 \qquad (7)$$

The problem is then formulated as follows:

$$
\begin{cases}
\text{Minimise} & \sum_{i=1}^{M} BG \\
\text{S.t.,} & \sum_{a=1}^{N} X_{aj} \times \left(\dfrac{QOF_a - \Delta QOF_a}{QPA_a} \right) \succ \\
 & \sum_{a=1}^{N} X_{aj} \times \left(E\!\left(\dfrac{QOF_a - \Delta QOF_a}{QPA_a} \right) \right), \ 1 \le j \le M \quad (8) \\
\text{and} & BG \le 5 \times QPA, \ 1 \le i \le M \\
\text{and} & \sum_{j=1}^{M} X_{ij} = 1, \ 1 \le i \le N \\
\text{and} & \left((X_{aj} + X_{bj}) \times \overline{y_{ab}} \right) \in \{0;1\}, \ 1 \le a \le N, \ 1 \le b \le N, \ 1 \le j \le M \\
\text{and} & X_{ij} \in \{0,1\}, \ 1 \le i \le N, \ 1 \le j \le M
\end{cases}
$$

5. PROPOSED SOLUTIONS

In order to solve complex optimisation problems, heuristics and metaheuristics are commonly used (Pirlot et Teghem, 2003; Hao et al., 2000). There exist several heuristics alternatives to solve such problem (Hao et al., 2000; Dear et Drezner, 2000; Van Breedam, 2001).

We have considered two heuristics. First, we use an heuristic, dedicated to this problem, that we have defined. Second, we use simulated annealing, which was first introduced by Kirkpatrick et al. (1983).

5.1 Dedicated heuristic

The principle of the dedicated heuristic we propose can be described as follows. The initial solution is the particular partition of {OF$_i$} in which each planned order constitutes a subset. Planned orders are then treated one after each other. For each planned order is tested the possibility to integrate it to an existing subset (that contains one or several orders). The most performing solution regarding our objective of minimising the upstream products requirements is selected. For equally performing solutions, the solution selected will be the one for which the fewest orders have to be manufactured together at the earliest stages of the production process. This cycle keeps going until it reaches the last planned order. The procedure can then be described as follows.

Step 1. The initial solution is defined as the particular partition in which each planned order constitutes a subset. The initial solution is the current solution.
Step 2. Select the first order from the list.
Step 3. Select the first subset from the list.
Step 4. Move the selected order to the selected subset.
Step 5. If the move is infeasible or does not improve the objective function value, then go to step 7.
Step 6. If neighbourhood solution is better than all previous solutions, then save the neighbourhood solution and go to step 7.
Step 7. If an entire evaluation cycle has not yet been completed for the order, which means that all the possible moves to subset have not yet been tested,

select the next subset. Go to step 4.
If an entire evaluation cycle has been completed for the order and the best neighbourhood solution has been saved, then this neighbourhood solution becomes the current solution. Select the next order. Go to step 3. If no improved neighbourhood solution has been saved during a complete evaluation cycle, then proceed with the next order. Go to step 3.

5.2 Simulated annealing

Principle. The principle of the SA metaheuristic is deduced from the physical annealing process of metals. It starts with a high temperature, gradually decreasing, and ends at cold temperatures. At each iteration a perturbation of the current solution is randomly generated. If the perturbation has a better objective function value, it is selected as the next current solution. If the selected perturbation does not provide a better result than the current one, it is selected with a probability depending both on the temperature and the two objective function values. At the beginning of the SA process, the temperature is high, and almost every non-improving perturbation is accepted. As the temperature goes down, the probability of selecting a non-improving perturbation decreases. At the end of the SA process, when the temperature is low, it is almost impossible to select a non-improving neighbour.

Implementation choices. Various implementations of SA are proposed in the literature. For our implementation, the following choices have been made.

Coding of the solution. A subset of n orders is represented by a vector $V(i)$ $(1 \le i \le n)$ where $v(i)$ indicate the number of the order labelled i in the subset. A bunch of subsets represents a solution.

Initial solution. The initial solution used is defined either as the worst solution (one subset per order) or as a good initial solution (resulting from the dedicated heuristic)

Neighbourhood. The heuristic generates a neighbourhood solution on a stochastic base. This implies that the orders selected to perform a move are selected at random. It is assumed that only feasible moves with respect to the combination rules are performed. Three types of moves are considered. They are respectively called the insertion, ejection and permutation moves.

- Insertion. The insertion move can be described as the move of an order from a randomly chosen subset to another (different) randomly chosen subset. This type of move is able to decrease the number of subsets.
- Ejection. The ejection moves generates a neighbourhood solution by creating a new subset containing a unique order taken away randomly from an existing subset. This type of move is able to increase the number of subsets.

- *Permutation*. The permutation move generates a neighbourhood solution by exchanging two orders between two subsets.

SA can be used only if the neighbourhood rules are such that the set of solutions is connex. Given the permutation moves we define, let us consider S^* the solution defined as the one for which there is one subset per order. Let us consider S and S' two randomly chosen solutions. By using a finite number of ejection moves, it is possible to pass from S to S^*. By using a finite number of insertion moves, it is possible to pass from S^* to S'. All the intermediary solutions are admissible, as they are less complex than S and S' that verify the problem constraints. Therefore, the connexity of the set of solutions is verified.

Cooling function. The cooling function we use for the reduction of the temperature is a simple geometric function. The temperature at iteration x, T_x, is obtained from the temperature of the previous iteration as $T_x = T_x.R$, where R represents the cooling rate. The number of iteration per step of temperature, K, specifies how many neighbour solutions will be tested for a given temperature. After this number of iteration is reached, the temperature will cool down.

Final solution. SA does not make any difference between two solutions that are equally performing regarding the objective function. Let us consider a set of 3 orders that verify the compatibility rules. A solution based on a partition in one subset, containing all of the 3 orders, and that consume 2 upstream products, is as performing as a solution proposing a partition in 2 subsets, one of them containing 1 order and the other one 2, each of those subsets consuming 1 upstream product. However, for such reasons as traceability, it is more difficult to manage products that have to be manufactured together at the earliest stages of the production process. In order to provide solutions that can be more easily used on the shop floor, the best solution provided by the SA process is modified in order to minimise the amount of orders that are planned to be produced along with others. In order to do so, ejection moves are tested on all the subsets obtained. If there is no difference in objective function value between the neighbour solution and the best solution, the neighbour solution (for which less orders have to be manufactured along with others) is conserved.

6. CASE STUDY AND DISCUSSION

6.1 Experimental parameters of the simulated annealing process.

The SA heuristics requires several parameters to be tuned, which are in our case study chosen as follows.

Initial temperature. The initial temperature is defined as a parameter that can be tuned. Values tested for this parameter are either 2000 (High temperature),

1000, 250, 20 or 1 (Cold temperature).

Cooling rate. The cooling rate is the fraction by which the temperature is reduced in the geometric temperature function. Values assigned to this parameter are either 0.99 (slow cooling), 0.95 or 0,90 (fast cooling).

Iteration per temperature step. Values assigned to the number of iteration per step of temperature, K, are either 100, 50 or 25.

Stopping criterion. The stopping criterion is satisfied if the value of the temperature is lower than a given parameter, ε. Values assigned to this parameter are either 0.1, 0.05 or 0.01.

6.2 Computational results

The proposed approaches have been tested and implemented in an Aluminium Conversion factory. This factory produces aluminium sheets. The amount of planned orders to be treated is around 2000 per run. Around 40% of those orders are so that the quantities of finished products needed do not match the largest quantity of finished product that can be generated from the upstream product initially required.

The dedicated heuristic provides interesting results. For a sample set of around 300 orders, compared to the requirements calculation performed by a classical MRP system, there is an average economy of 4% of metal (around 100 tons). Around 30% of the orders that may have initially generate remaining material are planned to be produced along with others. The calculation takes an average of half an hour for a whole set of orders on the machine used.

SA provides even better results than the heuristic we proposed. Figure 3 illustrates the evolution of the SA process on a sample of orders for different settings of the SA parameters. From the tests we have carried out, the quality of the initial solution seems to be important. Good initial solutions, as provided by the dedicated heuristic, do not always produce a significantly better final solution. However, with a good initial solution, it is possible to start with lower values of the initial temperature and to decrease the computing time. The results also reveal that higher temperatures do not provide significantly better solutions. Although it is not possible to obtain efficient solutions by starting with the worst solution as initial solution and low values of the initial temperature, the results that are obtained with an initial temperature of 2000 is not significantly better than those obtained with a value of 250. Last but not least, slow cooling does not provide significantly better results. However, slower cooling increases the computing times. Finally, the best performance has been found for low values of the temperature, and with the result of the dedicated heuristic as initial solution. For a sample set of around 300 orders, there is an average saving of 6 to 7% of metal.

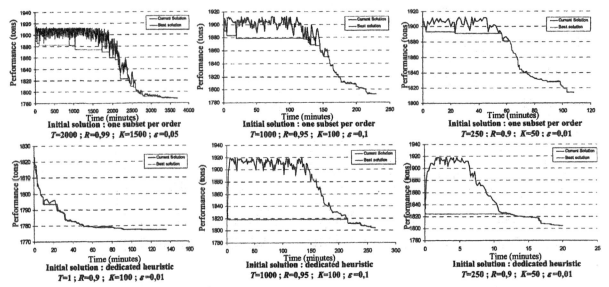

Fig. 3. Examples of the evolution of the simulated annealing process

Around 50% of the orders that may have initially generated remaining material are planned to be produced along with others. However, run lengths are long (usually more than a day for a whole set of order on the machine used).

7. CONCLUSIONS AND PERSPECTIVES

We present alternatives to compute the requirements in the case of a batch process industry in order to minimize the demand for upstream products. In order to do so, we present two heuristics approaches to solve the problem of the dependant demand calculation. Results from a case study show that, compared to a classical MRP approach, the consumption of upstream products slightly decreases, which may result in significant financial savings (those gains have not been detailed here for confidentiality reasons). However, the calculation length remains long.

Several research directions would need to be considered to better cope with the problem that has been highlighted in this article, which are as follows.

- More efficient approaches could be searched to solve the problem, both to minimise the demand for upstream products and to minimise the duration of the calculation. In this respect, other dedicated heuristics and metaheuristics approaches must be studied,
- The problem we have addressed in this article can probably be extended by taking into account existing stocks of finished products (that can eventually be cut to size to fit new customer requirements) and work in process. This necessitates taking into account new constraints (e.g. quantities required, compatibilities, etc.),
- Coping with the problem presented in this article will necessitate a software module to be developed. This module may have to be included or connected to the production planning software

the company uses. The coherence of this integration and its maintainability is a major concern. Indeed, modifying such systems as ERP is difficult and present industrial risks. Further research would be required to provide more adaptable and flexible systems. Research perspectives include the analysis of the possibilities offered by EAI (Enterprise Application Integration) systems.

REFERENCES

David, F., Pierreval, H., and Caux, C., 2003, ERP for Aluminium Conversion Industry: what is relevant?. *Proceedings of IEPM 2003, 6th International Conference on Industrial Engineering and Production Management*, May 26-28, Porto, Portugal.

Dear, R. G., and Drezner, Z., 2000, Applying combinatorial optimization metaheuristics to the golf scramble problem. *International Transactions in Operational Research*, **7**, pp. 331-347.

Hao, J.-K., Galinier, P., and Habib, M., 2000, Métaheuristiques pour l'optimisation combinatoire et l'affectation sous contraintes. *Revue d'intelligence artificielle*, **13** (2), pp. 283-324.

Kirkpatrick, S., Gelatt, C.D., and Vecchi, M.P., 1983, Optimization by Simulated Annealing. *Science*, **220** (4598), pp. 671-680.

Pirlot, M., and Teghem, J., 2003, *Résolution de problèmes de Recherche Opérationnelle par les métaheuristiques*. Editions Hermès, Paris, France.

Potts, C. N., and Kovalyov, M. Y., 2000, Scheduling with batching: A review. *European Journal of Operational Research*, **120**, pp. 228-249.

Van Breedam, A., 2001, Comparing descent heuristics and metaheuristics for the vehicle routing problem. *Computers & Operations Research*, **28**, pp. 289-315.

Ventola, D.P., 1991, Order Combination Methodology for short-term lot planning at an aluminium rolling facility. *Massachusetts Institute of Technology*, MsC Thesis.

PUBLICATIONS
www.elsevier.com/locate/ifac

Using Real-Time Model-Checking Tools in Agricultural Planning: Application to Livestock Waste Management

Arnaud HÉLIAS[a,b,c], François GUERRIN[a] and Jean-Philippe STEYER[b]

[a] *Organic Waste Management Team, CIRAD, Reunion Island, France*
[b] *Laboratory of Environmental Biotechnology, INRA, Narbonne, France*
[c] *Modelling and Control of Renewable Resources Team, INRIA, Sophia Antipolis,*
France, phone +33 492 387 684, fax +33 492 387 858, arnaud.helias@inria.fr

Abstract: This paper addresses the dynamical representation of a network made of a set of waste production units (*i.e.*, livestock farms) needing to transfer their wastes to a set of consumption units (*i.e.*, crops onto which wastes may be spread over). The dynamics of stocks (taken as continuous fluxes with imprecise parameters) should thus be coupled with management decisions or actions (taken as discrete events). Various temporal constraints determine the possibilities of waste transfers. These constraints, for each production or consumption unit, are modelled as a timed automaton. Possible allocation of wastes is then analysed by using model-checking techniques applied to the global timed automaton resulting from the product of all the elementary timed automata. For this, we used the Kronos software based on the Timed Computational Tree Logic (TCTL). Our approach is illustrated on the functioning of a typical farming system made of livestock and crop enterprises in the context of the Reunion Island.

Keywords: timed automata; model-checking; hybrid dynamical systems; imprecision; livestock waste management.

1. INTRODUCTION

In the Reunion Island, as well as in other regions worldwide, the management of livestock wastes (*i.e.*, the use of liquid or solid manure to fertilise crops) is hampered by the imbalance between the huge production of effluents by intensive livestock farms and too little cultivated area available for spreading. Hence, the problem of waste surpluses cannot be ignored by the farmers due to the increasing pressure of environmental and regulatory constraints. This is the case in the Reunion Island where pig and poultry farming intensely grew during the last 15 years.

With the perspective of farmers' decision support, our aim is to test and devise more accurate management strategies of livestock wastes. For this, we need to model a system made of a set of livestock farms (production units, PUs) that may transfer their production (*i.e.*, wastes) subject to constraints (namely temporal) towards a set of crops (consumption units, CUs). Among other modelling approaches (Guerrin and Paillat, [2003]), we use here timed automata as the modelling formalism of such PUs and CUs, and model-checking techniques to check the feasibility of waste spreading.

This paper is organised as follows: Section 2 gives an overview of the decision-making process in livestock waste management and the formalism of timed automata. Section 3 illustrates the use of timed automata to model the CUs and the PUs and derive a global model of the system. Finally, the use of model-checking to assess the spreading possibilities is presented and discussed in Section 4.

2. SYSTEM DESCRIPTION

2.1 Decision-making in livestock waste management

Some papers dealing with modelling livestock waste management can be found in the literature (see for example Polman and Thijssen [2002], Sorensen et al. [2003]). However, dynamical aspects are very seldom represented in these models even though the spreading decisions must be determined according to the evolution of the biophysical system as underlined by Thornton and Herrero [2001].

Aubry et al. [1998] listed the different aspects to be considered for representing decision-making in an agricultural context. These aspects are the following:

- *Structural parameters.* These parameters account for the overall farm structure determining the management strategies by farmers (*e.g.*, type of crops, size of livestock...).
- *Functioning indicators.* They represent information on the dynamic state of the biophysical system. The decision process is based on these indicators.
- *Decision rules.* A set of rules expressing the technical operations to be performed within specific periods and contexts as well as adaptations to compensate for disturbances (e.g., climatic events).
- *Management variables.* These elements correspond to the possible leverages to act upon the system.

Figure 1 summarises our conceptual view of livestock waste management highlighting stock interactions (i.e., manure storage facilities for the PUs and crop plots for the CUs).

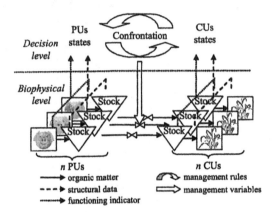

Figure 1. Sketch of livestock waste management.

A decision of spreading is based on the conjunction of two types of indicators according to the type of units (Fig. 1):
- PUs: Level of stocks, whose dynamics taken as continuous are classically modelled by ordinary differential equations based on mass balances.
- CUs: Agricultural events on crops (e.g., harvest, planting...) that are intuitively represented as discrete.

As a consequence, continuous and discrete subsystems must be combined to build an overall model of such a system. A classical approach is to represent the whole system within a unique formalism. To this end, we decided to model the different PUs and CUs using timed automata, one of the most popular timed discrete formalism, and to use associated model-checking tools to answer questions such as "Can this PU supply organic matter to this CU?".

2.2 Modelling methodology: timed automata and model-checking

Introduced by Alur and Dill [1994], a timed automaton is composed of a finite-state machine and a representation of continuous time. This formalism allows one to define temporal constraints using variables named *clocks*. These constraints are associated with the locations or the edges of the automaton. They are of the form $x \# c$ or $x - y \# c$ with x and y two clocks, $c \in \mathbb{Q}$ a constant and $\#$, a relation symbol from the set $\{<, \leq, =, \geq, >\}$.

The value of a clock grows linearly with time. A timed automaton is defined by a 6-tuple $G = (S, P, X, E, A, Inv)$ with:
- S a finite set of locations representing all the possible states of the system;
- P a mapping which associates to each location a set of atomic propositions valid at this location;
- X a finite set of clocks;
- E a finite set of labels;
- A a finite set of edges; each edge a joins two locations and has (i) an associated label, (ii) a clock constraint named *guard* that must be satisfied to trigger the discrete transition and (iii) a clock subset that must be reinitialised during the transition;
- Inv is a map that associates to each location a set of clock constraints, named *invariant*. The system can stay in this location as long as the invariant remains true.

Model-checking (*e.g.*, see Bérard et al. [2001]) is one of the most popular techniques to automatically check the properties of a system expressed as questions (*e.g.* which behaviour is expected from a given initial state?) formulated with a specific language. In the last decade, model-checkers have been extended to deal with real-time representations using the formalism of timed automata.

The timed computational tree logic (TCTL) defined by Henzinger et al. [1994] allows one to introduce temporal information into formulae expressed with the following grammar:
$$\phi := p \mid \neg\phi \mid x - y \# c \mid x \# c \mid \phi_1 \vee \phi_2 \mid \exists \Diamond_{\#c}\phi \mid \forall \Diamond_{\#c}\phi$$
with p an atomic proposition, x and y two clocks, $c \in \mathbb{N}$ a constant and $\#$ a relation symbol from the set $\{<, \leq, =, \geq, >\}$. "$\exists\Diamond$" means *"There exists at least one sequence of locations in which some property holds true"* and "$\forall\Diamond$" means *"For all sequences of locations, some property holds true"*.

Model-checking tools like *Kronos* [Yovine, 1997] have been developed for discrete-event systems.

This model-checker allows one to check TCTL formulae onto one or more timed automata. In our approach, *Kronos* is coupled with the Matlab® environment that we use to simulate the continuous system's components (i.e. stock dynamics).

3. MODELLING CONSUMPTION AND PRODUCTION UNITS

We model the PUs and the CUs with the timed automata formalism and then we use the *Kronos* software to check the possibility of spreading for each (PU, CU) pair.

3.1 Crops: the discrete part

From the CUs' point of view, spreading may occur in three main cases:
1. After a triggering event without fixed delay (*e.g.*, spreading at any time during the month after harvest);
2. After a triggering event with fixed delay (*e.g.*, spreading between one and three months after planting);
3. Previously planned at some period (*e.g.*, spreading between July and March because of the crop cycle constraints).

For each case, a time automaton can be defined. The first case is illustrated in Figure 2: when event e occurs, the system goes from location 1, where no spreading is possible, to location 2 and the clock x is reinitialised. The action can then be performed as long as the clock value remains less than or equal to δ (this is represented by location 3). When the clock value reaches δ the system goes back to location 1.

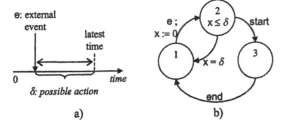

Figure 2. a) Triggering event without fixed delay; b) Resulting timed automaton.

The timed automaton corresponding to the second case is shown in Figure 3. The delay between the triggering event and the earliest time for action is represented by location 2 where the system must stay during δ^s time units.

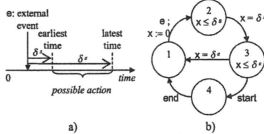

Figure 3. a) Triggering event with fixed delay; b) Resulting timed automaton.

Modelling the third case is illustrated by Figure 4. A clock y is reinitialised at the end of the cycle, when the clock value reaches Π (*e.g.* each year); the system goes then to location b where action is possible during the delay π.

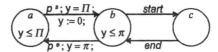

Figure 4. Timed automaton representing *a priori* planned actions.

Such triggering events can be either introduced by (i) one of the timed automata described by the Figures 2, 3 or 4; or (ii) a simple automaton describing a cycling loop with period μ (Figure 5).

Figure 5. Timed automaton modelling a cycling loop with period μ.

The overall spreading possibilities are then checked against the automaton resulting from the product of the various timed automata representing the constraints for each individual CU.

Using these generic timed automata, we modelled the main crops existing in the Reunion Island (*e.g.*, sugar cane, forage crop, market gardening, maize, fruit crops, etc.) with their specific parameters (*e.g.*, planting or harvest dates). For more details, see Hélias et al. [2003a].

For example, the spreading constraints holding for maize are modelled by combining: (i) the automaton of Figure 5 with a period of 182 days (in the Reunion Island, two maize harvests are obtained per year) and (ii) the automaton of Figure 2 with δ = 28 days (spreading should occur within a delay of four weeks after harvest).

A more complex example may be provided by sugar cane for which spreading is subject to the following constraints:

- *The period of harvest cycle is one year.* The automaton of Figure 5 is used with $\mu = 365$ days to generate a triggering event.
- *Spreading is possible within a delay of four weeks after harvest.* The automaton of Figure 2.b is used with $\delta = 28$ days.
- *A second manure application can be performed between one and three months after the first application.* The automaton of Figure 3.b is used with $\delta^s = 30$ and $\delta^e = 90$ days; here the triggering event e is not the harvest but the first manure application.
- *Spreading can occur between July and March next year.* The automaton of Figure 4 is used with $\Pi = 365$ and $\pi = 243$ days.

3.2 Livestock: the continuous part

At the PU level, spreading decisions are made according to stock levels which are generally represented by a mass-balance ODE system with real values. In order to integrate both the PUs' and CUs' representations within a unique formalism we devised an automatic procedure to abstract the continuous system's behaviour into a timed automaton. It proceeds as follows:

- Initial states and input parameters that are imprecisely known (e.g., the rate of waste production by livestock) may be estimated by numerical intervals;
- Using these intervals, two extremal ODE systems are derived from the single ODE system to simulate the lower and upper bounds of the system's behaviour;
- Outputs of simulation performed with the extremal ODE system are then translated into timed automata based on a state-space discretization by the means of thresholds partitioning the state-variable domains according to expert knowledge (e.g., minimal quantity for spreading, maximum stock capacity...);
- These automata may then be combined with other automata (e.g. representing crops) and analysed by model-checking techniques.

To cope with the risk of combinatorial explosion in the number of states, timed automata are purposely built to answer specific questions, according to some initial state and the properties that are to be checked.

From a simulation run of the extremal ODE system on a given time period the time point at which the state variables cross the thresholds are extracted.

The discretized dynamics of this continuous system is then approximated by a timed automaton where:

- a location and a proposition describing its characteristics (i.e., a threshold index and a trend symbolised by Δ or ∇) are assigned to each threshold crossing;
- invariants are defined by the latest time instants at which threshold crossing may occur;
- guards of edges are defined by the earliest time instants at which threshold crossing may occur.

Figure 3 gives a simple example of applying this procedure to a stock of pig slurry represented by the following system:

$$\begin{cases} \dot{x}_1 = p.z_1 \\ \dot{x}_2 = p.z_2 \end{cases} \quad (1)$$

with $p = 10$ breeding sows, x_1: volume of slurry, x_2: nitrogen amount, z_1: slurry production rate (between 0.035 and 0.04 $m^3 sow^{-1} day^{-1}$) and z_2: nitrogen production rate (between 0.015 and 0.018 kg $sow^{-1} day^{-1}$). We assume the initial state may be estimated by:

$x_1(t_0) \in [5 \ \ 15]$ (m^3) and $x_2(t_0) \in [22 \ \ 68]$ (kg).

Three thresholds relevant for the task of slurry management are defined for x_1: 0 m^3 (empty stock), 75 m^3 (minimal amount for spreading) and 100 m^3 (full stock capacity). The timed automaton obtained for x_1 is given in Figure 6.b (location s_0 is the initial state).

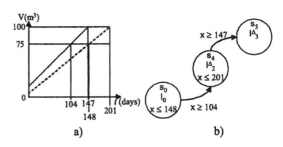

Figure 6. a) Simulation outputs of an extremal ODE system representing the dynamics of a pig slurry stock. b) Corresponding timed automaton.

Details about the full procedure can be found in Hélias [2003b] and Hélias et al. [2004a]; it is illustrated on a 4-dimensional non-linear ODE system in Hélias et al. [2004b].

This section has described the modelling approach to represent each PU or CU in the system. The next step is how to verify the feasibility of spreading for each (PU, CU) pair.

4. LIVESTOCK WASTE MANAGEMENT

4.1 Checking spreading possibilities

For each CU represented by a timed automaton, we assume that (i) at a given initial time, the location $init_{CU}$ and the associated clock values $V(X_{CU})$ are defined (this aspect is discussed in Hélias [2003b]) and (ii) there exists a set (possibly a singleton) of locations corresponding to spreading actions (*e.g.* location 3 of the automaton Figure 3). Let all the locations of this set have the associated proposition act_{CU}. The following TCTL formula:

$$init_{CU} \wedge V(X_{CU}) \Rightarrow \exists\Diamond(act_{CU}) \qquad (2)$$

can be checked using *Kronos* to answer the question: *"From the initial state, is there a sequence which leads to a state where a spreading action is possible?"*. The time instant t_{CU}^{max} at which the formula is true is taken as the latest date of a spreading action.

Similarly, abstracting a PU is performed with two thresholds: l_1, the minimal amount needed for a spreading action and l_2, the maximum stock capacity. By connecting a new location where the property act_{PU} holds true to each location where the stock level allows a spreading action (*i.e.*, locations with the associate property $l_1^{\wedge}$ or $l_2^{\vee}$), the following formulae can be checked:

$$s_0 \Rightarrow \exists\Diamond(act_{PU}) \qquad (3)$$

$$s_0 \Rightarrow \exists\Diamond(l_2^{\wedge}) \qquad (4)$$

Formula (3) checks the possibilities of spreading from the PU's point of view while (4) is related to stock overflowing.

Let t_{PU}^{max} be the earliest date where (4) is true. If there exists a location where the properties act_{PU} or act_{CU} can be reached in the PU and CU automata, then the interaction is checked on the single automaton resulting from the product of both automata by:

$$s_0 \wedge init_{CU} \wedge V(X_{CU}) \Rightarrow \exists\Diamond_{\leq t_{*U}^{max}}(act_{CU} \wedge act_{PU}) \quad (5)$$

Checking spreading possibilities is also performed according to the urgency of spreading. The operator constraint $\leq t_{*U}^{max}$ means that the property must be checked before a stock overflows (*i.e.*, t_{PU}^{max}) or before the latest date of the first occurrence of spreading (*i.e.*, t_{CU}^{max}).

Let n PUs and m CUs. By checking formulae (2) or (4) for all the units, the urgency indicators $t_{PU}^{max}(i)$ and $t_{CU}^{max}(j)$ are calculated ($i = 1,...,n$ and $j = 1,...,m$). The unit with the smallest urgency indicator (*i.e.*, having the closest deadline for spreading) should interact with another unit (i.e. try to spread or to be spread over). Based on expert priority rules, an ordered set associated to each unit lists the possible interactions with other units (*e.g.*, for a PU it provides the list of crops than may receive the kind of waste it produces ranked by preference order).

When (5) is verified for a (PU_i, CU_j) pair, spreading is performed: the ODE system describing each PU_i is reinitialised according to the volume of waste spread onto the CU_j and, for both units, the new initial time is taken as the spreading date. Then, the urgency indicators of both units are updated. Iterating the procedure allows one to explore the functioning of the system over a complete time period. This approach is summarized in Figure 7.

```
input { n PU, m CU, initial time}
//initialisation
for each UP
      discretisation procedure (§ 3.2) with the stock
         limit as a threshold
      t_PU^max(i) = latest date before overflowing
end
for each CU
      t_CU^max(j) = latest date for next spreading
end
//scenario test
while the time period is not totally explored
      let U* such as t_U*^max = min(t_CU^max, t_PU^max)
      while feasibility of spreading for pairs (U*,CU)
         or (PU,U*) is not checked
            test the pairs (Eq. 5) according to preferences
      end
      return spreading time
      evaluate new initial state for the PU
      update t_PU^max(i) and t_CU^max(j)
end
```

Figure 7. Procedure for testing a scenario.

4.2 Application to a typical farming system

To illustrate this approach, the functioning of a typical farming system in the context of the Reunion Island is presented. Production and consumption units are listed in Table 1. Table 2 gives for each PU the list of CUs ranked by preference order for spreading. Note that market gardening (CU_4) cannot receive slurry (PU_1) but only chicken manure (PU_2). CU_5, standing for fallow land, is given the least preference order as it is used (illegally) only in emergency cases when no other (legal) solution is found.

Table 1. Production and consumption units of the farm

Unit	Description
PU_1	20 sows, slurry stock capacity of 150 m^3
PU_2	3000 chickens, manure stock capacity of 38 T
CU_1	6.4 ha of sugar cane (production)
CU_2	1.6 ha of sugar cane (replanted)
CU_3	1 ha of maize
CU_4	2 ha of market gardening
CU_5	fallow land

Table 2. Preference order for spreading on the CUs

Unit	Order	Unit	Order
PU_1	$CU_1 > CU_3 > CU_2 > CU_5$		
PU_2	$CU_4 > CU_3 > CU_2 > CU_5$		
CU_1	PU_1	CU_4	PU_2
CU_2	$PU_2 > PU_1$	CU_5	-
CU_3	$PU_1 > PU_2$		

Figure 8 shows the result of ten years of waste management simulated for the studied farm:

- The normalised stock levels of PUs (volume of slurry or mass of solid manure) are on the left hand side with the vertical time axis labelled in days. For the sake of clarity, only the mean value lying between the lower and upper bounds of the double ODE system representing each stock is drawn.
- The right hand side of the Figure shows, for each CU, the time evolution of the normalised quantities of nitrogen spread per year.
- In the centre of the Figure, each black dot relates the PU_i, CU_j, and time points corresponding to each spreading action. A black star denotes spreading actions that have failed.

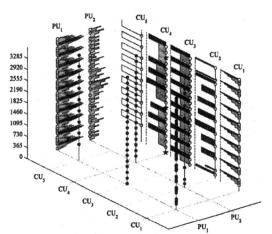

Figure 8. Time evolutions of production (PU) and consumption (CU) units for 10 years.

On Figure 8, we can see that (i) undesirable spreading on CU_5 is regularly performed because

of the pig slurry level of PU_1 and (ii) market gardening is under fertilised.

To improve management, we can check for new realistic rules: (i) a second spreading action, three months after the first, is allowed on CU_1; and (ii) two spreading actions are allowed on CU_2, one of slurry, one of poultry manure, instead of a sole manure or slurry fertilisation. The results of these changes are presented in Figure 9.

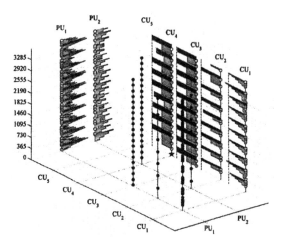

Figure 9. Time evolutions of the PUs and CUs after introducing new management rules.

We can note the better regulation of the stocks of PU_1 and PU_2, the correct fertilisation of CU_1-CU_4 and the (hopeful) absence of spreading on C_5.

For each scenario test, the units must be given initial states. In this example, initial slurry stock was set at 10% of the full stock capacity, and poultry manure 80%. Sensitivity of the system to initial conditions may be assessed for additional values for both PU_1 and PU_2 stocks: 0%, 33%, and 66% of the full stock capacity in addition to the others two. Figure 10 gives the maximal positive and negative difference with the mean value of both stock levels according to the sixteen scenarios (4 initial values for each PU). These results show that the influence of initial conditions vanishes after 2 years for PU_1 (pig slurry) and 8 years for PU_2 (poultry manure).

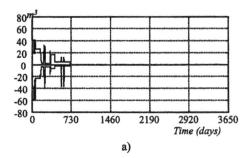

a)

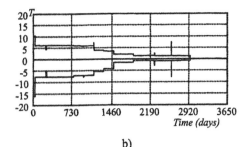

b)

Figure 10. Maximal positive and negative differences between stock levels simulated from various initial values and the mean value: a) for PU_1 (pig slurry) and b) for PU_2 (poultry manure).

This difference between both PUs can be partially explained by their spreading frequencies (on average 1/45 days for PU_1 and 1/125 days for PU_2). As consumption of the slurry stock occurs more often, this tends to make the stock level converge quicker to an equilibrium. As illustrated by this example, our approach to explore management options is based on iteratively refining actual or potential management strategies by altering the rules or some parameter values defined by agricultural expertise. This example is completely described in Hélias [2003b] in five iterations and the approach is also used to test spreading possibilities between a set of distinct farms (13 PUs and 35 CUs).

5. CONCLUSION

Our approach is based on modelling the temporal constraints related to each production or consumption unit by timed automata (a continuous-time discrete formalism). On the one hand, crop may be directly represented by timed automata accounting for agricultural events (*e.g.*, planting, harvest...) triggering their demand for waste application. On the other hand, the waste production by livestock, considered as a continuous process, is modelled as an extremal ODE system allowing one to handle the imprecision attached to its initial conditions and numerical parameters.

To handle the relationships between discrete crop representations and continuous stock models, we devised an automatic procedure. This procedure, first abstracts a simulated continuous evolution into a timed automaton; then, by making the product of this automaton with automata representing crops, it summarizes the whole system into a single automaton that may be analysed by model-checking tools. A discrete event system can thus be obtained from an hybrid dynamical system (i.e. embodying both discrete and continuous components).

In front of environmental risks generated by intensive livestock farming, modelling livestock waste management systems may be useful to help agricultural stakeholders improve their management strategies. We exemplified how to use our approach to improve management strategies by considering the possible waste transfers among a set of producers and a set of consumers.

ACKNOWLEDGEMENTS

This work was partially supported by the Region Réunion and the European Social Fund.

REFERENCES

Alur, R. and D. Dill, The theory of timed automata, *Theoretical Computer Science, 126*, 183-235, 1994.

Aubry, C., F. Papy and A. Capillon, Modelling Decision-Making Processes for Annual Crop Management, *Agricultural Systems, 56*(1), 45-65, 1998.

Bérard, B., M. Bidoit, A. Finkel, F. Laroussinie, A. Petit, L. Petrucci and P. Schnoebelen, *Systems and Software Verification - Model-Checking Techniques and Tools.* Springer-Verlag, 2001.

Guerrin, F. and J.-M. Paillat, Modelling biomass fluxes and fertility transfers: animal wastes management in Reunion Island, paper presented at International Congress on Modelling and Simulation (Modsim): Townsville, Australia, July 14-17, 2003.

Hélias, A., F. Guerrin and J.-P. Steyer, Représentation par automates temporisés de contraintes temporelles - Cas de la fertilisation organique des cultures de l'île de la Réunion, paper presented at 4ème Conférence francophone de modélisation et simulation (MOSIM): Toulouse, France, 2003a.

Hélias, A., Agrégation/abstraction de modèles pour l'analyse et l'organisation de réseaux de flux: application à la gestion des effluents d'élevage à la Réunion, Ph.D. thesis, ENSAM, Montpellier, France, 2003b.

Hélias, A., F. Guerrin and J.-P. Steyer, Abstraction of continuous system trajectories into timed automata, paper presented at IFAC Workshop on Discrete Event Systems (WODES 04): Reims (France), 6p., 2004a.

Hélias, A., F. Guerrin and J.-P. Steyer, Abstracting continuous system behaviours into timed automata: application to diagnosis of an anaerobic digestion process, paper presented at International Workshop on Principles of Diagnosis (DX 04), Carcassonne (France): 6p., 2004b.

Henzinger, T.A., X. Nicollin, J. Sifakis and S. Yovine, Symbolic Model Checking for Real-Time Systems, *Information and Computation, 111*(2), 193-244, 1994.

Polman, N.B. and G.J. Thijssen, Combining results of different models: the case of a levy on the Dutch nitrogen surplus, *Agricultural Economics*, *27*(1), 41-49, 2002.

Sorensen, C.G., B.H. Jacobsen and S.G. Sommer, An assessment tool applied to manure management systems using innovative technologies, *Biosystems Engineering*, *86*(3), 315-325, 2003.

Thornton, P.K. and M. Herrero, Integrated crop-livestock simulation models for scenario analysis and impact assessment, *Agricultural System*, 70, 581-602, 2001.

Yovine, S., Kronos: A verification tool for real-time systems, *Journal of Software Tools for Technology Transfer*, *1*, 123-133, 1997.

ELSEVIER
IFAC
PUBLICATIONS
www.elsevier.com/locate/ifac

MODELLING and CONTROL of PROCESSES in LOGISTICS – a GENERAL APPROACH

Petr Cenek

University of Žilina, Faculty of Management Science and Informatics
Department of Transportation Networks

Abstract: A logistic approach to problems should optimise sequence of production, storage and transportation processes. Management and control of these processes can be done on a macroscopic level using methods of operations research or on a microscopic level using simulations and control theory. Both types of models are fairly different in their approach and mathematical background even if they represent the same real system. Both model types should co-operate sharing their results and a general model can be possibly designed to serve various needs.

Keywords: Models, optimisation, logistics, transportation processes, control, hierarchical models

1 INTRODUCTION

There are many different processes in logistics, which serve a common goal – to supply a quality product to a customer in a desired time. A proper management and control of these processes can save costs and provide better quality of services to a customer. An optimisation on a general level uses methods of operations research and mathematical modelling, mostly methods of linear programming. Production, storage and transportation processes are then represented by commodity flows in a system. Such models can be referred to as macroscopic models.

Furthermore, various processes can also be analysed on a more detailed level. A distinction must be made among different processes in this case and we will focus on transportation processes. A detailed microscopic model of a transportation process will typically deal with individual vehicles/cars, with their kinetic and dynamic properties and finally with a concurrency or co-operation of processes. Microscopic models are frequently based on simulations, control theory or a traffic flow theory.

Custom models and optimisation methods can be used for models on a microscopic or macroscopic level while the real systems can remain unchanged,

which means that different models can represent the same vehicles and the same infrastructure. Two questions arise from this observation.

a. Is it necessary to use different models at different levels of detail?

b. How should such models co-operate for control and management of processes?

The rest of the paper is organized as follows: first, the basic characteristics of both types of models will be discussed, followed by the answers on the above stated questions, and conclusion will be drawn at the end of this paper.

2 MACROSCOPIC MODELS

There are many optimisation problems, which can be defined and solved on a macroscopic level. We can shortly list at least the important ones.

– *Location problem* which analyses the optimal capacity and location of production, storage or loading/unloading facilities.

– *Network design problem,* which makes an optimal choice of links (parts of infrastructure like roads or railroads or transportation services) to serve the transportation needs in a network.

- *Transportation/transhipment problems* that estimate the quantity of freight to be transported from sources (production plants) to sinks (customers).

- *Routing* should find an optimal path for vehicles serving several places of a distribution system.

- *Scheduling* makes a time-optimal planning of transportation services.

- *Shortest paths* problem seeks the least expensive path among the nodes in a transportation system.

The problems are ordered according to their importance and a scope of activities they influence. The location of centres and choice of links (network design) are typically strategic decisions, which demand a costly implementation and therefore are usually valid for a long time period. On the other hand finding the shortest path can be done efficiently and it demands little or no cost to apply the results in practice. The shortest paths can be thus estimated quite frequently according to the actual conditions in a system.

A mathematical model of linear programming can represent all of the above-described problems. The next section will show how it's done on an example of a transportation problem.

2.1 Model of a transportation problem

A simple model of a transportation problem will be used to show principles of building models on a macroscopic level. There are various approaches to modelling, but the most interesting ones are the mathematical models. These models are necessary to find a solution to an optimisation problem using a computer. The graphical representation is suitable for a problem presentation and results visualisation. A graphical model of a transportation problem is shown in Fig.1.

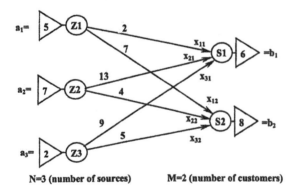

N=3 (number of sources)   M=2 (number of customers)

Fig.1: Graphical model of a transportation problem

The graphical model is easily understood and enables communication with users when creating a model and defining its parameters.

A solution to a transportation problem can be found using its mathematical model. How to create such model will be shown based on the example from Fig.1. The network consists of three sources Z1, Z2 a Z3 and two customers S1 a S2. Products can be transported from every source to any customer with defined unit costs along every edge (for instance $c_{11}=2$ stands for costs of traversal from a source Z1 to a customer S1).

The transportation problem is formulated by two disjoint subsets of nodes in a graph. One subset contains sources or nodes in which freight flows originate (for example production plants). The other subset is composed of sinks (or customers) with a defined demand on supplied freight. The capacity of sources is limited and unit costs describe the costs of traversal from source to sink. The solution to a transportation problem then estimates an optimal quantity of a commodity to be transported along each edge in a network, so that total costs are minimal, the production capacity in respected sources will not be exceeded and the demands of all customers will be fulfilled.

The optimisation problem must minimise total costs in the network. The total costs will be calculated as follows: x_{ij} describes a desired quantity to be transported along an edge (i,j) from a source Zi to a customer Sj and the unit costs of a transportation are c_{ij}. The total cost of transportation along the edge (i,j) can be calculated as $x_{ij} c_{ij}$. The total costs for all flows in the system will be a sum for all edges

$$C = x_{11} c_{11} + x_{21} c_{21} + x_{31} c_{31} + x_{12} c_{12} + x_{22} c_{22} + x_{32} c_{32}$$

or in a mathematical notation (for a general case with a number of sources N=3 and a number of customers M=2)

$$C = \sum_{i=1}^{3} \sum_{j=1}^{2} x_{ij} c_{ij}$$

The total cost formula above represents the minimization problem

$$\min \sum_{i=1}^{N} \sum_{j=1}^{M} x_{ij} c_{ij}$$

The solution must satisfy the constraints, which were stated in the transportation problem definition. Namely, the production capacity of sources must be respected and all customer demands must be fulfilled. These conditions can be formulated as equilibrium constraints for flows in nodes of a network.

Let's suppose that a flow equal to a source capacity enters a source node and quantities x_{ij} leave the node.

The incoming flows will be marked by a positive value and outgoing by a negative value. An equilibrium constraint must be fulfilled for all sources. For example, outgoing quantities from a source Z1 are x_{11} (to a customer S1) and x_{12} (to a customer S2) and the constraint can be written as:

$$(Z1:) \quad +5 - x_{11} - x_{12} = 0$$

or after moving the constant on the right hand side of the equation

$$(Z1:) \quad - x_{11} - x_{12} = - 5$$

From the perspective of the sinks the situation is similar. The flows leave the system (the quantity equal to a customer demand is consumed) while some flows enter the node. For a customer S1 there are incoming quantities x_{11} from a source Z1 and x_{21} from a source Z2. So the constraint will be

$$(S1:) \quad + x_{11} + x_{21} - 6 = 0$$

and after moving the constant on the right hand side of the equation

$$(S1:) \quad + x_{11} + x_{21} = + 6$$

Constraints respecting all flows in a network can be derived similarly and so a following system of constraints will be created (for all nodes in a model)

$$
\begin{array}{llllll}
(Z1:) & - x_{11} - x_{12} & & & = - 5 \\
(Z2:) & & - x_{21} - x_{22} & & = - 7 \\
(Z3:) & & & - x_{31} - x_{32} & = - 2 \\
(S1:) & + x_{11} & + x_{21} & + x_{31} & = + 6 \\
(S2:) & + x_{21} & + x_{22} & + x_{32} & = + 8
\end{array}
$$

All flows x_{ij} must be naturally nonnegative. A complete model will consist of costs function and a system of constraints

$$\min \quad \sum_{i=1}^{N}\sum_{j=1}^{M} x_{ij}\, c_{ij} \qquad (1)$$

subject to constraints

$$\sum_{j=1}^{M} x_{ij} = a_i \qquad \text{for } i = 1, 2, ..., N$$

(flow equilibrium at source nodes)

$$\sum_{i=1}^{N} x_{ij} = b_j \qquad \text{for } j = 1, 2, ..., M$$

(flow equilibrium at customer nodes)

$$x_{ij} \geq 0 \qquad \text{for } i = 1, 2, ..., N;\ j = 1, 2, ..., M$$

The above-derived mathematical model represents a typical macroscopic optimisation problem. We can briefly summarise, that an optimal solution can be found presuming that all necessary constants a_i, b_j and c_{ij} in a model are known. On the other hand, no assumptions were made on characteristics of

transported commodities, vehicles used or transportation modes. Some examples of macroscopic optimisation models are to be found in [3]-[6] and [8]-[10].

3 MICROSCOPIC MODELS

Microscopic models are built on more detailed level and they usually deal with individual vehicles, kinetic and dynamic properties of their motion and with concurrency of processes running simultaneously in a system.

There is an important difference in comparison to macroscopic models. The microscopic models must respect characteristics of a transportation mode, vehicles used and/or transported commodity. That is why the models cannot be created in a general way like macroscopic models without any regards to a transportation mode.

The differences between various transportation modes can be shown on vehicle kinematics and dynamics for two most frequent modes of transportation - road and railway. The railway transport is usually easier to describe.

3.1 Vehicle kinematics

The kinetic characteristics of a vehicle's motion is defined by geometric parameters of a road along which they move and by characteristics of a vehicle.

The railway vehicles are firmly bound to a rail track, which means that front wheels as well as rear wheels follow a trajectory described by a rail track. Therefore, there is no problem estimating the position of the front and rear wheels of a vehicle. The position of a vehicle is defined by two points, which correspond to a position of front and rear wheels of the vehicle. This also means that railway vehicles must follow a unique trajectory defined by the track geometry. Therefore the track geometry defines the maximum allowed speed on the track and a vehicle can be controlled only in one dimension (accelerate or decelerate). The kinetic behaviour of a railway vehicle is illustrated in Fig.2.

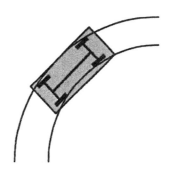

Fig.2: Railway vehicle following a track

The road vehicles behave differently. They can move freely on the road, which means that not only a longitudinal acceleration or deceleration can be controlled, but to a certain degree vehicle's trajectory can be changed as well. A shape of the trajectory of front wheels (a curve radius) will limit a vehicle's speed, so that the centrifugal acceleration will not surpass the maximum allowed value for actual road conditions and a vehicle type to result in a slide.

Rear wheels are "pulled" by front ones and they will just follow the direction defined by the longitudinal vehicle axis (a traktrix curve). Therefore, even if a correct trajectory for front wheels is chosen, the rear wheels may end up off the desired road boundaries. The position of front and rear wheels in a simple curve is shown in Fig.3.

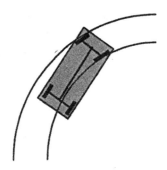

Fig.3: Trajectory of a road vehicle
(the rear wheels following the front wheels)

The mathematical description of rear wheels trajectory is not trivial and that is why only simple cases will be discussed. A vehicle motion will be described for a straight section and for a circular curve (a more complex trajectory can be approximated from these instances). The trajectory of rear wheels can be estimated for a steady state (after having run an infinite distance with constant curvature) and for a transient situation (after the trajectory curvature has changed).

The first case for a steady state on a straight section is rather simple. Rear wheels will simply follow the same line as front wheels. A case of steady state in a circular curve is slightly more complex as shown in

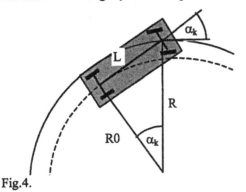

Fig.4.

Fig.4: Road vehicle in a curve with a fixed turning radius.

As shown in Fig.4 rear wheels at a steady state will move along a circle with a radius R0, which can be easily estimated as

$$R0 = \sqrt{R^2 - L^2} \qquad (2)$$

where

R – turning radius for front wheels [m],
L – length of a vehicle [m],
R0 – turning radius for rear wheels [m].

More information on road vehicle kinematics can be found in [1].

3.2 Vehicle dynamics

A general model of vehicle dynamics is based on the Newton's law

$$m.a = \sum_i F_i \qquad (3)$$

where

m – mass of a vehicle [kg],
a - acceleration [m.s^{-2}],
F_i – acting forces [N].

If acceleration *a* is substituted by a second derivative of distance against time, the following differential equation will describe the motion of a vehicle

$$\frac{d^2x}{dt^2} = \frac{1}{m} \sum_i F_i \qquad (4)$$

Solution to this differential equation will define the functions for speed, distance, acceleration and energy consumption against time. A time optimal or an energy optimal control can be derived for any vehicle using Pontrjagin's maximum principle, numerical methods or by experimenting with a simulation model. A real implementation seems to be reasonable only for railway vehicles, while they have an important consumption of energy and their trajectory is unambiguously defined by the track geometry.

Any optimal control must be feasible and so it must obey the speed limits of the track. Speed limits in railway transports can be defined according to a track geometry and to vehicle characteristics.

The maximum speed on the roadways will depend on a trajectory chosen by a driver and so it must be defined not only according to road geometry but also to an actual trajectory of a vehicle. The vehicle speed cannot surpass the maximum allowed centrifugal acceleration, otherwise the vehicle will slide off the road. The centrifugal acceleration should respect the following rule:

$$\frac{v_{max}^2}{R} \le a_C \qquad (5)$$

where

v_{max} – maximum applicable speed in a curve [m/s],
R - radius of a curve [m],
a_C – maximum permitted centrifugal acceleration
   [m.s^{-2}]

3.3 Concurrency of processes

Each vehicle in a transportation system represents one player, who tries to gain a part of limited shared resources of a system such as a space on a road or on a crossing or a service time of a loading/unloading device etc.

Modelling and control of such concurrent processes is probably the most difficult part of a control in transportation systems. These problems are solved by intelligence of a driver in real traffic situations. Rather difficult problems arise, when an automatic control of such a system is to be designed or when it should be simulated on a computer. The intelligence of a driver and his knowledge of traffic rules must be described formally and solution of all traffic situations must be implemented in a form of control subsystem in the overall model. These are actual research problems in the intelligent transportation systems and in simulation of transportation systems. A bit more detailed description of the problem can be found in [2].

A short description of microscopic models showed a model dependency on a mode of transportation. A mathematical formulation of a microscopic model is given by differential equations, which is quite different from a macroscopic model based mostly on linear programming.

4 CO-OPERATION OF MICROSCOPIC AND MACROSCOPIC MODELS

The basic characteristics and differences between microscopic and macroscopic models have been discussed in sections 2 and 3. Even if both types of models may differ substantially, they represent the same physical system and that is why certain common features can be expected. Now we can discuss the two questions stated in the introduction.

The first question can be answered if we realise that each model brings a certain abstraction and

simplification of a real system. It means that only some system characteristics are suitable for a model and for an actual optimisation task, while all other characteristics are unimportant and should be neglected. Every element may have some obligatory parameters and usually many more optional ones. A topology is an obligatory characteristic of a transportation system and all other parameters are optional and are to be used only in some models. A schematic organisation of data is shown in Fig.5.

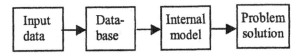

Fig.5: Scheme of data organisation

The raw data about a system will be stored in a database. The database may contain all available data, which will be used in either model. Internal model will be created "on demand" or according to an optimisation task to be solved. The only parameters necessary for a specific model will be extracted from a database and used in an internal model. Finally an optimisation problem can be solved using chosen model. All parameters have to be stored and available at a database but only obligatory parameters and chosen parameters for an actual optimisation problem will be used in an internal model. In this way, much more parameters such as a track geometry or mass, traction force and other dynamic parameters of a vehicle will be used in a microscopic model.

The second question was about co-operation or sharing results between both model types. One should recall the structural characteristics of the models from sections 2 and 3. A macroscopic model is suitable for finding paths in a network and for solving many other optimisation tasks. The validity of results will depend on the quality of input data (values of coefficients c_{ij} in our example model). The values can be simply estimated if they can be easily measured (as for example distances). The situation will be quite different if the coefficients stand for a travel time, energy consumption, loading/unloading time, time for sorting etc. A measurement of such values in a real environment would be costly and cannot be measured for a whole range of different traffic situations (as values of these parameters may depend strongly on a traffic load in the system). So we need data on traffic load as input data for running a microscopic model and results from this model to solve optimisation problems on a macroscopic level. This co-operation is schematically shown in Fig.6.

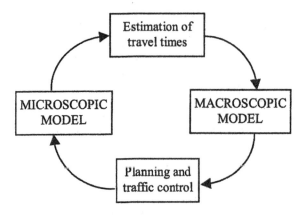

Fig.6: Co-operation of models

5 CONCLUSIONS

The paper presented two approaches to modelling of logistic and transportation processes. Macroscopic models serve mainly for planning and management, while microscopic models are used for detailed examination of processes usually through the simulations. Models may create a hierarchical system, where models on different levels share their results. An output from one model can serve as input value for the other one and vice versa. Both models represent processes in the same system and therefore they should share basic raw data from a common database.

Optimisation and simulation models of transportation systems are typically large and complex models, which demand significant computing power to run. Recent advancement of computer performance indicates that not only stand-alone applications but also complex hierarchical models can be run on high performance computers or using a distributed computing power of computer networks. Such distributed systems offer another advantage, as they reflect real life situations well. Individual computers can model traffic in logistic centres or nodes of a transportation system, other computers can model traffic on roads among them or traffic in the whole system.

The paper is based on experience with implementation of such models and it contains mostly reflections on how to build suitable models of logistics and transportation processes. Unfortunately, it is not easy to describe the models in all of their detail, so only basics necessary for the rudimentary implementation have been presented in the paper.

REFERENCES:

[1] Cenek P. (2003), Vehicle kinematics and micro-simulation models. Journal of Information, Control and Management Systems, Vol.1, No.1, FRI-ŽU, Žilina

[2] Cenek P. (2000), The simulation of transportation processes in logistics. IFAC MCPL 2000 Conference, Grenoble, 2000

[3] Janáček, J. (1999), Analysis and Structure Design of Distribution Systems. In Proceedings of the International Conference „Strategic Management and its Support by Information Systems", p. 120-124.

[4] Jánošíková, Ľ. (1997), An adaptation of the tabu search metaheuristic to the problem of transportation planning, In: Proceedings of the „IFAC/IFIP/IFORS Symposium - Transportation systems", Chania, Greece, **Vol. 2**, pp. 765-768

[5] Jánošíková Ľ., Sadloň Ľ., Cenek J. (2003), Model of Intelligent Transportation Infrastructure. Journal of Information, Control and Management Systems, Vol.1, No.1, FRI-ŽU, Žilina

[6] Kavička, A., Jánošíková, Ľ. (1999), Models of a trackage layout and estimation of a shortest path. Komunikácie, Scientific Letters of the University of Žilina, Žilina, **No. 2,** pp. 9-21

[7] Klima, V., Kavička, A. (1977), Virtual railway marshalling yard. In: Proceedings of IFAC/IFIP/IFORS Symposium – Transportation Systems, Ghania, Greece, **Vol. 2**, pp. 880-883

[8] Klingman, D. & Schneider, R.S., Microcomputer based Algorithms for Large Scale Shortest Path Problems. *Discrete applied mathematics and combinatorial operations research.* Vol.13, No.2, 3, 1986

[9] Magnanti, T.L. & Wong, R.T., Network and Transportation Planning: Models and algorithms. *Transportation Science*, 1984, pp.1-55

[10] Peško, Š. (1998), Multicommodity return bus scheduling problem. In Proceedings of the International Scientific Conference on Mathematics, University of Žilina, Žilina, p. 77-82.

[11] Die Traktrix, eine nicht algebraische Kurve. http://www.fh-lueneburg/u1/gym03/expo/jonatur/wisen/mahte/kurven/taktrix.htm.

The paper was prepared with a support of research grant VEGA 1/0498/03

PUBLICATIONS
www.elsevier.com/locate/ifac

USER INTERFACE FOR SIMULATORS OF THE NITRIFICATION PROCESS IN A ROTATING DISK REACTOR AND IN AN ACTIVATED SLUDGE SYSTEM

Carlos Muñoz, Christian Antileo, Cristian Bornhardt, Juan Carlos Araneda, Cesar Huiliñir, Martha T. Ramírez
e-mail: comunoz@ufro.cl
Universidad de La Frontera, Temuco, Chile

Abstract

The biological nitrification-denitrification process allows economic removal of high loads of nitrogen from industrial wastewaters in the fishing, chemical and agro-industrial areas. This work presents a user interface system developed in the GUIDE system of MATLAB for a set of computer simulators for the nitrification process. Simulators are based on models of an activated sludge process and a rotating disk reactor.

The result of this work is a user interface system that allows interactive operation of simulators by the scientific staff. The information is generated in a friendly way, making easy the data analysis and results interpretation in a rich graphic environment.

Keywords: user interface, computer simulator, bacterial nitrification

1. Introduction

An economic alternative to remove high nitrogen loads from industrial wastewater (IWW) generated by fishing, chemical and agro-industrial companies consists of the use of a biological Nitrification-Denitrification process, where autotrophic and heterotrophic bacteria convert ammonium into molecular nitrogen. In this biological process, the nitrification stage determines the efficiency of the general process; if this is carried out by means of high nitrite accumulation, significant savings on oxygen and organic consumption can be reached, as well as lower sludge production [1]. In this matter, the modeling and simulation of the biological nitrification process plays an important role, since it allows studying the process in lower time with respect to real time. Therefore, this allows determining with a high confidence degree which variables have greater effect to obtain a high degradation of the total ammonia nitrogen, and high nitrite accumulation. Considering this information, it is possible to design suitable control strategies for this process. The simulators used in this work are based on models developed by Pirsing [19] in the case of the activated sludge process, and Antileo [20] in the case of the rotating disk reactor (RDR) process.

It is desirable to develop a suitable system capable of simultaneously operate and analyze data generated by the nitrification process simulators, which can be accomplished using specialized routines of data processing with good user interfaces, graphical visualization and interaction, provided by the actual computer system. This new system is capable of performing a complex simulation, providing the scientific staff with valuable elements to achieve a high level of abstraction and phenomena comprehension [5]. Furthermore, it provides an interactive tool that simulates the process within a rich and interactive environment by displaying graphical information, making a positive contribution to the proper use of the simulators [4].

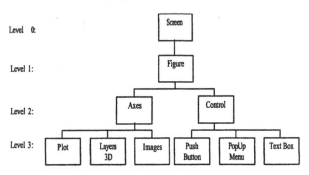

Figure 1. Hierarchy of graphical objects in Matlab

At the IWW Laboratory of the Department of Chemical Engineering of Universidad de La Frontera, two rotating disk reactors -RDR- and one activated sludge system was installed and instrumented. For each one of these reactors, a previously calibrated simulator was used. The development of graphically explicit user interfaces based on GUIDE objects, from Matlab v6.5 [7], provides a great capacity for managing 3D and 2D graphs and images through its father-son hierarchical structure (See Figure 1) [6]. This platform offers very good support, and considering the natural integration of simulators programmed using Simulink, GUIDE is the precise tool to use in order to generate this new system capable of providing the appropriate information to the scientific staff that will operate the simulators.

In section 2, the communication structure between the simulators and the user interface is described, as well as the technique used to capture, store and update the data generated by the simulators. In sections 3 and 4, the structure for designing the user interface for the nitrification process simulator in an activated sludge system and in a rotating disk reactor, RDR, respectively, is described. In section 5, the navigation structure for the user interface system using an opening interface is described. Finally, in section 6, the conclusions of this work are presented.

2. Simulator- User Interface Communication

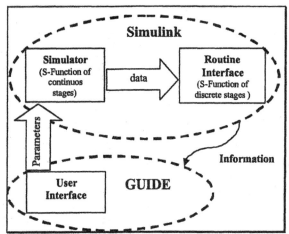

Figure 2. Scheme of Simulator-User Interface Communication

The communication between a nitrification process simulator and its user interface developed using Simulink and GUIDE of Matlab respectively, is shown in Figure 2. It can be seen that the Simulink model is made up of a process simulator represented by an S-function of continuous stages and of a specialized routine for processing data generated by the simulator, which is represented by an S-function of discrete stages. The routine that processes data generated by the simulator produces the information that would be visualized in the user interface through three-dimensional layers, tendency graphs, and present values. Furthermore, it allows the user interface operator to modify operational parameters on line.

2.1. Simulators' data gathering and storing

The routines in Figure 2 transform data from the process into information for the staff that operates the simulators. The data generated is captured by the simulators in a discrete way, storing it in vectors of constant length "m" for each variable of interest "x_n" provided by the simulators. Vectors are updated using a First In - Last Out rule approach as shown in Figure 3.

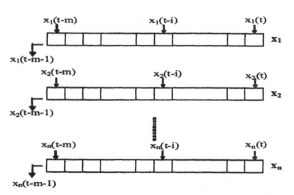

Figure 3. Framework used for storing data generated by the simulators.

3. User Interface for the simulator of nitrification process using an activated sludge system

The model of an activated sludge reactor (see Figure 4) was solved by a set of 8 coupled ordinary differential equations which are highly non-linear [14]. These equations are based on mass balance of the species (ammonium, nitrate, nitrite, oxygen and bacteria) and the proton balance [19].

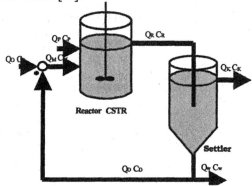

Figure 4. Framework of an activated sludge reactor

Figure 5 shows the user interface designed for the simulator of an activated sludge reactor. This user interface is divided into seven sections. Section 1 contains two buttons for navigating and a label that indicates the accumulated time simulated. Section 2 has three radio buttons which allows the user to choose the simulation mode between manual, semiautomatic, and automatic. Figure 6 shows the control scheme for the nitrification process.

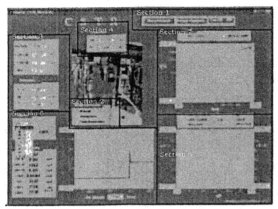

Figure 5. User interface for the simulator of nitrification process based in an activated sludge system.

Figure 6 illustrates three ways for operating the process. In the manual mode (process simulated without control loops) the user is allowed to manipulate all input variables of the process. In the semi-automatic mode (process simulated with regulatory control) the user can only manipulate the references of pH and dissolved oxygen in the reactor. In the automatic mode (self-control activated) the user can no longer manipulate the references of pH and oxygen dissolved in the reactor, because now the simulator is connected to a self-review system within the process, in charge of continuously generating new references in such a way, that the process always operates with maximum efficiency.

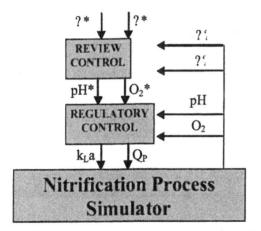

Figure 6. Control outline for the nitrification process.

Section 3 is made up of seven text boxes indicating the input variable values of the nitrification process using an activated sludge system; these values can be modified by the user. The user can modify the following variables: input flow Q_O, ammonia concentration $C_{O(N-3)}$, dissolved oxygen concentration $C_{O(O2)}$, and pH; additionally, some typical parameters such as the oxygen transfer coefficient k_La, the

recirculation rate of sludge N_R and the residence time of sludge in the settler can also be modified.

Section 4 has two text boxes where the user can adjust the flow, as well as the sodium carbonate concentration used to modify the pH in the reactor, Q_P and C_P, respectively. In section 5, a graph updates, from time to time, describes the dynamic behavior of the species in the reactor. To better visualize this, a pull-down menu allows the user to select the variable. The following variables can be observed: pH, ammonia concentration, nitrite concentration, nitrate concentration, dissolved oxygen concentration, sodium carbonate concentration, *Nitrosomonas* and *Nitrobacter* concentration.

Additionally, there is a set of eight labels showing the current value of each of the variables in the reactor. The text boxes that accompany the graph allows the user to change the time scale (x axis) and the range (y axis) in which is desired to observe the selected variables. On the other hand, the grid representing the time axis moves as the simulation progresses, to give the user the feeling of monitoring the simulator by him or her.

Section 6 contains a graph made up of two axes (one on top of another), six radio buttons and six text boxes. With the radio button the user can choose to observe the following combinations of variables: nitrite and nitrate concentration in the reactor, *Nitrosomonas* and *Nitrobacter* concentrations, alpha and beta (efficiency indices of the process in the reactor) versus pH or versus the dissolved oxygen concentration in the reactor. The text boxes allow the user to independently change the scales of the x-axis, as well as the y-axis.

Finally, section 7 contains a graph made up of two axes (one on top of another) similar to those in section 6, two checkboxes and six textboxes. With the selection of checkboxes the user can observe: the efficiency index of the alpha versus beta processes, or the concentration of *Nitrosomonas* versus concentration of *Nitrobacter* in the reactor; the user can also simultaneously observe in the reactor: alpha, beta, the concentration of *Nitrosomonas* and the concentration of *Nitrobacter*.

4. The user interface for the simulator of the nitrification process in a rotating disk reactor RDR

The nitrification process simulator in a rotating disk reactor, RDR (see Figure 7) developed by Huiliñir [2] uses a mathematical model made up of 214 coupled nonlinear ordinary differential equations; from those, 210 describe the behavior throughout the thickness of the biofilm, which are the result of quantization of seven partial differential equations, plus other four ordinary differential equations which describe the behavior of the system inside the fluid. The data generated by this simulator allows to observe how ammonia concentration, nitrite, and nitrate vary as function of time; these variables indeed can be measured, but not with the steadiness allowed by a simulator of the nitrification process.

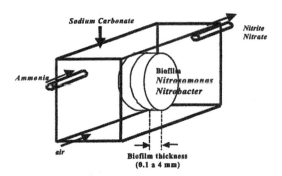

Figure 7. Outline of the nitrification biological process in a rotating disk reactor, RDR.

Besides, the simulator generates information of what happens inside the biofilm with the concentrations of ammonia, nitrite, nitrate, and the concentrations of the bacteria (*Nitrosomonas*, *Nitrobacter*, and inert biomass), and how the thickness of the biofilm varies. This information is complex to represent, since variables are a function of the biofilm thickness and time. Therefore, the simplest form to represent this is using graphs with three-dimensional layers, that allow the operator of the simulator to better understand the dynamic behaviour. It must be emphasized that information inside the biofilm is very difficult to obtain in a real process.

Figure 8 shows the design of the user interface for the nitrification process in a RDR made up by five sections: in Section 1, the user can stop the simulator at any time by pushing a button, and observe the simulated time in a text. In Section 2, the user can change operation parameters for the simulator in the text boxes designed for this purpose. The parameters that can be adjusted are: the input flow, ammonia concentration, dissolved oxygen, and pH in the inletflow; besides, the flow and the sodium carbonate concentration injected to the reactor to modify its pH can also be modified. Additionally, it is possible to adjust model parameters such as detachment of the biofilm, the fraction of liquid in the biofilm, the biomass density of the biofilm, the oxygen transfer coefficient, and the mass transfer coefficient.

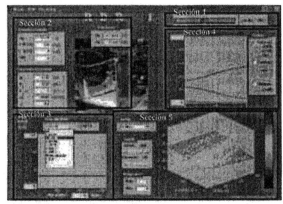

Figure 8. User interface of the nitrification process in a RDR.

In section 3, it is possible to visualize in a graph, how variables describing the behavior inside the fluid change in real time. For this, a pull-down menu gives the option to select the desired variable to observe in the graph. Among the variables that can be chosen are: the concentrations of nitrite, nitrate, ammonia, dissolved oxygen, sodium carbonate, and pH. In section 4, information showing what happens inside the biofilm at the last time simulated is updated according to how the simulation evolves. The data show the profile of the fractions of bacteria concentrations, substrates concentration (ammonia, nitrite and nitrate) and dissolved oxygen concentration. In this section, checkboxes to simultaneously activate biomass visualization, substrates, and dissolved oxygen are included.

In section 5, three-dimensional graphs are used to simultaneously represent from time to time the behavior and the evolution of the biofilm thickness. A pull-down menu is used to select the characteristic that is desired to visualize, among those are: the biomass fraction (*Nitrosomonas*, *Nitrobacter*, and inert biomass), substrates concentration (nitrite, nitrate, ammonia) or the concentration of dissolved oxygen in the biofilm.

On the other hand, considering that this simulator includes a high number of differentials equations that must be solved continuously, several yields?? of integration routines for stiff and non-stiff type problems were studied. The integration methods for stiff type problems managed to converge??; however, significant differences of yield among them were observed. For example, the Rosenbrock method manages to solve?? solved?? in 5 minutes the equivalent to 22 hours of simulation, whereas the routine *ode15s* was more efficient when using multiple integration steps, showing the capability to solve the equivalent to 1150 hours of simulation in the same 5 minutes.

5. The opening interface

The opening interface allows a fast and easy access to the set of user interfaces of the process simulator, allowing easy and interactive navigation between the user interfaces for the different simulators (see Figure 9).

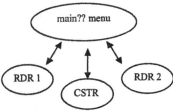

Figure 9. Scheme of navigation between user interfaces.

The opening interface is divided into 3 sections (see figure 10). Section 1 contains an image of a rotating disk reactor RDR, a button that allows simulating the nitrification process in the RDR 1, and one label indicating the current state of the simulator in two stages: simulation-on or simulation-off. Section 2 has an image of an activated sludge system, with a CSTR as main component, and similar to Section 1, it has a button that allows simulating the nitrification process, and one label indicating the current stage of the simulator. Section 3 is similar to section 1; the main difference consists of a button that allows simulating the nitrification process in the RDR 2.

Figure 10. Simulator opening interface.

6. Conclusions

The main result of this work is an interactive tool that allows operating the simulator with an appropriate and integrated routine. Additionally, it allows visualization of all the information generated by the simulators (which are executed with a suitable integrated routine) in a much easier way to interpret and to analyze the data with an enriched graphical environment. Working with this type of tools turn out to be an productive multidisciplinary work, in which scientists from the chemical engineering field interact and work with scientists from the modeling and simulation field, in order to study with more competence aspects such as what happens inside a biofilm in a RDR, or how to define and review control strategies to optimize the performance of an activated sludge system.

7. Acknowledgments

The authors acknowledge the financing granted by the research project FONDECYT 1030317 and the contribution made by the Center for Modeling and Scientific Computing of the Universidad de La Frontera.

8. References

[1] Antileo, C.; Roeckel, M.; Wiesman, U. (2003). "High Nitrite Buildup During Nitrification in a Rotating Disk Reactor". Water Environment Research **75**,151-162.

[2] Huiliñir, C.; Antileo, C.; Muñoz, C.; Araneda, J.C. (2004). "Modelación y simulación de un reactor de disco rotatorio aplicado a la nitrificación" (in spanish). MODOPT, Universidad de La Frontera, Temuco, Chile.

[3] Barmsnes, K.; Johnsen, T.; Sundling, C. (1997)."Implementation of Graphical User Interfaces in Nuclear Applications". ENS Topical Meeting on I&C of VVER, Prague, April 21-24.

[4] Riedl, M.; Amant, R. (2002). "Toward Automated Exploration of Interactive Systems". IUI'02, January 13-16, San Francisco, California, USA, 135-142.

[5] Chin, G.; Leung, L., Schuchardt, K.; Gracio, D. (2002). "New Paradigms in Problem Solving Environment for Scientific Computing". IUI'02, January 13-16, San Francisco, California, USA, 39-46.

[6] Atencia, J.; Nestar, R. (2001). "Aprenda Matlab 6.0 como si estuviera en primero". Escuela Superior de Ingenieros Industriales Industri Injineruen Goimailako Eskola. San Sebastián, España.

[7] The MathWorks. (2000) "Matlab. The Language of Technical Computing, Creating Graphical User Interfaces (Version 1)".

[8] Muñoz, C.; Contreras, A. (2003). "Simulación de Procesos con Matlab 5.3" Reporte Interno, Universidad de La Frontera, Temuco, Chile.

[9] Asmar, I. "Métodos Numéricos" http://www.unalmed.edu.co/~ifasmar, pp. 273-323.

[10] The MathWorks. (2000). "Matlab. The Language of Technical Computing. Using Matlab (Version 6)".

[11] The MathWorks. (2000). "Matlab. The Language of Technical Computing. Using Simulink (Version 6)".

[12] The MathWorks. (2000). "Matlab. The Language of Technical Computing. Using MATLAB Graphics (Version 6)".

[13] Nakamura, S. (1997). "Análisis Numérico y Visualización Gráfica con Matlab". Prentice-Hall Hispanoamericana,S.A., Primera Edición.

[14] Araneda, J.C. (2004). "Diseño y Simulación de un sistema control de un reactor biológico de flujo continuo CSTR" (in spanish). Trabajo de Título Ingeniería Civil Industrial mención Informática. Universidad de La Frontera, Temuco, Chile.

[15] Wanner, O.; Gujer, W. (1986). "A multispecies biofilm model". Biotechnology and Bioengineering **28**, 314-328.

[16] Okabe, S.; Hirata, K.; Watanabe, Y. (1995). "Dinamic changes in spatial microbial distribution in mixed population biofilms: Experimental results and model simulation". Water Science and Technology, **32**(8). (faltan páginas)

[17] Rittmann, B.; Stilwell, D.; Ohashi, A. (2002). "The transient-state, multiple-species biofilm model for biofiltration processes". Water Research **36**, 2342-2356.

[18] Morgenroth, E.; Wilderer, P. (2000). "Influence of detachment mechanism on competition in biofilm". Water Science and Technology **34**(2). (faltan páginas)

[19] Pirsing, A. (1996). "Reaktionstechnische Untersuchungen und mathematische Modellierung der Nitrifikation in hochbelasteten Abwässem". Reihe 15: Umwelttechnik Nr.156. VDI-Verlag. Düsseldorf, Germany.

[20] Antileo, C. (2003) "Oxidación biológica de amonio y acumulación de nitrito en bacterias inmovilizadas y suspendidas". Ph.D. Thesis, Department of Chemical Engineering. Universidad de Concepción.

Las referencias estaban muy "inhomogéneas"

ELSEVIER
IFAC
PUBLICATIONS
www.elsevier.com/locate/ifac

IMPLEMENTATION OF MASS CUSTOMIZATION
IN VIRTUAL FIRMS AND SUPPLY CHAINS

Julio Macedo

Institut Stratégies Industrielles
229 Forest, Pincourt, P.Q., J7V8E8, Canada

Abstract: An apparel virtual firm designs and distributes garments that are sewn by foreign contractors. Mass customization is a new strategy for making customized garments at low costs, similar to those obtained by applying mass production principles. Mass customization suggests making customized garments by combining common garment components and postponing the garments point of differentiation in the supply chain process. This paper presents a survey of the degree of implementation of mass customization principles by virtual firms in the apparel industry. The results show that most of these firms have not implemented mass customization principles.

Keywords: Manufacturing systems, Management systems, Model management, Process models, Production systems, Performance evaluation.

1. INTRODUCTION

A supply chain is a network of material suppliers, manufacturing centers, warehouses, distribution centers and retailers as well as the stocks that flow between these centers (Simchi-Levi et al., 2004). A virtual firm (VF) is a temporary, project-oriented alliance of a subset of firms of the supply chain (Martinez et al., 2001). In addition, a mass-customized firm makes products meeting the individual customer needs with near mass production efficiency, i.e. with low production costs (Tseng and Piller, 2003). Current literature has identified mass customization as a competitive strategy and the number of VF in the Canadian apparel industry is growing due to the transfer of the sewing operations to low-wage countries. However, research about the degree of implementation of mass customization principles in VF is lacking (Da Silveira et al., 2001). This paper presents a survey of the degree of implementation of mass customization principles in twelve VF firms of the Canadian apparel industry (Table 1).

2. METHODOLOGY

In this paper a supply chain consists of textile firms making fabrics, sewing firms assembling garments and the retailer stores selling garments to the customers. In this context, the VF activities are limited to the design and distribution of the garments to the retailers; the garments are assembled by sewing firms subcontracted by the VF that also buys the required fabrics. In practice, the sewing contractors are located in developing countries whereas the textile firms are located in technologically advanced countries around the world.

The methodology utilized to measure the degree of implementation of mass customization principles in VF is as follows. First, the expected characteristics of a VF designed according to mass customization principles are identified from current literature. Second, twelve VF of the Canadian apparel industry are visited. Third, the operations managers of these firms are asked to answer 24 open questions covering the implementation of mass customization principles. Fourth, the answers to these questions are aggregated

Table 1 Characteristics of analyzed virtual (VF) firms

VF name	Number of employees	Annual Income ($)	Types of garments sold	Percentage of VF income generated by sewing contractors	Location of contractors
B	26	3 millions	Ready to wear garments for women	100%	China
C	8	variable	Jeans for women	80%	China, India
R	?	variable	Ready to wear garments for women	?	China, India
CH	2000	226 millions	All types of garments for men and women	?	?
N	30	25 millions	All types of garments for men and women	100%	Canada
D	100	30 millions	High quality garments for women	80%	Canada, China, India
M	20	?	High quality garments for women	90%	Canada
S	1500	200 millions	High quality garments for women	80%	Canada, Europe, Asia
J	7	?	Jeans of high quality fabric	100%	Canada, Europe, Asia
Y	2000	?	Ready to wear garments for teenagers and women	100%	Asia
E	10	?	Uniforms for corporations	60%	China, Pakistan, Sri Lanka
A	100	30 millions	Street wear garments	80%	China

and compared to the expected characteristics above in order to formulate conclusions.

The expected characteristics of a VF and its supply chain designed according to mass customization principles are included in Figure 1 (Berman, 2002; Selladurai, 2004). As indicated, the customer is included in the VF supply chain: He participates, in real time, in the specification of the style and measurements of the garment he is buying. The garment style is specified using CAD databases and intelligent advice software available at the retailer (Anderson-Connell, 2002; Ulrich et al., 2003; Taylor et al., 2003) or by internet (Gurzki et al., 2003). The customer measurements are obtained with 3D scanning equipments available at the retailer (Ashdown et al., 2004). Another characteristic of a VF is that it can deliver customized garments quickly because the physical structures of the firms and flows included in the VF supply chain are well integrated, enough to support this type of response. For example, the raw materials are standardized and some production activities outsourced (Anderson, 2004).

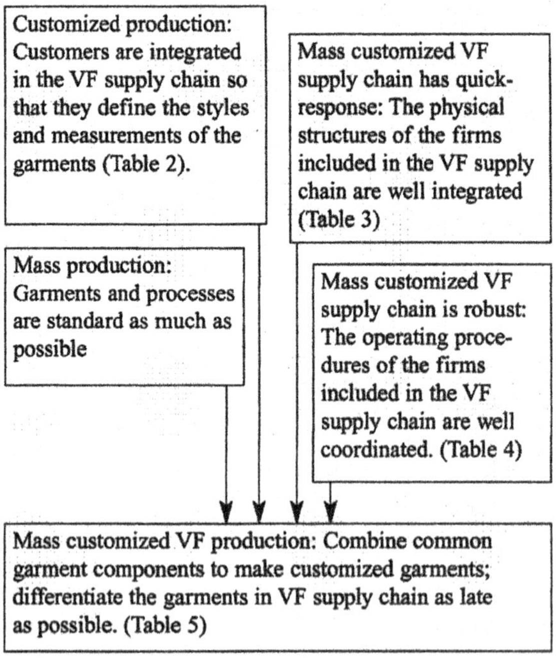

As indicated in Figure 1, a VF supply chain designed according to mass customization principles is robust: The firms included in this supply chain react locally, but in an integrated way, to sudden events because their working procedures are coordinated. This coordination is supported by Internet-based software (Ghiassi and Spera, 2003) that allows the retailers to communicate the quantities of garments sold to the upstream firms of the VF supply chain. Hence, the

Customized production: Customers are integrated in the VF supply chain so that they define the styles and measurements of the garments (Table 2).

Mass customized VF supply chain has quick-response: The physical structures of the firms included in the VF supply chain are well integrated (Table 3)

Mass production: Garments and processes are standard as much as possible

Mass customized VF supply chain is robust: The operating procedures of the firms included in the VF supply chain are well coordinated. (Table 4)

Mass customized VF production: Combine common garment components to make customized garments; differentiate the garments in VF supply chain as late as possible. (Table 5)

Fig. 1. Expected characteristics of a mass customized virtual firm (VF) and its supply chain.

emergence of large security stocks at different parts of the supply chain to prevent sales increases is avoided (Dejonckheere et al., 2003; Sheu, 2003). This software can include operations research models that calculate numeric values for the parameters of the supply chain operating procedures: Sizes of lots to produce (Stadtler, 2004; Dong and Chen, 2004), sizes of lots of raw materials to purchase (Crama et al., 2004),

Table 2 Degree of integration of customers in virtual firms (VF) supply chains

Degree of customer integration		Firms	Percentage of firms
Time required by VF designers to develop a new collection of garments until the prototypes are ready	1 week	CH	8.30%
	2 weeks	E	8.30%
	3 weeks	N, Y	16.70%
	4 weeks	R, D, M, A	33.30%
	5 weeks	S	8.30%
	8 weeks	B	8.30%
	10 weeks	J	8.30%
	24 weeks	C	8.30%
Sewing contractor time to approve a prototype	1 week	CH	9.10%
	2 weeks	Y, A	18.20%
	3 weeks	E	9.10%
	4 weeks	C, R, J	27.30%
	7 weeks	M	9.10%
	8 weeks	D	9.10%
	10 weeks	B, N	18.20%
Software utilized by VF designers to sketch new product styles	Illustrator with sewing contractor participation	C, CH, J, Y, A	45.50%
	Illustrator without sewing contractor participation	R, B	18.20%
	Lectra's Modaris without sewing contractor	N	9.10%
	Sketch done manually without sewing contractor participation	D, M, E	27.30%
Number of apparel shows attended annually by VF designers	2 shows	C, M, S, E	36.40%
	4 shows	B, R, D, J, A	36.40%
	10 shows	Y	9.10%
	30 shows	CH	9.10%

sizes of lots to transport between firms (Wang and Sarker, 2004), location and sizes of supply chain buffer stocks (Lee and Chu, 2004), allocation (Ma et al., 2002) and scheduling (Kolisch, 2000) of the activities of the firms in the supply chain.

Figure 1 also indicates that a VF designed according to mass production principles respects the two following principles. First, the customers design the customized garments by combining available common product components and modules (Ernst and Kamrad, 2000; Lee and Chen, 1999). Second, the point for differentiating the garments in the production process (and in the supply chain process where it is possible) is delayed as much as possible to have a standard production process (Partanen and Haapasalo, 2004; Svensson and Barfod, 2002; Ma et al., 2002). In this way, the individual needs of the customers are respected partially to preserve some garment and process standardization as suggested by mass production efficiency principles.

3. SURVEY RESULTS

The results in Table 2 indicate that the supply chains of the analyzed VF cannot integrate the customer in the design of the garment he is buying. Effectively, Table 2 indicates that 50% of VF require between 3 and 4 weeks for developing a new collection of garments, and, most of the time the sewing contractor requires 4 weeks to approve the garment prototype for serial production (27.3% of cases). Therefore, the current process for designing new garments in VF is too long to allow the real time participation of the customers. In addition, Table 2 shows that 45.5% of VF designers utilize Illustrator software for sketching new garment styles instead of CAD databases of allowed garment components. Hence, VF designers create customized garments from zero as in craft production and not from a pool of garment components that the VF can make efficiently.

Table 3 Degree of integration of structures of firms included in virtual firms (VF) supply chains

Integration of firms strucures			Firms	Percentage of firms
Sewing contractor delay to deliver an urgent small order (100 garments) to a Canadian VF	2 weeks		R, CH, D	27%
	4 weeks		B, N, M, J	36%
	6 weeks		E	9%
	8 weeks		A	9%
	12 weeks		C, Y	18%
Compatibility of pattern software utilized by VF designers with the one utilized by sewing contractors	Yes		R, CH, D	25%
	No		B, C, N, M, S, J, Y, E, A	75%
Frequency of communication between VF designers and sewing contractors for designing an easy to assemble garment	0 times		S, J	20%
	1 times		N, E	20%
	3 times		R	10%
	Frequently		B, C, D, Y, A	50%
Frequency of communication between VF designers and fabric designers during the design of a new garments	0 times		C, R, N, S, D	50%
	2 times		B	10%
	3 times		CH, E	20%
	Frequently		D, A	20%
ISO certification of sewing contractor	Yes		B, D	17%
	No		C, R, CH, N, M, S, J, Y, E, A	83%
Sewing contractor equipments can package garments as required by VF retailers	Yes		12	100%
	No		0	0%
Fabric manufacturer equipments can package fabrics as required by VF sewing contractor	Yes		R, CH, N, D, M, J, E, A	67%
	No		B, C, S, Y	33%
Transportation system utilized to move garments from sewing contractor factory to Canadian VF facilities (average delivery delay)	Ship		B, C, R CH, N, D, S, Y, E, A (4 weeks)	83%
	Air freight		M, J (2 days)	17%

The results in Table 3 suggest that the supply chains of the analyzed VF do not have quick responses: A small urgent order can be delivered in 2 to 4 weeks by 63% of VF. This situation is the result of poor integration of the structures of the firms included in the VF supply chain. For example, most of time the garment pattern software utilized by the VF designers and the sewing contractors is not compatible (75% of cases). Hence, they are forced to communicate frequently (50% of cases) during the development of new garments. In addition, the majority of the sewing contractors are not ISO certified (83% of cases) so that the number of defective garments produced can be high, lengthening the production delay. Finally, the transportation of the garments from the sewing contractors' factories to Canada takes 4 weeks (83% of cases) because it is done by sea (airfreight is too expensive). At this point note that the situation above (software incompatibili-

ty, frequent communication, lack of ISO certificates) increases the overhead costs of the garments offered by the VF. Hence, sewing garments in low-wage countries does not always minimize the total (direct and overhead) costs of the garments.

The results of Table 4 point out that the supply chains of the analyzed VF are not robust. The firms included in a VF supply chain cannot react coherently to unexpected events. In effect, most VFs have neither software to plan and control the development of new garments (75% of cases) nor software to coordinate their operating activities (67% of cases). In addition, 33% of VFs do not collect information regarding the amounts and styles of garments sold at the retailer and another 33% of VF collect this information by telephone. The latter is neither accurate, nor reliable.

Table 4 Degree of coordination of operating procedures of firms included in virtual firms (VF) supply chains

Coordination of firms procedures		Firms	Percentage of firms
Software utilized by VF to plan and control the activities of developing a new garment	Excel	C, CH, M	25%
	Nothing	B, R, N, D, S, J, Y, E, A	75%
Software utilized by VF to coordinate the activities of firms included in VF supply chain	Excel with internet	CH, Y	17%
	Mytech with internet	N	8%
	E-mail	B	8%
	Nothing	C, R, D, M, S, J, E, A	67%
Software utilized by VF to collect retailer sales data and transfer it to sewing contractor	Store 21 with internet	CH	8%
	Retail Plus with internet	Y	8%
	Mytech with internet	N	8%
	Momentis with internet	D	8%
	Telephone	R, S, J, E	33%
	Nothing	B, C, M, A	33%

The results of Table 5 show that the analyzed VF do not apply product modularity and do not postpone the point for differentiating the products in the supply chain process. Effectively, most of VF designers do not know the principles to design easy to assemble garments (75% of cases). Hence, these designers probably do not know the principle of product modularity, i.e. designing garments by combining common garment components that are made efficiently by the VF. In addition, Table 5 shows that 58% of VF designers communicate 0 or 1 time with the sewing contractor in order to define the garment production process. This communication is so infrequent that the best point of the supply chain process for differentiating the garments is probably not identified by these people.

4. CONCLUSIONS

In summary, most of VFs in the Canadian apparel industry have not implemented the principles of mass-customization (Figure 1). The lack of integration of the physical structures of the firms included in the VF supply chains (Table 3) seems difficult to be solved in the short term because this integration requires the improvement of the infrastructure of the countries of the firms included in the VF supply chains. This improvement is out of the control of the Canadian VF.

REFERENCES

Ashdown S., S. Loker, K. Schoenfelder K., L. Lyman-Clarke (2004), Using 3D scans for fit analysis, *Journal of Textile and Apparel, Technology and Management*, 4(1), p.1-12.

Anderson-Connell L. J., P. Ulrich, E. Brannon (2002), A consumer-driven model for mass customization in the apparel market, *Journal of Fashion Marketing and Management*, 6(3), p.240-258.

Anderson D. M. (2004), *Build to Order and Mass Customization*, CIM Press, Cambria: CA.

Berman B. (2002), Should your firm adopt a mass customization strategy?, *Business Horizons*, July-August, p.51-60.

Crama Y., R. Pascual, A. Torres (2004), Optimal procurement decisions in the presence of total quantity discounts and alternative product recipes, *European Journal of Operational Research*, 159, p.364-378.

Da Silveira G., D. Borenstein, F. Fogliatto (2001), Mass customization: Literature review and research directions, *International Journal of Production Economics*, 72, p.1-13.

Dong M., F. Chen (2004), Performance modeling and analysis of integrated logistics chain: Analytic framework, *European Journal of Operational Research*, in press.

Dejonckheere J., S. Disney, M. Lambrecht, D. Towill (2003), Measuring and avoiding the bullwhip effect: A control theoretic approach, *European Journal of Operational Research*, 147, p.567-590.

Ernst R., B. Kamrad (2000), Evaluation of supply chain structures through modularization and postponement, *European Journal of Operational Research*, 124, p.495-510.

Gurzki Th., H. Hinderer, U. Rotter (2003), Individualized avatars and personalized consulting. In: *The Customer Centric Enterprise* (M. Tseng, F. Piller (Ed.)), p.477-489. Springer Verlag, Berlin.

Table 5 Degree of implementation of mass customization principles by virtual firms (VF)

Implementations related to mass customization		Firms	Percentage of firms
Principles applied by VF designers to develop varied but easy to assemble garments	Garments made of standard fabrics	CH, D	17%
	Garments made of common components	M	8%
	No principles	B, C, R, N, S, J, Y, E, A	75%
Frequency of communication between VF designers and sewing contractors to define garment production process	0 times	D, M, S, Y	33%
	1 time	R, N, J	25%
	2 times	CH, A	17%
	4 times	B	8%
	Frequently	C, E	17%

Ghiassi M., C. Spera (2003), Defining the Internet-based supply chain system for mass customized markets, *Computers & Industrial Engineering,* **45,** p.17-41.

Kolisch R. (2000), Integration of assembly and fabrication for make to order production, *International Journal of Production Economics,* **68,** p.287-306.

Lee C. C., W. Chu (2004), Who should control inventory in a supply chain? *European Journal of Operational Research,* in press.

Lee S. E., J. C. Chen (1999), Mass-customization methodology for an apparel with a future, *Journal of Industrial Technology,* **16(1),** p.2-8.

Martinez M. T., P. Fouletier, K. H. Park, J. Favrel (2001), Virtual enterprise organization, evolution and control, *International Journal of Production Economics,* **74,** p.225-238.

Ma S., W. Wang, L. Liu (2002), Commonality and postponement in multistage assembly systems, *European Journal of Operational Research,* **142,** p.523-538.

Partanen J., H. Haapasalo (2004), Fast production for order fulfillment: Implementing mass customization in electronics industry, *International Journal of Production Economics,* **90,** p.213-222.

Svensson C., A Barford (2002), Limits and opportunities in mass customization for build to order SMEs, *Computers in Industry,* **49,** p.77-89.

Sheu J. (2003), A multi-layer demand responsive logistics control methodology for alleviating the bullwhip effect of supply chains, *European Journal of Operational Research,* in press.

Stadtler H. (2004), Supply chain management and advanced planning basics, overview and challenges, *European Journal of Operational Research,* in press.

Selladurai R. S. (2004), Mass customization in operations management: Oxymoron or reality?, *OMEGA,* **32,** p.295-300.

Simchi-Levi D., Ph. Kaminsky, E. Simchi-Levi (2004), *Managing the Supply Chain.* Mc Graw Hill, New York: NY.

Taylor C. R. Harwood, J. Wyatt, M. Rouse (2003), Implementing a mass customized clothing service. In: *The Customer Centric Enterprise* (M. Tseng, F. Piller (Ed.)), p.465-475. Springer Verlag, Berlin.

Tseng M. M., F. Piller (2003), The customer centric enterprise. In: *The Customer Centric Enterprise* (M. Tseng, F. Piller (Ed.)), p.3-16. Springer Verlag, Berlin.

Ulrich P., L. J. Anderson-Connell, W. Wu (2003), Consumer co-design of apparel for mass customization, *Journal of Fashion Marketing and Management,* **7(4),** p.398-412.

Wang S., B. Sarker (2004), An assembly type supply chain system controlled by kanbans under a just in time delivery policy, *European Journal of Operational Research,* in press.

ELSEVIER

IFAC

PUBLICATIONS
www.elsevier.com/locate/ifac

Voice Compression Model by means application of Fast Fourier Transform and Technical Digital for the Processing of Signals and Formats of Compression of Images.

Dr. Héctor Kaschel C Dr Francisco Watkins Dr (c) Enrique San Juan U.

Electric Engineering department, Engineering Faculty

Industrial Technologies Department, Technological Faculty

University of Santiago from Chile.

Av. Ecuador #3519 Estación Central. Santiago Chile.

Phons: (56) 2 -7762963 - (56) 2 2396 fax: (56) 2 - 6819079

hkaschel@lauca.usach.cl.   fwatkins@lauca.usach.cl   esanjuan@lauca.usach.cl

Summary

The present work shows the implementation of a model for voice compression through the use of digital technics for processing of signals and of the application of images compression formats . This work is based on the hypothesis that the compression of the voice under this scheme achieves a significant reduction of the figure of bytes used and the reduction of the speed in the voice transmission. The philosophy of this model is centered in which the voice signal is splited by a filter bank and then each band is converted into a compressed images. These signal are later transmitted through the channel. In the receptor the voice frame is recover through a synthesis process. For the voice processig techniques in the time domain and frequency domain are used, such like the application of digital filters IIR (Infinite Impulse Response), the FFT (Fast Fourier Transform), and different formats of images compression.

Keywords: Compression; Filters IIR; FFT; JPEG; PNG; TIFF; MOS.

1. Introduction

It is highly unlikely actually   of carrying out transmission of multimedial information in a format without compression. Mainly because day by day are more applications are created in this context, that require of a wide range of quality and performance according to the requirements of heterogeneous users. The alternative now, is that it is possible the massive compression of the data before making its transmission. Fortunately, during the last decades, a great number of investigations has led to many techniques of compression algorithms, that make feasible the multimedia transmission. All the compression systems require two algorithms, one for the compression of the data at the origin and another for the decompression in the destination. In the literature these algorithms are known respectively as encode and decoding algorithms. For many applications a multimedia document will only be coded once when it being stored in the server, but you can decode thousands of times when being seen by the clients. This asymmetry allows that the encode algorithm be slow and require expensive hardware, provided the decoding algorithm is quick and don't require a hardware of high cost. On the other hand, for the real time multimedia, as the videoconferences and the voice over IP, slowly encode it is

not acceptable. Another asymmetry arises to that is not necessary that the coding/decoding process you can inverse. For example, when compressing, to transmit and to decompress a file of data the user hopes to receive in correct way until the last bit of the original information be processed. In multimedia this requirement doesn't exist. In general, it is acceptable that the signal after coding and o decoding is lightly different from the original one. The encoding systems with losses are important, because to accept a small loss of information can offer enormous advantages in the relationship of possible compression. For example, the compression algorithm of images JPEG and voice compression LPC [1][2][3].

2. Formulation of the Model

To make the model formulation it is needed to explain differents aspects of its performance. Such aspects are describing as follows.

2.1 Voice conversion in Image

This technique has as objective to transform voice frames into compressed images which are transmitted, then to make in the receiver the synthesis process to recover the voice frame. To get it, the voice frame is processed through a bank of a band pass filters, which eliminate redundancy in the information. The bank of filters is inside the range of frequencies where the main formants of the voice are located. The main formants are detected dynamically by the software system or they are given by the users in arbitrary form (according to studies carried out by diverse investigations). In this way, the frequencies in which these formants are located as the central frequency of each band pass filter of the filters bank.. The signals filtered by the bank of filters is processed later by means of an FFT algorithm. As the Fourier transform gives complex samples, the real and imaginary part of the signal filtered in the frequency dominion. Each of them becomes an separated image at which is applied a image compression format   like the well-known JPEG, TIFF and PNG. Getting two compressed format of the voice frame as images. Both images are are transmitted and they are received in the receiver, in where the inverse process is made.In this way, to obtain the signal in the domain of the time. The model described for the transmission stages and reception are shown in figures 1, 2 and 3 respectively.

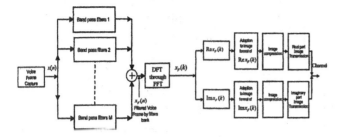

Figure 1: Compression Model at the Transmitter

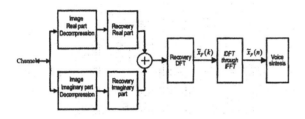

Figure 2: Decompression Model at the Receiver

2.2 WAV file Structure

In Table 1 the structure of a file WAV is described. Its knowledge and handling is key for the implementation of the program of simulation of the outlined compression model. This format is one of the most utilized to store sounds, it is to store the samples one after other (after the head of the file that indicates the sampling frequency among other things), without any type of compression of data, with uniform quantification. The simplicity of this format makes it ideal for the digital treatment of the sound, in fact Matlab has functions to work with this type of files [5].

Table 1: Format of the files WAV

Bytes	Usual content	Description
00 - 03	"RIFF"	Identification block (without quotation marks).
04 - 07	Data	Long integer. Size of the file in bytes, including head.
08 - 11	"WAVE"	Another badge.
12 - 15	"fmt "	Another badge.
16 - 19	16, 0, 0, 0	Size of the head until this point.
20 - 21	1, 0	Format label. .
22 - 23	1, 0	Number of channels (2 if it is stereo).
24 - 27	Data	Sampling frequency (samples/s).
28 - 31	Data	Half number of bytes/s.
32 - 33	1, 0	Block alignment.
34 - 35	8, 0	Number of Bits for sample (usually 8, 16 or 32).
36 - 39	Data	Marker that indicates the beginning of the data of the samples.
40 - 43	Data	Number of bytes sampling.
rest	Data	You show (uniform quantification)

3. Implementation of Model Description

In this section it corresponds to introduce the outlined model description for implementing. This is shown infigures 1 and 2. The simulation is carried out by means of Matlab. A flow diagram associated to this implementation is also shown (figure 3). In synthesis the model allows to transform a voice frame into a compressed image by means of the formats JPEG, TIFF and PNG and in this way, to carry out the transmission of a compressed image through the network.This image contains the the voice information. In in the receiver the process of voice recovery is carry out

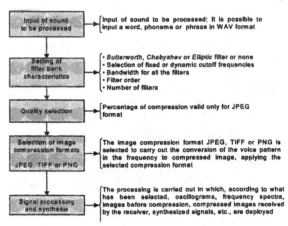

Figures 3: Diagram of flow for the simulation of the model

The voice frames are in WAV format. The steps that the simulation program does to implement the model of figures 1 and 2 are the following :

3.1 Voice frame Selection

This option allows to select the voice frame to be This files are in WAV format.

3.2 Filters Bank Selection

There are two types filters bank to select. In one of them the characteristic are given by the user.The other the characteristic are given by the system dynamicway. In the fixed filters the locations of the cut frequencies and bandwidths for each one of them it is determined arbitrarily. This is done according to the location of the first two formants for the five Spanish vowels. For the four pass band filters fixed incorporated in the program, the locations of the formants are specified in the Table 2. The bank of dynamic filters consists on determining the cut frequencies of the amount of filters of user-defined dynamically.The users can choose Butterworth, Chebyshev or Elliptic filter, in a range between 0 and 4 kHz.Also should be selected, the order of the filters and the bandwidth that will have got. The methodology to select the location of the filters inside the range 0 to 4 kHz, is centered in an analysis of the frequencies spectrum of the complete frame of the voice signal. by means of the determination of the peaks of amplitude of the spectrum. It is observed that the multiplication among the selected bandwidth for the filters and the quantity of selected filters doesn't overcome the 4 KHZ, that is to say that it completes the following relationship *Filters numbers* x *BW*≤4kHz. The program doesn't consider another peak inside oneself bandwidth where has already been a peak [5].

Filter No.	Lower cutoff frecuency (Hz)	Upper cutoff frecuency (Hz)	Bandwidth BW
1	200	300	100
2	500	800	300
3	1150	1500	350
4	2000	2500	500

In this way, the program finds the amount of filters selected with a certain BW.

Another important parameter is the order of the filter which is selectionable as much for the fixed ones as for the dynamic ones. The voice frame is processed by each filter by means of time convolution, this way an important saving of time with respect to the operation over the filters were done in the frequency.

3.3 Reduction of redundancy by means of bank filters

The application of the bank of filters is very important , since it is possible to remove redundancy to the signals. this also allows that for many frequencies the levels of amplitude of the spectrum are very small or zero. That means when a filtered voice frame is converted in an image, these levels are presented like a very similar gray levels. Which it allows to achieve a bigger compression of the data when a selected images compression format is applied.

3.4 Conversion of the voice signal in image

When the FFT algorithm is applied to the output filtered signal of the filter pass band bank an M, N conjugated complex samples are obtained. So for transmission purposes it is enough to process one of the samples of each conjugated pair, i.e. a total of N/2 samples, thereby producing important BW savings in the channel. Then it is convenient to process separately the real and the imaginary parts of the samples. Thus, N/2 real and N/2 imaginary numbers are respectively converted into image format in a gray scale. Obtaining in this way an image format of the real part and another one of the imaginary part.

3.5 Application of images compression formats

When the voice frame information in the frequency domain is obtained in two vectors in image format, it is possible apply to them some image compression format such as JPEG. Which has been proven to have a good quality/compression ratio.Which is a good alternative for the proposed model.That also makes it possible to decide the quality of the compression to be used. In order to make a qualitative and quantitative measurement of the efficiency of the JPEG format in this model, it is interesting to compare with others formats. The TIFF and PNG formats are also incorporated. As seen in Figure 1, the compression format is applied separately to the real and the imaginary parts of the DFT, carried into image format, thereby

creating two compressed images which are transmitted through the channel.

3.6 Voice synthesis in the receiver

The compressed images of the real and imaginary parts of the DFT reaching the receiver are converted into the frequency domain, obtaining N/2 complex samples of type $a+jb$. Conversion of the compressed image of the real part gives the real component a, and conversion of the compressed image of the imaginary part gives the imaginary part b. Then their conjugates ($a-jb$) must be constructed in order to get the N complex samples that constitute the DFT. If we designate as DFT_T the DFT of the signal filtered by the filter bank in the transmitter, and as DFT_R the DFT in the receiver, it is seen that both are not equal because DFT_R comes from an image compression process that causes losses. The synthesis process is achieved when the IDFT is applied to the DFT_R, and in this way the synthesized voice pattern is obtained.

4. Develop of the simulation

An example for a particular simulation is given below by processing the sentence "El perro olfatea la comida de su amo" ("The dog smells his master's food") spoken by a woman. It has a duration of 3.573 ms, and its samples have a weight of 28,485 kbytes. The data delivered to the simulation program are given in Table 3.

Table 3: Entry data for the simulation

Setting	Selection
Compression quality	50
Type of filter for the bank: Butterworth, Chebyshev, Elliptic or none	Butterworth
Selection of filter bank: fixed or dynamic	Fixed
Band width BW for all the filters in the bank, 200, 300, 400, 500 or according to Table 2	According to table 2
Filter order 2 or 4	4
Number of filters composing the bank	According to table 2
Name or type of WAV file to be processed	Phrase
Selection of compression format: JPEG, TIFF or PNG	JPEG

Figure 4 shows the oscillogram or time graph of the original signal before processing.

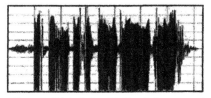

Figure 4: Original signal in time before processing

In the figure 5 are shown the real spectrum of frequencies and the imaginary spectrum of frequencies after applying the FFT to the output of filters bank, in both graphics the spectra are inside a compressed and normalized range between 0 and 1, for their later conversion in image. Figure 6 the equivalent images are shown for these spectra before their compression in the transmitter. In the figure 7 you can appreciate the real and imaginary images in the decompressed receiver. In the figure 8 the synthesized voice is shown starting from the conversion from these images to the real and imaginary parts of the FFT in the receiver one and then to calculate the IFFT.

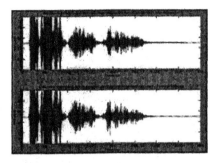

Figures 5: Real and imaginary spectra of frequencies.

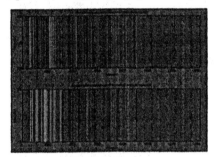

Figure 6: Real and imaginary images in the transmitter before compressing.

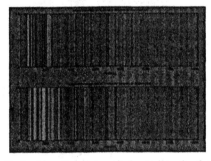

Figure 7: Real and imaginary images in the receiver after decompressing.

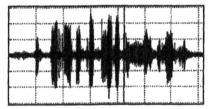

Figure 8: Oscillogram of the signal synthesized in the receiver.

4.1 Filters Performance

In relation to the performance of the pass band filters bank, it is obvious that decreasing the number of filters makes up the bank favors compression, however a difference appears in the voice audible quality. In this case the difference is not due mainly to noise, but rather to the application of different numbers of the filters that compose the bank, which results in a gradual alteration of the voice characteristics of the speaker as the number of filters in the bank is decreased. The extreme case occurs when the bank is taylored with just only one pass band filter whose bandwidth of less than 400Hz . That it makes the synthesized voice unintelligible. The use of different numbers of filters in the filter bank will depend on the characteristics of the particular test and of the compression levels that it is desired to achieve. Application of the filter bank provides up to 60% more compression compared to not applying the filter bank. With respect to which it is the filter more adapted to use in the model, it is found that the differences between the three types of filter banks are very subtle and the results obtained indicate that the *Butterworth* filter bank shows the best performance in terms of compression, while the *Elliptic* filter bank shows the lowest performance. Since the differences are minimal, it is recommended to choose the *Butterworth*, which is the most common and simple to implement. It is important to make some comments on dynamic and fixed filter banks. In general, fixed filter banks achieve a slightly lower compression level than dynamic filter banks, since the fixed Bank cover a greater frequency range than the dynamicfilter bank. This also implies that the audible quality obtained is slightly higher than if the process is carried out with dynamic filter banks.

4.2 Image compression format performance

In relation to which image compression format is more adequate or efficient, it will obviously depend on the particular application that is required. Based on the results, the most efficient performance is achieved with the JPEG and PNG formats, followed by the TIFF format. The JPEG format has the advantage of allowing the quality of the percentage compression to be selected, leading to great flexibility and making it a very convenient format for this voice compression model. In addition to making it possible to work with qualities that reach 1%, and even then in some cases to obtain a signal that have poor quality but intelligible. With this format, even with 25% quality a synthesized voice with a MOS rating of 4.5 can be obtained, achieving in this case image compression for transmission between approximately 54% and 71%, depending on the number of filters used in the bank. For the PNG format the compression percentage in general is comparable to those of the JPEG between 25% and 50% in quality. With respect to the MOS, PNG is comparable to JPEG in the quality range between 75% and 25%. These characteristics make PNG a highly convenient format to be used in similar way as JPEG. Even though it does not have the possibility of selecting the compression quality. Finally, the format that provides the least is TIFF, with about 50% of the compression achieved with the PNG format.

5. Conclusions

After evaluating the results, it can be concluded that the objective of this work was satisfactory, since it is possible to convert voice patterns into images which are compressed by means with an image compression format and then are transmitted through the channel. The voice pattern is recovered in the receiver through a synthesis or decoder process, therefore confirming the proposed hypothesis, which states that voice compression under this scheme achieves a significant reduction of the number of bytes and in consequense decrease of the speed in bit/s needed for transmitting the information. It has been shown that the voice compression model proposed here is viable and it is possible to achieve important voice compressions, in some cases reaching values of about 60% compression (MOS 4.5, fixed filter bank) with reasonable quality depending on the particular application. The compression values obtained are varied and depend largely on the setting of the filters, the compression formats selected and the desired audible quality level. But for a MOS quality between approximately 4.0 and 4.7 speeds between 15.5 kbit/s and 42.7kbit/s, respectively, can be achieved. The model developed is innovative, since at first the conversion of voice patterns into equivalent images does not seem obvious. Under this scheme it might be possible to transmit audio images together with video images properly. Both through the same channel in the appropriate manner, in this way probably avoiding synchronization problems between audio and video, with the corresponding saving of one channel and improvement of the service. They could be continued giving this model's possible applications. Depending in great measure of the particular necessities that are presented. It is necessary to highlight that the model provides very good expectations in what concerns to compression of the voice. It is interesting to continue considering it in experimental applications and in the future to constitute a good alternative of outline of voice compression in real applications.

References

[1] Andrew S Tanenbaum. "Computer Networks" Pearson, 3° edition 1997

[2] Scott Keagy et.al "Integration Voice and Networks, Cisco Dates System, Pearson , 2001

[3] Jonathan Davidson, James Peters, "Voice over IP Fundamentals", Cisco System, Pearson , 2001

[4] Format of the files wav. http://www.upv.es/protel/usr/jotrofer/sonido/sound.htm

[5] Marcos Faúdez Zanuy, digital "Treatment of voice and image", Marcombo , 2000

[6] Arturo of the Stairway, "Vision for computer", Prentice Hall, 2001

[7] Castleman, K.R.: "Digital Image Processing", Prentice-Hall, Englewood Cliffs, New Jersey 07632, 1996

[8] Usevitch, B.E. "To tutorial on modern lossy wavelet image compression: foundations of JPEG 2000", IEEE Signal Processing Magazine, Volume: 18 Issue: 5, Sept. 2001 Page(s): 22 -35

[9] Skodras, A.; Christopoulos, C.; Ebrahimi, T. "The JPEG 2000 still image standard" compression, IEEE Signal Processing Magazine, Volume: 18 Issue: 5, Sept. 2001, Page(s): 36 -58

[10] Deborah Olive tree. "JPEG: Method per the compressione gave immagini." Maggio 1998 http://utenti.lycos.it/debolivo/jpeg/indexj.html

[11] JPEG. http://coco.ccu.uniovi.es/immed/compresion/ descripcion/jpeg/jpeg.htm

[12] Proakis, J.G. and Manolakis D.G. "Digital Signal Processing. Principles, Algorithms and Applications" Prentice Hall, INC. 1998

[13] Vinay K. Groin, John G. "Digital Proakis Signal Processing. Using Matlab V.4." PWS Publishing Company.1997

[14] Maurice Bellanger. "Digital Processing of Signals. Theory and Practice." John Wiley & Sons. 1986

[15] "Signal Processing Toolbox. For uses with Matlab." User's Guide version 4.2. The MathWorks Inc. 1999. http://www.mathworks.com

[16] Schefer and Rabiner. "Digital Representation of Speech Signals", IEEE 1975

[17] J.Makhoul. To "line Predictions: To Tutorial Review" Speech and Signal Processing, IEEE 1975

[18] Frédéric J.Harris. " On the Uses of Windows for Harmonic Analysis with the Discrete Fourier Transform." Proceeding of the IEEE, 1978

[19] Luis Alfaro Ferreres; María José Rock Estellés. Systems of diffusion of audiovisual information in Internet: Graphic format PNG. http://www.conganat.org/iicongreso/comunic/008/png.htm

[20] Tutorial of the SPRING. The format TIFF: http://www.dpi.inpe.br/spring/usuario_spa/tiff_format.htm

ELSEVIER
IFAC
PUBLICATIONS
www.elsevier.com/locate/ifac

Optimal Reliability Allocation: Simplification and Fallacy

Yu Haiyang[*], Chu Chengbin, Yalaoui Farouk, Châtelet Éric

Fax: +33-03 25 71 56 49, Phone: +33-03 25 71 84 34, Email: Haiyang.YU@utt.fr

Institute of Information Science and Technology of Troyes (ISTIT), University of Technology of Troyes, France

ABSTRACT—Reliability is a principal measure of system performance, which anticipates the uncertainty of system failures. Nevertheless, high reliability often indicates huge system cost, rendering the problem of system reliability optimization. This paper emphasizes an insight of modelling the system reliability for optimization purposes. The omission of the system lifetime in allocating the redundancies is proved to be improper. Besides, the dependences within systems are found to be essential. That is, dependencies are inherent for the reliability evaluation. Unfortunately, the independency is generally utilized in the literature. With the dependencies concerns, our study breaks down one popular view in the reliability engineering, which believes that cold-standby systems are often more 'reliable' than warm-standby ones. This paper is hence contributed to the right conception on the optimization of system reliability.

KEYWORDS— system *lifetime, system reliability, dependency, optimization*

1. INTRODUCTION

Things fail. Many failures are significant in both their economic and safety effects [1]. That is where the study of system reliability goes. The probability that a system successfully performs as designed is called 'system reliability', or 'probability of survival' [2].

Generally, for a system without maintenance, its reliability will consume up after a long-run,

$$\lim_{T \to \infty} R_S(T) = 0 \qquad (1)$$

Consequently, if the reliability level of a system is required, there must be a given lifetime, i.e.

$$R_S(T) \geq R_0 \Rightarrow T < +\infty \qquad (2)$$

In the case of no maintenance, this finite lifetime is often omitted in the studies of optimal reliability allocation. The reliability is then taken as a real number [3]-[7]. The object of these studies is to manipulate the system structure, e.g. the number of redundancies and their reliability levels, so that the system reliability fulfils the reliability demand and the system cost is minimized,

$$\min C_S(k, r_1, r_2, ..., r_k)$$
$$s.t. \ R_S(k, r_1, r_2, ..., r_k) \geq R_0 \qquad (3)$$

Divers optimal algorithms, e.g. exact methods [3][4], genetic algorithms [5][6], and recently the Ant Colony Algorithm [7] are implemented, respectively, to obtain a favourable system structure.

However, the implicit omission of the lifetime excludes a promising way to optimize the system reliability

during the time horizon, as shown by R. Zhao & B. Liu [8] and David [9]. In fact, this omission has excised the stochastic distinctions from the reliability disciplines.

For an insight of this finite lifetime, taking an outcome [4] of the optimal reliability allocation in the literature, this paper carries out a comparison study in Section 2, the result of which questions the basis of optimal reliability allocation. Then, a further discussion on parallel redundancies is performed in Section 3, pointing out the fallacy of independence supposition in the redundancy studies. Section 4 shows the significance of the dependence, by reviewing the reliability of two-component standby systems. Our results rebut the popular view that, cold-standby systems are more 'reliable' than warm-standby ones. Finally, a concluding remark is addressed.

2. SPECIFIED LIFETIME OF A SYSTEM

2.1. An Optimal Solution

Recently, a basic problem of optimal reliability allocation has been achieved by C. Elegbede, et el [4]. One of the sub-problems is to minimize the cost of the system, whose components are in parallel,

$$\min C_S(k, r_1, r_2, ..., r_k) = \sum_{i=1}^{k} C(r_i) \qquad (4)$$

subject to the system reliability demand,

$$R_S(k, r_1, r_2, ..., r_k) = 1 - \prod_{i=1}^{k}(1 - r_i) \geq R_0 \qquad (5)$$

If the cost function $C(r_i)$ takes the same form for each component, and the function $h(x) = C(1 - e^x)$ is convex, the optimal solution holds that the reliability levels of components must be identical:

$$r_i = 1 - \sqrt[k]{1 - R_0}, i = 1, 2, ..., k \qquad (6)$$

where k is the optimal number of the components.

2.2. Another Way to Construct the System

Now recalling (1)(2), a finite lifetime of the system is actually connoted in (6):

$$r_i(T) = 1 - \sqrt[k]{1 - R_0}, i = 1, 2, ..., k \qquad (7)$$

As no maintenance is concerned, a component deteriorates during the system lifetime. The formula (7) means that each component is required to meet the same reliability level at the end of system lifetime.

On the other hand, given a time $\tau < T$, it holds for any of the components,

$$r(\tau) > r(T) \qquad (8)$$

In particular, we can observe such a τ so that

$$r(\tau) = R_0 \qquad (9)$$

Here τ notes the moment that the reliability of a single component provides the system reliability demand. Now our question comes naturally: what would happen if we add the parallel components one by one every the duration τ? Is this planning better than the former one? Readily, the system reliability of the new planning at the end of $n\tau$ duration can be derived,

$$R_S(n\tau) = r(\tau) + r(2\tau)[1 - r(\tau)]$$
$$+ r(3\tau)[1 - r(2\tau)][1 - r(2\tau)] + \dots \qquad (10)$$
$$+ r(n\tau)\prod_{i=1}^{n}[1 - r(i\tau)]$$

With (9), it is apparent that the system reliability is no less than the reliability demand,

$$R_S(k\tau) > R_S[(k-1)\tau] > R_S[(k-2)\tau] > \dots$$
$$> R_S(2\tau) > R_S(\tau) = r(\tau) = R_0 \qquad (11)$$

The system reliability within the duration $k\tau$ is illustrated in Fig 1.

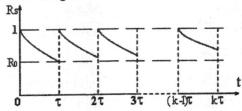

Figure 1. System reliability during time horizon

The new planning of the system satisfies the demand (5). In the following context, our interest is to compare the total 'reliable' duration $k\tau$ with the former system lifetime T.

2.3. The Comparison

In general, $k\tau/T$ depends on the number of redundancies k and the reliability demand R_0. Besides, given the probability distribution of $r(t)$, T and τ can be obtained from (7) and (9), respectively. The failure distributions of the component also affect $k\tau/T$. For example, given the two-parameter Weibull distribution as the component failure distribution,

$$r(t) = \exp\left[-\left(\frac{t}{\theta}\right)^{\beta}\right] \qquad \theta, \beta > 0 \qquad (12)$$

the $k\tau/T$ can be estimated through (13) and illustrated in Fig 2.

$$k\tau/T = k\left[\frac{\ln R_0}{\ln\left(1 - \sqrt[k]{1 - R_0}\right)}\right]^{\frac{1}{\beta}} \qquad (13)$$

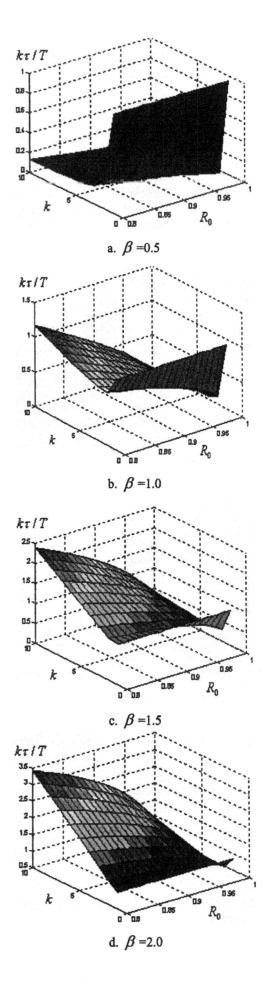

a. $\beta = 0.5$

b. $\beta = 1.0$

c. $\beta = 1.5$

d. $\beta = 2.0$

76

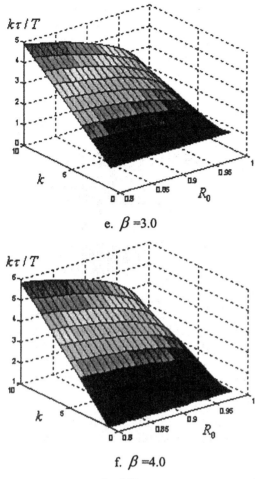

e. $\beta = 3.0$

f. $\beta = 4.0$

Figure 2. $k\tau / T$ with respect to
$k = 1,2,...,10$ and $R_0 \in [0.8, 0.99]$

Fig 2 shows that the 'reliable' duration $k^*\tau$ can be longer than the former system lifetime T. Note, such an improvement is only induced by the finite lifetime of systems. This implies that, the omission of the lifetime does underestimate the optimal solutions.

3. THE FALLACY OF INDEPENDENCY

In the problem (5), it is implicitly supposed that adding components gives no effects to the failures of components that to be added. It is kind of fallacy, because it indicates that *the failure distinction of the system is independent of the system loading*. Actually, as emphasized by Misra [10], in the case of electronic devices and in the case of mechanical components, the independence assumption is almost always incorrect.

The simplification of independency obviously distorts the truth of reliability, because it negates the importance of dependences in a system. Fricks and Trivedi [11] point out that the dependences should never be ignored in reliability models. The failure dependences are commonly studied by many scholars, e.g. [12][13].

In addition, besides the failure dependences, the dependences still exist. Parallel components share the system loading. The failure of the system always depends on the system structure because it operates against the loading under certain conditions [14]. And the given system loading will be redistributed due to the

mutation of system structure. Such redistribution is intrinsic, e.g. the asymmetric components induce an inhomogeneous loading on components, which reacts on the failure rate of each component, which is an underlying truth in the reliability engineering.

The following section is devoted to realizing the fallacy in current studies on optimal reliability allocation. Hence, a comparative study is developed to show the indispensability of dependency.

4. TWO-COMPONENT STANDBY SYSTEMS

To make our comparison simpler, the systems are composed only by identical components in the following discussions. In addition, the time to failure of each component is exponentially distributed.

4.1. Warm-standby

We are given two identical components in parallel. The system fails only if both of the components fail. Its reliability can be obtained directly:

$$R_S^W(t,\lambda) = 1 - \left[1 - e^{-\lambda t}\right]^2 = 2e^{-\lambda t} - e^{-2\lambda t} \quad (14)$$

where $\lambda > 0$ is the failure rate of the components and R_S^W is the system reliability.

4.2. Cold-standby

We are given two identical components. Firstly, one component is put into the system and the other component is put into as soon as the primal one fails; the system fails if both of the components have failed. Suppose x, y is the time to failure of the components, the probability of the sum $x + y$ can be calculated through its joint p.d.f,

$$P(x+y<t) = \int_0^t \left(\int_{-\infty}^{\infty} f_{X,Y}(x, s-x) dx \right) ds \quad (15)$$
$$= \int_0^t \lambda^2 s e^{-\lambda s} ds = 1 - (1 + \lambda t) e^{-\lambda t}$$

Then we obtain the system reliability as

$$R_S^C(t,\lambda) = P(x+y \geq t)$$
$$= 1 - P(x+y<t) = (1 + \lambda t) e^{-\lambda t} \quad (16)$$

where $\lambda > 0$ is the failure rate of the components and R_S^C is the system reliability.

In addition, the reliability differences between cold-standby and warm-standby systems can be illustrated in the Fig 3:

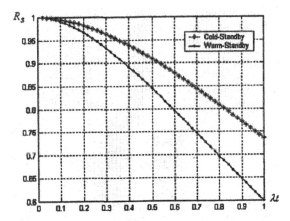

Figure 3. Reliability comparison of a cold-standby system and a warm-standby system

4.3. Load-sharing

We are given two identical components in parallel. The system fails only if both of the components fail. But there is now a dependency between the two components. If one component fails, the failure rate of the other λ component increases as a result of the additional load on it. Markov analysis is adequate to determine the system reliability.

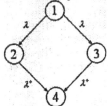

Figure 4. Rate diagram for a two-component load-sharing system

The rate diagram and the four states of the system are shown in Fig. 4, where $\lambda^+ > \lambda$ represents the increased failure rates of components. The probability transition equations of the system are:

$$
\begin{cases}
\dfrac{dP_1(t)}{dt} = -2\lambda P_1(t) \\[2mm]
\dfrac{dP_2(t)}{dt} = \lambda P_1(t) - \lambda^+ P_2(t) \\[2mm]
\dfrac{dP_3(t)}{dt} = \lambda P_1(t) - \lambda^+ P_3(t)
\end{cases}
\tag{17}
$$

With the initial condition $P_1 = 1$, $P_2 = P_3 = P_4 = 0$, the solution to these equations can be derived:

$$P_1(t) = e^{-2\lambda t}$$

$$P_2(t) = P_3(t) = \frac{\lambda}{2\lambda - \lambda^+}\left[e^{-\lambda^+ t} - e^{-2\lambda t}\right] \tag{18}$$

And the system reliability will be:

$$
R_S^L(t, \lambda, \lambda^+) = P_1(t) + P_2(t) + P_3(t)
$$
$$
= \frac{1}{2\lambda - \lambda^+}\left[2\lambda e^{-\lambda^+ t} - \lambda^+ e^{-2\lambda t}\right] \tag{19}
$$

where R_S^L is the reliability of the load-sharing system and is shown in Fig5, compared to the reliability of cold-standby system.

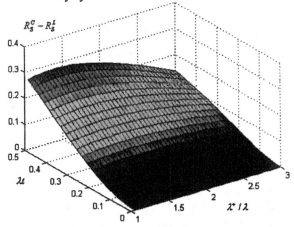

Figure 5. Reliability comparisons of a cold-standby system and a load-sharing system

Fig 3 brings the classic view that a cold-standby system is often much 'reliable' than a warm-standby system. Nevertheless, Fig 5 reaches the same conclusion even the failure dependencies have been concerned.

4.4. True Cold-standby

In the load-sharing system, it is often supposed that, the failure rate of the remaining component increases as a result of the additional load placed on it. In the same way, for the cold-standby system, each component in fact has actually already shared the additional load because only one component takes the whole system loading. That is,

$$
R_S^{C^*}(t, \lambda^+) = P(x + y \geq t)
$$
$$
= 1 - P(x + y < t) = (1 + \lambda^+ t)e^{-\lambda^+ t} \tag{20}
$$

where $R_S^{C^*}$ is the reliability of the true cold-standby system

Similarly, the reliability differences between the true cold-standby system and the load-sharing system are addressed in Fig 6.

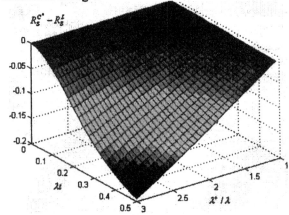

Figure 6. Reliability comparisons of the true cold-standby system and the load-sharing system

Because of the general dependence, Fig 6 offers a totally different observation: true cold-standby systems are absolutely not more 'reliable' than a true warm-standby (load-sharing) system. This result directly puts down the popular belief that, the cold-standby systems are generally more 'reliable' than warm-standby ones.

5. CONCLUDING REMARKS

A finite lifetime should not be omitted in system reliability optimization, if no maintenance is concerned. Omission of such a lifetime excises the stochastic characters of the system reliability, which excludes many promising ways to improve the system reliability. Recent studies in optimal reliability allocation should be reviewed to reflect this fact. Otherwise, the optimal solutions could be underestimated.

If the system reliability is independent of the system loading, the warm-standby system with components in parallel is not better than the cold-standby system that puts the components online in sequences. However, in the formulation of system reliability optimization, a finite system loading is often supposed implicitly. This fact makes it indispensable to concern the general dependencies in system reliability. Our presented paper shows that, the dependent warm-standby system is more reliable than a cold-standby one, which absolutely rebuts the classical belief.

In a word, those studies that optimize the system reliability in non-stochastic scenery, must realize their fatal errors at the very beginning.

REFERENCES

[1] E.E. Charles, *An introduction to reliability and maintainability engineering.* (International Editions 1997) New York: McGraw-Hill Companies, Inc., 1997, ch. 9, 11.

[2] W. Kuo, V.R. Prasad, F.A. Tillman, C.L. Hwang, *Optimal Reliability Design: Fundamentals and Applications*, Cambridge University Press, UK, 2001.

[3] G. Levitin, "Redundancy optimization for multi-state system with fixed resource-requirements and unreliable sources", *IEEE. Trans. Reliability*, vol. 50, March 2001, pp. 52-59.

[4] C. Elegbede, C. Chu, Kondo H, F. Yalaoui, "Reliability allocation through cost minimization", *IEEE. Trans. Reliability*, vol 52, March 2003, pp. 106-111.

[5] V. R. Prasad, W. Kuo, "Reliability optimization of coherent systems", *IEEE. Trans. Reliability*, vol. 49, no. 3, September 2000, pp. 323-330.

[6] David W. Coit, Alice E. Smith, "Reliability optimization of series-parallel systems using a genetic algorithm", *IEEE. Trans. Reliability*, vol. 45, June 1996, pp. 254-260.

[7] Yun-Chia Liang, Alice E. Smith, An ant colony approach to redundancy allocation, *submitted* to IEEE Transaction on Reliability.

[8] R. Zhao, B. Liu, "Stochastic programming models for general redundancy-optimization problems", *IEEE. Trans. Reliability*, vol. 52, June 2003, pp. 181-191.

[9] W. C. David, "Cold-standby redundancy optimization for non-repairable systems", *IIE Trans.*, vol. 33, 2001, pp 471-478.

[10] K. B. Misra, *New trends in system reliability evaluation*, The Netherlands: Elsevier Science Publishers, Amsterdam, 1993

[11] R. Fricks and K. S. Trivedi, "Modeling failure dependencies in reliability analysis using stochastic petri nets", *Proc. European Simulation Multi-conference (ESM '97)*, Istanbul, June 1997.

[12] J. K. Vaurio, "Treatment of general dependencies in system fault-tree and risk analysis", *IEEE. Trans. Reliability*, vol. 51, September 2002, pp. 278-288.

[13] A. Bobbio, L. Portinale, M. Minichino, E. Ciancamerla, "Improving the analysis of dependable systems by mapping fault trees into Bayesian networks", *Reliability Engineering & System Safety*, vol. 71, March 2001, pp. 249-260.

[14] G. A. Klutke, Y. Yang, "The availability of inspected systems subject to shocks and graceful degradation", *IEEE Trans. Reliability*, vol. 51, September 2002, pp. 371-374.

ELSEVIER

IFAC

PUBLICATIONS
www.elsevier.com/locate/ifac

ON DIAGNOSIS OF MACHINING PROCESSES: A FUZZY APPROACH

Orlando Durán[1*], Luiz A. Consalter[2], Ricardo Soto[3]

[1] Pontificia Universidad Católica de Valparaíso, Esc. Ing. Mecánica, Phone: 56.32.274471, Fax
56.32.274471 Av.Los Carrera 1567, Quilpué,CHILE, orlando.duran@ucv.cl * Corresponding Author
[2] Universidade de Passo Fundo, FEAR Passo Fundo, RS, Brasil, lac@upf.br
[3] Pontificia Universidad Católica de Valparaíso, Esc. Ing. Informática, ricardo.soto@ucv.cl

ABSTRACT

Manufacturing operations have empiric nature. Under any kind of problem during the execution of a given process, practical knowledge, experience and intuition have been used in solving these problems. However, many firms face problems since they do not have an adequate knowledge base to maintain and represent skills and expertise. Therefore, the importance of developing intelligent systems that assist the users in generating solutions for given problems. This paper presents a fuzzy based approach for a diagnosis system for solving machining problems. Because of the probability of having a great number of fuzzy rules, a rule navigation procedure, which aim is accelerating the searching process enhancing the system performance is proposed.

KEYWORDS: Diagnosis, Machining, Fuzzy Logic

INTRODUCTION

Manufacturing processes, specially, machining process, have essentially an empiric character. Thus, any initiative to solve practical problems or to carry out optimizations in the production will depend on experimental results, empiric appreciations and mainly, on technical experience. The great quantity of cutting parameters and variables involved in a manufacturing process (with a strong relationship among them) increases the difficult of optimization efforts. To ignore the fact that modification of given cutting parameter, serious prejudices can be caused to the manufacturing process. This fact leads to higher costs, low productivity indexes and low quality response. As an example for better understanding, consider the case of the intent to solve the problem of successive crashes of the cutting tool. To avoid tool breakages the technician or specialist could adopt tool geometry with a negative angle. However, if the cause of the problem were the resulting vibration caused by precarious rigidity of the machine-tool-fixture-piece system and, therefore, the solution adopted by the technician would increase the forces on tool surfaces causing a higher vibration in the operation, and the final collapse of the tool. An appropriate solution for this case would be, indeed, the simple selection of another tool material with a higher toughness. Many situations as the described here can be found within national and foreign industry.

In spite of the availability of information within catalogs and charts, plus the information that comes from the technical support that certain suppliers offer to manufacturers, the knowledge that professionals that act within the shop floor, is not properly systematized for the correct application and use in industry. This can lead to mistaken decisions and loss of time in the solution of problems. As a consequence, one will have an unnecessary elevation of the production costs, fall of the productivity index and a low quality performance, what can be reflected, in definitive, by the loss of competitiveness within a demanding market. All of this could be avoided or quickly overcomed if the technical knowledge were available in a structured manner. In addition, if one considers the fact that systematized knowledge doesn't exist in a company, or there is a lack of feasible records and technical documentation, this can lead to erroneous decisions in process planning tasks. This might occur mainly when the professionals are inexperienced or when the resources and technological procedures evolve and people don't accompany this evolution.

During operation execution, some problems can arise. Many of those problems have to be recovered to avoid superior consequences. Among the problems that could exist during operations, we can highlight tool breakage, vibration, deficient quality, low surface integrity, chip control and many more. In the tentative of solving those problems, many sources can be used, such as machining handbooks, tool manufacturers catalogues, etc. However, almost all the information contained in these sources has a quantitative character with nominal values. To use this information and advices for solving problems, a sensors system is needed. In other hand, many manufacturers do not have enough instrumentation to monitoring or controlling operation's behavior. Thus, any possibility of analyzing operation's performance concentrates on visual and auditive observations made by humans, such as operators and/or technicians. Almost all the judgments made by technicians and operators when trying to solve any operations problems, is made through imprecise and qualitative information. Therefore, nominal and quantitative values and recommended actions in the literature can be considered as a good starting point only in trying to solve the operations problems.

Another factor that should be taken into account is the knowledge escape from the firm. It occurs when skilled professionals migrate to other companies. If the company wants to maintain its level of quality and competitiveness, it has to recompose the lost knowledge with new investments in training or incorporation of new professionals with similar skills and experience in the area. This can consume time, during which the company is exposed to new errors and risks, proper of any learning curve.

This article attempts to describe a fuzzy based approach as a basis for diagnosis in cutting operations. In the next sections, we will review the literature about topics such as cutting operations, expert systems and fuzzy logic. Finally, we describe the proposed approach. In the conclusions, we present the application potential and next actions pointing to implement a tele-diagnosis expert system for machining operations.

BACKGROUND

Among the manufacturing processes, machining has the major presence in industrial applications. It is known that approximately 80% of manufactured products have, directly or indirectly some machining operation. In addition, it is known that now the relation between effective time and downtime in machining corresponds to 70:30, the opposite of what happened before the 70s. That change occurred in function of new technological development and new machine tools. According to the paradigms of modern production, the economic and technical results of machining operations much depend on what happens during the effective cutting time (time where the tool is in effective contact with the piece) rather

than the benefits obtained from improving unproductive phases.

This leads to a great responsibility in the process-planning phase. At this point, it is evident the necessity of applied knowledge to avoid problems and to generate optimized solutions aiming at technical and economically satisfactory processes. This is possible through at least three types of resources: hardware, tooling and knowledge. Great part of the companies have hardware and cutting tools with high performance to face the markets demands, The opening markets and globalization guarantees the access to the most modern resources of production to anyone that wants it. Therefore, that is not the problem, but the absence or lack of systematization of machining knowledge, as mentioned previously, has been the great barrier to continuous improvement in machining operations.

The solution would be the systematization of machining knowledge in the form of decisions rules, forming a knowledge base within the machining domain. This knowledge base would be available to end-users and would be expandable and updateable.

The availability of low costs of software dedicated to the knowledge systematization and specifically, Experts Systems platforms, allows users create their own knowledge bases. Therefore, within environments where knowledge is available, it is no more admissible to use the information in an unstructured manner, however is an imperative the creation of systems able to supporting and managing company knowledge.

EXPERTS SYSTEMS AND MANUFACTURING

Expert Systems (ES) in the area of Manufacturing have been investigated and applied recurrently the last years. A recent survey in the literature made by Pham and Pham (1999) shows how different tools of artificial intelligence (AI) are applied in manufacturing environments. Diverse experiences point to the implementation of control systems, pattern recognition, production, and process planning combining the use of neural networks, fuzzy logic and genetic algorithms. Monostori (2003) comments the obtained advances with applications of artificial intelligence in the production, and concludes that the most promissory application in this area are those related with Agents and Holonic Systems. AI applications that are related with implementations within the factory are vision, natural language recognition and the diagnosis. AI applications may go through several stages of the product life cycle, from the Conceptual Design (Durán and Zanoni, 2001; Durán, 1999) until the programming in high-level languages of programmable devices (Durán and Batocchio, 1997a; 1997b; 1998), covering also the production programming, simulation, and training.

Considering machining operations, few have been published in applications of Experts Systems. This is, we believe, because a great quantity of variables associated in these processes. As Wong et alli. (1999) pointed, the selection of cutting data is a complex task and it cannot be easily formulated through deterministic models. According to the mentioned authors, optimized information on appropriate cutting conditions is obtained basically, from the experience of operators and technicians, as well as, based on human intuition. Most of the taken decisions are made based on incomplete and approximate information, situation that leads to think of tools of AI to achieve correct computational modeling. Also according to Wong et alli. (1999), Fuzzy Logic is an appropriate methodology for representing the strategy and actions that a specialist executes when select cutting tools and their cutting conditions. A similar work was presented by Almeshai and Oraby (2003) who proposed a solution based on AI to evaluate selected cutting parameters, comparing them with information of the available resources, production objectives and restrictions. The developed system by the authors seeks to predict the performance of operations through certain factors as tool wear, cutting forces, superficial quality, roughness and even vibration caused by the operation. The solution proposed by the authors consists on two fundamental steps, application of mathematical fundamental model, such as Taylor´s equations and other, and in a second stage, the use of logical algorithms for the implementation of the evaluation task of the conditions suggested by the user.

In spite of the few work found in the literature about AI techniques and processes planning in machining operations, we can conclude that for the type of information and the nature of the process, Fuzzy Logic is the most appropriate approach for automating the process planning process and representing machining decisions.

FUZZY LOGIC AND MACHINING

We can concentrate this discussion on three main and recent articles in the literature. Chen et alli. (1995) presented a Fuzzy Logic based Expert System called SAM (Smart Assistant to Machinist) that basically aids the specialists in the selection of a tool and it establishes the using conditions, starting from imprecise and approximate knowledge. Fang (1995) proposed a system that allows diagnosing of turning operations, using fuzzy logic. The methodology used by the author introduces the concept of fuzzy relationship matrix to quantify the real threat that exists starting from a series of coming signals of behavior (cutting forces, vibrations and other parameters of the machining operation). The aforementioned article points to the use of the proposed system with an on-line machine tool monitoring module. More recently, Hasmi et alli. (2000) presented a system of selection of cutting data for drilling operations based on a Fuzzy approach. The system allows selecting the cutting speed for this operation. According to the authors, the relationship among the hardness of a material and the cutting speed may be set as being of a Fuzzy type. Evaluating imprecise information starting from similar cases of drilling operations, materials hardness, different holes diameters and feed values, the system suggests the appropriate value of cutting parameters for the operation that is being planned.

DIAGNOSIS EXPERT SYSTEMS AND INTERNET

The Internet technology is presented as a powerful tool for quick dissemination of information and for taking decisions in distributed atmospheres. Maybe for that reason, this technology is beginning to be used for the diagnosis in different knowledge domains. For the same reason, the Internet technology has won significant space as a tool for the teaching-learning process, mainly when allied with multimedia resources. Ong et alli. (2001) present a prototype that carries out the remote diagnosis of fault caused by tool wear in numeric controlled machine-tools. For such an objective, the authors have based the work on the technology of agents. They use specifically, two agents, an agent of learning and another agent whose function is to carry out the diagnostic of the operation. The breakthrough particularity of this proposal, it is that when the fault cannot be solved by the prototype, the same one has the capacity of acquiring knowledge, translating it into new rules using the called "rule builder." Other examples that deserve to be mentioned are: Duan et alli. (2003) who developed a system to diagnose problems in aquaculture using the web, specifically, looking for the causes of deaths in fish through the participation of an expert system that interacts with remote users through the internet. Another initiative that uses Internet to perform diagnosis functions is that related by Wang et alli. (2003) that developed a system for analysis of metallographies. The proposed system points to detect and to analyze particles of waste through samples. The system, called RIESFD is based on a hybrid solution combining Neuronal Networks and Fuzzy Logic to carry out the diagnostics. The main output of the system is the level of probability (fuzzy term) that each one of the faults (in a set of five possibilities) has, that is: high temperature, contamination for water, fatigue, and presence of wear particles or overloads.

In the articles mentioned previously and in others that we will mention to continue, is important to mention the didactic aspects in the use of these systems. For example, Parkinson and Hudson (2002) highlighted that multimedia techniques offer great potential benefits like teaching tools. Therefore, these techniques can be used to extend considerably the experience of learning. The authors show this potential through a study of applied cases to the teaching of the design in engineering. In this case, the authors emulate the knowledge of diverse experts in design and make the students interact with this knowledge as if they were part of a team of specialist participating in a collaborative project experience. A similar reference is Pooley and Wilcox (2000) who described a prototype of a multi-user system based on Internet for applications in a factory atmosphere. The system, entirely written in Java, allows experimenting with the relative knowledge about production lines and their fine programming and planning. Accordingly Cao and Bengu (2000), the technology of Agents is the one that has the biggest potential for effectives environment of teaching through the Internet.

PROPOSED APPROACH

The approach is divided into two main parts. The first part points to structuring a knowledge base where intelligence on machining conditions is stored in a set of fuzzy rules. This fuzzy knowledge base is defined by machining specialists through the definition of a set of linguistic terms and its correspondent membership functions. Decisions rules are written using these terms and membership functions. The fuzzy rules will assist the user in resolving machining problems through the recommendation of cutting conditions and alterations in tool geometry or type.

Because machining operations involve a great number of variables and conditions, this approach could generate a great number of decisions rules. This constitutes a well-known

limitation of rule-based expert systems affecting the implementation's performance. Some authors have been investigating to develop efficient mechanisms for accelerating the search process within a rule based expert system. We suggest here the utilization of an interactive searching algorithm similar to the one proposed by Liu and Liu (2003). The authors created a meta-programming algorithm that builds a rule base in an automatic way. In other words, the propose is to implement a meta-mechanism that allows the system to select subsets of rules dedicated to certain symptoms and/or problems, concentrating the search process only on the rules associated with a particular element of the system Machine-Fixture-Part-Tool (MFPT). The selected elements are those related with the specific problem detected or declared by the user.

The proposed mechanism of navigation within the rule set can better understood through the following example (figure 1):

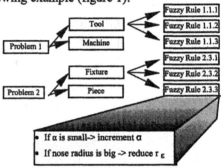

Figure 1. Navigation algorithm for enhancing the search process

As several problems or symptoms can arise simultaneously, it is necessary that the user identifies or qualifies the degree of intensity that each problem/symptom have. With such an objective, a scale of linguistic terms that represents the degree of intensity of each type of problem is manifested in a given application (Table 1) is defined.

Table 1.

PROBLEM			Linguistic	Terms	
Vibrations	V.high	High	Medium	Low	Null
Roughness	High	High	Medium	Low	Very low
Power Consumption	High	High	Medium	Low	Very low
Tool Wear	High ..etc	Medium	Normal		

Then, through a defined set of membership functions corresponding to the linguistic terms that represent the intensity which each one of the problems or symptoms are arising during a given application. For example, for the term of "high" roughness, it can be considered a membership function delimited by the following limits (2.0, 10.0, 12.0, 20.0) as is shown in figure 2.

The problems can be associated, with different levels of intensity, to anyone of the elements of the system Machine-Fixture-Part-Tool (MFPT), this in the sense of expressing the impact that each one of those elements may have in the generation of the problem declared by the user.

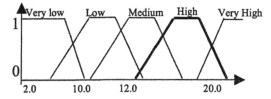

Figure 2. Membership function for the terms associated to the symptom roughness.

That association can also be made using certain linguistic terms to represent the degree of influence among each one of the elements of the MFPT system and the probable problems. The terms are obtained from agreement or consensus of a set of judgments of one or more machining specialists. According to aforementioned example, the attributed linguistic terms could be as follows in Table 2.

Analyzing the vibration problem, table 2 shows that the elements that exert main influence in generating that problem are: the machine and the tool. From this the system will analyze just the subset of rules that are related to these elements and the mentioned problem. Through this strategy, the inference time will be reduced significantly and the system performance will be enhanced.

Table 2.

PROBLEM	Machine	Fixture	Part	Tooling
Vibration	High	Medium	Medium	High
Roughness	Low	Low	Medium	High
Power Consumption	Etc.	...	...	...
Chip Type	...	...	...	...

Suppose that, after the user inputs the information about a given cutting operation, plus the cutting data and the description of prevailing problem that is presenting. For instance, suppose that cutting tool vibration is the main symptom that manifests during the operation. Also, suppose that it is known that the element that influences operation stability, and therefore causes the existence of vibrations, is the cutting tool. The system will sweep the rules associated to that element (the tool), as observed in figure 1, and extract a series of recommendations to solve that problem.

Suppose that the relief angle in the described situation is 4°. The membership functions for the relief angle (α) are shown in figure 2. In this example, the discrete value of α (4°) crosses two membership functions associated to respective linguistic terms. Therefore, angle α can be partially considered as a small angle and partially as a medium angle. The term "partially" can be represented by a percentage or weight that will be used to evaluate the fuzzy rules and to extract the recommendations from the knowledge base. For example, and observing the figure 2, the relief angle of 4° can be considered in 75% as medium sized angle or 60% as a small sized angle. That is, the fuzzy rules that incorporate the term "small relief angle" will participate with an equivalent weight of 60% in the recommendations extracted from the knowledge base. Nevertheless, at the other hand, the rules that incorporate the term "medium relief angle" will have a weight of 75% in recommended actions. This supposes a combination of two or more recommended actions and these actions will be declared by the system with their respective weights. In a similar manner, when evaluating the rules that regard the cutting speed we use the membership functions shown in figure 4.

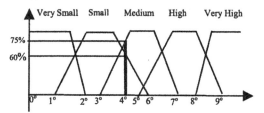

Figure 3. Membership Function associated to the relief angle.

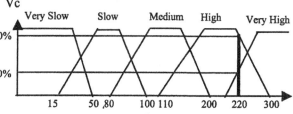

Figure 4. Membership functions of cutting speed.

If the speed used in the situation described before were 220m/min. This value will be fuzzified to be used in the evaluation of the following rule:

If Vc is very high - > reduce Vc

What will give a certain percentage of weight to the concept of very high Vc (20%). In function of such a value, the system will emit a given recommendation, in the case of to reduce the speed with a given weight.

CONCLUSIONS

A Fuzzy Based - Tele Diagnosis system was proposed. The system aims at resolving cutting operations problems from the description of a given cutting situation and its cutting conditions. The central hypothesis is that the knowledge about machining operations is unstructured, approximated and imprecise. This leads to the use of fuzzy logic techniques. When the user describes a given cutting situation, he/she does it using imprecise judgments and qualitative values. Through the use of fuzzy approach, the proposed methodology allows user to describe a given cutting condition, enumerating the resources that he is using and the problem that are arising during the operation with fuzzy terms. These terms are the same that those will be used to sweep the set of rules, searching for the recommendations for solving the described problem.

REFERENCES

CAO LL., BENGU G., 2000; Web-based for reengineering engineering education. Journal of Educational Computing Research. 23 (4) : 421-430, 2000.

CHEN, Y., HUI, A. DU.R., 1995, A fuzzy Expert System for the Design of Machining Operations, Int.J.Mach.Tools Manufact. Vol.35, No. 12 pp.1605-1621, 1995.

DUAN YQ., FU ZT., LI DL., 2003; Toward developing and using Web-based tele-diagnosis in aquaculture. Expert Systems with Applications, 25(2):247-254 aug. 2003

DURÁN, O, R.ZANONI, 2001; Evaluación de Alternativas de Diseño usando Lógica Difusa, Revista de la Facultad de Ingeniería U.T.A. Vol.9, pp.:43-51.

DURÁN, O. 1999; Definición de una metodología de evaluación de alternativas de diseño usando Inteligencia Artificial In: Congreso Iberoamericano de Ingeniería Mecánica, Santiago do Chile. 1999.

DURÁN O., A.BATOCCHIO 1997a; "Programmable Controllers Programme Generation using a Textual Interface", Proceedings of the 7th. International FAIM Conference Flexible Automation and Intelligent Manufacturing. United Kingdom, 1997.

DURÁN O., A.BATOCCHIO, 1997b; Gerador Automático de Software para CP: um enfoque orientado a objeto, Revista Máquinas e Metais Oct.1997.

DURÁN O., A.BATOCCHIO, 1998, Automatic PLC Software Generation with a Natural Interface. International Journal of Production Research,. UK,1998.

FANG, X.D., 1995, Expert System-supported Fuzzy Diagnosis of Finish Turning Process States. Int.J.Mach.Tools Manufact. Vol.35, No. 6 pp.913-924, 1995.

HASHMI,K. GRAHAM, I.D., MILLS, B., 2000, Fuzzy logic based data selection for the drilling process, Journal of Materials Processing Technology, 108, pp.55-61, 2000

LIU, S.C., LIU,S.Y., 2003, An Efficient Expert System for machine Fault Diagnosis, Int. J. Adv. Manuf. Technol. 21:691-698, 2003

MONOSTORI, L., 2003; AI and Machine Learning Techniques for Managing Complexity, changes and uncertinties in Manufacturing. Engineering Applications of Artificial Intelligence. 16(4): 277-291, 2003

ONG, S.K., AN,N., NEE, C.Y., 2001, Web-based Fault Diagnosis and Learning System, Int. J. Adv. Manuf. Technol. 18:502-511, 2001.

PARKINSON B., HUDSON P., 2002; Extending the learning experience using the Web and a knowledge-based virtual environment. Computers & Education 38(1-3): 95-102 Jan-Apr 2002

PHAM DT., PHAM PTN, 1999; Artificial Intelligence in Engineering, International Journal of Machine Tools & Manufacture, 39(6): 937-949, Junio, 1999.

POOLEY R., WILCOX P. 2000; Distributing decision making using Java simulation across the World Wide Web. Journal of the Operational Research Society, 51(4): 395-404, Apr.2000

WANG JD., CHEN DR. KONG XM., 2003; a Web-based remote intelligent expert system for ferrography diagnosis. Damage Assessment of Structures, Proceedings Key Engineering Materials, 245-2:367-372, 2003

WONG SV, HAMOUDA AMS, EL BARADIE MA, 1999; Generalized fuzzy model for metal cutting data selection Journal Of Materials Processing Technology 90: 310-317 Sp. Iss. SI MAY 19 1999

SUPERVISION OF COMPLEX PROCESSES: STRATEGY FOR FAULT DETECTION AND DIAGNOSIS

T. Kempowsky1, A. Subias1,2, J. Aguilar-Martin1,3

1LAAS-CNRS, 7 avenue du Colonel Roche, 31077 Toulouse, France
2 INSA/DGEI, 135 Avenue de Rangueil 31077 Toulouse, France
3ESAII-UPC, Rbla. St. Nebridi,10. 08222 Terrassa (Catalunya), Spain

Abstract: the present work proposes a fault detection and diagnosis strategy using a classification tool in order to identify process situations from measured data, with a direct participation of the expert. Using this strategy, the supervision may be developed though different processes: tracking, detection of behavioural deviations, diagnosis, preventive and corrective actions.

Keywords: Fault detection, process monitoring, data mining.

1. INTRODUCTION

Complex process industries, such as the chemical industry, invest large amounts of money in the automation of their plants in order to increase productivity taking environmental and safety constraints for the operator as for process equipment into account. For these reasons industries must develop supervision system to perform failure detection, diagnosis and to take decisions for corrective actions or reconfigurations. These problems of failure detection and diagnosis are considered from several years in both the control community through what is called Fault Detection and Isolation, and the Artificial Intelligence community, with specific methods and tools as shown in table 1 (Dash et al,2000). The presented work concerns a history based approach, which is well suited for industrial continuous processes as they are dynamic, usually with a great number of components and a noisy environment, and where a mathematical or quantitative model is not always available. Additionally, for this kind of processes, the number of sensors is determined in advance and they are placed for security, optimization and/or control purposes, but not for failure detection and diagnosis purposes. Moreover, advances in sensor technologies, control devices and computer technologies have increased the amount of data collected making it more difficult to analyze.

This paper is organized as follows: section 2 is dedicated to the strategy for fault detection and diagnosis. Section 3 focuses on the method used to implement the strategy. All the results are given on an illustrative example in section 4. Concluding remarks are drawn in section 5.

Table 1. Classification of diagnosis methods

Process based		Process History based	
Qualitative	Quantitative	Quantitative	Qualitative
• Residual • Observers • Parity-space • Assumption	• Causal model • Signed digraph	• Neural Nets • Statistical	•Trend analysis •Rule-based

2. PROPOSED STRATEGY FOR FAULT DETECTION AND DIAGNOSIS

The objective of the proposed strategy is to perform situation assessment; this means to supply to the process operator or to the expert relevant information every time a new situation arises on the process and also to identify this new situation. A situation results from an evolution of the process and corresponds to an operational state of the studied process. As a failure can occur at any time on the process, these states can either be normal, that means expected states, or abnormal states i.e. unexpected states. Therefore, situation assessment induces the capability to detect these abnormal states. Moreover, this general objective can be completed with the causes identification of the abnormal situations which corresponds to fault diagnosis.

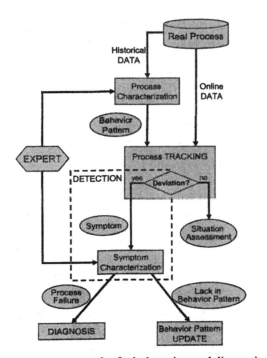

Fig. 1. The strategy for fault detection and diagnosis

Figure 1 presents a general diagram of the situation assessment strategy proposed. The entire strategy is based on the analysis of available information issued from process sensors. Two different but non-independent steps can be pointed out: an offline step in which historical data are analyzed and processed to characterize the known behavior of the system, and an online step, during which the obtained behavior and the online data are used to determine the current expected state of the process. Both stages are fully described as follows.

2.1 Process characterization

For complex processes, the generation of a suitable model is an actual challenge. One solution is to use the large amount of process-history data. But, this solution implies others difficulties to manage. Indeed, most of the time these data are heterogeneous and in great number as they result from several recordings. The data analysis is then a very hard task for an expert. As all the future decisions of the expert about the process depend on this analysis, it is then necessary to assist the expert during this analysis step. By means of machine learning techniques significant information can be extracted from historical data. Data mining is one way to help the expert. From this relevant information a "reference model" of the process behavior can be obtained. This model is a *behavior pattern* elaborated offline. This pattern obviously does not include modeling errors; nevertheless it is not exhaustive as the historical data cannot cover the entire "life" of the process. This lack of completeness implies that an abnormal situation can either characterize a failure situation or a situation not present in the behavior pattern. In some cases, this pattern can include situations

identified with precision that correspond to critical states of the process in terms of equipments or operators safety.

2.2 Process Tracking

Once the process characterization is performed, online situation assessment is possible. This consists in a process tracking phase. During this stage, the "incoming" data are analyzed to decide if the current observed situation is normal. Situations recognition is then based on the comparison between the different possible states of the process recorded in the behavior pattern and the observed situations. Any deviation from the expected behavior leads to a symptom detection. This detection procedure based on such a discrepancy principle allows taking into account abnormal situations due to real failures on the process sensors and/or actuators but also unexpected situations which correspond to a normal operating of the process not considered in the elaboration of the behavior pattern. At this stage, the knowledge of the symptom is not sufficient to describe what happened that means what is the nature of the observed deviation. A characterization of the symptom is then necessary.

2.3 Symptom Characterization

This symptom characterization is performed offline. The aim is to refine the abnormal situation detection making a distinction between failure situation and unexpected normal situation. For this, it is first necessary to isolate the data associated to the observed deviation, received from the process. Then, several analyses on these data are provided to the expert in such a way that he can interpret the meaning of the symptom. This characterization step is crucial in the decision making of future actions. A symptom corresponding to a real process failure must activate a diagnosis procedure. The aim of the diagnosis is to find the failure origin i.e. to find the cause of the failure situation. The set of detectable patterns is updated whenever a new failure, or previously unobserved normal behavior, is detected. In this way, the construction of a process behavior model as well as the identification of possible faults, are performed iteratively.

3. PROPOSITION AND IMPLEMENTATION OF A STRATEGY

By considering Data Mining research and terminology, the process characterization step, which leads to establish a pattern, corresponds to a *learning stage*. In a similar way, the tracking step is equivalent to a *recognition stage*. And finally, the symptom characterization can be associated to a new clustering procedure called *active supervised learning*. Several Data Mining modeling techniques allow constructing the behavior pattern via a learning stage. A complete state of the art is beyond the scope of this work (Duda et al, 2001; Michie et al, 1994). We give now a brief presentation of the main approaches that we consider convenient for the attempted result.

<u>Methods based on clustering</u>: they attempt to find natural groups of data, according to the similarities among them. Typically the "similarity" concept is defined as the distance between a data vector and the cluster prototype (center). The characteristics of the prototypes are not usually known beforehand, they are chosen randomly and updated at the same time as the partitioning of the data is made. The *K-means* (Jain et al, 1999) and *Fuzzy C-means (FCM)* (Bedzek, 1981) algorithms are based on an iterative optimization of an objective function (e.g. variability within clusters). A drawback of these techniques is that they are sensitive to the selection of the initial partition since it is made in a random way. Moreover, the number of clusters must be given at the start.

<u>Statistical techniques</u>: they cover "classical" algorithms that in some way are derived from Fisher's *Linear Discriminants (LD)*. The "modern" statistical techniques provide an estimate of the joint distribution of the features within each class (Michie et al,1994). For Fisher's LD approach, the space of attributes can be partitioned by a set of hyperplanes, each one defined by a linear combination of the attribute variables. It requires no probability assumptions. Modern statistical techniques are distribution-free classification procedures. The *K-nearest neighbor* for instance, is based on the idea that a new object is most likely to be near to observations from its own proper population. A major difficulty of this method is the *proper scaling* of the observations. Some standardization is usually required. For large dataset, this method is very time consuming since all the training data must be stored and examined for each classification.

<u>Decision Trees</u>: are classifiers in the form of a tree structure. Each node represents a decision or question for a given characteristic of the objects. From each node two or more branches may be attached depending if the decision is binary or not. The decision for the assignment of an object to a given class is taken in the terminal nodes R. Quinlan's ID3 and CART (Duda et al, 2001) are some of the most known decision tree algorithms. ID3 algorithm, using the entropy measure, searches through the attributes space to extract the attribute that best separates the given examples. There are no restrictions for the nature of the data (they can handle both continuous and symbolic variables) and understandable rules can be easily generated. A drawback of decision trees is that most of them only examine a single field at a time and they are not well suited for non-linear domains. Moreover, they can be computationally expensive in the growing combinatorial process of a decision tree.

<u>Neural Networks based methods</u>: are used to predict outputs from a set of inputs by taking linear combinations of the inputs and then making nonlinear transformations of the linear combinations using an activation function. In supervised learning networks (Perceptron, the Multi Layer Perceptron (MLP) (Mange et al, 1998)), measurements in the training data set are accompanied by labels indicating the class where they belong. In the unsupervised learning case, the network adapts purely in response to its inputs. Neural nets perform very well on difficult non-linear domains, where it becomes more and more difficult to use Decision trees, or Rule induction systems, which cut the space of examples parallel to attribute axes. They also perform slightly well on noisy domains. One of the drawbacks is a slow learning process. Moreover neural networks do not give explicit knowledge representation in the form of rules, or some other easily interpretable form.

<u>LAMDA:</u> is a fuzzy methodology of conceptual clustering and classification. It is based in finding the global membership degree of an individual to an existing class, considering all the contributions of each of its attributes. This contribution is called the *marginal adequacy degree (MAD)*. MADs are combined using "fuzzy mixed connectives" as aggregation operators in order to obtain the *global adequacy degree (GAD)* of an individual to a class (Aguilar et al,1982). LAMDA methodology enables classifying individuals represented with quantitative and/or qualitative attributes (i.e. descriptors). When the descriptor is a numerical type, the *MAD* is calculated by selecting one of the different function possibilities. When the descriptor is a qualitative one, the observed frequency of its attribute is used to evaluate the *MAD*. To avoid the assignment of a not very representative object to any class, that is an object with a small membership, a minimum global adequacy threshold is employed. This threshold corresponds to the *GAD* given by the Non-Informative Class (NIC). Therefore, if the maximum value of all the GADs of an individual is less than the NIC threshold, it will not be classified to any existing class. The LAMDA method allows working with multiple variables, corresponding with both numerical information and qualitative information (Piera et al,1989). Moreover, the characteristics of the resulting classes after a learning stage are interpretable by the operators. This method allows mixing different learning strategies as well as combining recognition. LAMDA accepts the creation of new classes given a number of observations that have been rejected (unrecognized), keeping the previously created classes. It also accepts the evolution of the existing classes. For these reasons, the proposed strategy described in this paper has been implemented using LAMDA method via the development of a software tool called SALSA: "Situation Assesment using LAMDA claSsification Algorithm". In the remainder, the implementation of each step of the situation assessment strategy is detailed.

In a general way, the learning stage consists in finding groups of objects, which have similar characteristics. This learning can be done with *a priori* knowledge of the different possible categories of objects. The objective is then to establish a rule by which a new observation can be classified into one of the existing groups. In this case the learning is referred as "supervised learning". If the categories are unspecified (no previous knowledge), then from a set of observations the objective is to establish the existence of groups in the data. This is called unsupervised learning or "self-learning". With SALSA the operator can use different learning strategies and is a real actor during each step (Kempowsky et al, 2003). Indeed, SALSA has been designed to enhance the knowledge of the expert. If supervised learning is used, the expert, according to some criteria, pre-defines a number of significant groups or classes to a set of measured data from the process. Then, SALSA generates each class profile according to the type of variables (numerical or symbolic). These pre-defined classes can be validated by means of a passive recognition procedure. Because it may occur that some pre-assigned individuals do not fit to any existing class, SALSA gives the possibility to the user to create new classes from the available information, or to modify and adjust the characteristics of the given classes. If a great amount of data is available and the expert is not able to identify the different situations, an unsupervised learning strategy is possible. SALSA makes a partition of the individuals in the data set, according to the adequacy concept, generating clusters or classes. In the offline stage, the user can obtain different partitions with the same data set, by changing LAMDA parameters in order to improve the quality of the final classification. The process expert proceeds to a knowledge-based interpretation of the resulting classes and maps them into significant process situations or states. SALSA provides different possibilities for the expert to map the resulting classes into significant states. Once a suitable classification has been found a reference model may be constructed as the classification is made in a sequential way. This model is given to the expert via a graphical representation of a supervision automaton. The automaton states are issued from the classes or from the states in case of mapping. The transitions from one class to another are given by the adequacy degrees of an element to the different classes. The expert is assumed to validate this model and to complete it for instance by adding other possible transitions between states.

3.2 Recognition stage

Once the behavior pattern is elaborated, the next step is to recognize what is the current functional state of the monitored process when a new observation has been made. Online normal or abnormal situation assessment is a process of pattern recognition. For this the different adequacies (GAD) of the object to each of the classes are produced and the object will be assigned to the class with the highest adequacy. Of course, the tracking process can only confirm an existing (expected) state or reject a new observation that is too different and cannot be assigned to any class. Symptoms are identified by deviations observed during the online recognition process. These deviations are detected in two ways: when an object is assigned to the NIC, that means that the system is not able to assign the online observations to any of the existing classes, i.e. they do not correspond to any of the expected behaviors. The other possible type of deviation is when a new situation is identified but the transition between the previous and the new situation is not an expected one, i.e. the transition between two classes has not been learned in the behavior pattern These two kind of deviations are integrated in the supervision automata to alert the expert.

3.3 Active supervised learning

When a symptom has been detected through the assignation of elements to the NIC, the aim is to identify and characterize this symptom, by launching a new learning procedure. New classes are generated from the characteristics of the unrecognized elements, keeping unmodified the existing classes. Therefore, this stage constitutes an active supervised learning stage. The expert will next interpret the new classes in order to characterize the new behavior. The result of this stage leads to update the automata by adding states for instance.

4. ILLUSTRATIVE EXAMPLE

A pilot process of a section of a nuclear power plant has been chosen to illustrate the fault detection and diagnosis strategy. The installation is constituted of four subsystems: a *receiver* with the water supply system, a *boiler* heated by a thermal resistor, a *steam* flow system and a complex *condenser* coupled with a heat exchanger. This illustration focuses on the boiler subsystem (figure 2). The feed water flow $F3$ is pumped to the boiler via a pump $P1$. An on-off controller, to maintain a constant water level $L8$ in the boiler, controls this pump. The heat power value $Q4$ depends on the available accumulator pressure $P7$. When the accumulator pressure drops below a minimum value, the heat resistance delivers maximum power, and when the accumulator reaches a maximum pressure the heat resistance is cut off, in order to keep the pressure within ±0.2bars of the set-point.

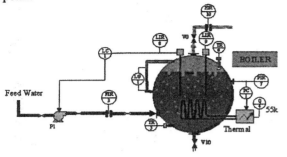

Fig.2. Steam generator subsystem:boiler

The steam flow is measured with *F10*. Two scenarios where exploited to illustrate the different stages of the strategy: process tracking, symptom detection and characterization, diagnosis and behavior pattern update. The first scenario corresponds to the detection of a normal situation that was not learned during the process characterization. The second is a fault situation; it corresponds to a steam pipe blockage at the outlet of the boiler. The descriptors used are five sensor measures: feed water flow (*F3*), heat power (*Q4*), boiler pressure (*P7*), boiler level (*L8*) and steam flow (*F10*).

4.1 Process Characterization: Learning stage

The behavior pattern is obtained from a training data set of 1800 samples (individuals) representing the normal behavior. Unsupervised learning is chosen since no prior knowledge about the possible situations is given. Because all descriptors are quantitative SALSA needs to normalize them to the unit interval [0,1] in such a way that they can be treated simultaneously. Figure 3 shows the behavior of the different variables (descriptors) chosen for the construction of the behavior pattern with the regulation of the level (*L8*) and the pressure (*P7*).

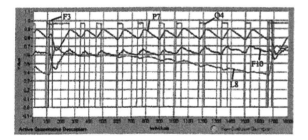

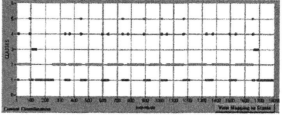

Fig. 3. Training data: Normal behavior

Fig. 4. Classification results for Normal operation

SALSA initially identifies five different classes (figure 4) that best represent the operation of the boiler in normal conditions. These classes are obtained using the distance to a center for the *MAD* function and Zadeh's Fuzzy Logic Minimum-Maximum for the connective family.

$$MAD(x_j, \rho_{kj}) = \rho_{kj}^{1-d_{kj}} (1 - \rho_{kj})^{d_{kj}}, \ d_{kj} = \left| x_j - c_{kj} \right|$$

To validate the resulting classification another data set is used. Even if all the classes are correctly identified, there are some unrecognized individuals. In order to identify and characterize the unrecognized situation a new learning procedure is launched. This induces the creation of new classes keeping unmodified the existing classes (Figure 5). From the profile of the classes, the expert is able to interpret and identify the situations (table 2). The two different regulations are identified independently (class 1 and class 6) as well as it is possible that they occur simultaneously (class 3). Class 2 and 5 characterize the case where there is no regulation and class 4 represents the transition from the no regulation situation to the pressure regulation. Because it is possible that several classes may characterize the same situation, the expert can map them into a single state. Here, classes 2 and 5 are mapped into one single state, the no regulation state.

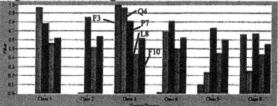

Fig. 5. Class profile Normal operation

Table 2. Identified situations in normal operation

Class	Situation	D1: F3	D2: Q4	D3: P7	D4: L8	D5: F10
1	Pressure reg.	0.0	0.96	0.78	0.55	0.02
2	No regulation	0.0	0.0	0.84	0.51	0.63
3	Press & Level Regulation	0.99	0.95	0.80	0.43	0.61
4	Transition	0.01	0.69	0.80	0.50	0.62
5	No regulation	0.10	0.23	0.72	0.44	0.59
6	Level Reg.	0.65	0.25	0.66	0.43	0.55

4.2 Process Tracking: Recognition

The resulting classification as well as the behavior pattern where then used to perform the online situation assessment. Figure 6 illustrates the online recognition performed for a period of 900 seconds. During the first 800 samples the different normal situations are correctly detected

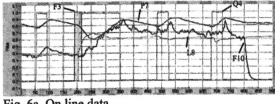

Fig. 6a. On line data

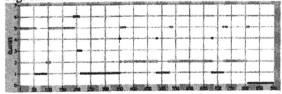

Fig. 6b. Online Process Tracking

4.3 Symptom identification and diagnosis

After simple 901, a deviation is detected i.e. samples are not classified to any existing class. In order to characterize the observed symptom, a new learning procedure is launched. With the same parameters SALSA, automatically generates new classes from the unrecognised individuals without modifying the existing ones. After this learning procedure a new class has been created (class 7) (Figure7). From the new class profile the user can identify that it corresponds to a very small value for the steam flow (F10) while the pressure (P7) has not decreased. So, according to the expert knowledge and the characteristics of the descriptors of class 7 (Table 3), the detected deviation is interpreted as a blockage in the outlet of the boiler. After this diagnosis step the behavior pattern is updated. Figure 8 shows the resulting discrete event model for the boiler operation including the steam pipe outlet blockage.

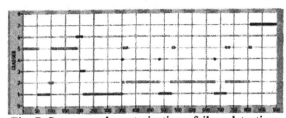

Fig. 7. Symptom characterization - failure detection

Table 3. Failure profile - Steam pipe blockage

Class	Situation	D1: F3	D2: Q4	D3: P7	D4: L8	D5: F10
7	Steam pipe Blockage	0.00	0.01	0.82	0.62	0.02

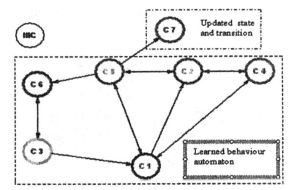

Fig. 8 Updated behaviour pattern

CONCLUSION

Supervision of petrochemical processes is an actual challenge. This paper presents a strategy for fault detection and diagnosis implemented via a software tool: SALSA (Situation Assessment using LAMDA claSsification Algorithm). The strategy is shown on the example of a steam generator subsystem, it is mainly based on different stages. First, a learning offline stage, in order to create the behaviour pattern. Then this pattern is used during the online recognition stage for fault detection. An active supervised leaning stage is added to characterize any new situation. During each stage, all the functionalities of SALSA have been designed to enhance the expert knowledge A discrete event model of the tracking of the process evolutions can be built. Current research focuses on the development of the discrete event model, especially on the identification of a state change according to the descriptors values.

ACKNOWLEDGEMENT

The authors would like to thank the laboratory of Automatic Control and Computers for Industry of the University of Lille, for the use of the plant pilot and also the CHEM European project within SALSA has been developed www.chem-dss.org.

REFERENCES

Aguilar J., López de Mántaras R., "The process of classification and learning the meaning of linguistic descriptors of concepts". Approximate Reasoning in Decision Analysis p. 165-175. Nor1th Holland, 1982.

Bezdek, J., "Pattern Recognition with Fuzzy Objective Function", Plenum Press, USA,1981

Dash S. and Venkatasubramanian V., "Challenges in the industrial applications of Fault Diagnostic Systems", *Computers & Chemical Eng.*, 24(2-7), 785-791, 2000.

Duda R., Hart P., Stork D., "Pattern Classification", Second Edition, Wiley-Interscience, 2001.

Jain A.K., Murty M. N., and Flynn P. J. "Data clustering: a review", *ACM Computing Surveys*, 31(3):264-323, 1999.

Kempowsky T., Aguilar-Martin J., Subias A., Le Lann M.V., "Classification Tool based on Interactivity between Expertise and Self-Learning Techniques", IFAC-Safeprocess 2003, Washington D.C. (USA), June 2003.

Mange D., Tomassini M., "Bio-Inspired Computing Machines", presses polytechniques , 1998

Michie D., Spiegelhalter D.J., Taylor C.C., "Machine Learning, Neural and Statistical Classification", Ellis Horwood series in AI, February, 1994.

Piera N., Desroches P., Aguilar J., "LAMDA: An incremental Conceptual Clustering System", Report 89420 LAAS-CNRS, 1989.

ELSEVIER

IFAC

PUBLICATIONS
www.elsevier.com/locate/ifac

DESCRIPTION AND OPTIMIZATION OF DISCRETE AND CONTINUOUS HIERARCHICAL ENVIRONMENT

Tiit Riismaa

Institute of Cybernetics at Tallinn University of Technology
Akadeemia tee 21, Tallinn 12618, Estonia
Phone: +372 6204192 Fax: +372 6204151
tiitr@ioc.ee

Abstract: Two different approaches, a discrete and continuous one, of description and optimization of the structure of hierarchical production system are presented and compared. For discrete statement the feasible set of hierarchies is a set of oriented trees. Connections between neighboring levels are presented by adjacency matrixes. The graph theory description of feasible set of hierarchies is given. For continuous statement a feasible set of hierarchies is an ordered set of sizes of levels. Connections between the adjacent levels are described by density function of distribution. This density function describes how the size of each level is distributed between the sizes of next level. Two different types of variable parameters are defined: parameters for the level size and parameters for connections between adjacent levels. For described variable parameters the structure optimization problem is a discrete-convex programming problem. The given choice of variable parameters and the statement of the both integer and continuous optimization problem as a double minimum problem enable to construct methods for finding a global optimum of goal function and construct the corresponding structure.

Keywords: hierarchical systems, continuous hierarchical environment, structure optimization, mathematical programming, convex programming.

1. INTRODUCTION

For very complicated systems the hierarchical (multi-level) partitioning procedure is useful. The qualities of overall system depend on this multi-level partitioning.

Consider the processing (aggregation, packing, clustering etc.) of n parts. In case of one processing unit the overall processing and waiting time for all n parts is proportional to n^2 and is a quickly increasing function. For this reason the hierarchical system of processing can be suitable. From zero-level (level of object) the units will be distributed between p_1 first-level processing units (robots) and processed (aggregated, packed etc.) by these processing units (robots). After that the parts will be distributed between p_2 second-level processing units and processed further and so on. From p_{s-1} $(s - 1)$-level

the parts will be sent to the unique s-level processing unit and processed (aggregated, packed etc.) finally. The number of processing units on each level is a variable parameter. A problem of optimal multiple borders inside processing system arises.

In modern processing technology (information technology, robotics etc.) the Euclidean metrics is pressed to the background and the number of connections between the subsystems and elements of the overall system are important.

The problem of structuring of processing system can be stated mathematically as multi-level optimal decomposition problem. Some multi-level classification methods for conceptual structures are based on multi-level partitioning of the given finite set to subsets.
In multi-level assembling process the choice of optimal number of workstations on each level is mathematically

complicated problem what must be largely simplified. The assembling problem as well as a broad class of design and implementation problems, such as component selection in production systems, reconfiguration of manufacturing structures, optimization of the hierarchy of decision making systems, multi-level aggregation, etc. could be mathematically stated as a multi-level selection problem (Mesarovic et al, 1970; Riismaa, 1973; Laslier, 1997).

A broad class of design and implementation problems such as component selection in production systems, reconfiguration of manufacturing structures, optimization of the hierarchy of decision making systems etc. could be mathematically stated as multi-level partitioning problems, where the qualities of the system depend on this partitioning.

In earlier papers (Littover et al, 2001; Riismaa and Randvee, 2002; Riismaa et al, 2003; Riismaa, 2003; Riismaa and Randvee, 2003) a method for description and optimization of the structure of multi-parameter multi-level selection procedure often to be dealt with in automated production systems is introduced. The feasible set of structures is a set of oriented trees corresponding to the full set of multi-level partitioning of given finite set (Part 2.1). Each tree from this set is represented by sequence of Boolean matrices, where each of thus matrices is an adjacency matrix of neighboring levels. To guarantee the feasibility of the representation, the sequence of Boolean matrices must satisfy some conditions (Part 2.1, Part 2.2). Described formalism enables to state the problem as a double-step discrete optimization problem (Part 2.3) and construct some classes of solution methods (Riismaa, 2003). Variable parameters of the inner minimization are used for the description of connections between adjacency levels. Variable parameters of the outer minimization problem are used for the representation of the number of elements at each level. This double-step minimization approach guarantees that the involved mathematical properties – possibility to continue the objective function to the convex function - enable us to construct an effective algorithm for finding the global optimum (Riismaa and Randvee, 2003; Riismaa, 2003).

The structural selection procedure considered in earlier papers is based on the full set of hierarchical trees of feasible structure what could be composed from the given set of elements (Riismaa, 1993). In the context of this approach an element were considered as a logical part of the production system that is carrying out an identifiable mission and obeys the necessary functionality and autonomy (Berio and Vernadat, 1999).

In this paper the continuous statement of structure optimization problem of hierarchies is considered. In the continuous case each level is characterized by nonnegative real valued parameter called as size.

Connections between the levels of the system are defined as a finite sequence of density functions of distributions. The i-th distribution will correspond to the connections between the levels i and $i-1$. The i-th distribution shows how the size z_{i-1} of level i-1 is distributed between the size z_i of level i.

2. THE DISCRETE APPROACH

2.1 The Feasible set of Discrete Structures

Consider s-level hierarchies, where nodes on level i are selected from the given nonempty and disjoint sets and all selected nodes are connected with selected nodes on adjacent levels. All oriented trees of this kind form the feasible set of hierarchies (Littover et al, 2001).

The illustration of this formalism is given in Fig.1

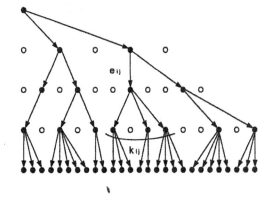

Figure 1 - Feasible set of structures (discrete statement)

Suppose Y_i is an adjacent matrix of levels i and $i-1$ $(i=1,...,s)$. Suppose m_0 is the number of 0-level elements (level of object).

Theorem 1. All hierarchies with adjacent matrixes $\{Y_1,...,Y_s\}$ from the described set of hierarchies satisfy the condition

$$Y_s \cdot .... \cdot Y_1 = \underbrace{(1,...,1)}_{m_0} , \qquad (1)$$

The assertion of this theorem is determined directly.

2.2 The Graph Theory Statement of General Problem of Structure Optimization

The general optimization problem is stated as a problem of selecting the feasible structure which correspond to the minimum of total loss given in the separable-additive form:

$$\min\left\{\sum_{i=1}^{s}\sum_{j=1}^{m_i}\mathrm{h}_{ij}\left(\sum_{r=1}^{m_{i-1}}d^i_{jr}y^i_{jr}\right)\middle|\ Y_s\cdot\ldots\cdot Y_1=\underbrace{(1,\ldots,1)}_{m_0}\right\}$$

over $Y_1,\ldots,Y_s,$ (2)

where $h_{i,j}(\ \cdot\ )$ is an increasing loss function of j-th element on i-th level and d^i_{jr} is the element of $m_i\times m_{i-1}$ matrix D_i for the cost of connection between the i-th and $(i$-1)-th level (Littover et al, 2001; Riismaa and Randvee, 2002).

The meaning of functions $h_{i,j}(\ k\ )$ depends on the type of the particular system.

Mathematically this problem is an integer-programming problem with a non-continuous objective function and with a finite feasible set. For solving this kind of problems only ineffective methods are known.

For solving the problem (2) in (Riismaa and Randvee, 2002; Riismaa, 2003) a class of recursive algorithms, where the recursion index is the level number, is constructed.

2.3 The Reduced Optimization Problem Statement

In this part we consider an important special case where the connection cost between the adjacent levels is the property of supreme level. In this case the total loss depends only on the sums (Figure 1) $\sum_{r=1}^{m_{i-1}}y^i_{jr}=k_{ij}$.

Recall that

$$\sum_{j=1}^{p_i}k_{ij}=p_{i-1},\quad (i=1,\ldots,s),$$

where p_i is the number of nodes on i-th level and $k_{i,j}$ is the number of edges outgoing from the j-th node on i-th level.

If we additionally suppose that $h_{i1}(k)\le\ldots\le h_{im}(k)$ for each integer k, it is possible to reorganize the optimization procedure into two sequential stages:

$$\min\left\{\sum_{i=1}^{s}g_i(p_{i-1},p_i)\middle|\ (p_1,\ldots,p_{s-1},1)\in W^s\right\},$$

over $p_1,\ldots,p_{s-1}$ (3)

where

$$g_i(p_{i-1},p_i)=\min\left\{\sum_{j=1}^{p_i}h_{ij}(k_{ij})\middle|\sum_{j=1}^{p_i}k_{ij}=p_{i-1}\right\}$$

over $k_{i1},\ldots,k_{ip_i}$ (4)

and

$$W^s=\left\{(p_1,\ldots,p_{s-1},1)\middle|1\le p_i\le p_{i-1}\right\}.$$

Free variables of the inner minimization are used to describe the connections between the adjacency levels. Free variables of the outer minimization are used for the representation of the number of elements at each level.

For solving the optimization problem (3), (4) two classes of methods are constructed. The first class of methods is recursive algorithms where in each step the minimization information obtained only on previous step is used (Riismaa et al. 2003). The second class of methods is based on the known algorithms of convex programming (Rockafellar, 1970) adapted for the integer programming. They are so called local search methods (Riismaa and Randvee, 2002; Riismaa 2003; Riismaa et al. 2003).

2.4 Analytical Method of Solving Reduced Problem for Horizontal Homogeneous Hierarchies

The hierarchy is called horizontal homogeneous if

$$h_{ij}(k)=h_i(k)\quad (j=1,\ldots,m_i;i=1,\ldots,s). \quad (5)$$

Now it is possible to solve the inner minimization problem (4) (to find the optimal connections between the adjacent levels) analytically:

$$g_i(p_{i-1},p_i)=$$

$$=\left(p_i\cdot\left(\left[\frac{p_{i-1}}{p_i}\right]+1\right)-p_{i-1}\right)\cdot h_i\left(\left[\frac{p_{i-1}}{p_i}\right]\right)+$$

$$+\left(p_{i-1}-p_i\cdot\left[\frac{p_{i-1}}{p_i}\right]\right)\cdot h_i\left(\left[\frac{p_{i-1}}{p_i}\right]+1\right)$$

(6)

$$(i=1,\ldots,s-1),$$

$$g_s(p_{s-1},1)=h_s(p_{s-1}),\ [p]\text{-integer part of }p.$$

Denote $\quad p_i\cdot\left(\left[\dfrac{p_{i-1}}{p_i}\right]+1\right)-p_{i-1}=p_i^*$

and $\quad p_{i-1}-p_i\cdot\left[\dfrac{p_{i-1}}{p_i}\right]=p_i^{**}.$

Recall

$$k_{ij} = \left[\frac{p_{i-1}}{p_i}\right] \left(j = 1,...,p_i^*\right),$$

$$k_{ij} = \left[\frac{p_{i-1}}{p_i}\right] + 1 \quad \left(j = p_i^* + 1,...,p_i^* + p_i^{**}\right) \quad (7)$$

and $(i = 1,..,s)$.

Certainly $p_i^* + p_i^{**} = p_i$.

To complete the solving of problem (3), (4) it is enough to use (6) for outer optimization problem (3).

3. THE CONTINUOUS APPROACH

In the continuous case the structure of the system is defined as a finite sequence density functions of distributions

$$\{f_1(x),..., f_s(x)\}, \; f_i : R^+ \to R^+ (i = 1,...,s).$$

The i-th distribution will correspond to the connections between the levels i and $i-1$. Here the parameter $z_i > 0 (i = 0,...,s)$ is the size of level i.

The function $f_i(x)$ describe how the size z_{i-1} of level i-1 is distributed between the size z_i of level i (the condition in (9)).

The statement of the problem is as follows:

$$\min_{z_1,...,z_{s-1}} \left\{\sum_{i=1}^{s} q_i(z_{i-1}, z_i) \mid z_i \in R^+ (i = 1,...,s-1), z_s = 0\right\}$$

$$(8)$$

where

$$q_i(z_{i-1}, z_i) = \min_{f_i(z)} \left\{\int_0^{z_i} h_i(x, f_i(x)) dx \middle| \int_0^{z_i} f_i(x) dx = z_{i-1}\right\}$$

$$(i = 1,...,s-1) \qquad (9)$$

and $q_s(z_{s-1}, 0) = h_s(0, z_{s-1})$.

Here $h_i(x, z)$ describe connections between the elements of level i and of level i-1. The meaning of these functions depends on the type of the particular system.

4. THE CONTINUOUS HORIZONTALLY HOMOGENEOUS HIERARCHIES

For continuous statement the hierarchy is called horizontally homogeneous if

$$h_i(x, z) = h_i(z) \; (i = 1,...,s) . \qquad (10)$$

This condition substantially simplify the problem (8), (9) and enables to find the solution for inner optimization problem (9) analytically:

$$f_i^*(x) = \frac{z_{i-1}}{z_i} , \; x \in [0, z_i] \; (i = 1,...,s). \quad (11)$$

This relation will mean, that the size z_{i-1} of level i-1 is distributed between the size z_i of level i uniformly.

To use (11) the outer problem (8) transform to continuous programming problem

$$\min_{z_1,...,z_{s-1}} \left\{\sum_{i=1}^{s} z_i \cdot h_i\left(\frac{z_{i-1}}{z_i}\right) \middle| z_i \in R^+ (i = 1,...,s-1)\right\}.$$

$$(12)$$

Lemma 1. If $h_i(x) \; (i = 1,...,s)$ are convex functions, then (12) is convex programming problem.

The convexity of goal function of problem (12) is controllable immediately.

Conditions (7) and (11) shows the relation between solutions between discrete statement and continuous statements:

$$k_{ij} = \left[\frac{p_{i-1}}{p_i}\right] \leq \frac{z_{i-1}}{z_i} \; \left(j = 1,..., p_i^*; i = 1,...,s\right) ,$$

$$\frac{z_{i-1}}{z_i} \leq \left[\frac{p_{i-1}}{p_i}\right] + 1 = k_{ij} \; \left(j = p_i^* + 1,..., p_i; i = 1,...,s\right)$$

5. OPTIMALITY OF HORIZONTALLY HOMOGENEOUS HIERARCHIES

Recall z_i is the size of level i. Now to change the

variable $\frac{z_{i-1}}{z_i} = w_i \; (i = 1,...,s),$ $\qquad (13)$

from (13) follows:

$$\min_{w_1,...,w_s} \left\{ \sum_{i=1}^{s} w_{i+1} \cdot ... \cdot w_s \cdot x_s \cdot h_i(w_i) \middle| w_1 \cdot ... \cdot w_s \cdot x_s = x_0 \right\}$$

(14)

,

or

$$\min_{w_1,...,w_s} \left\{ \sum_{i=1}^{s} \frac{x_0}{w_1 \cdot ... \cdot w_i} h_i(w_i) \middle| w_1 \cdot ... \cdot w_s \cdot x_s = x_0 \right\}$$

(15).

The (13) mean that the size z_{i-1} of level i-1 is distributed between the size z_i of level i uniformly and the proportion for each unit of size z_i is equal to w_i.

The Lagrangian function for minimization problem (15) is

$$L(w,\lambda) = \sum_{i=1}^{s} \frac{x_0}{w_1 \cdot ... \cdot w_i} h_i(w_i) + \lambda(w_1 \cdot ... \cdot w_s - x_0)$$

(16).

To calculate the derivatives and equalizing these to zero (well-known saddle point conditions) the conditions of optimality are as follows:

$$\frac{dh_i}{dw_{i+1}} + h_i(w_i) - w_i \cdot \frac{dh_i}{dw_i} = 0 \ (i=1,...,s-1)$$

(17)

$$w_1 \cdot ... \cdot w_s \cdot x_s = x_0.$$

Recall x_0 and x_s are fixed.

6. OPTIMALITY OF HOMOGENEOUS HIERARCHIES

For continuous statement the hierarchy is called homogeneous if

$$h_i(x,z) = h(z) \ (i=1,...,s).$$

In this case (12) transforms to

$$\min_{z_1,...,z_{s-1}} \left\{ \sum_{i=1}^{s} z_i \cdot h\left(\frac{z_{i-1}}{z_i}\right) \middle| z_i \in R^+ (i=1,...,s-1) \right\}$$

(18)

and (17) transforms to

$$\frac{dh}{dw_{i+1}} + h(w_i) - w_i \cdot \frac{dh}{dw_i} = 0 \ (i=1,...,s-1)$$
$$w_1 \cdot ... \cdot w_s \cdot x_s = x_0$$

. (19)

Recall x_0 and x_s are fixed.

The condition (19) is a nonlinear system of equations and can be solved easily.

6. CONCLUDING REMARKS

Many discrete or finite hierarchical structuring problems can be formulated mathematically as multi-level partitioning procedure of a finite set of nonempty subsets. This procedure may be considered as a hierarchy where to the subsets of partitioning correspond nodes of hierarchy and the relation of containing of subsets defines arcs of hierarchy. The feasible set of structures is a set of hierarchies (oriented trees) corresponding to the full set of multi-level partitioning of given finite set. Examples of problems of this class are aggregation problems, structuring of decision-making systems, database structuring, and the problems of multiple distribution or centralization. Described formalism enables to state the problem as a double-step discrete optimization problem and construct some classes of solution methods (Littover et al, 2001; Riismaa and Randvee, 2002). Variable parameters of the inner minimization are used for the description of connections between adjacency levels. Variable parameters of the outer minimization problem are used for the representation of the number of elements at each level. This double-step minimization approach guarantees that the involved mathematical properties – possibility to extend the objective function to the convex function - enable us to construct algorithms for finding the global optimum (Littover et al, 2001; Riismaa and Randvee, 2002; Riismaa et al, 2003).

The main difficulty from point of view of optimization is that the number of subsets of partitioning is a variable parameter. The finding of solution considered here nonlinear integer programming problem is complicated mathematically. This is a reason why continuous statements are more useful.

In this paper a continuous or infinite method for description and optimization of the structure of multi-parameter multi-level selection procedure is introduced. For this continuous statement the size of level may have non-integer values too and connections between the adjacent levels are described by density function of distribution. This density function describes how the size of level is distributed between the sizes of next level. Considered statement is a problem of conditional calculus of variation. But some reduced variants of this problem have good mathematical properties (convexity of objective function) and may be solved analytically.

7. ACKNOWLEDGEMENT

This work has been supported by the Estonian Science Foundation.

REFEENCES

Berio, G., F. B. Vernadat (1999). New Developments in Enterprise Modeling Using CIM-OSA. *Computers in Industry*, **40**, 99-114.

Laslier, J.-F. (1997). *Tournament Solutions and Majority Voting*. Springer, Berlin, Germany.

Littover, M., I. Randvee, T. Riismaa, J. Vain (1999). Recursive Integer Programming Technique for Structural Optimization. In: *Proceedings of the 4-th International Scientific Colloquium on CAx Techniques. September 13-15, Bielefeld, Germany* (H. Ostholt. (Ed)), 95-102. Druckerei Robert Bechauf, Bielefeld, Germany.

Littover, M., I. Randvee, T. Riismaa, J. Vain (2001). Optimization of the Structure of Multi-Parameter Multi-Level Selection. In: *Proceedings of the 17th International Conference on CAD/CAM, Robotics and Factories of the Future, CARS&FOF, July 10-12, Durban, South Africa* (G. Bright, W. Jannsens. (Eds)), 317-322. University of Natal, Durban, South Africa.

Mesarovic, M. D., D. Macko and Y. Takahara (1970). *Theory of Hierarchical Multi-Level Systems*. Academic Press, New York.

Mesarovic, M. D.,Y. Takahara (1975). *General System Theory: Mathematical Foundations*. Academic Press, New York.

Riismaa, T. (2004). Description and optimization of discrete and continuous hierarchical structures. In: *Proceedings of the 20th International Conference on CAD/CAM, Robotics and Factories of the Future, July 21-23*, (M. Marquez. (Ed)), 282-289. Nadie Nos Edita Editores, San Cristobal, Venezuela.

Riismaa, T. (2003). Optimization of the structure of fuzzy multi-level decision-making system. In: *Proceedings of the International Conference on Modelling and Simulation of Business Systems MOSIBUS 2003, May 13-14, Vilnius, Lithuania* (H. Pranevicius[et al.]. (Eds)), 31-34. Technologija, Kaunas, Lithuania.

Riismaa, T. (2002). Optimization of the Structure of Multi-Level Parallel Processing. In:, *Proceedings of the IASTED International Conference in Applied Informatics, February 18-21, Innsbruck, Austria* (M. H. Hamza. (Ed)), 77-80. ACTA Press, Anaheim, Calgary, Zurich.

Riismaa, T. (1993). Description and Optimization of Hierarchical Systems. *Remote Control*, **12**, 146-152.

Riismaa, T., I. Randvee (2003). Description and optimization of the structure of multi-level parallel selection. In: *Proceedings of the 19th International Conference on CAD/CAM, Robotics and Factories of the Future, July 2-24, Kuala Lumpur, Malaysia* (Wan Abdul Rahman Wan Harun. (Ed)), 423-429. SIRIM Berhad, Selangor, Malaysia.

Riismaa, T.,I. Randvee (2002*). Recursive Algorithms for Optimization of Multi Level- Selection. In: Proceedings of the 18th International Conference on CAD/CAM, Robotics and Factories of the Future, July 3-5, Porto, Portugal* (J. J. Pinto Ferreira. (Ed)), 333-340. INESC Porto, Porto, Portugal.

Riismaa, T., I. Randvee, J. Vain (2003). Optimization of the structure of multi-level parallel assembling. In: *Proceedings of the 7th IFAC Workshop on Intelligent Manufacturing Systems, 6-8 April, Budapest, Hungary* (L.Monostori [et al.]. (Ed)), 235-238. Elsevier, Oxford, England.

Riismaa, T., O. Vaarmann (2003). Optimal decomposition of large-scale systems. In: *Proceedings of the International Workshop on Harbour, Maritime and Multimodal Logistics Modelling & Simulation HMS 2003, September 18-20, Riga, Latvia* (Y. Merkuryev [et al.]. (Eds)), 385-394. Riga Technical University, Riga, Latvia.

Randvee, I., T. Riismaa, J. Vain (2000). Optimization of Holonic Structures. In: *Proceedings of the Workshop on Production Planning and Control, WPPC'2000. October 2-4, Mons, Belgium* (A. Artiba. (Ed)), 61-66. Ateliers de la FUCaM, Mons, Belgium.

Riismaa, T. (2000). Optimization of the Structure of Multi-Level Processing. In: *Proceedings of the Second International Conference on Simulation, Gaming, Training and Business Process Reengineering in Operations. September 8-9, Riga, Latvia* (Yuri Merkurjev [et al.]. (Eds)), 43-45. Riga, Latvia.

Riismaa T., I.Randvee (1997). Restructuration of Control Scenarios. In: *Proceedings of 23rd EUROMICRO Conference New Frontiers of Information Technology. September 1-4, Budapest, Hungary* (P. Milligan, P. Corr. (Eds)), 281-287. IEEE Computer Society, Washington, Brussels, Tokyo.

PUBLICATIONS
www.elsevier.com/locate/ifac

DIAGNOSIS METHODS USING ARTIFICIAL INTELLIGENCE. APPLICATION OF FUZZY PETRI NETS AND NEURO-FUZZY SYSTEMS

Maxime Monnin, Nicolas Palluat, Daniel Racoceanu, Noureddine Zerhouni

Laboratoire d'Automatique de Besançon
UMR CNRS 6596
24, rue Alain Savary 25000 Besançon, France
Tel.: +33 (0) 3 81 40 27 92   Fax: +33 (0) 3 81 40 28 09
Email: {mmonnin , npalluat, daniel.racoceanu , zerhouni}@ens2m.fr

Abstract: In this paper an overview of the most important artificial intelligence diagnosis tools is given. For each tool, we focus on diagnosis principles and on their advantages and disadvantages. That allows us to extract four important points that a diagnosis tool should fulfilled. Using these results, we propose a tool based on fuzzy Petri nets and a tool based on neuro-fuzzy systems, which allow to make a diagnosis using a model easy to build and that take into account the uncertainties of maintenance knowledge. These tools provide abductive approaches of fault propagations research with an efficient localization and a characterization of the fault origin. At the end, we apply our tools on a comparative example of a flexible system diagnosis.

Keywords: Monitoring, Diagnosis aid systems, Fuzzy Petri nets, Neuro-Fuzzy systems.

1. INTRODUCTION

The monitoring of industrial systems may be split into two phases: the first phase is the faults detection and the second one is the diagnosis. The diagnosis must allow the localization and the identification of the causes of these faults: it should make it possible, starting from the observation of symptoms, to go back to the causes explaining these symptoms. That amounts to put forth an assumption on the causes according to the observation of symptoms. The realization of such a diagnosis is difficult because the information available is usually incomplete, vague and dubious. Such a diagnosis system must imperatively take into account these uncertainties so as to make relevant assertions.

This paper is organized as follows: In the first section, a classification of diagnosis methods is given. A definition of each method and the related tools coming from Artificial Intelligence is given. Then, in the second section, we focus on the principles behind each method and their application to the diagnosis. This study allows giving four important points to build a diagnosis tool. Finally, we propose two diagnosis aid tools which take into account the conclusion of our work.

2. DIAGNOSIS AND ARTIFICIAL INTELLIGENCE

2.1. Diagnosis definitions

In (Dubuisson, 2001) we can find the next definition:
A diagnosis problem can be defined as a problem of pattern recognition. The set of states is the counterpart of a set of classes and the pattern vector is the vector of observed parameters.

This definition is valid in diagnosis cases in which the modes are well identified, and the associated faults origins and localization well known. In those cases, pattern recognition methods are quite efficient.

There are other approaches of diagnosis. Peng, (Peng, 1990) defines diagnosis as follows:
Given a set of observed manifestations (symptoms sss...), the goal is to explain their occurence, to go back to the cause using knowledge on the considered system.

This definition will be adopted in our study. Indeed, this approach is common to many authors, (Zwingelstein, 1995), (Grosclaude, 2000), (Bouchon-Meunier, 2003) and better put in evidence the real challenge of the diagnosis, like it is seen by our industrial partners.

Close to this definition, diagnosis as monitoring, takes into account numeric and symbolic data. Moreover, to make a diagnosis possible, the diagnosis needs causal knowledge on the system. Indeed, a default is easily described by the relation between its causes and its effects. The diagnosis problem deals with finding explanations for the observed symptoms. The inference used to make this reasoning possible, which allows to "go back to the causes" is called *Abductive Inference*:
Given a fact "B" and the association (causal relation) "A→B" (A implies B), infer "A" possible.

Such a diagnosis is called "Abductive Diagnosis". Given the complexity of diagnosis problems, many methods have been developed using different tools. The next section gives a classification of one of the most efficient one based on the Artificial Intelligence methods, in order to see which kind of diagnosis is possible with each method.

2.2. Classification of symbolic methods

We choose to work on symbolic methods which seem to be more adapted to deal with numeric and symbolic data for diagnosis. Indeed the complexity of industrial system does usually not allow making a numeric model. "Symbolic methods" is a wide term to define methods without models. The classification proposed here joins the classifications in (Basseville, 1996) and (Aghasaryan, 1998). It gives three classes of methods: the methods based on behavioural models, the recognition methods and the methods based on explicative models.

Methods using behavioural models are characterised by the possibility to simulate the behaviour of the system. They are generally based on tools like finite state machines or Petri nets.

Recognition methods (like patterns recognition systems and rules based systems) work in two phases: learning and recognition.

Methods using explicative models are based on models which give a representation of a causal analysis of the system to diagnose.

The next section exposes principles and applications for each method of diagnosis.

3. SYMBOLIC METHODS FOR THE DIAGNOSIS

3.1 Methods based on behavioural models

A method using finite state machines, presented in (Aghasaryan, 1998) is described in (Sampath, 1996). It is characterized by two phases to carry out the diagnosis. Initially, a model of the system with a total automaton is developed by composition of elementary automata corresponding to local systems. Then, in the second phase, a diagnosis assistant also corresponding to an automaton is built starting from the total model. This last one carries out a diagnosis by observing on-line a sequence of events. For each consecutive event, the diagnosis assistant provides an estimation of the state of the system and of events not observed, from where the occurrences of the breakdown are deduced.

The second main approach uses Petri Nets. A interesting use of Petri nets is described in (Anglano, 1994). The Petri nets considered are a model of behaviour of the system to be diagnosed. This model is built with a BPN (Behavioral Petri Net) which includes possibly states of breakdowns. The authors introduce two types of tokens; a normal token and an inhibiting token. These two types of tokens allow to introduce three types of markings, namely: { true, false, unknown }. The BPN are safe and deterministic networks. Backward firing rules are defined and make it possible to discover possible inconsistencies in the search of causes. In this approach the diagnosis is carried out by a backward analysis of the networks which makes it possible to go back to the initial marking (causes) starting from the marking of the related state observed.

Finally, finite state machines and Petri nets constitute tools relatively well adapted to build mechanisms of detection when the normal operation of the system is described by these formalisms. On the other hand, their uses in diagnosis are still limited. For the automata, the main difficulties are the significant size of the space of state, this leads to problems of memory and speed of execution for the diagnosis. However, as it is underlined in (Valette, 1994), Petri nets are a powerful tool of modeling and may be seen as a tool among other to describe knowledge necessary to the diagnosis.

3.2 Recognition Methods

These methods assume that no model is available to describe the relations between causes and effects. Only the knowledge relying on the human expertise consolidated by a solid feedback is considered, as in (Zwingelstein, 1995). Most of these methods are based on the Artificial Intelligence with particular tools such as case based reasoning, neural networks, fuzzy logic and neuro-fuzzy networks.

The use of CBR in diagnosis begun about fifteen years ago, one can quote as an example system MOLTKE and PATDEX, (Bergmann, 1998). For an application in diagnosis, the principal singularity of CBR is due to the definition of the cases structure. A new diagnosis problem is solved by *retrieving* one or more previously known cases, *reusing* the known case in one way or another, *revising* the solution based on reusing a previous case, and *retaining* the new experience by storing it into the existing knowledge-base (case-base), (Aamodt, 1994). It makes it possible starting from the description of a breakdown to find the causes and to propose an action for a possible intervention of maintenance. Adapted to the diagnosis, the structure of the cases is thus the following one:

Problem ↔ Symptoms (description of the particular situation of diagnosis)

Solution ↔ Origins (several possible origins)

Conclusion ↔ Actions (strategy of maintenance).

Thus the use of CBR in diagnosis seems relatively easy, with the suitable structure for cases. However, the difficulty is precisely due to this structure of case and information which it must contain. Indeed, the extraction of knowledge and its representation are of primary importance in this type of application.

In the same way, one can find neural networks in diagnosis systems. Many architectures have been developed but we focus on the RRBF (Recurrent Radial Basis Function) proposed in (Zemouri, 2002). Indeed, this architecture seems best suited for applications of forecast and dynamic monitoring. Moreover thanks to its simplicity (separation of the dynamic memory and the static memory) and to the relatively short training time, it allows the on-line training, which makes it particularly interesting within the framework of the dynamic monitoring. When applied to the diagnosis, it provides powerful classification means. Indeed, the neural networks belong to the recognition pattern methods of diagnosis. However, even if the neuronal tool is able to identify modes of failures of a system, it does not explain really the causes at the origin of these modes of failures. For an application in diagnosis, the neural networks and more specially the RRBF belong to detection methods which come before the diagnosis in the monitoring of the systems. The results obtained with temporal neural networks make them particularly interesting tools for industrial applications of monitoring. Their capacities of detection and classification would make possible to use them within the framework of an application of diagnosis for the generation of intelligent alarms because the detection that they carry out is followed by a treatment and allows a classification. Indeed they could constitute a good tool which would make possible to obtain symptoms associated with a failure that a system of diagnosis would use to carry out the localization and the identification of the causes of this failure

Introduced by Zadeh in 1965, fuzzy logic as it is described in (Bouchon-Meunier, 1995) allows formalizing the representation and the treatment of vague or approximate knowledge. It allows handling systems of a great complexity with for example human's factors. Its use in fields such as the decision-making aid or the diagnosis thus seems natural because it provides a powerful tool to assist in an automatic way the human actions which are usually inaccurate.

In these various contexts (diagnosis, decision-making aid), knowledge or data given by human expert are incomplete, imprecise and informal. Thus fuzzy logic makes it possible on one hand to take into account the data inaccuracies and on the other hand to specify rules to diagnose or to find the appropriate action. An example of such architecture is given in (Rahamani, 1998), in which fuzzy logic is used on three different levels (a fuzzy expert system and two classification stages). Fuzzy logic is also associated with other tools like Petri Nets for example. In (Minca, 2002), Petri Nets are used to model fault trees and logical expressions of fault trees are translated into fuzzy rules. So, starting from a fuzzy expression of symptoms, the tool gives a constant dynamic analysis of the state of degradation of the system. In these various applications, fuzzy logic is rather natural to treat the inaccuracy, uncertainty related to knowledge of the field. However one cannot consider these applications as real applications of diagnosis because these various tools do not localize nor identify causes of failures. Used with fault trees, fuzzy logic should provide an evaluation on the occurrence or the presence of the basic events of the fault tree which are at the origin of the top event. One would thus obtain the evaluation of the causes at the origin of a dysfunction.

In applications of diagnosis, one finds mainly hybrid neuro-fuzzy models for which neural networks and fuzzy systems are combined in a homogeneous way. The majority of the applications met in (Palade, 2002), (Uppal, 2002), are based on the establishment of a diagnosis starting from the classification of residues, they thus require to be able to establish a numeric model of the system. Possibilities of establishing models with neuro-fuzzy techniques have been developed but their applications remain limited. It would be thus interesting to use these techniques taking into account their capacities while being completely free from a model of the system to diagnose.

We have seen in this part the various techniques of the Artificial Intelligence which make it possible to carry out a diagnosis based on recognition methods. Their applications are numerous and for some, the results are overall satisfactory. However, most of these methods carry out a classification by pattern recognition. The diagnosis thus amounts to identify an operating mode of the process which reflects the state of breakdown. In this direction, the diagnosis carried out does not make it possible to identify the causes of the dysfunction unless if they are explicitly described in the identified mode or case as in the CBR. For the other tools, the applications are connected more with "intelligent detection", for which the output of the system of diagnosis is carrying information on the state of the system, but does not give the causes of them. These tools thus seem better suited for the modules of detection in a complete architecture of monitoring.

3.3 Methods based on explicative models

These methods are mainly based on the representation of the relations between the various states of breakdowns and their possibly observable effects. They thus rely on a major analysis of the system, so as to have sufficient knoledges on these relations of cause to effect. Some models allow to use an abductive approach which consists in going back to the causes of the breakdowns starting from the observations corresponding to the symptoms.

The causal graphs, as those described in (Brusoni, 1995), (Brusoni, 1997) (Grosclaude, 2000), are a tool particularly interesting for the diagnosis because they can bring a justification of the diagnosis suggested through causal way followed in the graph. Moreover, the algorithms of abductive diagnosis make it possible starting from the observation of symptoms to seek possible causes. Lastly the introduction of temporal constraints, contradictory effects and interactions between the breakdowns gives better approaches of the physical reality of the system to be diagnosed. The causal graphs rely on the formalization of the causal bonds which govern the states of breakdowns and require an important knowledge of the system to establish these causal bonds as well as the temporal constraints. This knowledge can be extract from diagnosis tools like fault tree or FMECA (Failure Modes, Effects and Criticality Analysis). Moreover, the algorithms of temporal abductive diagnosis are relatively complex and impose long computing times. A diagnosis on-line is thus not easily realizable by using the causal model directly.

The contextual graphs were introduced starting from the decision trees. The representation based on the context of the event is turn into a representation based on the context of resolution of the event. The branches of the tree which lead to the same final action are gathered in the graph and a temporal connection is introduced into the graph to account for the actions and the decisions which can be carried out in parallel. They allow the representation of multiple actions depending on the context. Moreover this representation takes into account the dynamics of the context in its evolution. They also present the advantage to be able to handle great structures such as industrial applications. This representation is comprehensible by operators since it is similar to their mode of reasoning. Lastly, their flexibility and their modularity allow incremental acquisition of knowledge so as to integrate new practices. The contextual graphs thus seem to be a tool adapted for modelling activities which comprising a procedure/practices duality. They are applicable in fields where an interpretation or an adaptation of general rules is necessary to take into account the richness of the real context. Within the framework of an application of supervision, they could be applied in cases where the context takes a significant place between diagnosis of defects and the actions of recovery.

Petri nets also allow within the framework of the diagnosis an approach in term of fault model. The places constitute states of breakdowns and the architecture of the network makes it possible to account for the relations existing between these breakdowns. Several techniques are based on a fault model, with in particular, the use of stochastic Petri nets. In (Aghasaryan, 1997), (Aghasaryan, 1998), (Tromp, 2000) and (Fabre, 2001), an approach by Petri nets to the problems of fault detection and diagnosis is used. The partially stochastic Petri nets are presented through two original approaches for the diagnosis. The first relates to an application to the management of the faults in the telecommunications networks, the second is applied to complex industrial systems. The use of partially stochastic Petri nets models makes it possible to preserve the probabilistic independence between competitor events, and does not require the exploration of the whole state space because only the local contexts are necessary to the propagation of the events or to the estimation of probabilities. Moreover this aspect of local context makes the method more robust, the tool is to a certain extent relatively quite evolutionary. On the other hand, results in (Tromp, 2000), tend to make partially stochastic Petri Nets a tool not very adapted to the diagnosis of complex industrial process.

Finally, fuzzy logic can also be applied in order to work out an explicative model as in (Bouchon-Meunier, 2003). It deals with an explanation oriented diagnosis which makes it possible to explain the presence of a set of symptoms and to go back to the origin of these observations. With this aim one uses the modelling of a rule of causes to effects between the dysfunctions and the symptoms. Moreover one also takes into account the uncertainty of the observation of certain symptoms. According to this approach, the diagnosis is carried out thanks to the confrontation between knowledge and the observations. The diagnosis is carried in two stages : it uses an index of coherence which allows to eliminate dysfunctions which are incoherent with the observations and it is refined by an index of relevance which select and sort by order of suspicion remaining dysfunctions. Other works (Mellouli, 2000), on fuzzy logic study the realization of an abductive reasoning by the inversion of the modus ponens. Being given a fuzzy rule of the type "if U is A then V is B", and an observation of the type V', the authors seek to characterize the assumptions of the type U is A' answering the question "why V is B' "? This abductive approach based on the inversion of the modus ponens must make it possible to find the assumptions satisfying a given observation. If the starting rule models a causal relation for which premise A is a cause and the conclusion B is an effect the abductive approach on the basis of the B' observation of the effect makes it possible to characterize the cause by the A' assumption and thus carries out a diagnosis.

For all these methods, four important points can resume the suitable properties for a diagnosis tool. For the beginning, the first step in order to make a diagnosis system is the model acquisition. As we have seen, making a diagnosis uses the system knowledge. Thus explicative models seem to be the most adapted to express the causal knowledge of a system, which are essential to carry out a diagnosis. In the industrial maintenance practice, this knowledge – based on human expertise - is often uncertain. Fuzzy logic is the best tool to express and take into account these uncertainties. Moreover, a diagnostic tool has to be robust and must use a generic method. At last whatever tool we use, results have to be validated by an expert. The tools proposed in the next section try to take into account all these points.

4. FUZZY PETRI NETS AND NEURO-FUZZY SYSTEMS FOR THE DIAGNOSIS

Taking into account these requirements for the diagnosis, we focus on methods based on explicative models. They indeed are adapted best for the modelling of the causal relations essential to the diagnosis. A fault model seems more suited for a diagnosis tool and Petri Net are chosen in a first time for their modelling capabilities. Given the difficulties of the model acquisition, expertises already available in a company are invaluable sources of information. The Petri net fault model is based on the tools used for diagnosis which are fault trees and FMECA. As in (Tromp, 2000) and (Minca, 2002) Petri nets model fault trees and provide powerful tools, able to represent *AND* and *OR* influences. The fault tree is a formalization of the causal bonds which govern the states of breakdowns. Modelling such analysis with Petri net allows to give a justification of the diagnosis by the way followed in the net as in causal graphs. Instead of probabilistic approaches like Bayesian Networks or Stochastic Petri nets, which lead to tools hardly relevant for on-line diagnosis, the fuzzy approach allows reasoning close to the human one. The association of the Fuzzy logic with the Petri net's computing abilities, allows an abductive approach for an on-line diagnosis without increasing the complexity of the tool. Fuzzy logic is introduced in order to take into account uncertainties linked to the maintenance knowledge and represents a formalization of the important knowledge given in the FMECA. It makes it possible to give fuzzy degrees of credibility as in (Looney, 2003), associated to the states of breakdowns. Lastly, we use an abductive approach which makes it possible to characterize the causes starting from the observations. The algorithm is a downward approach in the fault tree. Given a top event observed this observation is translated into a fuzzy degree of credibility and propagated in the net. So, each fault which could be at the origin of the top event is characterized by its own fuzzy degree.

In the same time we work on neuro-fuzzy systems as described in the next. Our method is divided into several points, (Palluat, 2004) :
- Acquisition of relevant information of the system. Using studies carried out on the system (Failure Modes Effects and Critical Analysis – FMECA, fault tree, functional analyzes …), and with the help of the operators and the experts, it is necessary to extract the critical zones to supervise, as well as information available on these zones: static (fault tree, functional analyzes …) and dynamic information (CMMS – historic, given sensors …).

- Application of the detection system based on the dynamic neural networks.

On input of the detection system, we find the information given by the sensors; it can be a binary or a real value. On output, the experts identify the operating mode (symptom) of the supervised element. The use of neural networks is justified by their training ability, their parallel computation ability, their capacity to solve problems inherent to the system non-linearity and their computation speed when implemented in an integrated circuit.

- Application of the diagnosis system based on a fuzzy neural network.

On input of the diagnosis system, we find the degree of membership of each operating mode given by the detection system. We find also external qualitative or quantitative inputs like information given by operators to improve diagnosis. On output, we find a list of possible causes ordered by degree of credibility, and as optional information: the degree of gravity. This diagnosis aid system can be used for real-time monitoring and it improves its accuracy by on-line learning. Moreover, the natural formulation often available in CMMS can be integrated thanks to the fuzzy logic procedure. The diagnosis aid system gives also to the operator a fuzzy interpretation of all possible causes and origins linked to a given symptom and may help the maintenance manager to plan the maintenance.

5. EXAMPLE

Let us consider the following fault tree of a flexible manufacturing sub-system from the "Institut de Productique" of Besançon (Fig. 1)

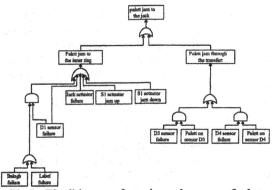

Fig.1 : Flexible manufacturing sub-system fault tree

Given the *AND* and *OR* influences we obtain the Fuzzy Petri Net shown in Fig.2. The different places in the net are the different events of the fault tree and the transitions are the different *AND* and *OR* dependencies. In this example we consider that events corresponding to the places P_2 and P_3 are the observable symptoms. We suppose that we have a detection system which provides us the fuzzy credibility degree corresponding to both symptoms. Information given by the FMECA are introduced in order to parameterize the backward propagation of the degree in *AND/OR* dependencies. This propagation depends on the fault frequency of the corresponding event which is one of the characteristic given in the FMECA. This frequency determines the center of a sigmoid function which allows the backward propagation, (Looney, 2003). For each causal relation, the degree corresponding to the

consequent is transformed through the sigmoid function in order to provide the degree for the antecedent.

The same fault tree is used in order to build the fuzzy neural network (Fig 3.). AND/OR dependencies are given with only two types of neurons: one with linear activation and one with semi-sigmoid activation.

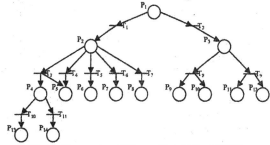

Fig.2: Example of Fuzzy Petri Net (FPN)

As for the fuzzy Petri net, the FMECA allow to configure the fuzzy neural network. A factor α for semi-sigmoid neurons represents the frequency of the fault. For basic events, this factor follows the following rule:

IF *this event is in the FMECA,*
THEN *the factor is found directly by a fuzzyfication of the frequency of the fault*
ELSE *the factor is the minimum that we fixed in preliminaries.*

For upper level, this factor is the maximum of factors that are below.

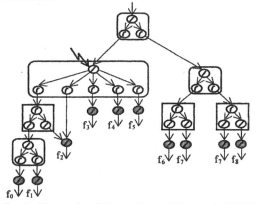

Fig. 3 : Example of Fuzzy Neural Network (FNN)

We assume that we observed the event *"Pallet jam to the inner ring"* with the fuzzy degree 0.46, results for both tools is given in the table 1.

Table 1.

Events	FPN	FNN
P_{13}/f_0	0.018	0.449
P_{14}/f_1	0.018	0.449
P_5/f_2	*0.574*	*0.967*
P_6/f_3	0.231	0.430
P_7/f_4	0.098	0.226
P_8/f_5	0.45	0.725
P_9/f_6	0	0
P_{10}/f_7	0	0
P_{11}/f_8	0	0
P_{12}/f_9	0	0

This degree is propagated through the Petri Net in places which could lead to P_2 in a forward propagation. The maintenance knowledge of the FMECA and the causal relations of the fault tree used in the fuzzy Petri net allow suspecting the event corresponding to P_5 as the fault origin with a fuzzy credibility of 0.574. The same example has run trough the FNN which were parameterize with the same FMECA. The results are quite similar and the same event: *"D1 sensor failure"* is suspected. This information can held a maintenance operator in his diagnosis. The FPN and the FNN provide an abductive diagnosis: given a symptom, it gives a set of fault origin directly linked by causal relations to the observed event with a backward analysis of the fault tree. Moreover it takes into account the uncertainties on the maintenance knowledge by giving a fuzzy characterisation of each fault origin. So, localization and identification of the fault origin are implicitly given by the descriptions in the fault tree.

6. CONCLUSION

We have studied the problem of diagnosis aid systems using Artificial Intelligence methods. We focused on methods based on symbolic models. Most of these models handle at the same time numerical and symbolic data which are essential to the realization of a diagnosis. After having drawn up a panorama of the various tools met in the literature, we proposed the significant points necessary to the realization of a diagnosis. Finally, we sketch tools for diagnosis based on an explanatory model carried out with the fuzzy Petri nets and the fuzzy neural network.

Further work will investigate methods to better exploit the FMECA in these tools, include feedbacks as a training capability and link the tools to a dynamic neural network which perform a dynamic detection, in order to have a complete, dynamic monitoring system.

REFERENCES

Aamodt, A. , Plaza, E. (1994), *Case-Based Reasoning : Foundational Issues, Methodological Variations and Systems Approaches*, AI Communication. IOS Press, Vol. 7: 1, pp. 39-59.

Aghasaryan, A. *(1998), Formalisme HMM pour les réseaux de Pétri partiellement stochastiques : Application au diagnostic de pannes dans les systèmes répartis*, Thèse de doctorat université de Rennes1.

Aghasaryan, A., Boubour, R., Fabre, E., Jard, C., Benveniste, A (1997), *A Petri Net Approach to fault detection and diagnosis in distributed systems*, Armen., Publication interne n°1117 IRISA.

Anglano, C., Luigi Portinale, L. (1994), *B-W Analysis : a Backward Reachability Analysis for Diagnostic Problem Solving Suitable to Parallele Implementation*, Proceeding of the 15th International Conference on Application and Theory of Petri Nets, Zaragoza Spain.

Basseville, M., Cordier, M-O. (1996), *Surveillance et diagnostic de systèmes dynamiques : approches complémentaires du traitement du signal et de l'intelligence artificielle*, Rapport INRIA n°2861.

Bergmann R. (1998), *Introduction To Case-Based Reasoning*, University of Kaiserslautern, http://www.dfki.unikl.de/~aabecker/Mosbach/Bergmann-CBR-Survey.pdf

Brusoni, V., Console, L., Terenziani, P., Theseider Dupré, D. (1995), *Characterizing temporal abductive diagnosis,* , Proc. of International Workshop on Principles of Diagnosis pp. 34-40.

Brusoni, V., Console, L., Terenziani, P., Theseider Dupré, D. (1997), *An Efficient Algorithm for Temporal Abduction*, Lecture notes in Artificial Intelligence 1321 pp. 195-206.

Bouchon-Meunier B. (1995), *La Logique Floue et ses applications*, Edition Addison-Wesley.

Bouchon-Meunier B., Marsala C. (2003), *Logique floue, principes, aide à la décision*, Ed. HERMES.

Dubuisson, B. (2001), *Diagnostic, Intelligence Artificielle et reconnaissance de formes*, Ed. HERMES.

Fabre, E., Benveniste, A., Jard, C., Smith, M. (2001), *Diagnosis of distributed discrete event systems, a net unfolding approach*, Publication Interne IRISA.

Grosclaude, I., Quiniou, R. (2000), *Dealing with Interacting Faults in Temporal Abductive Diagnosis*, Proc. of the 8th International Workshop on Principles of Diagnosis.

Looney, Carl G., Liang, Lily R., (2003) *Inference via Fussy Bielef Petri Nets*, , 15th IEEE International Conference on Tools with AI (ICTAI' 03), Sacramento, California, USA.

Mellouli, N., Bouchon-Meunier, B. (2000), *Fuzzy Approaches to Abductive Inference*, Conference on Non-Monotonic reasoning, Berckenridge, Colorado .

Minca E., Racoceanu D., Zerhouni N. (2002), *Monitoring Systems Modelling and Analysis Using Fuzzy Petri Nets*, Studies in Informatics and Control, Vol. 11, No. 4.

Palade, V., Patton, R. J., Uppal1, F. J., Quevedo, J., Daley, S. (2002), *Fault diagnosis of an industrial gas turbine using neuro-fuzzy methods*, IFAC.

Palluat, N., Racoceanu, D., Zerhouni, N. (2004), *Diagnosis Aid System using a Neuro-Fuzzy Approach*, In "Advances in Maintenance and Modelling, Simulation and Intelligent Monitoring of Degradation", IMS'2004, July 15-17, 2004 Arles, France.

Peng Y., Reggia J. A. (1990), *Abductive Inference Models For Diagnostic Problem-Solving* Springer-Verlag, New York.

Rahamani K., Arabshahi P., Yan T.-Y., Pham T., and Finley S. G. (1998), *An Intelligent Fault Detection and Isolation Architecture For Antenna Arrays*, TDA Progress Report 42-132.

Sampath M., Sengupta R., Lafortune S., Sinnamohideen K., and Teneketzis D. C. (1996), *Failure Diagnosis Using Discrete Event Models*, IEEE Transactions On Control Systems Technology, Vol. 4, No. 2.

Tromp L. (2000), *Surveillance et diagnostic de systèmes industriels complexes : une approche hybride numérique/symbolique*, Thèse de doctorat université de Rennes 1.

Uppal F. J., Patton R. J. and Palade V. (2002), *Neuro-fuzzy based Fault Diagnosis applied to an Electro-Pneumatic Valve*, IFAC.

Valette R., Künzle L. A. (1994), *Réseaux de Petri pour la détection et le diagnostic*, Journées Nationales : Sûreté, Surveillance, Supervision. GDR Automatique Paris.

Zemouri, R., Racoceanu, D., Zerhouni, N. (2002) *Application of the dynamic RBF network in a monitoring problem of the production systems*, 15e IFAC World Congress on Automatic Control, sur CD ROM, 6 p., Barcelone, Espagne, juillet 2002.

Zwingezlstein G. (1995), *Diagnostic des défaillances : théorie et pratique pour les systèmes industriels* Ed. HERMES.

OVERVIEW ON DIAGNOSIS METHODS USING ARTIFICIAL INTELLIGENCE APPLICATION OF FUZZY PETRI NETS

Maxime Monnin, Daniel Racoceanu, Noureddine Zerhouni

Laboratoire d'Automatique de Besançon
UMR CNRS 6596
24, rue Alain Savary 25000 Besançon, France
Tel.: +33 (0) 3 81 40 27 97   Fax: +33 (0) 3 81 40 28 09
Email: {mmonnin, daniel.racoceanu , zerhouni}@ens2m.fr

Abstract: This paper studies diagnosis-aid systems that use Artificial Intelligence tools. This kind of systems become very interesting in an uncertain industrial environment especially for flexible production systems. An overview of the most important artificial intelligence diagnosis tools is given. For each tool, we focus on diagnosis principles and on its advantages and disadvantages. That allows us to extract four important points that a diagnosis tool should fulfilled. Using these results, we propose a tool based on fuzzy Petri nets which allows to make a diagnosis using a model easy to build and that take into account the uncertainties of maintenance knowledges. This tool provides abductive approaches of fault propagations system with an efficient localization and a characterization of the fault origin. At the end, we apply our tool on an illustrative example of a flexible system diagnosis is presented.

Keywords: Monitoring, Diagnosis, Diagnosis aid systems, Fuzzy logic, Fuzzy Petri nets.

1. INTRODUCTION

The monitoring of industrial systems may be split into two phases: the first phase is the faults detection and the second one is the diagnosis. The diagnosis must allow the localization and the identification of the causes of these faults : it should make it possible, starting from the observation of symptoms, to go back to the causes explaining these symptoms. It is thus assumed that there is a causal relation of implication between the causes and the effects observed: causes ⇒ effects (symptoms). However, logic does not make it possible to provide information on the antecedent from the consequent. The basic idea developed here is thus to put forth an explanatory assumption on the antecedent in comparison with the observations relative to the consequent (Mellouli, 2000). Applied to the diagnosis, that amounts to put forth an assumption on the causes according to the observation of symptoms. The realization of such a diagnosis is difficult because the information available is usually incomplete, vague and dubious. The system of diagnosis must imperatively take into account these uncertainties so as to make

relevant assertions. This paper is organized as follows. In the first section, a classification of diagnosis methods is given. A definition of each method and the related tools coming from Artificial Intelligence is given. Then, in the second section, we compare the principles behind each method and how it applies to diagnosis. This comparative study allows to give four important points to build a diagnosis tool. Finally, we propose a diagnosis aid tool, which takes into account the conclusion of our work.

2. DIAGNOSIS AND ARTIFICIAL INTELLIGENCE

2.1. Diagnosis definitions

A first definition is given by (Dubuisson, 2001), in which diagnosis is considered as follows :
A diagnosis problem can be defined as a problem of pattern recognition. The set of states is the counterpart of a set of classes and the pattern vector is the vector of observed parameters.
This definition can be applied to many applications of diagnosis in which pattern recognition methods are used.

There are other approaches of diagnosis Peng, (Peng, 1990) defines diagnosis as follows :

Given a set of observed manifestations (symptoms, ...),the goal is to explain their occurence, to go back to the cause using knowledges on the considered system.

We will rely on this definition in the rest of the paper. Indeed, this approach is common to many authors, (Zwingelstein, 1995), (Grosclaude, 2000), (Bouchon-Meunier, 2003).

Closed to this definition, diagnosis as monitoring, takes into account numeric and symbolic data. Moreover, to make a diagnosis possible, it needs causal knowledges on the system. Indeed, a default is easily described by the relation between its causes and its effects. When we have a theory modeling such relations, a diagnosis problem deals with finding explaniations for the observed symptoms. The inference used to make this reasoning possible, which allows to "go back to the causes" is called *Abductive Inference* :

Given a fact "B" and the association (causal relation) "A→B" (A implies B), infer "A" possible.

Such a diagnosis is called "Abductive Diagnosis". Given the complexity of diagnosis problems, many methods have been developed using different tools, more specifically tools of Artificial Intelligence. The next section gives a classification of these methods in order to see which kind of diagnosis is possible with each method.

2.2. Classification of symbolic methods

We choose to work on symbolic models which seem to be more adapted to deal with numeric and symbolic data for diagnosis. Indeed the complexity of industrial system does usually not allow to make a numeric model. "Symbolic methods" is a wide term to define methods without models. The classification proposed here joins the classifications in (Basseville, 1996) and (Aghasaryan, 1998). It gives three classes of methods : the methods based on behavioural models, the recognition methods and the methods based on explicative models.

Methods using behavioural models are characterised by the possibility to simulate the behaviour of the system. They are generally based on tools like finite state machines or Petri nets.

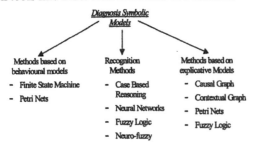

Fig. 1 : Classification of diagnosis symbolic models

Recognition methods (like patterns recognition systems and rules based systems) work in two phases : learning and recognition.

Methods using explicative models are based on models which give a representation of a causal

analysis of the system to diagnose. The next section exposes principles and applications for each methods of diagnosis.

3. SYMBOLIC METHODS FOR THE DIAGNOSIS

3.1 Methods based on behavioural models

A method using finite state machines, presented in (Aghasaryan, 1998) is described in (Sampath, 1996). It is characterized by two phases to carry out the diagnosis. Initially, a model of the system with a total automaton is developed by composition of elementary automata corresponding to local systems. Then, in the second phase, a diagnosis assistant also corresponding to an automaton is built starting from the total model. This last one carries out a diagnosis by observing on-line a sequence of events. For each consecutive event, the diagnosis assistant provides an estimation of the state of the system and of events not observed, from where the occurrences of the breakdown are deduced.

The second main approach uses Petri Nets. A interesting use of Petri nets is described in (Anglano, 1994). The Petri nets considered are a model of behaviour of the system to be diagnosed. This model is built with a BPN (Behavioral Petri Net) which includes possibly states of breakdowns. The authors introduce two types of tokens; a normal token and an inhibiting token. These two types of tokens allow to introduce three types of markings, namely: { true, false, unknown }. The BPN are safe and deterministic networks. Backward firing rules are defined and make it possible to discover possible inconsistencies in the search of causes. In this approach the diagnosis is carried out by a backward analysis of the networks which makes it possible to go back to the initial marking (causes) starting from the marking of the related state observed.

Finally, finite state machines and Petri nets constitute tools relatively well adapted to build mechanisms of detection when the normal operation of the system is described by these formalisms. On the other hand, their uses in diagnosis are still limited. For the automata, the main difficulties is the significant size of the space of state, this leads to problems of memory and speed of execution for the diagnosis. However, as it is underlined in (Valette, 1994), Petri nets are a powerful tool of modeling and may be seen as a tool among other to describe knowledges necessary to the diagnosis.

3.2 Recognition Methods

These methods assume that no model is available to describe the relations between causes and effects. Only the knowledge relying on the human expertise consolidated by a solid feedback is considered, as in (Zwingelstein, 1995). Most of these methods are based on the Artificial Intelligence with particular tools such as case based reasoning, neural networks, fuzzy logic and neuro-fuzzy networks.

The use of CBR in diagnosis begun about fifteen years ago, one can quote as an example systems MOLTKE and PATDEX, (Bergmann, 1998). For an application in diagnosis, the principal singularity of CBR is due to the definition of the cases structure. A new diagnosis problem is solved by *retrieving* one or more previously known cases, *reusing* the known case in one way or another, *revising* the solution based on reusing a previous case, and *retaining* the new experience by storing it into the existing knowledge-base (case-base), (Aamodt, 1994). It makes it possible starting from the description of a breakdown to find the causes and to propose an action for a possible intervention of maintenance. Adapted to the diagnosis, the structure of the cases is thus the following one:

Problem ↔ Symptoms (description of the particular situation of diagnosis)

Solution ↔ Origins (several possible origins)

Conclusion ↔ Actions (strategy of maintenance).

Thus the use of CBR in diagnosis seems relatively easy, with the suitable structure for cases. However, the difficulty is precisely due to this structure of case and information which it must contain. Indeed, the extraction of knowledge and its representation are of primary importance in this type of application.

In the same way, one can find neural networks in diagnosis systems. Many architectures have been developed but we focus on the RRBF (Recurrent Radial Basis Function) proposed in (Zemouri, 2003). Indeed, this architecture seems best suited for applications of forecast and dynamic monitoring. Moreover thanks to its simplicity (separation of the dynamic memory and the static memory) and to the relatively short training time, it allows the on-line training, which makes it particularly interesting within the framework of the dynamic monitoring. When applied to the diagnosis, it provides powerful classification means. Indeed, the neural networks belong to the recognition pattern methods of diagnosis. However, even if the neuronal tool is able to identify modes of failures of a system, it does not explain really the causes at the origin of these modes of failures. For an application in diagnosis, the neural networks and more specially the RRBF belong to detection methods which comes before the diagnosis in the monitoring of the systems. The results obtained with temporal neural networks make them particularly interesting tools for industrial applications of monitoring. Their capacities of detection and classification would make possible to use them within the framework of an application of diagnosis for the generation of intelligent alarms because the detection that they carry out is followed by a treatment and allows a classification. Indeed they could constitute a good tool which would make possible to obtain symptoms associated with a failure that a system of diagnosis would use to carry out the localization and the identification of the causes of this failure

Introduced by Zadeh in 1965, fuzzy logic as it is described in (Bouchon-Meunier, 1995) allows to formalize the representation and the treatment of vague or approximate knowledges. It allows to handle systems of a great complexity with for example humans factors. Its use in fields such as the decision-making aid or the diagnosis thus seems natural because it provides a powerful tool to assist in an automatic way the human actions which are usually inaccurate.

In these various contexts (diagnosis, decision-making aid), knowledges or data given by human expert are incomplete, imprecise and informal. Thus fuzzy logic makes it possible on one hand to take into account the data inaccuracies and on the other hand to specify rules to diagnose or to find the appropriate action. An example of such an architecture is given in (Rahamani, 1998), in which fuzzy logic is used on three different levels (a fuzzy expert system and two classification stages). Fuzzy logic is also associated with other tools like Petri Nets for example. In (Minca, 2002), Petri Nets are used to model fault trees and logical expressions of fault trees are translated into fuzzy rules. So, starting from a fuzzy expression of symptoms, the tool gives a constant dynamic analysis of the state of degradation of the system. In these various applications, fuzzy logic is rather natural to treat the inaccuracy, uncertainty related to knowledge of the field. However one cannot consider these applications as real applications of diagnosis because these various tools do not localize nor identify causes of failures. Used with fault trees, fuzzy logic should provide an evaluation on the occurrence or the presence of the basic events of the fault tree which are at the origin of the top event. One would thus obtain the evaluation of the causes at the origin of a dysfunction.

In applications of diagnosis, one finds mainly hybrid neuro-fuzzy models for which neural networks and fuzzy systems are combined in a homogeneous way. The majority of the applications met in (Palade, 2002), (Uppal, 2002), are based on the establishment of a diagnosis starting from the classification of residues, they thus require to be able to establish a numeric model of the system. Possibilities of establishing models with neuro-fuzzy techniques have been developed but their applications remain limited. It would be thus interesting to use these techniques taking into account their capacities while being completely free from a model of the system to diagnose.

We have seen in this part the various techniques of the Artificial Intelligence which make it possible to carry out a diagnosis based on recognition methods. Their applications are numerous and for some, the results are overall satisfactory. However, most of these methods carry out a classification by pattern recognition. The diagnosis thus amounts to identify an operating mode of the process which reflects the state of breakdown. In this direction, the diagnosis carried out does not make it possible to identify the causes of the dysfunction unless if they are explicitly described in the identified mode or case as in the CBR. For the other tools, the applications are connected more with "intelligent detection", for which the output of the system of diagnosis is carrying information on the state of the system, but does not give the causes of them. These tools thus seem better suited for the modules of detection in a complete architecture of monitoring.

3.3 Methods based on explicative models

These methods are mainly based on the representation of the relations between the various states of breakdowns and their possibly observable effects. They thus rely on a major analysis of the system, so as to have sufficient knoledges on these relations of cause to effect. Some models allow to use an abductive approach which consists in going back to the causes of the breakdowns starting from the observations corresponding to the symptoms.

The causal graphs, as those described in (Brusoni, 1995), (Brusoni, 1997) (Grosclaude, 2000), are a tool particularly interesting for the diagnosis because they can bring a justification of the diagnosis suggested through causal way followed in the graph. Moreover, the algorithms of abductive diagnosis make it possible starting from the observation of symptoms to seek possible causes. Lastly the introduction of temporal constraints, contradictory effects and interactions between the breakdowns gives better approaches of the physical reality of the system to be diagnosed. The causal graphs rely on the formalization of the causal bonds which govern the states of breakdowns and requires of this fact a important knowledge of the system to establish these causal bonds as well as the temporal constraints. This knowledge can be extract from diagnosis tools like fault tree or FMECA (Failure Modes, Effects and Criticality Analysis). Moreover, the algorithms of temporal abductive diagnosis are relatively complex and impose long computing times. A diagnosis on-line is thus not easily realizable by using the causal model directly.

The contextual graphs were introduced starting from the decision trees. The representation based on the context of the event is turn into a representation based on the context of resolution of the event. The branches of the tree which lead to the same final action are gathered in the graph and a temporal connection is introduced into the graph to account for the actions and the decisions which can be carried out in parallel. They allow the representation of multiple actions depending on the context. Moreover this representation takes into account the dynamics of the context in its evolution. They also present the advantage to be able to handle great structures such as industrial applications. This representation is comprehensible by operators since it is similar to their mode of reasoning. Lastly, their flexibility and their modularity allow incremental acquisition of knowledges so as to integrate new practices. The contextual graphs thus seem to be a tool adapted for modelling activities which comprising a procedure/practices duality. They are applicable in fields where an interpretation or an adaptation of general rules is necessary to take into account the richness of the real context. Within the framework of an application of supervision, they could be applied in cases where the context takes a significant place between diagnosis of defects and the actions of recovery.

Petri nets also allow within the framework of the diagnosis an approach in term of fault model. The places constitute states of breakdowns and the architecture of the network makes it possible to account for the relations existing between these breakdowns. Several techniques are based on a fault model, with in particular, the use of stochastic Petri nets. In (Aghasaryan, 1997), (Aghasaryan, 1998), (Tromp, 2000) and (Fabre, 2001), an approach by Petri nets to the problems of fault detection and diagnosis is used. The partially stochastic Petri nets are presented through two original approaches for the diagnosis. The first relates to an application to the management of the faults in the telecommunications networks, the second is applied to complex industrial systems. The use of partially stochastic Petri nets models makes it possible to preserve the probabilistic independence between competitor events, and does not require the exploration of the whole state space because only the local contexts are necessary to the propagation of the events or to the estimation of probabilities. Moreover this aspect of local context makes the method more robust, the tool is to a certain extent relatively quite evolutionary. On the other hand, results in (Tromp, 2000), tend to make partially stochastic Petri Nets a tool not very adapted to the diagnosis of complex industrial process.

Finally, fuzzy logic can also be applied in order to work out an explicative model as in (Bouchon-Meunier, 2003). It deals with an explanation oriented diagnosis which makes it possible to explain the presence of a set of symptoms and to go back to the causes to the origin of these observations. With this aim one uses the modelling of a rule of causes to effects between the dysfunctions and the symptoms. Moreover one also takes into account the uncertainty of the observation of certain symptoms. According to this approach, the diagnosis is carried out thanks to the confrontation between knowledge and the observations. The diagnosis is carried in two stages : it uses an index of coherence which allows to eliminate dysfunctions which are incoherent with the observations and it is refined by an index of relevance which select and sort by order of suspicion remaining dysfunctions. Other works in (Mellouli, 2000), on fuzzy logic study the realization of an abductive reasoning by the inversion of the modus ponens. Being given a fuzzy rule of the type "if U is A then V is B", and an observation of the type V', the authors seek to characterize the assumptions of the type U is A' answering the question "why V is B' "? This abductive approach based on the inversion of the modus ponens must make it possible to find the assumptions satisfying a given observation. If the starting rule models a causal relation for which premise A is a cause and the conclusion B is an effect the abductive approach on the basis of the B' observation of the effect makes it possible to characterize the cause by the A' assumption and thus carries out a diagnosis.

For all these methods, four important points can resume the properties for a diagnosis tool. To begin with, the acquisition of the model is the first step to make a diagnosis system. As we have seen, to make a diagnosis possible we need knowledges on system to diagnose. Thus explicative models seem to be the best adapted to express the causal knowledges on a system, which are essential to carry out a diagnosis. These knowledges based on human expertise are uncertain. Fuzzy logic is the best tool to express and take into account these uncertainties. Moreover, a diagnostic tool has to be robust and relatively generic but diagnosis is intimately linked to the system so, a generic method is better adapted and would be applied on different systems. At last whatever tool we use, results have to be validated by an expert. The tool proposed in the next section tries to take into account all these points.

4. FUZZY PETRI NETS FOR THE DIAGNOSIS

Taking into account these requirements for the diagnosis, we focus on methods that are based on explicative models. They indeed are adapted best for the modelling of the causal relations essential to the diagnosis. A fault model seems more suited for a diagnosis tool and Petri Net are chosen for their modelling capabilities. Given the difficulties of models acquisition, expertises already available in a company are invaluable sources of information, the fault model realized with the Petri nets is based on the tools running for diagnosis which are fault trees and FMECA. As in (Tromp, 2000) and (Minca, 2002) Petri nets model fault trees, they provide powerful tools, able to represent *AND* and *OR* influences between the faults. Then the model is completed with information of the FMECA. Moreover in order to take into account uncertainties linked to the maintenance knowledges, we introduce the Fuzzy logic, which makes it possible to give fuzzy degrees of credibility as in (Looney, 2003), associated to the states of breakdowns. Lastly, we use an abductive approach which makes it possible to characterize the causes starting from the observations. The algorithm is a downward approach in the fault tree. Given a top event observed, this observation is translated into a fuzzy degree of credibility and propagated in the net. So, each fault which could be at the origin of the top event is characterized by its own fuzzy degree.

5. EXAMPLE

Let us consider the following fault tree of a flexible manufacturing sub-system from the "Institut de Productique" of Besançon (Fig. 2)

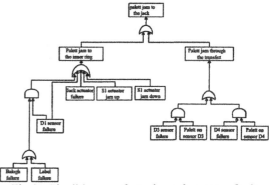

Fig.2 : Flexible manufacturing sub-system fault tree

Given the *AND* and *OR* influences we obtain the following model using Fuzzy Petri Net (Fig.3) :

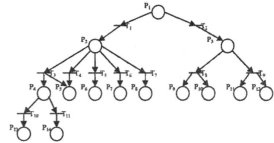

Fig.3 : Example of Fuzzy Petri Net

The different places in the net are the different events of the fault tree and the transition are the different *AND* and *OR* dependencies. In this example we consider that events corresponding to the places P_2 and P_3 are the observable symptoms. We suppose that we have a detection system which provide us the fuzzy credibility degree corresponding to both symptoms. Informations given by the FMECA are introduced in order to parameterize the backward propagation of the degree in *AND/OR* dependencies. This propagation depends on the fault frequency of the corresponding event which is one of the characteristic given in the FMECA. This frequency determine the center of a sigmoid function which allows the backward propagation, (Looney, 2003). For each causal relation, the degree corresponding to the consequent is transformed through the sigmoid function in order to provide the degree for the antecedent.

Simulations have been done with the software MATLAB and give us different results resumed in table 1 : *Fuzzy marking for a propagation from P_2.*

Places	Degree
P1	
P2	0.46
P3	0.0958
P4	0.574
P5	0.2915
P6	0.0958
P7	0
P8	0
P9	0
P10	0
P11	0
P12	0
P13	0.0180
P14	0.0180

Table 1

We assume that we observed the event of the place P_2 with the fuzzy degree 0.46. This degree is propagated through the Petri Net in places which could lead to P_2 in a forward propagation. The maintenance knowledges of the FMECA and the causal relations of the fault tree used in the fuzzy Petri net allow to suspect the event corresponding to P_4 as the fault origin with a fuzzy credibility of 0.574. This information can held a maintenance operator in his diagnosis. The Fuzzy Petri Net provides an abductive diagnosis : given a symptom, it gives a set of fault origin directly linked by causal relations to the observed event with a backward analysis of the fault tree. Moreover it takes into account the uncertainties on the maintenance knowledge by giving a fuzzy characterisation of each fault origin. So, localization and identification of the fault origin are implicitly given by the descriptions in the fault tree.

6. CONCLUSION

We have studied the problem of decision-making aids in diagnosis and more particularly the methods of diagnosis based on Artificial Intelligence tools. We focused on methods based on symbolic models. Most of these models handle at the same time numerical and symbolic data which are essential to the realization of a diagnosis. After having drawn up a panorama of the various tools met in the literature, we proposed the significant points necessary to the realization of a diagnosis. Lastly, we sketch a tool for diagnosis based on an explanatory model carried out with the fuzzy Petri nets.

Further work will investigate methods to better exploit the FMECA in the Petri net and include feedbacks as a training capability.

REFERENCES

Aamodt, A. , Plaza, E. (1994), *Case-Based Reasoning : Foundational Issues, Methodological Variations and Systems Approaches,* AI Communication. IOS Press, Vol. 7: 1, pp. 39-59.

Aghasaryan, A. *(1998), Formalisme HMM pour les réseaux de Pétri partiellement stochastiques : Application au diagnostic de pannes dans les systèmes répartis,* Thèse de doctorat université de Rennes1.

Aghasaryan, A., Boubour, R., Fabre, E., Jard, C., Benveniste, A (1997), *A Petri Net Approach to fault detection and diagnosis in distributed systems,* Armen., Publication interne n°1117 IRISA.

Anglano, C., Luigi Portinale, L. (1994), *B-W Analysis : a Backward Reachability Analysis for Diagnostic Problem Solving Suitable to Parallele Implementation,* Proceeding of the 15th International Conference on Application and Theory of Petri Nets, Zaragoza Spain.

Basseville, M., Cordier, M-O. (1996), *Surveillance et diagnostic de systèmes dynamiques : approches complémentaires du traitement du signal et de l'intelligence artificielle,* , Rapport INRIA n°2861.

Bergmann R. (1998), *Introduction To Case-Based Reasoning,* University of Kaiserslautern, http://www.dfki.unikl.de/~aabecker/Mosbach/Bergmann-CBR-Survey.pdf

Brusoni, V., Console, L., Terenziani, P., Theseider Dupré, D. (1995), *Characterizing temporal abductive diagnosis,* , Proc. of International Workshop on Principles of Diagnosis pp. 34-40.

Brusoni, V., Console, L., Terenziani, P., Theseider Dupré, D. (1997), *An Efficient Algorithm for*

Temporal Abduction, Lecture notes in Artificial Intelligence 1321 pp. 195-206.

Bouchon-Meunier B. (1995), *La Logique Floue et ses applications,* Edition Addison-Wesley.

Bouchon-Meunier B., Marsala C. (2003), *Logique floue, principes, aide à la décision,* Ed. HERMES.

Dubuisson, B. (2001), *Diagnostic, Intelligence Artificielle et reconnaissance de formes,* Ed. HERMES.

Fabre, E., Benveniste, A., Jard, C., Smith, M. (2001), *Diagnosis of distributed discrete event systems, a net unfolding approach,* Publication Interne IRISA.

Grosclaude, I., Quiniou, R. (2000), *Dealing with Interacting Faults in Temporal Abductive Diagnosis,* Proc. of the 8th International Workshop on Principles of Diagnosis.

Looney, Carl G., Liang, Lily R., (2003) *Inference via Fussy Bielef Petri Nets,* , 15th IEEE International Conference on Tools with AI (ICTAI' 03), Sacramento, California, USA.

Mellouli, N., Bouchon-Meunier, B. (2000), *Fuzzy Approaches to Abductive Inference,* Conference on Non-Monotonic reasoning, Berckenridge, Colorado .

Minca E., Racoceanu D., Zerhouni N. (2002), *Monitoring Systems Modelling and Analysis Using Fuzzy Petri Nets,* Studies in Informatics and Control, Vol. 11, No. 4.

Palade, V., Patton, R. J., Uppal1, F. J., Quevedo, J., Daley, S. (2002), *Fault diagnosis of an industrial gas turbine using neuro-fuzzy methods,* IFAC.

Peng Y., Reggia J. A. (1990), *Abductive Inference Models For Diagnostic Problem-Solving* Springer-Verlag, New York.

Rahamani K., Arabshahi P., Yan T.-Y., Pham T., and Finley S. G. (1998), *An Intelligent Fault Detection and Isolation Architecture For Antenna Arrays,* TDA Progress Report 42-132.

Sampath M., Sengupta R., Lafortune S., Sinnamohideen K., and Teneketzis D. C. (1996), *Failure Diagnosis Using Discrete Event Models,* IEEE Transactions On Control Systems Technology, Vol. 4, No. 2.

Tromp L. (2000), *Surveillance et diagnostic de systèmes industriels complexes : une approche hybride numérique/symbolique,* Thèse de doctorat université de Rennes 1.

Uppal F. J., Patton R. J. and Palade V. (2002), *Neuro-fuzzy based Fault Diagnosis applied to an Electro-Pneumatic Valve,,* IFAC.

Valette R., Künzle L. A. (1994), *Réseaux de Petri pour la détection et le diagnostic,* Journées Nationales : Sûreté, Surveillance, Supervision. GDR Automatique Paris.

Zemouri M-R. (2003), *Contribution à la surveillance des systèmes de production à l'aide des réseaux de neurones dynamiques : Application à la e-maintenance,* , Thèse de Doctorat, Université de Franche Comté.

Zwingezlstein G. (1995), *Diagnostic des défaillances : théorie et pratique pour les systèmes industriels* Ed. HERMES.

ELSEVIER
IFAC
PUBLICATIONS
www.elsevier.com/locate/ifac

MODELING AND ANALYSIS OF TIME CONSTRAINTS USING P-TIME PETRI NETS FOR A MULTI-HOIST ELECTROPLATING LINE

Fatah CHETOUANE [(1)], **Simon COLLART DUTILLEUL** [(2)], **Jean-Paul DENAT** [(3)]

[(1)]*Faculty of Engineering, Université de Moncton, Moncton, New-Brunswick, Canada E1A 3E9*
Phone: +1 (506) 863-2074, Fax: +1 (506) 858-4082, E-mail: chetouf@umoncton.ca

[(2)]*LAIL, École Centrale de Lille, Cité Scientifique, BP48, France 59651 Villeneuve d'Ascq*

[(3)]*LAMII-CESALP, ESIA, Université de Savoie, 41 avenue de la Plaine, BP 806, France 74016 Annecy*

Abstract — The aim of this paper is to propose an original analysis approach for multi-hoists surfaces treatment lines using P-time Petri nets. A decomposition approach of the multi-hoist line into several single hoist sub-lines is presented. P-time Petri-nets are used for modeling and calculation of time constraints on the sub-lines. It is assumed that all hoists move on the same track and production run in a cyclical mode.

Keywords: Control, Decomposition, Hoist scheduling Problem, P-time Petri nets, Surfaces treatment line, Time delay, Time window constraints.

1. INTRODUCTION

Surface treatment lines are composed of a set of tanks containing chemical or plating agents such as H_2SO_4 activating or Nickel plating. After a part is manufactured, some of its surfaces may have to undergo further processing in order to improve certain technical properties. Treating a part consists of immersing it sequentially in a set of chemical tanks according to its treatment routing using material handling hoist. A line can be hundreds of meters long. Several computer-controlled hoists are then used on the same track (see Fig. 1). The parts enter the workshop in large batch sizes. For a given lot, the part mix changes infrequently thus the production regime tends to be repetitive. In this case, the hoists are controlled to perform cyclical movement sequences. For a given batch, the purpose of the control is to determine the hoist transport sequence that ensures the required throughput and respects all process constraints.

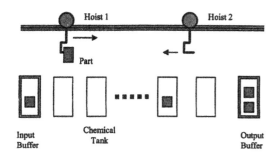

Fig. 1. Multi-hoist surfaces treatment line

2. THE HOIST SCHEDULING PROBLEM

Let us consider a multi-hoists line, composed by NT chemical tank TA_i ($i = 1, 2, ... , NT$), where input and output buffer are designated as TA_0 and TA_{NT+1}. The treatment duration for a part in a tank TA_i is defined by an interval $[a_i, b_i]$. If the staying duration of a part in a tank doesn't respect its duration limits the part is considered as defective. A tank can treat only one part at a time and a hoist can carry only one part at a time. The hoist traveling times between two tank TA_i and TA_j is not negligible and it is designated as $h_{i,j}$ ($i \neq j$ and $i\ j = 1, 2, ... , NT$). After pickup a part must be transferred immediately (no waiting) to the next treatment to avoid any oxidation by air. There must be no collision between hoists sharing the same track. The problem of finding a schedule for hoists operation that will satisfy all process constraints is known as the Hoist Scheduling Problem (HSP) and was proven to be NP-complete (Lei and Wang, 1989). The single-hoist case has attracted the interest of researchers for several decades (Ge and Yih, 1995; Phillips and Unger, 1976; Yih, 1994). Little research has been suggested, however, for the multi-hoist case (Armstrong et *al.*, 1996; Lei and Wang, 1991; Thesen and Lei, 1990). Among existing research, the robustness aspect of hoist control faced with disturbances has been considered in (Chetouane et *al.*, 1999; Collart-Dutilleul and Denat, 1997, Chetouane et *al.* 2001).

In the following section 3 of the paper a static decomposition approach for the multi-hoist case will be presented and. In section 4, P-time Petri nets modeling tools will be used to analyze the scheduling sequence for the hoist in order to deduce all time windows constraints involved in the control of such process.

3. STATIC DECOMPOSITION APPROACH

A static decomposition of a two-hoist line means that the exchange of parts between hoists will take place in a fixed tank TA_m of the line. This is done by separating the line into two independent sub-lines with one hoist each: $\{SLN_1, Hoist_1\}$ and $\{SLN_2, Hoist_2\}$ as shown in Fig. 2. Tank TA_m is considered to be a dummy output buffer for SLN_1 and a dummy input buffer for SLN_2. To avoid collision, only one hoist can enter the exchange zone at a time to serve TA_m. The decomposition of the two-hoist line and the determination of the tank TA_m are performed according to the following two steps (Chetouane et al. 2001).

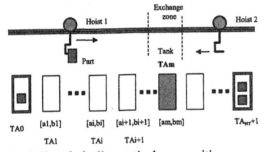

Fig. 2. Two-hoist line static decomposition

STEP 1: Hoist movement sequence calculation

For a two-hoist line $\{TA_0, TA_1, TA_2, ..., TA_{NT-1}, TA_{NT}, TA_{NT+1}\} \cup \{Hoist_1, Hoist_2\}$, we can build NT pairs of sub-lines with one hoist each: $SLN_1(v) = \{TA_0, TA_1, TA_2, ... TA_{v-1}\} \cup \{Hoist_1\}$ and $SLN_2(v) = \{TA_{v+1}, TA_{v+2}, ... TA_{NT+1}\} \cup \{Hoist_2\}$ for $v = 1, 2, ...$, NT. Using the algorithm of Yih (1994), the desired cyclical hoist movement sequences $SEQ_1(v)$ and $SEQ_2(v)$ can be computed for each of the sub-lines $SLN_1(v)$ and $SLN_2(v)$. Although any algorithm can be used for this purpose, the algorithm of Yih was used here because of its simplicity and efficiency.

STEP 2: Cycle time evaluation and TAm determination

The cycle time for $Hoist_j$ is denoted by C_j ($j =1$ or 2). For any cyclical functioning, each sub-line cycle time $C_1(v)$ and $C_2(v)$ varies in an interval (Collart-Dutilleul and Denat, 1997). This interval can be evaluated using P-time Petri nets modeling and evaluation technique for each sequence $SEQ_j(v)$ (j

=1 or 2) computed in step 1 (The P-time Petri net modeling and analysis technique used for this purpose is detailed later). Thus: $C_1(v) \in [C_{1min}(v), C_{1max}(v)]$ and $C_2(v) \in [C_{2min}(v), C_{2max}(v)]$.

There must exist at least one common value between cycle time intervals of sub-lines $SLN_1(v)$ and $SLN_2(v)$. None respect of this condition leads to a violation of capacity constraint for the common tank TA_v. In such a case, the decomposition is not feasible and another $SEQ_j(v)$ or/and TA_v would have to be selected. For $v \in \{1, 2, ... , NT\}$, the feasibility condition for the decomposition $SLN_1(v)$ and $SLN_2(v)$ can be expressed as: $I(v) \neq \varnothing$, where $I(v)=[C_{1min}(v), C_{1max}(v)] \cap [C_{2min}(v), C_{2max}(v)]$. The search process for TA_m is completed when the value of v is varied all over the set $\{1, 2, ... , NT\}$.

Application example

Let us consider a two-hoist line defined by Table 1. The input and output buffers are respectively TA0 and TA8. Because there is no constraints on the staying time in these buffers, they can be assimilated to a current chemical tank with processing time interval $[a_i, b_i] = [0, \infty[$.

i	1	2	3	4	5	6	7
a_i	25	35	30	1	20	25	11
b_i	30	40	35	∞	25	31	∞
travel time $h_{i,j} = $	$|i\text{-}j|$	time unit					

Table 1. Processing time for two-hoist line with 7 tanks

Let us assume that the tank TA_4 is the actual tank TA_v tested by the subroutine.

By using the algorithm of (Yih, 1994) a desired cyclical hoist moves sequences SEQ_1 and SEQ_2 can be computed for each sub-line $SLN_1 = \{TA_0, TA_1, TA_2, TA_3\}$ and $SLN_2=\{TA_5, TA_6, TA_7, TA_8\}$ respectively. The tank TA_4 is considered as an unloading station for SLN_1 and a loading station for SLN_2. Between two successive loaded moves $TA_i \rightarrow TA_j$ and $TA_k \rightarrow TA_l$, hoists perform an empty move from $TA_j \rightarrow TA_k$.

These sequences SEQ_1 ($Hoist_1$) and SEQ_2 ($Hoist_2$) are represented using P-time Petri net model in order to calculate the cycle time interval $[C_{1min}, C_{1max}]$ and $[C_{2min}, C_{2max}]$ respectively. Cycle time intervals are needed in Step 2 of the decomposition subroutine. The model of each sequence is represented in Fig. 3.

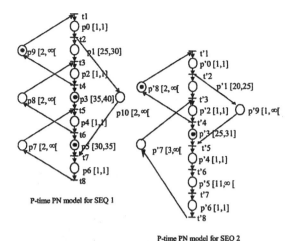

P-time PN model for SEQ 1

P-time PN model for SEQ 2

Fig. 3. P-time Petri nets model for hoist sequences

The two set $\{p_0, p_2, p_4, p_6\}$, and $\{p'_0, p'_2, p'_4, p'_6\}$ represent loaded movements performed between adjacent tanks by *Hoist1* and *Hoist2* respectively. The two set $\{p_1, p_3, p_5\}$, and $\{p'_1, p'_3, p'_5\}$ represent treatment operations in sub-lines *SLN1* and *SLN2* respectively. Places on the left side (resp. right side) for each P-time PN model represent hoist backward (resp. forward) empty movements.

4. P-TIME PETRI NET MODELING APPROACH

Khansa (1997) defined a P-time PN as a pair $<R, I_0>$ where R is a marked PN and I_0 is a function defined as:

$I_0: P \rightarrow Q+ \cup \infty \times Q+ \cup \infty$

$p_i \rightarrow I_i = [a_i, bi]$ with $0 \leq a_i \leq b_i$, $\forall p_i \in P$, where: P is the set of places on the PN, I_i defines the interval of staying times of a mark in place p_i, and $Q+$ is the set of non-negative rational numbers.

A mark in the place p_i is taken into account for transition validations when it has stayed in p_i more than a duration a_i and no longer than b_i. After the b_i duration, the mark is said to be dead. From this definition, if $St_i(n)$ is the nth firing instant of transition t_i, the positivity constraints between the nth firing instant of the input transition t_i of place p_i containing m_i tokens (marks) and its output transition t_{i+1}, for each place of a P-time Petri net are:

$$St_i(n) + a_i \leq St_{i+1}(n + m_i) \leq St_i(n) + b_i \qquad (1)$$

In a 1-periodic functioning mode with a functioning cycle value C, and $\forall n \in \aleph$ set of natural numbers we have:

$$St_i(n) = St_i(1) + (n-1).C \qquad (2)$$

equation (1) can be rewritten as:

$$St_i(n) + a_i \leq St_{i+1}(n) + m_i.C \leq St_i(n) + b_i \qquad (3)$$

In the following section we will present an analysis approach for the P-time PN model based on its critical structure.

5. PRIMARY CRITICAL STRUCTURE CONSTRAINTS

We define a *critical structure* as an elementary circuit or structure that contains parallel elementary path between a parallelism transition and a synchronization transition of a PN model. Critical structures, which contain only one place that models an empty hoist movement, are called *primary critical structures*. The PN model for SEQ_1 of Fig. 3 contains four primary critical structures: the first one is made up of two directed elementary paths between transitions t_2 and t_7 (in SEQ_1 PN model). This is called *critical structure of type 1* (parallelism-synchronization). The three remaining structures are called *critical structure of type 2* (circuit).

In a critical structure of *type 1*, t_i denotes a parallelism transition and t_j denotes a synchronization transition. Let us introduce the following notations:

$h_{i,j}$: time duration associated with the path from t_i to t_j. Physically, it represents the hoist travel time between tanks TA_i and TA_j;

$w_{i,j}$: hoist waiting time on its empty movement represented by the path from t_i to t_j;

$\sum\limits_{\alpha=i}^{\alpha=j} q_\alpha$: sum of time duration of all places

between t_i and t_j in the routing model;

m_i : number of tokens in place p_i;

$m_{i,j}$: number of tokens in all places belonging to the path from transition t_i to transition t_j;

$m_{i,j,i}$: number of tokens existing in the circuit crossing transitions t_i, t_j and back to t_i;

C : functioning cycle time for the PN model;

$St_j(n)$: the nth firing instant of t_i;

Now, the positivity constraints that give the portion of the path that models the routing on the critical structure of *type 1* are:

$$St_j(n + m_{i,j}) = St_i(n) + \sum\limits_{\alpha=i}^{\alpha=j} q_\alpha \qquad (4)$$

The path that models hoist empty movement gives:
$$St_j(n + m_{i,j}) = St_i(n + m_{i,j}) + h_{i,j} + w_{i,j} \qquad (5)$$

From constraints (4) and (5), we have:

$$St_i(n + m_{i,j}) - St_i(n) = - h_{i,j} - w_{i,j} + \sum\limits_{\alpha=i}^{\alpha=j} q_\alpha \qquad (6)$$

In a critical structure of *type 2*, where t_i and t_j are respectively the input and the output transition of the routing, positivity constraints give:

$$St_i(n + m_{i,j,i}) - St_i(n) = h_{j,i} + w_{j,i} + \sum_{\alpha=i}^{\alpha=j} q_\alpha \qquad (7)$$

6. CYCLE INTERVAL CONSTRAINTS CALCULATION BASED ON PRIMARY STRUCTURE ANALYSIS

Once the sequence is determined using scheduling algorithm, the problem of finding cycle constraints for each sequence is treated using the primary structure defined above. The technique is to set the value of the staying duration q_α of places p_α in the critical structures of *type 1* (resp. of *type 2*) to their maximum value b_α (resp. to their minimal value a_α). From equation (6), for the *type 1* structures, this leads to the following equation (8):

$$St_i(n) - St_i(n - m_{i,j}) = - h_{i,j} - w_{i,j} + \sum_{\alpha=i}^{\alpha=j} b_\alpha \qquad (8)$$

The *maximal gap* between entry transition firing instants for this structure type is obtained for a functioning mode with no waiting time: $w_{i,j} = 0$. From (2) and (8), this 1-periodic functioning mode can be described by:

$$h_{i,j} + w_{i,j} = \sum_{\alpha=i}^{\alpha=j} b_\alpha - m_{i,j}.C \qquad (9)$$

The maximal period C_{max} is satisfied for a zero waiting time on one of the hoist forward empty movements:

$$C_{max} = (\sum_{\alpha=i}^{\alpha=j} b_\alpha - h_{i,j})/m_{i,j} \qquad (10)$$

For the structures of *type 2*, we have the following equation:

$$St_i(n) - St_i(n - m_{i,j,i}) = h_{j,i} + w_{j,i} + \sum_{\alpha=i}^{\alpha=j} a_\alpha \qquad (11)$$

From equation (11), the *minimal gap* between input transition firing instants for this structure type is obtained for a functioning mode with no waiting time: $w_{j,i} = 0$. Equations (2) and (11), lead us to equation (12) below:

$$h_{j,i} + w_{j,i} + \sum_{\alpha=i}^{\alpha=j} a_\alpha = m_{i,j,i}.C \qquad (12)$$

The minimal period C_{min} is satisfied for a zero waiting time on one of the hoist backward empty movement:

$$C_{min} = (\sum_{\alpha=i}^{\alpha=j} a_\alpha + h_{j,i})/m_{i,j,i} \qquad (13)$$

Equations (10) and (13) allow us to assign a cycle value interval $[C_{min}, C_{max}]$ to any hoist movement sequence. This interval represents the cycle time constraint for the related sequence. If the cycle time falls out of this interval, its related sequence is no longer valid. Therefore, new hoist movement sequences must be computed.

For each P-time PN model of Fig. 3, this calculation allow us to obtain for SEQ_1, $[C_{1min}, C_{1max}] = [39, 52.5]$; and for SEQ_2, $[C_{2min}, C_{2max}] = [42, 56]$. Thus, $[C_{1min}, C_{1max}] \cap [C_{2min}, C_{2max}] = [42, 52.5]$ is not an empty interval. Accordingly, the previous decomposition using TA_4 as an exchange area is feasible for SEQ_1 and SEQ_2.

7. TIME CONSTRAINTS FOR THE MULTI-HOIST LINE CONTROL

The multi-hoist line control consists in determining the firing instants of the transitions (or the staying times of the tokens in the places of the net). According to the cycle time intervals $[C_{min}, C_{max}]$ for each sequence, a common functioning cycle must be chosen for both hoists to avoid any capacity constraint violation for tank TA_4.

If C_j is the cycle time chosen for sub-line SLN_j, the nth firing instant for a transition t_i from sequence SEQ_j will be at $St_i(n) = St_i(1)+(n-1).C_j$, (j=1, 2). For instance, for $C_1 = 46$ time unit (t.u), $[St_1(1), St_2(1), St_3(1), St_4(1), St_5(1), St_6(1), St_7(1), St_8(1)] = [0, 1, 27, 28, 17, 18, 11, 12]$ for SEQ_1 of Fig.3.

From this sub-line control, the global control of the two-hoist line must include additional security zone time constraint linked to hoist traffic, and to the respect of the treatment duration in the exchanging tank. For the application example of section 3, TA_4 $[a_4, b_4]$ is the tank where the exchange of product take place. The constraints can be expressed using the nth firing instant of each sub-line as follow (if there is no part in TA_4):

$$St_8(n) + a_4 \leq St'_1(n) \leq St_8(n) + b_4 \qquad (14)$$

Assuming that TA_4 has to be loaded and unloaded at the same position, to avoid that one hoist enter the security zone, while another hoist still in there we must add the constraint:

$$St_8(n) + t_f \leq St'1(n) - t_o \qquad (15)$$

The variable t_f is the time needed by $Hoist_1$ to free the security zone, and t_o is the time needed by $Hoist_2$ to occupy the security zone. Generally, we have $t_f \leq h_{4,3}$ and $t_o \leq h_{5,4}$.

For the application example, by considering $t_f = h_{4,3} = 1$ t.u, and $t_o = h_{5,4} = 1$ t.u, using constraints (14) and (15), we can deduce that:

$$St'_1(n) - St_8(n) \geq 2 \text{ t.u.}$$

By taking the lowest value of this gap (2 t.u), the global control of the two-hoist line is defined by the following firing time setting, according to the same time origin:

$$[St_1(1), St_2(1), St_3(1), St_4(1), St_5(1), St_6(1), St_7(1), St_8(1)] = [0, 1, 27, 28, 17, 18, 11, 12], C_1 = 46 \text{ t.u.}$$

$$[St'_1(1), St'_2(1), St'_3(1), St'_4(1), St'_5(1), St'_6(1), St'_7(1), St'_8(1)] = [14, 15, 38, 39, 20, 21, 32, 33], C_2 = 46 \text{ t.u.}$$

This allows the respect of collision constraint, capacity and time windows constraints for TA_4, and keep the two hoist functioning at the same cycle time as if they are totally independent. Such control can be implemented on the hoists programmable logical controller with a scan cycle C_{nom} such as:
$$C_{nom} = [C_{1min}, C_{1max}] \cap [C_{2min}, C_{2max}].$$

8. CONCLUSION AND PERSPECTIVES

The P-time Petri nets approach presented in this paper allows a process based representation for the multi-hoist line. The hoist movement sequences and treatment time windows can be represented in a simple way.
Constraints on cycle time and hoist collision are evaluated using an original approach which is only valid in the case of a repetitive functioning mode for the P-time Petri Nets.

The validity of the presented approach is affected by any change in the hoist movement sequences, since we assumed the sequences unchangeable, which is compatible with the static decomposition of the multi-hoist line. Nevertheless, hoists cycle time can be adjusted in their intervals without changing the pre-defined hoist movement sequences.
This gives considerable flexibility for real time control, especially in front of time disturbances such as delays, without dealing with the NP complexity of the sequencing problem.

A future research is needed to focus on dynamic decomposition of the multi-hoist line in which the movement sequences can be adapted to ensure part exchange in a tank that has to be located in real-time by the multi-hoist line control system. This could be investigated using simulation techniques.

9. REFERENCES

Armstrong, R., S. Gu and L. Lei (1996). A greedy algorithm to determine the number of transporters in a cyclic electroplating process. *Process IEE transactions*, **289**(5), 347-355.

Chetouane, F., Z. Binder, and L. Little (1999). Control and robustness for production systems: application to electroplating systems. In: *Proceedings of the 14th World Congress IFAC, Beijing, China*, **A**, 331-335.

Chetouane F., J.P. Denat and S. Collart-Dutilleul (2001). « Robustness of cyclical control for material handling hoists in surface treatment workshop with time windows». Computer & Industrial Engineering an International Journal, Accepted for publication, Ref. 01-132.

Collart Dutilleul, S. and J.P. Denat (1997). P-time Petri nets and robust control of electroplating lines using several hoists. In: *Proceedings of the IFAC-IFIP-IMACS Conference on Control of Industrial Systems*, Belfort, France, 3, 501-506.

Ge, Y. and Y. Yih (1995). Crane scheduling with time windows in circuit board production lines. *International Journal of Production Research*, **33**(5), 1187-1199.

Khansa, W. (1997). *Réseaux de Petri P-temporels: Contribution à l'étude des systèmes à événements discrets.* Doctorat Thesis, Université de Savoie, France.

Lei, L. and T.J. Wang (1989). A proof: the cyclic hoist scheduling problem is NP-complete. *Working paper. Rutgers University and Bell Communication Research.* USA.

Lei, L. and T-J. Wang (1991). The minimum common-cycle algorithm for cyclic scheduling of two material handling hoists with time window constraints. *Management Science,* **37**(12), 1629-1639.

Phillips, L.W. and P.S. Unger (1976). Mathematical programming solution of a hoist scheduling program. *AIIE Transactions,* **8**(2), 219-225.

Thesen, A. and L. Lei (1990). An expert scheduling system for material handling hoists. *Journal of Manufacturing Systems,* **9**(3), 247-252.

Yih (1994). An algorithm for hoist scheduling problems. *International Journal of Production Research*, **32**(3), 501-516.

www.elsevier.com/locate/ifac

Control System for a Tentacle Manipulator with Variable Length

Mircea Ivanescu
Department of Mechatronics
University of Craiova
Craiova, Romania
E-mail: ivanescu@robotics.ucv.ro

Abstract — The paper presents the control problem of a class of tentacle arms, with variable length, that can achieve any position and orientation in 3D space and can increase the length in order to get a better control in the constraint operator space. First, the dynamic model of the system is inferred. In order to avoid the difficulties generated by the complexity of the nonlinear integral - differential model, the control problem is based on the artificial potential method. Then, the method is used for constrained motion in an environment with obstacles.

Numerical simulations for spatial and planar tentacle models are presented in order to illustrate the efficiency of the method.

Keywords — potential, control, hyper-redundant system.

I. INTRODUCTION

Tentacle manipulators are hyper-redundant or hyper-degree-of-freedom manipulators and there has been a rapidly expanding interest in their study and construction lately.

The control of these systems is very complex and a great number of researchers have tried to offer solutions for this difficult problem. In [1] the control by cables or tendons meant to transmit forces to the elements of the arm in order to closely approximate the arm was analysed as a truly continuous backbone. In [2], Gravagne analysed the kinematic model of "hyper-redundant" robots, known as "continuum" robots. Robinson and Davies [7] present the "state of art" of continuum robots, outlining their areas of application and introducing some control issues.

In other papers [8-9] several technological solutions for actuators used in hyper-redundant structures are presented and conventional control systems are introduced.

The difficulty of the dynamic control is determined by integral-partial-differential models with high nonlinearities that characterise the dynamic of these systems.

In [11] the dynamic model for 3D space is inferred and a control law based on the energy of the system is analysed.

This paper discusses a class of tentacle arms of variable length. First, the dynamic model of the system is inferred.

The method of artificial potential is developed for these infinite dimensional systems. In order to avoid the difficulties associated with the dynamic model, the control law is based only on the gravitational potential and a new artificial potential. It is shown that it is possible to drive the tentacle robot to a desired position if the artificial potential is a potential functional whose point of minimum is an attractor for this dissipative controlled system. Also, this method is used for constrained motion in an environment with obstacles.

II. BACKGROUND

2.1. Technological model

We will study a class of tentacle arms, of variable length, that can achieve any position and orientation in 3D space and that can increase their length in order to get a better control in the operator space with a constraint area.

We will consider a class of tentacle arms based on the use of flexible composite materials in conjunction with active controllable electro-rheological (ER) fluids that can change their mechanical characteristics in the presence of electrical fields.

The general form of a tentacle model is shown in Figure 2. It consists of a number of elements, cylinders made of fiber-reinforced rubber. There are four internal chambers in the cylinder, each of them containing the ER fluid with an individual control circuit. The deformation in each cylinder is controlled by an independent electrohydraulic pressure control system combined with the distributed viscosity control of the ER fluid.

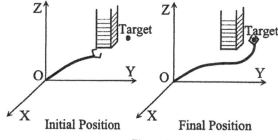

Figure 1.

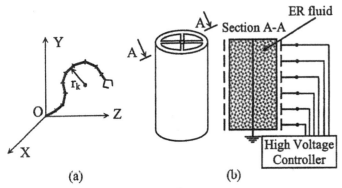

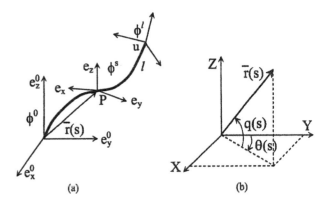

Figure 2.

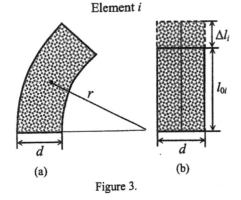

Figure 3.

Figure 4. Tentacle system parameters.

$$u = \sum_{i=1}^{N} \Delta l_i \qquad (2.3)$$

determines the control variable of the arm length.

The position of a point s on curve C is defined by the position vector,

$$\bar{r} = r(s) \qquad (2.4)$$

when $s \in [0, l]$. For a dynamic motion, the time variable will be introduced, $\bar{r} = \bar{r}(s, t)$ (Figure 4).

$$\bar{r}(s, t) = [x(s,t) \quad y(s,t) \quad z(s,t)]^T \qquad (2.5)$$

For an element dm, kinetic and gravitational potential energy will be

$$dT = \frac{1}{2} dm \left(v_x^2 + v_y^2 + v_z^2 + v_u^2 \right) \qquad (2.6)$$

$$dV = dm \cdot g \cdot z \qquad (2.7)$$

where

$$dm = \rho ds \qquad (2.8)$$

From (2.6)-(2.8) we obtain

$$T = \frac{1}{2} \rho \int_0^l \left(\left(\int_0^s \left(-\dot{q} \sin q' \sin \theta' + \dot{\theta}' \cos q' \cos \theta' \right) ds' \right)^2 + \right.$$

$$+ \left(\int_0^s \left(-\dot{q}' \sin q' \cos \theta' - \dot{\theta}' \cos q' \sin \theta' \right) ds' \right)^2 +$$

$$+ \left. \left(\int_0^s \dot{q}' \cos q' ds' \right)^2 \right) ds + \frac{1}{2} \rho \int_0^l \dot{u}^2 ds \qquad (2.9)$$

$$V = \rho g \int_0^l \int_0^s \sin q' ds' ds \qquad (2.10)$$

The elastic potential energy will be approximated by two components, one determined by the bending of the element [3-5]

The length of the element can also be increased by the pressure control in all the chambers (Figure 3) and a new variable Δl_i is used in order to define the position of each element.

Each element of the arm has two types of deformations: bending and axial tension/compression. Thus, we will use an extensible model that closely matches the characteristics of the prototype arm. Also, the system is frictionless and any other damping and frictions are neglected.

2.2. Theoretical model

The essence of the tentacle model is a 3-dimensional backbone curve C that is parametrically described by a vector $r(s) \in \mathbf{R}^3$ and an associated frame $\phi(s) \in \mathbf{R}^{3 \times 3}$ whose columns create the frame bases (Figure 4a).

The independent parameter s is related to the arc-length from the origin of the curve C, a variable parameter, where (Figure 3b)

$$l = \sum_{i=1}^{N} (l_{0i} + \Delta l_i) \qquad (2.1)$$

or

$$l = l_0 + u \qquad (2.2)$$

where l_0 represents the length of the N elements of the arm in the initial position and

$$V_{eb} = k\frac{d^2}{4}\sum_{i=1}^{N}\left(q_i^2 + \theta_i^2\right) \qquad (2.11)$$

and the other given by the axial tension/compression energy component

$$V_{ea} = \frac{1}{2}ku^2 \qquad (2.12)$$

where we assumed that each element has a constant curvature and a uniform equivalent elasticity coefficient k, assumed constant for the entire length of the arm.

We will consider $F_\theta(s,t)$, $F_q(s,t)$ the distributed forces on the length that determine motion and orientation in the θ-plane, q-plane and $F_u(t)$, the force that determines axial motion, assumed constant along the length of the arm.

III. DYNAMIC MODEL

In this paper, the manipulator model is considered a distributed parameter system defined on a variable spatial domain $\Omega = [0,l]$; the spatial coordinate is denoted by s [10, 15].

The state of this system at any fixed time t is specified by the set $(\omega(t,s), v(t,s))$, where $\omega = \begin{bmatrix} \theta & q & u \end{bmatrix}^T$ and v represents momentum densities. The set of all functions of $s \in \Omega$ that ω, v can take on at any time is a state function space $\Gamma(\Omega)$. We will consider that $\Gamma(\Omega) \subset L_2(\Omega)$.

The control forces have the distributed components along the arm, $F_\theta(t,s)$, $F_q(t,s)$, $s \in [0,l]$ and a lumped component $F_u(t)$.

A practical form of the dynamic model expressed only as a function of generalised coordinates is derived by using Lagrange equations developed for infinite dimensional systems,

$$\frac{\partial}{\partial t}\left(\frac{\delta T}{\delta\dot{\theta}(t,s)}\right) - \frac{\delta T}{\delta\theta(t,s)} + \frac{\delta V}{\delta\theta(t,s)} + \frac{\delta V_e}{\delta\theta(t,s)} = F_\theta \qquad (3.1)$$

$$\frac{\partial}{\partial t}\left(\frac{\delta T}{\delta\dot{q}(t,s)}\right) - \frac{\delta T}{\delta q(t,s)} + \frac{\delta V}{\delta q(t,s)} + \frac{\delta V_e}{\delta q(t,s)} = F_q \qquad (3.2)$$

$$\frac{\partial}{\partial t}\left(\frac{\partial T}{\partial\dot{u}}\right) - \frac{\partial T}{\partial u} + \frac{\partial V}{\partial u} + \frac{\partial V_e}{\partial u} = F_u \qquad (3.3)$$

where $\partial/\partial(\cdot)$, $\delta/\delta(\cdot)$ denote classical and functional partial derivatives (in Gateaux sense [15]), respectively.

In Appendix 1 the dynamic model of this ideal spatial tentacle manipulator will be developed and the difficulties to obtain a control law will be easily inferred.

IV. UNCONSTRAINED CONTROL

The artificial potential is a potential function whose points of minimum are attractors for a dissipative controlled system. It was shown [10, 13, 14] that the control of robot motion to a desired point is possible if the function has a minimum in the desired point. In this section we will extend this result for the infinite dimensional model of the tentacle manipulator with variable length.

We consider that the initial state of the system is given by

$$\omega_0 = \omega(0,s) = [\theta_0, q_0, l_0]^T \qquad (4.1)$$

$$v_0 = v(0,s) = [0, 0, 0]^T \qquad (4.2)$$

where

$$\theta_0 = \theta(0,s), \quad q_0 = q(0,s), \quad s \in [0,l_0] \qquad (4.3)$$

$$l_0 = l(0) \qquad (4.4)$$

corresponding to the initial position of the manipulator defined by the curve C_0

$$C_0 : (\theta_0(s), q_0(s), l_0), \quad s \in [0,l_0] \qquad (4.5)$$

The desired point in $\Gamma(\Omega)$ is represented by a desired position of the arm, the curve C_d,

$$\omega_d = [\theta_d, q_d, l_d]^T, \quad v_d = [0, 0, 0]^T$$
$$C_d : (\theta_d(s), q_d(s), l_d), \quad s \in [0,l_d] \qquad (4.6)$$

The system motion (3.3)-(3.5) corresponding to a given initial state (ω_0, v_0) defines a trajectory in the state function space $\Gamma(\Omega)$.

The control problem of the manipulator means the motion control by the forces F_θ, F_q, F_u, from the initial position C_0 to the desired position C_d.

From the viewpoint of mechanics, the desired position (ω_d, v_d) is asymptotically stable if the potential function of the system has a minimum at

$$(\omega, v)(s) = (\omega_d, v_d)(s), \quad s \in [0,l]$$

and the system is completely damped [10, 13].

As a control problem, the results of [10, 13] will be extended in this paper for the infinite dynamic systems.

We will consider the control forces,

$$F_\theta(t,s) = \frac{\delta V}{\delta\theta(t,s)} + \frac{\delta V_e}{\delta\theta(t,s)} - F_{\theta d} - \frac{\delta\Pi}{\delta\theta(t,s)} \qquad (4.7)$$

$$F_q(t,s) = \frac{\delta V}{\delta q(t,s)} + \frac{\delta V_e}{\delta q(t,s)} - F_{qd} - \frac{\delta\Pi}{\delta q(t,s)} \qquad (4.8)$$

$$F_u(t) = \frac{\partial V}{\partial u(t)} + \frac{\partial V_e}{\partial u(t)} - F_{ud} - \frac{\partial\Pi}{\partial u(t)} \qquad (4.9)$$

The first two terms compensate the gravitational and elastic potential, the third components assure the damping control and the last terms define the new artificial potential introduced in order to assure the motion to the desired position. The minimum points of this potential must be identical with the desired positions of the manipulator, as attractors of its motion. For example, the potential Π can be selected as a functional of generalised coordinates,

$$\Pi(\theta, q, u) = \int_0^l \left((\theta - \theta_d(s))^2 + (q - q_d(s))^2 \right) ds + (l_0 + u - l_d)^2 \tag{4.10}$$

The control law (4.7)-(4.9) modifies the system potential and the Lagrange equations (3.3)-(3.5) become

$$\frac{\partial}{\partial t}\left(\frac{\delta T}{\delta \dot{\theta}(t,s)} \right) - \frac{\delta T}{\delta \theta(t,s)} + \frac{\delta \Pi}{\delta \theta(t,s)} = F_{\theta_d} \tag{4.11}$$

$$\frac{\partial}{\partial t}\left(\frac{\delta T}{\delta \dot{q}(t,s)} \right) - \frac{\delta T}{\delta q(t,s)} + \frac{\delta \Pi}{\delta q(t,s)} = F_{q_d} \tag{4.12}$$

$$\frac{\partial}{\partial t}\left(\frac{\partial T}{\partial \dot{u}} \right) - \frac{\partial T}{\partial u} + \frac{\partial \Pi}{\partial u} = F_{u_d} \tag{4.13}$$

The force components F_{θ_d}, F_{q_d}, F_{u_d} represent the damping components of the control [9, 10, 15], and have the form

$$F_{\theta_d}(s,t) = -\int_0^l K_\theta(s,s')\dot{\theta}(s',t)ds' \tag{4.14}$$

$$F_{q_d}(s,t) = -\int_0^l K_q(s,s')\dot{q}(s',t)ds' \tag{4.15}$$

$$F_{u_d}(t) = -K_u \dot{u}(t) \tag{4.16}$$

where $K_\theta(s,s')$, $K_q(s,s')$ are positive definite specified spatial weighting functions on $(\Omega \times \Omega)$ and K_u is a positive constant. For practical reasons, the derivative components of the control have the form

$$K_\theta(s,s') = \delta(s - s') \cdot k_\theta(s) \tag{4.17}$$

$$K_q(s,s') = \delta(s - s') \cdot k_q(s) \tag{4.18}$$

V. CONSTRAINED CONTROL

Let B be the region of the state space where the mechanical system motion is not admissible, its complement $\bar{B}$ is the region of admissible movements and ∂B is the boundary of B. The control problem is to determine the potential function $\Pi(\theta, q, u)$ which would determine the motion to the desired position $(\omega_d(s), v_d(s))$, $s \in [0, l]$ without penetrating the constrained area B. In terms of the artificial potential, this means that this functional should have a single stationary point in $\bar{B}$ and that it grows without limit when the system penetrates the boundary ∂B.

We will consider the following artificial potential [13],

$$\Pi(\theta, q, u) = max\left\{ \Pi_1(\theta, q, u), \Pi_2(\theta, q, u) \right\} \tag{5.1}$$

where $\Pi_1(\theta, q, u)$ is the artificial potential for unconstrained problem and $\Pi_2(\theta, q, u)$ is the potential for constrained control problem.

$\Pi_2(\theta, q, u)$ is a non-negative, continuous functional defined in $\bar{B}$ and

$$\lim_{d \to 0} \Pi_2(\theta, q, u) = \infty \tag{5.2}$$

where d is the distance between the current state (θ, q, u) and the boundary ∂B.

VI. SIMULATIONS

In this section, some numerical simulations are carried out on 3D and 2D tentacle manipulators.

Example 1. We consider a spatial tentacle manipulator that operates in OXYZ space. The mechanical parameters are: linear density $\rho = 2.2$ kg/m and the length of the arm $l = 0.03$ m. is assumed constant.

The initial position of the arm is assumed to be horizontal (0Y-axis),

$$\theta(s, 0) = 0; \quad q(s, 0) = 0; \quad s \in [0, 0.03] \tag{6.1}$$

and the desired position is represented by a line in OXYZ frame that is defined in terms of motion parameters as

$$\theta_d(s) = \frac{\pi}{5}; \quad q_d(s) = \frac{\pi}{5}; \quad s \in [0, 0.3] \tag{6.2}$$

The unconstrained control law is given as (4.7)-(4.9) where the gravitational potential V has the form

$$V = \rho g \int_0^{l_0} \int_0^s \sin q' ds' ds \tag{6.3}$$

where $q' = q(s')$, $s' \in [0, s]$, the artificial potential is obtained from (4.10)

$$\Pi(\theta, q_0) = \int_0^{l_0} \left(\left(\theta(s) - \frac{\pi}{5} \right)^2 + \left(q(s) - \frac{\pi}{5} \right)^2 \right) ds \tag{6.4}$$

and the damping control components have the form (4.14), (4.15), (4.17), (4.18) with

$$k_\theta(s) = k_q(s) = 1.015, \quad s \in [0, l] \tag{6.5}$$

To simulate the dynamic model the integral-differential model discussed in Appendix is used. A discretization of the Ω – space with an increment $\Delta = 0.05$ m is used

$$s_i = i \cdot \Delta \quad i = 1, 2, ..., 6$$

and a MATLAB system is applied. The result is presented in Fig. 5. We see the initial position (on the OY axis), the final position as well as several intermediary positions.

The phase portrait of the evolution is plotted in Fig. 6 where the error for the global system is defined as

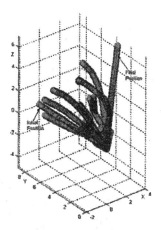

Figure 5. 3D model motion.

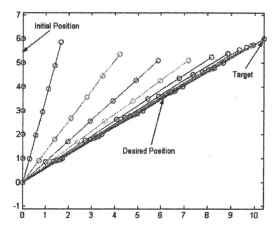

Figure 7. 2D nonconstrained variable length motion

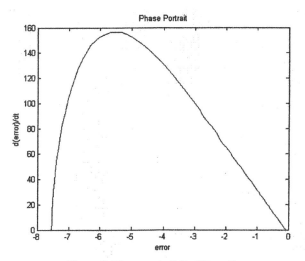

Figure 6. Phase portrait for 3D-motion.

$$e(t) = \int_0^l \left((q(s,t) - q_d(s))^2 + (\theta(s,t) - \theta_d(s))^2 \right) ds \qquad (6.6)$$

We see the stability of motion and error convergence to zero.

Example 2. A better understanding of the control can be obtained for 2D-tentacle arm.

We will analyse now the case of a planar tentacle model in OXZ plane. The dynamic model is obtained for the case $\theta = 0$. Also, in order to get a better image of the motion, the initial length of the arm is $l_0 = 0.03$ m. but the lengthening of the arm is not limited. The variable u is not constrained. The initial position is determined by the vertical line (OZ-axis),

$$q(s,0) = \frac{\pi}{2}, \quad s \in [0,\ 0.03] \qquad (6.7)$$

and the motion objective is to achieve the fixed target,

$$x_T = 0.06\,\text{m}, \quad y_T = 0.03\,\text{m}. \qquad (6.8)$$

The artificial potential is selected as

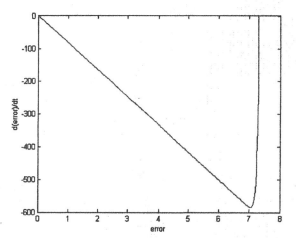

Figure 8. Phase portrait for 2D-motion

$$\Pi(q,u) = \frac{1}{2}\left(\left(\int_0^{l_0+u} \cos q\, ds - x_T \right)^2 + \left(\int_0^{l_0+u} \sin q\, ds - y_T \right)^2 \right) \qquad (6.9)$$

and the control laws (4.7)-(4.9) are used with

$$k_{q_i} = 0.152, \quad i = 1,2,\dots,6 \qquad (6.10)$$

$$k_u = 5.6. \qquad (6.11)$$

The phase portrait and the arm motion are presented in Figure 7 and Figure 8. We observe the convergence of the motion to the target and the lengthening of the arm from the initial value 0.03 m to the final length 0.078 m.

Example 3. A new constrained control problem is simulated by using the variable length of the arm. The constrained area is defined by the circle

$$\partial B: (x - 0.09)^2 + (y - 0.05)^2 = 5 \cdot 10^{-4} \qquad (6.12)$$

and the target of the motion is

$$x_T = 0.14\,\text{m}, \quad y_T = 0.075\,\text{m}. \qquad (6.13)$$

The initial position is the vertical like OY.

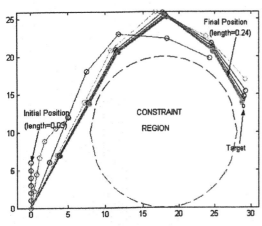

Figure 9. 2D constrained variable length motion

The artificial potential for the constrained area has the form (5.1). The arm motion is plotted in Figure 9. We observe the increasing of the arm from the initial value $l_0 = 0.03$ m to the final value $l_f = 0.24$ m and a very good interception of the target, avoiding the constraint area.

APPENDIX

We will consider a spatial tentacle model, an ideal system, neglecting friction and structural damping. We assume a uniformly distributed mass with a linear density ρ [kg/m].

We will use the notations:

$$q = q(s,t), \quad s \in [0,l], \quad t \in [0,t_f]$$

$$\dot{q} = \frac{\partial q(s,t)}{\partial t}, \quad s \in [0,l], \quad t \in [0,t_f]$$

$$F_q = F_q(s,t), \quad s \in [0,l], \quad t \in [0,t_f]$$

From (4.11)-(4.13), it results,

$$\rho \int_0^s \int_0^s (\ddot{q}'(\sin q' \sin q'' \cos(q'-q'') + \cos q' \cos q'') -$$

$$- \ddot{\theta}' \cos q' \sin q'' \sin(\theta''-\theta') +$$

$$+ \dot{q}'^2 (\cos q' \sin q'' \cos(\theta'-\theta'') - \sin q' \cos q'') +$$

$$+ \dot{\theta}'^2 \cos q' \sin q'' \cos(\theta'-\theta'') -$$

$$- \dot{q}' \dot{q}'' \sin(q''-q')) ds' ds'' + \rho g \int_0^s \cos q' ds' + \frac{1}{2} k d^2 q = F_q$$

$$\rho \int_0^s \int_0^s (\ddot{q}' \sin q' \cos q'' \sin(\theta''-\theta') + \ddot{\theta}' \cos q' \cos q'' \cos(\theta''-\theta') -$$

$$- \dot{q}'^2 \cos q' \cos q'' \sin(\theta''-\theta') +$$

$$+ \dot{\theta}' \cos q' \cos q'' \sin(\theta''-\theta') -$$

$$- \dot{\theta}' \dot{q}' \sin q' \cos q'' \cos(\theta''-\theta')) ds' ds'' + \frac{1}{2} k d^2 \theta = F_\theta$$

$$\rho \ddot{u} + \frac{1}{2} \rho \dot{u}^2 + k u = F_u$$

VII. CONCLUSIONS

The paper treats the control problem of a tentacle manipulator of variable length. In order to avoid the difficulties generated by the complexity of the nonlinear integral-differential equations that define the dynamic model of this system, the control problem is based on the artificial potential method. This method is developed for these infinite dimensional systems of unknown dynamic model except for the gravitational component.

It is shown that the control of a tentacle robot to a desired position is possible if the artificial potential is a potential functional whose point of minimum is an attractor of this dissipative controlled system. This method is also used for constrained motion in an environment with obstacles.

These results are illustrated by the simulation of several tentacle models in 2D and 3D space.

REFERENCES

[1] A. Hemami, "Design of light weight flexible robot arm", Robots 8 Conference Proceedings, Detroit, USA, June 1984, pp. 1623-1640.

[2] Ian A. Gravagne, Ian D. Walker, "On the kinematics of remotely - actuated continuum robots", Proc. 2000 IEEE Int. Conf. on Robotics and Automation, San Francisco, April 2000, pp. 2544-2550.

[3] G. S. Chirikjian, J. W. Burdick, "An obstacle avoidance algorithm for hyper-redundant manipulators", Proc. IEEE Int. Conf. on Robotics and Automation, Cincinnati, Ohio, May 1990, pp. 625 - 631.

[4] G.S. Chirikjian, J. W. Burdick "Kinematically optimal hyper-redundant manipulator configurations", Proc. IEEE Int. Conf. on Robotics and Automation, Nice, May 1992, pp. 415 - 420.

[5] G.S. Chirikjian, J. W. Burdick "Kinematically optimal hyper-redundant manipulator configurations", IEEE Trans. Robotics and Automation, vol. 11, no. 6, Dec. 1995, pp. 794 - 798.

[6] H. Mochiyama, H. Kobayashi, "The shape Jacobian of a manipulator with hyper degrees of freedom" Proc. 1999 IEEE Int. Conf. on Robotics and Automation, Detroit, May 1999, pp. 2837-2842.

[7] G. Robinson, J.B.C. Davies, "Continuum robots – a state of the art", Proc. 1999 IEEE Int. Conf. on Robotics and Automation, Detroit, Michigan, May 1999, pp. 2849-2854.

[8] K. Suzumori, S. Iikura, H. Tanaka, "Development of flexible micro-actuator and its application to robot mechanisms", IEEE Conference on Robotics and Automation, Sacramento CA, April 1991, pp 1564-1573.

[9] G. S. Chirikjian, "A continuum approach to hyper-redundant manipulator dynamics", Proc. 1993 Int. Conf. on Intelligent Robots and Systems, Yokohama, Japan, July 1993, pp. 1059 - 1066.

[10] M. Takegaki, S. Arimoto, "A new feedback methods for dynamic control of manipulators", Journal of Dynamic Systems, Measurement and Control, June 1981, pp. 119-125.

[11] M. Ivanescu, "Dynamic control for a tentacle manipulator", Proc. Int. Conf. on Robotics and Factories of the Future, Charlotte, 1984, pp. 315-327.

[12] M. Ivanescu, V. Stoian, "A variable structure controller for a tentacle manipulator", Proc. IEEE Int. Conf. on Robotics and Automation, Nagoya, May 1995, pp. 3155-3160.

[13] N. V. Douskaia, "Artificial potential method for control of constrained robot motion", IEEE Trans. on Systems, Man and Cybernetics, part B, vol. 28, June 1998, pp. 447-453.

[14] S. A. Masoud, A. A. Masoud, "Constrained motion control using vector potential fields ", IEEE Trans. on Systems, Man and Cybernetics, part A, vol. 30, May 2000, pp. 251-272.

[15] P. K. C. Wang "Control of distributed parameter systems", in Advance in Control Systems, by C.T. Leondes, Academic Press, 1965.

ELSEVIER

IFAC
PUBLICATIONS
www.elsevier.com/locate/ifac

ON MONOTONIC CONVERGENCE OF ILC / RC SYSTEMS
A WAY TO FIGHT AGAINST OUTPUT'S OVERSHOOTS [1]

Francis Gérardin, Daniel Brun-Picard

Laboratoire des Sciences de l'Information et des Systèmes, LSIS - UMR CNRS 6168 - Equipe IMS
2, cours des Arts et Métiers - AIX-EN-PROVENCE cedex 1, F-13617, FRANCE
E-mail: Francis.Gerardin@aix.ensam.fr, Daniel.Brun-Picard@aix.ensam.fr - http://www.lsis.org

Abstract: Industrial machines can get better performances when operated with an "iterative learning controller" (ILC) or a "repetitive controller" (RC), because the best command law will be produced for a specific and repetitive task after some initial iterations. A well-designed controller aims to guarantee for monotonic decay of the norm of the trajectory error. But even in that case, it is often observed that the output response exhibits an overshoot in the repetition domain. Unsatisfactory solutions exist to overcome this problem, from reduction of learning gain to switching the learning unit off. This paper presents a new solution, based on observations of the behavior in the frequency domain. Vector analysis shows that phase lag accumulation in the learning loop is the key reason of the output's overshoot. A well-designed phase lead operator may cancel this phase rotation and help to reduce the overshoots.

Keywords: Robot Control, Machine control, Iterative Learning control, Repetitive Control, Vector Analysis, Phase rotation, Overshoot Cancellation.

1. INTRODUCTION

Learning control endows machines with the ability to reduce their trajectory error over and over, by repetitively reinforcing its knowledge. In practice, the tasks submitted to the machine are mainly continuous repetitive or iterative with idle time. Therefore, the corresponding two classes of learning controller are called repetitive controller (RC) and iterative learning controller (ILC). They share almost the same theoretical basis, whose first contribution was attributed to Arimoto [1] and which is quite well summarized in [10].

Recently, works have focused on monotonic convergence of the error [6] [8] [9] [11]. The objective was to understand the phenomenon of pseudo-divergence of the sup-norm of the error. Some new theory and proposals where derived to get a better behavior of the controlled process, mainly by the use of a PD type ILC controller, which is named *"higher order in time"* in [7]. But monotonic convergence of the error doesn't imply that the trajectory of the output is free from overshoots, and there are cases where overshoots must be avoided.

Investigation in the time domain gives poor information on the reasons for which overshoots come up. In fact, dealing with norm leaves aside phase rotation of the signals; therefore detailed analysis must be done in the frequency domain. In this paper, it will be shown that this phenomenon is mainly due to the excess phase rotation of the error signal, an auxiliary effect of the learning controller. This rotation increases at each repetition, due to accumulation of phase shift by the learning controller.

The first section of this paper gives basic relations obtained from a simplified ILC/RC scheme, which is normally suitable for many plants. The second section looks at the reasons for which the system's output may exhibit some overshoots. The third section gives an example of simulation which shows the effectiveness of the proposed solution.

2. SPECIAL SETUP FOR THE ANALYSIS

Figure 1 shows the setup used for that study. Such a scheme has been extensively studied and simulated in [2] [3] [4] [5]. His main properties are fast convergence and ability to work properly in iterative learning control mode or in repetitive control mode.

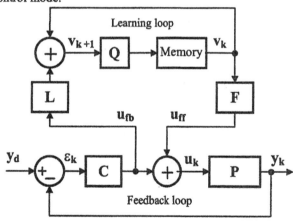

Figure 1: Scheme of the system

The learning loop consist of a (possibly) infinite-dimensional memory, a (zero-phase) filter **Q** requested for robustness purpose, and the **L** and **F** operators whose usefulness will be considered later. The group delay T_d of the loop is set equal to the cycle time Tc of the desired motion law $y_d(t)$.

[1] 3rd International Conference on Management and Control of Production and Logistics MCPL2004, Santiago, 3-5 Nov 2004

If the convergence conditions are satisfied, the learnt action v_k will rise from $\mathbf{0}$ (initial value at k=0 or start up cycle) to the fixed-point v_∞ according to the equation (1). This lets one think that ε_k vanishes as (subscript) iteration number k rises. Of course, the learning loop takes direct advantage of the benefits (error amplification, noise filtering, etc) given by the C controller.

$$v_{k+1} = Q(v_k + LC\varepsilon_k), \qquad L > 0 \quad (1)$$

A. Scheme relations and behavior

The main loop (or feedback loop) consist of the plant $\mathbf{P}$ and the controller $\mathbf{C}$, whose behavior is governed by the basic equations (2) of his nodes. Let $\mathbf{Q=I}$ to simplify the equations:

$$\begin{array}{c|c} y_k = P\,u_k & \varepsilon_k = y_d - y_k \\ u_k = F\,v_k + C\,\varepsilon_k & v_k = v_{k-1} + LC\,\varepsilon_{k-1} \end{array} \quad (2)$$

The resulting command law (3) is easily obtained and shows that the corrective action of the previous cycle feedback error ε_{k-1} may be canceled if the auxiliary operators are chosen with default values such that $\mathbf{LF = I}$. This *first-order ILC* scheme becomes then a less efficient *zero-order* one.

$$u_k = u_{k-1} + C\ \varepsilon_k + F\left(LF - I\right)F^{-1}C\ \varepsilon_{k-1} \quad (3)$$

Let the sensibility function $\mathbf{S}$ and the closed loop transfer function $\mathbf{T}$ (complementary function of $\mathbf{S}$) be defined by (4).

$$S = \left(I + P\,C\right)^{-1}; \qquad T = C\left(I + P\,C\right)^{-1}P \quad (4)$$

For that learning scheme, the convergence condition (5) is derived from the recurrence equation (6). One may think that a well-tuned feedback controller yields a quasi unity closed-loop transfer function $\mathbf{T}$ on a sufficiently wide band of frequencies. In consequence, the norm (5) is close to zero and implies a fast convergence of the learnt action v_k.

$$\left\| I - L\,T\,F \right\|_\infty < 1 \qquad \forall \omega \quad (5)$$

$$v_{k+1} = \left(I - L\,T\,F\right)v_k + LCS\,y_d \quad (6)$$

In fact, most real systems behave like low pass filters and the previous assumption fails in the high frequency band. This is a reason for which a robustness filter $\mathbf{Q}$ is necessary. Finally, it appears with equation (7) that the speeds of convergence of both the error ε_k and the learnt action v_k are similar.

$$\varepsilon_k = \left[I - C^{-1}TFLC\right]\varepsilon_{k-1} \quad (7)$$

B. About the L and F operators

Assume the system is working well below its cut-off frequency ω_{-3dB}. Whatever its order is, the transfer function $\mathbf{T}$ can be seen in the time domain as a weakening delay line with a constant gain $\rho \le 1$ and a constant delay $\tau > 0$, according to equations (8). This delay corresponds typically to the time constant of the closed-loop system. Referring to equation (5), it is clear that the $\mathbf{L}$ and $\mathbf{F}$ operators must compensate for the attenuation and the delay brought by $\mathbf{T}$.

$$\rho = |T_\omega|; \qquad \tau = -\frac{d}{d\omega}Arg(T_\omega); \qquad \omega \ll \omega_{-3dB} \quad (8)$$

Operator $\mathbf{L}$ may be a constant (or adaptive) gain, for example $\mathbf{L = I/\rho}$, or an input filter for the learning loop, in which case an additional delay exists. Thus, there is an non-causal operator $\mathbf{F}$ that must compensate for the delays. Fortunately, it can be easily implemented inside the learning loop, by adding an appropriate output out of the delay line (memory). Note that $\mathbf{F}\,v_k$ defines the feed forward signal u_{ff}.

C. Defining additional functions

Let $\mathbf{W}_k$ and $\mathbf{S}_k$ be the two cross-coupled transfer functions at iteration k, obtained from the compounded learning and feedback loop, as defined by (9), with $\mathbf{S}$ defined in (4). These new functions will help to better analyze the convergence phenomenon in the next section. The command law in (2) can then be rewritten as (10). Since $\varepsilon_\infty \approx 0$ when the convergence is established, $\mathbf{S}_\infty \approx \mathbf{0}$ and we get $\mathbf{PFW}_\infty \approx \mathbf{I}$.

$$W_{k+1} = \frac{v_{k+1}}{y_d} = Q\sum_{i=0}^{k}\left(Q^{k-i}LCS_i\right) \quad (9)$$

$$S_k = \frac{\varepsilon_k}{y_d} = S\left(I - PFW_k\right), \qquad S_0 = S$$

$$u_k = F\,v_k + C\,\varepsilon_k = \left(F\,W_k + C\,S_k\right)y_d \quad (10)$$

3. VECTOR ANALYSIS OF THE PHENOMENON

When working with a learning controller, we can observe that the trajectory output magnitude often exhibits one or more overshoots in the repetition domain, a kind of pseudo-divergence that may be dangerous for an industrial process. To understand this phenomenon, the idea is to drive the system with a pure sinusoidal signal $y_d(t)$. For the sake of simplicity, the operators $\mathbf{Q}$, $\mathbf{C}$, $\mathbf{L}$ and $\mathbf{F}$ are set to unity. This will slow down the convergence process and give us a simple and clear vector diagram.

Let $\mathbf{P}$ be a plant producing almost no harmonic at the selected working frequency (a linear system with delay can be a good candidate) and let the system work in *repetitive control* mode. The frequency of the desired output trajectory $y_d(t) = \sin(\omega_c.t)$ is chosen in the vicinity of the cut-off frequency of the open loop plant. This is because stronger overshoots appear when the input / output phase difference of the process grows, assuming the closed loop gain is not too weak. Since all signals are mastered, there will be no problem of convergence as long as the working frequency is not too high.

A. First trial (k=0)

The memory being reset at start up time, there is no initial learnt action available, so $v_0 = \mathbf{0}$ and $u_0 = C\,\varepsilon_0 = \varepsilon_0$. Figure 2 corresponds to the initial vector diagram which shows that, in the complex plane, the error signal ε_0 is the difference between the set point y_d and the output signal y_0. This diagram is the starting point to explain how the whole system works from cycle to cycle.

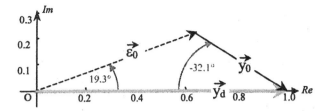

Figure 2: Vector diagram of the first trial (k = 0)

B. First iteration (k=1)

As seen on Figure 3, the first learnt action v_1 is a true copy of the error signal ε_0 recorded during the initial trial. The corresponding vector is slipped axially with its tip stuck at the origin of the diagram. The command law becomes $u_1 = \varepsilon_1 + v_1$ with the resulting effect that the phase and magnitude differences of the two vectors y_1 and y_d are lessened.

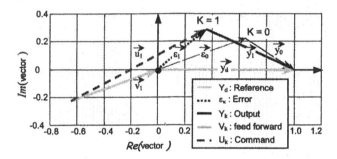

Figure 3: Vector diagram of the second trial (k = 1)

C. Further iterations (k→ ∞)

The learnt feed forward action v_k evolves according to equation (11) and converges to the fixed point v_∞ only if the error vanishes as the number of iteration grows to infinity. This is the case with the selected periodic time Tc, since the convergence condition (12) is satisfied.

$$\mathbf{v}_k = \varepsilon_0 + \varepsilon_1 + ... + \varepsilon_{k-1}, \quad \varepsilon_k = \big[\mathbf{I} - \mathbf{T}\big]_\infty \varepsilon_{k-1} \quad (11)$$

$$\big\|\mathbf{I} - \mathbf{T}\big\|_\infty < 1, \quad @\omega_c \approx \omega_{-3dB} \quad (12)$$

Writing some of the first error equations gives the following recurrent relations (13), which show again that the learning controller transfer function realizes the best inverse model of the plant for the specific motion law y_d. As defined by (9), $\mathbf{W}_k$ features the equivalent transfer function of the compound *feedback and learning loop* at iteration k.

$$\varepsilon_0 = \mathbf{S}\big[\mathbf{y}_d - \mathbf{P}\,\mathbf{v}_0\big] = \mathbf{S}\big[\mathbf{I} - \mathbf{P}\,\mathbf{W}_0\big]\mathbf{y}_d = \mathbf{S}_0\,\mathbf{y}_d$$

$$\varepsilon_1 = \mathbf{S}\big[\mathbf{y}_d - \mathbf{P}\,\mathbf{v}_1\big] = \mathbf{S}\big[\mathbf{I} - \mathbf{P}\,\mathbf{W}_1\big]\mathbf{y}_d = \mathbf{S}_1\,\mathbf{y}_d$$

$$\varepsilon_2 = \mathbf{S}\big[\mathbf{y}_d - \mathbf{P}\,\mathbf{v}_2\big] = \mathbf{S}\big[\mathbf{I} - \mathbf{P}\,\mathbf{W}_2\big]\mathbf{y}_d = \mathbf{S}_2\,\mathbf{y}_d \quad (13)$$

$$\mathbf{S}_k = \mathbf{S}\big[\mathbf{I} - \mathbf{P}\,\mathbf{W}_k\big], \quad \mathbf{W}_k = \sum_{i=0}^{k-1}\mathbf{S}_i \xrightarrow{k\to\infty} \mathbf{P}^{-1}$$

Finally, equation (14) shows that the effect of the learning process is to slide the closed loop transfer function $\mathbf{T}_k$ gradually from the classical transfer function $\mathbf{PW}_0 = \mathbf{T}$ to the unity one $\mathbf{PW}_\infty = \mathbf{I}$ as iterations continue over and over.

$$\mathbf{y}_k = \mathbf{T}_k\,\mathbf{y}_d = \mathbf{P}\big(\mathbf{C}\,\mathbf{S}_k + \mathbf{FW}_k\big)\mathbf{y}_d \xrightarrow{k\to\infty} \mathbf{y}_d \quad (14)$$

Figure 4 presents the evolution of the phases and magnitudes of the different signals appearing on Figure 1 for the six first trials. From that diagram, it is obvious that the overshoot in magnitude of y_k is mostly due to the phase rotation of the error ε_k. Depending on the frequency ω_c, multiple revolutions of vector ε_k around the origin may be observed. Its magnitude has small effect and this phenomenon cannot be highlighted by calculus made with $\|\varepsilon_k\|_\infty$. Also, this diagram suggests that the rotation of vector ε_k should be strongly reduced if the angular position of the vector v_k could be clamped down.

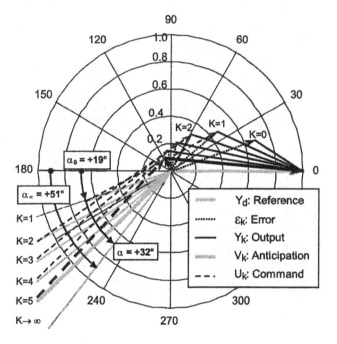

Figure 4: Phase rotation after multiple trials (k → ∞)

4. COUNTERACTING THE PHENOMENON

Assume $v_1 = v_\infty$ (since $v_0 = 0$) directly in the first iteration, which needs to pick the rights operators L and F. This implies $\varepsilon_k = 0$ for all k > 0, so the learnt action v_k will not change according to equation (11). But the question is how to pick L and F?

A. Evaluation of the requested phase lead α

Examining Figure 4 for k = 1 and k = ∞, we observe that v_1 has an initial angle of 19° (relative to the reference y_d), and v_∞ has a final one of 51°. The purpose of operator F is to give the learnt action v_k a phase lead of $\alpha = 32°$, and produce the best feed forward $\mathbf{F}\,\mathbf{v}_k$ so that the command node can build the best law $\mathbf{u}_k$ according to equation (2).

$$\alpha_0 = Arg\left(\varepsilon_0\right) = Arg\left(\mathbf{S}\right)_{\omega_c} = 19° \quad (15)$$
$$\alpha_\infty = Arg\left(\mathbf{u}_\infty\right) = -Arg\left(\mathbf{P}\right)_{\omega_c} = 51°$$

The initial angle α_0 is given by relative phase of the closed loop system without the learning controller (15) when the final angle α_∞ is due to the plant alone. In practice, picking a value for α (16) is quite simple, even when no information on the plant is available, provided that it is possible to physically measure the requested angles α_0 and α_∞.

$$\alpha = \alpha_\infty - \alpha_0 = -\left(Arg\left(\mathbf{P}\right) + Arg\left(\mathbf{S}\right)\right)_{\omega_c} = 32° \quad (16)$$

Formally, an equation (17) may be derived for a specific plant $\mathbf{P}$ in the frequency domain, with $\mathbf{F}$ being replaced by the phase shift operator $e^{i.\alpha}$.

$$\min_\alpha \left\| \frac{\varepsilon_1(\alpha)}{\mathbf{y}_d} \right\| \Rightarrow \frac{d}{d\alpha}\left\| \left(\frac{\varepsilon_1(\alpha)}{\mathbf{y}_d}\right)_{j\omega_c} \right\| = 0 \quad (17)$$

B. Design of a simplified learning controller

For a particular cycle pulsation $\omega_c = 2\pi/T_c$, such a phase lead is equivalent to the time lead: $\theta = \alpha/\omega_c$, which defines where to place the intermediate outlet out of the learning memory (or equivalent delay line). Normally, θ should be set for each new operating frequency ω_c, but fortunately θ leads toward the time constant of the closed loop as the working frequency decreases. Finally, the learning gain $\mathbf{L}$ is set to the inverse DC gain of the closed loop system $\mathbf{T}_0 = \mathbf{T}$ so that a simplified learning controller may look like the one of Figure 5. Note that we get $\mathbf{v}_k = \mathbf{F}^{-1}.\mathbf{u}_{ff.k}$ from a causal point of view.

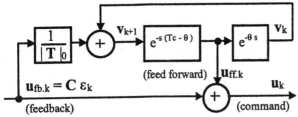

Figure 5: Scheme of the simplified learning controller

C. Effect of non-nominal phase lead α

Assume $\alpha = arg(\mathbf{F})$ and $\varphi = arg(\mathbf{T}) < 0$, the phase lag of the output signal $\mathbf{y}$ referenced to $\mathbf{y}_d$, with $\mathbf{T}$ the closed loop classical transfer function defined by (4). Taking $\mathbf{C} = \mathbf{I}$, equation (7) can be rewritten as (18) in the frequency domain.

$$\varepsilon_k = \left[1 - e^{i(\alpha+\varphi)}\right]\varepsilon_{k-1}, \quad \forall\omega, \ \alpha > 0, \ \varphi < 0 \quad (18)$$

It is clear from equation (19), valid for angular mismatch less than 0.35 rd (i.e. 20°), that the speed of convergence is a linear function of the difference between α and $|\varphi|$. The error still converges monotonically, even with $\alpha = 0$. This fact may be considered as robustness.

$$\|\varepsilon_k\|_\infty \le \sqrt{2\left(1 - \cos(\alpha+\varphi)\right)}\|\varepsilon_{k-1}\|_\infty \approx |\alpha + \varphi|\|\varepsilon_{k-1}\|_\infty \quad (19)$$

Plainly, we get overshoots on the output signals, what we try to avoid. In fact, it is not clear how large that mismatch can be so that no overshoot appears on the output signals.

5. SYSTEM SIMULATION IN RC MODE

To illustrate the general results obtained with ILC / RC controllers and to simplify the following calculus, the simplest example of a first order system is used for this simulation. But since the used method is bound to the phase lag and the attenuation brought by the system, there is obviously no restriction on the plant, whatever is its order, as long as there are almost no harmonics.

Lets take the first order simple plant $P(s) = 1/(s+1)$ and apply a sinusoidal law for forty cycles to the input of the controlled loop. The cut-off frequency of the closed loop system is then 0.16Hz ($\omega_{-3dB} = 1$ rd/s). We set $\omega_c = \pi$ rd/s (period Tc = 2s), to get visible overshoots. This implies from (20) a leading phase of 57.5°. Equation (21) gives the magnitude and the angle of the vector corresponding to the error signal at the end of the first trial, after transients have elapsed.

$$\alpha = arctg\left(\frac{\omega_c}{2}\right) = 57.5°, \quad \theta = \frac{\alpha_{rd/s}}{\omega_c} = 0.32s \quad (20)$$

$$\frac{\varepsilon_0}{y_d} = \left(\frac{i\omega+1}{i\omega+2}\right)_{\omega=\pi} = 0.88\underline{|14.8°} \quad (21)$$

Simulation is more efficient if equations (13) are rewritten in the non-recursive style of (22), since gain $\mathbf{C} = \mathbf{I}$ is constant.

$$\mathbf{S}_k = (\mathbf{I}+\mathbf{P})^{-(k+1)} \quad \mathbf{W}_k = \mathbf{P}^{-1}\left(\mathbf{I}+\mathbf{S}_{k-1}\right) \quad (22)$$

The scheme of Figure 5 is used in repetitive control mode, with no resetting of the initial conditions at the beginning of each new cycle. Figure 6 shows the chronograms of the signals recorded on two similar systems; a reference one ($\alpha = 0°$) and an improved one ($\alpha = 57°$).

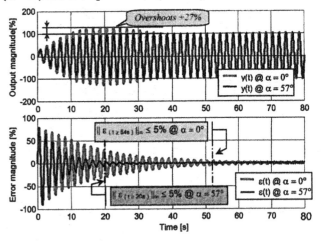

Figure 6: Chronograms of signals in the time domain

During the first cycle (0 to 2s), the output waves are identical. Then the learning controller starts to produce the feed forward signal, which forces the error signal to vanish after some iteration. Considering a relative magnitude level of 5%, ten iterations are necessary for the improved system to converge (t ≥ 20s), and at least twenty-seven for the reference one (t ≥ 54s).

The magnitude of the output signal of the reference system

crosses over the level 100% for some cycles when the magnitude of the error always decreases. This points out the fact that the monotonic convergence of the error does not imply the monotonic convergence of the output signal.

First important point, overshoots are canceled if the **F** operator is correctly set. Second one, convergence of the error is faster with the **F** operator than without it.

Figure 7 gives the best insight on the output convergence problem. The desired profile of convergence may be adjusted in the repetition domain, as a pseudo-periodic response may be in the time domain. An equivalent *response time* in the repetition domain can then be defined, by picking a value for α in the range 0 to -φ. This gives robustness to this setting. Simulations show that the best result is obtained with α set to a mid range value when the motion law y_d is no more a pure sinusoid, with plants of different order.

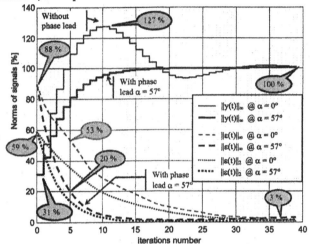

Figure 7: Norms of signals in the repetition domain

5. GENERALISATION

The objective set out is to cancel the error signal ε_1 from the very first iteration $k = 1$, whatever the order of the plant is, whatever the **Q** and **C** operators are. The (non restrictive) assumption is that **L** and **F** are chosen according to equation (8), the closed loop system T_ω being equivalent to a delay and an attenuation at the specified working frequency ω_c.

From equation (9), we get: $PFW_1 \approx I$ since we have only an approximate model of the plant **P**. Finally, the operators **L** and **F** are simply defined by equations (23) which show that the best approximation for these operators is (one more time) the inverse model of the closed loop system, and that **Q** can be a filter of any kind (zero-phase type or not).

$$PFQLCS \approx I, \quad \Rightarrow \quad FQL \approx T^{-1}$$

$$T_\omega = \rho \cdot e^{j\varphi}, \quad F_\omega = e^{j\alpha}, \quad L_\omega = k \qquad (23)$$

$$\alpha \approx -\arg(QT), \quad k \approx \left\| T_\omega Q_\omega \right\|_\infty^{-1}$$

6. CONCLUSIONS

Learning controllers may give a closed loop system the ability to strongly reduce its trajectory error when working with repetitive tasks. Specifications to ensure monotonic convergence of the error may be found in many publications, but there still has been nothing concerning the output trajectory.

This paper focuses on the outline of the output norm in the repetition domain (can we say convergence?) because this one often exhibits overshoots that may be unacceptable. An auxiliary operator named **F** (a kind of phase shifter) is proposed, which produces a feed forward signal from the learnt action. This results in the speed up of the error convergence and in the smoothing of the output overshoots.

The behavior of the whole system has been analyzed in the frequency domain, because the key reason of overshoots emergence is the continuous phase shift of the error. It has been observed in the vector diagram (Figure 4) that the error vector wind up can be thwarted with an appropriately wound up feed forward vector. As a result, the proposed **F** operator will help to fight against output overshoots.

Simulations made with the RC scheme of Figure 5 give encouraging results and generalization follows easily. But this scheme cannot be used as is for ILC structures, since there will be a lack of signal at the end of the forwarded action u_{ff}. To overcome this problem, a solution is to wind up v_k it-self inside the learning memory. This can be done in batch mode and only once at the end of the first trial. The continuity of v_k will certainly be lost and exhibit a leap between the main shifted part and the one moved to the end. This last one corresponds to the transient part of the signal and cannot be considered as reliable.

Further works are needed to improve ILC schemes including phase shifter, and to obtain analytic relations to control overshoots in the repetition domain.

REFERENCES

[1] S. Arimoto. S. Kawamura and F. Miyazaki. *Bettering operation of robots by learning*. Journal of Robotic Systems, 1984, vol. 1, pp. 123-140.

[2] D. Brun-Picard, Aline Cauvin, *About "Learning by being told" applied to multivariable systems : Stability, precision and rapidity of convergence* , "IEEE/SMC International Conference on Systems, Man and Cybernetics ", *System Engineering in the Service of Humans* Vol. 4, Num. *nc*, pp 143-146, 1993

[3] D. Brun-Picard, P. Mitrouchev, MA. Faure, JSS. Sousa, *Autonomous learning control for robot axis: efficiency and robustness in presence of interactions and vibrations.* 2nd Japan-France Congress on Mechatronics, Takamatsu, Kagawa, Nov. 1-3, 1994.

[4] D. Brun-Picard, João S. S. Sousa, *Design of machines and robots endowed with a permanent learning ability.* Control Engineering Practice, n° 7, pp. 565-571, Pergamon, 1999.

[5] F. Gerardin and D. Brun-Picard, *From iterative learning control to Analogy learning control* (in french), 4th international conference on industrial automation, Montreal, 2003.

[6] R. W. Longman and Y.-C. Huang, *The Phenomenon of Apparent Convergence Followed by Divergence in Learning and Repetitive Control*, Intelligent Automation and Soft Computing, the Journal of the World Automation Congress,, Special Issue on Intelligent Control, Vol. 8, N° 2, pp. 107-128, 2002.

[7] KL. Moore and YQ. Chen, *On Monotonic Convergence of High-Order Iterative Learning Update Laws,* IFAC World Congress, Barcelona, Spain, July 21-26, 2002.

[8] KL. Moore and YQ. Chen, *A Separative High-Order Framework for Monotonic Convergent Iterative Learning Controller Design*, in Proceedings of the 2003 American Control Conference, June, 2003.

[9] M. Norrlöf and S. Gunnarsson, *Time and frequency domain convergence properties in Iterative Learning Control,* International Journal of Control, Vol 75, N° 14, pp. 1114-1126, 2002.

[10] MQ. Phan, RW. Longman, KL. Moore. *Unified Formulation of Linear Iterative Learning Control.* AAS/AIAA Space Flight Mechanics Meeting, Clearwater, Florida, 2000, Paper No. AAS 00-106.

[11] Y. Wang and R. W. Longman, *Monotonic Vibration Suppression using Repetitive Control and Windowing Techniques*, Advances in the Astronautical Sciences, Vol. 93, pp. 945-965, 1996.

ELSEVIER
IFAC
PUBLICATIONS
www.elsevier.com/locate/ifac

OPTIMAL MOTION PLANNING OF COOPERATIVE MOBILE ROBOTS IN 2D ENVIRONMENTS

Antoneta Iuliana BRATCU*, Dan DULMAN*, Iulian MUNTEANU* and Alexandre DOLGUI**

*Advanced Control Systems Research Centre
"Dunărea de Jos" University of Galaţi
111, Domnească, 800021, Galaţi, Romania
Phone/Fax: (+40) 2 36 46 01 82
E-mail: {Antoneta.Bratcu, Dan.Dulman, Iulian.Munteanu}@ugal.ro

**Centre "Génie Industriel et Informatique" (G2I)
École Nationale Supérieure des Mines de Saint–Étienne
158, Cours Fauriel, 42023 Saint–Étienne cedex 2, France
Phone: +33 (0) 4 77 42 01 66; Fax: +33 (0) 4 77 42 66 66
E-mail: dolgui@emse.fr

Abstract: This paper deals with the optimal motion planning of two robots manipulating a rectangular object in a known 2D environment, with obstacle avoidance. It is proposed that a lowest cost path algorithm (the so-called A* algorithm) be used to this end. Its cost evaluation function takes into account the position changes, the orientation changes and the changes of robots arrangement around the object, while taking into account the manipulation constraints induced by the presence of obstacles. Some results of this algorithm performing on a given map are presented, which are intended to guide the effective control of mobile robots under position control.

Keywords: mobile robots, optimal search techniques, cooperation, obstacle avoidance, heuristics.

1. INTRODUCTION

The concept of cooperation by multiple mobile robots was proposed as an inspiration from the human actions, to improve task flexibility and fault tolerance (Arai and Ota, 1996), wherever the mobile robots are expected to undertake various tasks (including difficult environments). The robots carrying out transport tasks are particularly interesting. Many studies have been conducted on cooperative transportation and manipulation by multiple robot systems, either mobile autonomous, decentralized controlled (Rus, et al., 1995; Sawasaki and Inoue, 1996; Kosuge, et al., 1997), which process their sensors' information, or wheeled vehicles (Hashimoto, et al., 1995), manipulators (Sugar and Kumar, 2002), etc.

The difficulty of the motion planning problem results from the large dimensions of the configuration space (C-space). Different methods of operations research have been used to cope with this aspect (Barraquand and Latombe, 1991; Gupta and Guo, 1992), sometimes combined with multi-heuristics (Kondo, 1991). The C-space dimensions increase even more if considering the constraints of an object's motion (that is, the explicit computation of the contact and friction forces). The problem of motion planning with collision avoidance for the transfer of a large object in 2D by a team of mobile robots is discussed by Ota, et al., (1995). A method of reducing the dimension of freedom (DOF) of the C-space for motion planning in 3D environments is proposed by Yamashita, et al., (2003).

In this paper a motion planning method for the cooperative transportation of a large object by two small mobile robots in a 2D environment is proposed. The motion planner must be conceived to ensure that the robots transport the object as quickly as possible, while avoiding obstacles (whose configuration is a priori known and supposed fixed) and decide the motions that robots will undertake. As these aspects cannot be treated at once, because of the inadmissible increase of the computational time, the problem has been split in two (Yamashita, et al., 2003): 1) obtaining an optimal path by considering only the geometrical and topological conditions (shapes of the objects involved), and 2) determining the effective control actions by considering the objects static (contact forces). The planners corresponding to the two phases are called the *global (rough)* and the *local (fine)* one, respectively.

These planners interact in the following way: the rough one decides the moment, the place and the manner of manipulation, based upon the result of the fine planner. The first planner provides the path of the object and the robots, which is optimal from the point of view of the *manipulation cost*. As motion errors by robots can occur, it is highly probable that robots cannot move according to precise orders following exactly the trajectory found by the global planner. Hence, the fine planner outputs a method for manipulation and information about it. This paper concerns the design of the global planner, using the A* search algorithm.

This paper is organized as follows. In the next section the problem to solve is presented, together with the assumptions needed for formalization. Section 3 presents the steps of the A* search adapted to the presented problem. The cost evaluation function used is detailed in Section 4. Some results obtained by running the proposed algorithm are discussed in Section 5. Finally, Section 6 concludes this paper.

2. PROBLEM DESCRIPTION AND ASSUMPTIONS

The solution proposed for the path planning has to be both computational cost-effective and feasible too. The robot motions must be in agreement with their real surroundings (collision avoidance). The following assumptions are adopted in the motion planning:
- related to the *environment* (the "map"):
1) 2D environment with obstacles represented as polygons;
2) the shape, position and orientation of each obstacle is known and supposed fixed;
- related to the *object*:
3) the object is supposed rectangular;
4) the friction coefficient between the object and the floor is known;
- related to the *mobile robots*:
5) the robots are taken as rectangular and can move in any direction;
6) the robots always keep in contact with the object when they transport it, by using a pushing action;

- related to *manipulation*:
7) all motions are quasi-static;
8) all frictional interactions of the object are described by Coulomb's law of friction.

The role of the global planner is to continuously plan the positions of the robots and the position and the orientation of the object from a given start configuration to a desired goal configuration, provided the geometry of the obstacles (the "map") and the centre of mass of the object. Therefore, the result of the global planner is the solution of an optimization problem subject to constraints: *minimizing the manipulation cost, i.e., the time needed to finish the manipulation, such to avoid obstacles.* This is a lowest cost path graph search problem, which can be solved to optimality, for example, by the Dijkstra's algorithm (Miller, 1999). But when computation time is restrictive, one would rather prefer to quickly find a solution, even if this is not the optimal one. Heuristic methods are then used. A classical, but meanwhile efficient way of heuristic search is the use of the A* search algorithm (Pearl, 1984). At each node n, the search is guided by the cost evaluation function, denoted by f, having two components:

$$f(n) = g(n) + \alpha \cdot h(n) , \qquad (1)$$

where $g(n)$ contains the minimal cost of the path *from* the start obtained until that moment, $h(n)$ stands for the heuristic part of f, representing an estimation of the cost *to* the goal and α is the heuristic's parameter. A large value of α corresponds to a strongly oriented to the goal search; thus, even if sacrificing optimality, a good, but non-optimal solution can be quickly found. On the contrary, a small value of this parameter gives more importance to the optimality ($\alpha = 0$ reduces the A* algorithm to the Dijkstra's algorithm).

In the sequel, some important results on the heuristic search are listed. It is said that any algorithm that is guaranteed to find an optimal path to the goal is *admissible*. A well-known result states that there are conditions on graphs and on the h function guaranteeing the admissibility of the A* algorithm (Miller, 1999):
- each node in the graph has a finite number of successors (if any);
- all arcs have positive costs;
- for all nodes in the search graph, $h(n) \leq H(n)$, where $H(n)$ is the *real* value of the best path to the goal.

The last condition means that the h function never overestimates the actual value, H; for this reason, h is called an *optimistic estimator*. Ideally, h would be set to the maximal lower bound of H. Another useful result on h concerns the *consistency condition*. This may be defined both locally (Miller, 1999), and globally too in an equivalent manner (Pearl, 1984). The local definition is more intuitive: considering two nodes in the search graph such that n_j is one of the successors of n_i, it is said that h is consistent (or *monotone*) if the following condition is met:

$$h(n_i) \le h(n_j) + c(n_i, n_j), \qquad (2)$$

Condition (2) is suggestively represented in figure 1 and states that along any path in the search graph, the estimate of the optimal (remaining) cost to the goal cannot decrease by more than the arc cost along the path.

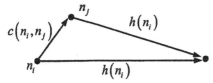

Fig. 1. The consistency condition is similar to the triangle inequality.

The choice of the h function is, therefore, very important for the effectiveness of the search. The h function chosen for the presented problem is admissible, but it is not consistent, as it will be shown in Section 4.

3. STEPS OF THE A* ALGORITHM

The steps of the A* algorithm are listed here below.

1. Create a search graph, G, consisting only of the start node, n_0, and put this node on a list called OPEN.

2. Create a list called CLOSED, which is initially empty.

3. If OPEN is empty, exit with failure.

4. Select the first node on OPEN, let it be n, remove it from OPEN and put it on CLOSED.

5. If n is the goal node, exit successfully: the solution is obtained by tracing a path along the pointers from n to n_0 in G (these pointers define a search tree, as established in step 7).

6. Expand node n, generating the set, S, of its successors *that are not already ancestors* of n in G.

7. Establish a pointer to n from each of those members of S that were not already on either OPEN or CLOSED. Let M be the set of all these nodes. Add M to OPEN. For each member, m, of $S - M$, redirect its pointer to n if the best path to m found so far is through n.

For each member of S already on CLOSED, redirect the pointers of each of its descendants in G so that they point backward along the best paths found so far to these descendants.

8. Reorder the list OPEN in increasing order of the f values.

9. Go to step 3.

Remarks: The second part of the redirection from step 7 might save subsequent search effort, but with a possible exponential computation effort. An important result states that, if the consistency condition on h is satisfied, then when A* expands a node n, it has already found an optimal path to n. Therefore, in this case, A* never has to redirect pointers. However, even if the consistency condition is not met, the second part of step 7 is often not implemented.

4. DESIGN OF THE COST EVALUATION FUNCTION

The cost evaluation function associated to a search node, n, must reflect the manipulation cost, divided into two components, as to relation (1): *from the start* – that is, $g(n)$ – and *to the goal* – that is, $h(n)$. Obviously, these two functions will have almost the same structure.

4.1 Reduction of the C-space DOF

In order to give the detailed expressions of g and h, let us analyze the possible ways of manipulation. Note that the transportation task's features can be used to reduce the dimension of freedom of the C-space, as the original DOF becomes too large (the object moving in 2D has a DOF of 3 and the two robots have each a DOF of 3). In fact, the possible paths and motions are limited, being different from the paths (even if constrained) of free flying objects. Robots can transport the object only in some specific configurations. Thus, the positions and orientations of the robots can be taken as *arrangements* in relation to the object coordinates and codified accordingly (figure 2).

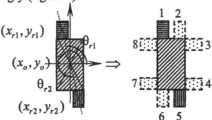

Fig. 2. Codification of the robots' arrangements around the object.

It is considered that robots are always placed at the object corners, in order to use their entire pushing force when they execute a rotation of the object. Instead of characterizing each robot independently (by the coordinates of their mass centres and the angles related to the object's mass centre, $(x_{r1}, y_{r1}, \theta_{r1})$ and $(x_{r2}, y_{r2}, \theta_{r2})$ respectively, as in the left side of figure 2), the arrangement of the robots is globally characterized by codifying each possible position of robots around the object (as in the right side of figure 2). It is also considered that robots can be placed either on the same side of the object, either on opposite sides at opposite corners; thus, some combinations are infeasible (for example, [1,3] or [7,4]).

Apart from arrangement changes, position changes and orientation changes of the object are also possible. Each general motion of the object and the robots can thus be decomposed into three *independent* motions. Correspondingly, three primitive operations may be defined (figure 3): a) the robots change the object's position in the absolute coordinate through a *position change* operation; b) the robots change the object's orientation through an *orientation change* operation; c) the robots change their arrangement around the object while handling it.

Each of the above operation is independent from the

other two (that is, for example, in the position change, the object's orientation and robots arrangement are maintained, etc.).

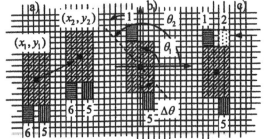

Fig. 3. Three primitive operations of a rectangular object by two robots in 2D: a) position change; b) orientation change; c) arrangement change.

As a conclusion, the DOF of the C-space has been reduced from 3+2·3=9 to only 3, from which two are continuous (the position of the mass centre and the orientation angle) and one is discrete (the arrangement, constrained to represent 8 feasible combinations of robots' positions around the object). Each continuous search coordinate is conveniently discretized, such that the motion takes place on a map of pixels (as suggested in figure 3).

4.2 The cost evaluation function

As it is intended to guide the search in order to minimize the manipulation cost (that is, the manipulation time), while avoiding the obstacles, it is proposed that the two components of the cost evaluation function be defined as follows:

$$\begin{cases} g(n) = g_t(n) \\ h(n) = h_t(n) + h_{obst}(n) \end{cases}, \quad (3)$$

where the components indexed with "*t*" measure the time cost. Component $h_{obst}(n)$ of $h(n)$ takes into account the possibility of colliding with the obstacles. This component represents the *repulsion force* by the obstacles and has been modelled as a potential field, as shown in the sequel.

Function $g_t(n)$ must contain the cumulated cost (time) of a path from the start to the current node, n, that is, the sum of the costs of all arcs of this path. The cost of an arc from node n_1 to node n_2 is defined using the notations from figure 3:

$$c(n_1, n_2) = c_p(n_1, n_2) + c_o(n_1, n_2) + c_a(n_1, n_2), \quad (4)$$

where the three components measure respectively the cost of the position change (the Euclidean distance):

$$c_p(n_1, n_2) = \sqrt{(x_2 - x_1)^2 + (y_2 - y_1)^2}, \quad (5)$$

the cost of the orientation change:

$$c_o(n_1, n_2) \approx |\Delta\theta| = |\theta_2 - \theta_1|, \quad (6)$$

and the cost of the arrangement change, expressed in terms of distance that robots have to walk for changing their positions (as suggested in figure 3c)). Note that the orientation change must also be

expressed as a distance, namely the length of the circle segment corresponding to the angle change, $\Delta\theta$. The orientation changes larger than π are modulo π reduced. Thus, an approximate value of the orientation cost may be:

$$c_o(n_1, n_2) = \frac{L}{\pi} \cdot |\Delta\theta|, \quad (6')$$

where L is the length of the object.

Function $h_t(n)$ has the same structure as the cost of an arc, considering as final node the goal node, and represents an estimation of the manipulation cost to the goal. Concerning the repulsion force by the obstacles, $h_{obst}(n)$, there are many ways of defining potential fields (Latombe, 1991; Hwang and Ahuja, 1992). The function used in this paper is:

$$P_{pot}(d, obstacle) = \begin{cases} \dfrac{d^2}{r^2} - \dfrac{2 \cdot d}{r} + 1, & \text{if } d \le r \\ 0, & \text{if } d < r \end{cases}, \quad (7)$$

where d is the distance of the considered point to the obstacle and r is some fixed positive distance (for example, $r = \max(a,b)$, where a and b are the robot's dimensions). Irrespectively from the form of the potential field, it is proposed that the *total repulsion force* exerted by an obstacle on the object and the robots be the sum of the potential fields in all the edges of the resulted object (there are 8 edges, no matter of the robots' arrangement – see figure 2 – whose coordinates can be computed, given the configuration of the associated node). Thus:

$$h_{obst}(n) = \sum_{j=1}^{nb\_obst} \sum_{i=1}^{8} P_{pot}(d_i, obstacle_j), \quad (8)$$

where nb_obst is the number of obstacles.

One can conclude that function $h(n)$ represents a trade-off between the attraction force by the goal and the repulsion force by the obstacles. On the other hand, it can be easily verified that function $h(n)$ is not consistent (because of its component measuring the orientation change, which depends on modulo π reduced angles – see relation (6')). This is the reason for which a supplementary method of collision avoidance was necessary when effectively running the algorithm. This method is based on deciding whether a given point is inside or outside a polygon.

5. DISCUSSION OF THE RESULTS

In this section the results of the proposed A* search algorithm performance on a small scale example are presented. The dimensions of the map in pixels are 70x50 and it contains 4 obstacles, the dimensions of the object are 9x5 and of the robots are 3x2 (figure 4). The start node is characterized by the coordinates of the mass centre, (9,6), and the orientation angle, $\theta_{start} = 0$. The goal node is represented by (65,26) and $\theta_{goal} = 3\pi/4$. The algorithm made 194 expansions of nodes and took 73.882 seconds when run on a Pentium II MMX at 266 MHz.

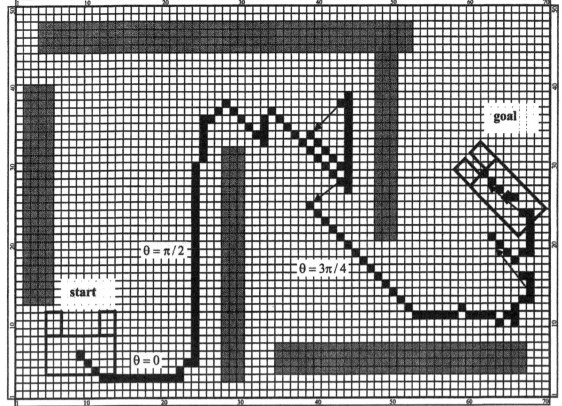

Fig. 4. The path planned for a small scale case. The arrows indicate the zones of local minima.

One can note that the studied case, although small, has a quite restrictive feature, namely the dimensions of the map in relation to those of the object. For legibility sake, in figure 4 the arrangement of the robots has not been plotted. Because the heuristic function, h, is not monotone, some cyclic behaviour has been noted in the performance of the algorithm, corresponding to local minima. The arrows on figure 4 mark the empirically performed "jumps" necessary to escape these local minima.

The trajectory illustrated in figure 4 roughly corresponds to the expectations. Obviously, such a trajectory needs to be filtered before being transformed into control sequences and transmitted to the robots. This problem concerns the design of the local planner and the interaction between the two planners, which is a point of future work.

6. CONCLUSION. FUTURE WORK

This paper presented a method of planning the operation of an object by two robots in 2D known and fixed environments, which is optimal from the point of view of the time manipulation (this can be easily converted into energy consumption), with the avoidance of obstacles. The designed planner is a rough (global) one and outputs the moment, the place and the manner of manipulation, for instant considering that robots can follow ideally the ordered trajectory. The motion errors by the robots, as well as their explicit static, will be considered in the future design of the fine (local) planner.

Trying to define heuristic functions which obey the consistency condition is an immediate investigation task. Furthermore, combining these heuristic functions with other types of potential fields, in order to achieve a monotone search map, is an useful research direction.

The proposed planning method is finally intended to be embedded in a real time application, eventually with changing environments. In this case, the representation of the map is continuously obtained by video feedback processing from a camera, which must be both quick and accurate. The planning method needs then to be globally robustified to errors, delays and environment changes.

ACKNOWLEDGMENT

This work has been partially supported by the National Scientific Research Council of Romania, under Grant no. 3334/2004, AT type, code 80.

REFERENCES

Arai, T. and J. Ota (1996). Let us work together – Task planning of multiple mobile robots. In: *Proceedings of the IEEE/RSJ International Conference on Intelligent Robots and Systems*, pp. 298-303.

Barraquand, J. and J.-C. Latombe (1991). Robot motion planning: A distributed representation approach. *Int. J. Robot. Res.*, **10(6)**, pp. 628-649.

Gupta, K.K. and Z. Guo (1992). Motion planning for many degrees of freedom: Sequential search with backtracking. In: *Proceedings of the IEEE International Conference of Robotics and Automation*, pp. 2328-2333.

Hashimoto, M., F. Oba and S. Zenitani (1995). Object transportation control by multiple wheeled vehicle planar Cartesian manipulator systems. In: *Proceedings of the IEEE/RSJ International Conference on Robotics and Automation*, pp. 2267-2272.

Hwang, Y.K. and N. Ahuja (1992). A potential field approach to path planning. *IEEE Transactions on Robotics and Automation*, **8(1)**, pp. 23-32.

Kondo, K. (1991). Motion planning with six degrees of freedom by multi-strategic bidirectional heuristic free-space enumeration. *IEEE Transactions on Robotics and Automation*, **7(3)**, pp. 267-277.

Kosuge, K., T. Osumi and K. Chiba (1997). Load sharing of decentralized controlled multiple mobile robots handling a single object. In: *Proceedings of the IEEE/RSJ International Conference on Robotics and Automation*, pp. 3373-3378.

Latombe, J.-C. (1991). *Robot motion planning*, Kluwer, Norwell MA, U.S.A.

Miller, R.E. (1999). *Optimization: Foundations and Applications*, John Wiley and Sons.

Pearl, J. (1984). *Heuristics: intelligent search strategies for computer problem solving*, Addison-Wesley Longman Publishing Co., Inc., Boston, MA, U.S.A.

Rus, D, B. Donald and J. Jennings (1995). Moving furniture with teams of autonomous robots. In: *Proceedings of the IEEE/RSJ International Conference on Intelligent Robots and Systems*, pp. 235-242.

Sawasaki, N. and H. Inoue (1996). Cooperative manipulation by autonomous intelligent robots. *JSME International Journal*, C, **39(2)**, pp. 286-293.

Sugar, T.G. and V. Kumar (2002). Control of cooperating mobile manipulators. *IEEE Transactions on Robotics and Automation*, **18(1)**, pp. 94-103.

Yamashita, A., T. Arai, J. Ota and H. Asama (2003). Motion Planning of Multiple Mobile Robots for Cooperative Manipulation and Transportation. *IEEE Transactions on Robotics and Automation*, **19(2)**, pp. 223-237.

PUBLICATIONS
www.elsevier.com/locate/ifac

Control of Assembly Systems for Automated Threaded Fastenings

Mongkorn Klingajay, Lakmal D. Seneviratne, Kaspar Althoefer

Centre of Mechatronics, Division of Engineering, King's College London
London, WC2R 2LS, United Kingdom
{mongkorn.klingajay, lakmal.seneviratne, k.althoefer}@kcl.ac.uk

ABSTRACT: Assembly is a common manufacturing procedure aiming to establish a rigid connection between two or more components. Threaded fastenings are a common assembly method, accounting for over a quarter of all assembly operations. Screw fastenings permit easy disassembly for maintenance, repair, and relocation. Screw insertions are typically carried out manually, since it is a difficult operation to automate. The focus of this paper is on real-time parameter estimation during threaded assembly. The Newton Raphson estimation technique is employed to estimate one parameter during the insertion of a self-tapping screw. Experimental results are presented to show that the parameter can be reliably estimated in real time, based on an analytical model and measured torque signature signals. The estimated parameter can be used to develop automated monitoring strategies for threaded assembly systems.

Keywords: Production management, Automated Threaded Fastenings, Screw insertions, Parameter estimation, Newton Raphson Method.

1   INTRODUCTION

Automation and condition monitoring are two key issues in implementing intelligent manufacturing systems. This paper deals with the problem of real-time monitoring of the threaded fastening assembly operation.

Human operators are particularly good at threaded fastenings in manual assembly, employing past experience and force sensing capabilities. However, the increased speeds when using electrically powered screwdrivers, reduces the human ability to monitor the insertions on-line. Thus, an automated on-line monitoring strategy for the screw fastening process is highly desirable.

The torque Vs insertion depth signature signal has a unique identity for a given screw fastening operation. One automated on-line monitoring strategy is based on the "Torque Vs Insertion Depth" curve measured during screw insertion; if this signal is within a pre-defined bound of the correct insertion signal, then the insertion is considered to be satisfactory. The torque signature signal for a correct insertion is either taught or predicted using an analytical algorithm [1, 2, 4]. However, determining the ideal signature signal prior to production is a difficult task. For large scale or high unit cost production, it is feasible to teach the ideal signal prior to production as a set up procedure. However this approach is not feasible in a flexible manufacturing system.

The approach adopted in this study is to use an analytical model of the process together with estimated screw insertion parameters, to estimate ideal signature signals. An analytical model for a general self-tapping screw insertion operation is presented in [1]. This model requires various process parameters as input, and it is not always possible to know these parameters in advance with sufficient accuracy. Hence the focus of this paper is to investigate the feasibility of real-time parameter identification.

A methodology for estimating unknown parameters for a general self-tapping screw insertion was proposed in [6]. The approach is based on the Newton Raphson Method (NRM) [10]. The paper is based on a computer simulation study and showed that the parameters required by this model can be reliably estimated during screw insertion process. This is very useful since parameters required by the model, such as friction and hole diameter, maybe difficult to measure.

This paper extends the above work, by employing an experimental study. Experimental results are presented to show that the parameter can be reliably estimated in real time, based on an analytical model and measured torque signature signals. The estimated parameter can be used to develop automated monitoring strategies for threaded assembly systems.

2 THE SCREW INSERTION PROCESS

2.1 General Screw Insertions

The screw insertion process aims to fasten together two (or more) parts with a predetermined clamping force, Figure 1. Screw insertions are especially popular in assemblies that need to be reassembled or disassembled for repair, maintenance or relocation. A standard self-tapping screw is shown in Figure 2.

In [1, 2], a general self-tapping screw insertion process is described by a torque signature signal. The insertion signal is a two dimensional curve of "Axial torque Vs insertion depth". The axial torque is necessary to drive the screw through the hole as the insertion develops [6, 7, 8, 9, 10].

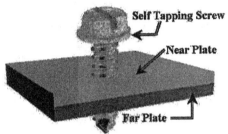

Figure 1: A typical self-tapping screw insertion.

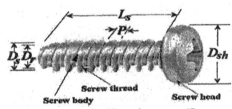

Figure 2: A typical self-tapping screw (Screw geometry).

2.2 Torque Signature Signals

The screw insertion process consists of the five main stages. These stages are shown in Figure 3 and described by the points T_0, T_b, T_c, T_{fb}, T_{nb}, and T_{ft} :

a) T_0-T_b – Stage 1 - "Screw Engagement" - where the screw gets into contacts with the hole wall and continues until thread engagement is completed.
b) T_b -T_c - Stage 2 - "Screw Breakthrough" - ends when the screw first breaks through the hole.
c) T_c-T_{fb} – Stage 3 - "Screw Full-Breakthrough" – ends when thread forming is completed.
d) T_{fb}-T_{nb} - Stage4 - "Screw Sliding"- the screw advances without thread cutting until the screw head contacts the top plate.
e) T_{nb}-T_{ft} - Stage5 - "Screw Tightening" - the final tightening torque is applied.

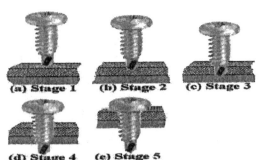

Figure 3: The 5 stages of screw insertion

An ideal screw insertion signal (torque Vs. insertion depth), and its 5 main stages are shown in Figure 4 [8, 9, and 10].

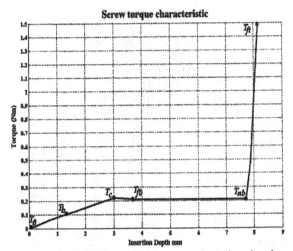

Figure 4: An ideal self-tapping screw insertion signal

3 MATHEMATICAL MODEL

The mathematical model presented in [1, 4, 5] is used in this study. This model consists of analytical equations for each of the five stages of the self-tapping screw fastening process, Appendix 1.

Stage 2 gives the torque required to drive the screw from screw engagement until initial breakthrough. This is the most critical stage of the entire screw fastening process and is employed in this study for parameter estimation:

$$\tau_2 = R_s A_c\, \sigma_{UTS}\, cos\, \theta + 2\mu\, R_f\, K_f\, \sigma_f cos\theta\, (\phi\text{-}\alpha). \qquad (1)$$

Equation (1) can be written as

$$\tau_2 = f(\mu, \theta, \sigma_{UTS}, \beta, \phi, D_s, D_h, L_b, P\,), \qquad (2)$$
where,

R_s= Position from the centre of the tap-hole to the point of action of shear forces.

A_c= The full-thread cross-sectional area.

σ_{UTS} = Ultimate tensile stress of tap-plate.

θ = Thread helix angle.

μ = Friction coefficient.

R_f =Radial distance to center of screw thread cross-section area.

K_f= Geometric parameter relating to friction force.

σ_f= The yield stress for the material of the hole wall.

ϕ = Angular rotation of screw.

α =Angular position value corresponding to the length of the screw cutting-portion.

β =Screw thread crest half-angle.

τ = Insertion torque.

4   MODEL PARAMETERS

Equation (1) contains a number of model parameters, which can be divided into 3 groups: Screw properties, plate properties, and material properties. The main screw properties consist of the screw geometry, shown in Figure 2. The plate properties consist of the hole diameter and the thicknesses of the two plates. The material properties included the yield stress, ultimate tensile stress, modulus of elasticity and the friction coefficient. These can be obtained from published information and databases, provided the material type is known.

Table 1 summarises the screw, plate and material properties for the experimental study reported in this paper:

Screw type : AB4	
Screw major diameter (D_s)	2.87 mm
Screw head diameter (D_{sh})	5.03 mm
Screw root diameter (D_r)	2.02 mm
Screw thread pitch (P)	1.10 mm
Screw taper-length (L_t)	2.79 mm
Screw total threaded-length (L_s)	9.69 mm
Material type : Polypropylene	
Tap Hole Diameter (D_h)	2.0 mm
Tensile Strength (σ_{UTS})	51.1 Mpa
Yield Strength (σ_Y)	45 Mpa
Elastic Modulus (E)	2.35 Gpa
Coefficient of Friction (μ)	0.24
Tap (far) plate thickness (T2)	4.46 mm
Tap (near) plate thickness (T1)	0

Table 1: The screw, plate, and material properties.

The parameters shown in Table 1 may vary depending on manufacturing tolerances, measurement uncertainties and wear. Generally, the cheaper the screw, the larger are these variations. Also, the friction coefficient is often difficult to measure. Due to these uncertainties and measurement difficulties, parameter estimation becomes necessary.

5   ESTIMATION STRATEGY

In this study, the aim is to estimate the unknown friction coefficient μ, based on the analytical model in [1] and measured torque signature signals.

An analytical model for estimating torque signals with a general self-tapping screw insertion had been described above. This model requires various parameters as input, before the torque signals can be predicted. Most of them can be measured or are known, but the friction coefficient is difficult to obtain and is estimated here.

Based on Equations (1, 2), the torque signature of stage 2 can be expressed as a function of the angular rotation of the screw and the unknown parameter, μ:

$$\tau_2 = F(\phi, \mu) \qquad (3)$$

Let τ_{2i} be the torque value sampled at insertion angle ϕ_i. Then equation (2) can be written as

$$\hat{\tau}_{2i} = F(\phi_i, \hat{\mu}_i), \qquad (4)$$

where $\hat{\mu}_i$ is the ith estimate of the friction coefficient and $\hat{\tau}_{2i}$ is the estimated torque at ϕ_i. Let the estimation error at ϕ_i be

$$e_i = \tau_{2i} - \hat{\tau}_{2i}, \qquad (5)$$

where τ_{2i} is the measured value of torque sampled at angular position ϕ_i. The Newton-Raphson method is used to find the next update value,

$$\hat{\mu}_{i+1} = \hat{\mu}_i - e_i. \qquad (6)$$

The NRM improves the estimate iteratively until the error e_i is reduced to a small, predefined threshold:

$$e_i = \tau_{2i} - \hat{\tau}_{2i} \rightarrow 0 \qquad (7)$$

This iterative process is depicted in Figure 5.

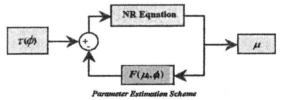

Parameter Estimation Scheme

Figure 5: Parameter estimation using NRM

6   EXPERIMENTAL PROCEDURE

The screw insertions are performed using an industrial electric powered screwdriver manufactured by Desoutter Ltd. The experimental system consists of a screwdriver, a torque sensor, a torque meter and an optical encoder, Figure 6 and 7. The screwdriver system is connected to a PC, which contains a data acquisition card for recording the screw insertion signals.

In order to perform the estimation task, the torque vs insertion-depth signature signals have to be measured. A rotary strain-gauge based torque sensor, with a range between 0 and 8 Nm and with an accuracy of 1% of full scale deflection, is used. The torque sensor is attached to the output shaft of the screwdriver to measure torque directly at the tip of screwdriver. This direct measurement approach avoids hysteresis and friction effects which would be introduced if the torque was measured at a point closer to the drive motor.

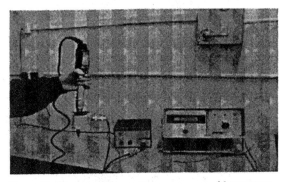

Figure 6: The screwdriver, torque sensor and instrumentation used in the experimental study.

An optical encoder with a resolution of 60° is used to measure the rotation angle of the screwdriver, Figure 7, and hence the insertion-depth is found from *a priori* knowledge of the screw pitch. For the experiments performed in this study, the angular measurements range from 0 to 3000 degrees.

Figure 7: Optical shaft encoder

Data from the torque sensor and the shaft encoder was recorded using a National Instruments multifunction data acquisition (DAQ) card. The Analog-Digital converters on-board the card have a 12 bit resolution.

Figure 8: The screw insertion test rig.

Screw insertion experiments were performed using the electric screwdriver, Figure 8, and the results are presented in Section 7.

7   TEST RESULTS

The estimation algorithm based on the NRM is tested using the screw insertion system described in the previous section. The screw and material properties shown in Table 1 are used. It is noted that the friction coefficient, μ, for the plate material Polypropylene, is 0.24 (Table1).

The torque and angular positions were measured during the screw insertion. Using the knowledge of the screw pitch the angular position is converted into insertion depth. An example of an acquired torque vs. insertion depth signal is shown in Figure 109.

Before the acquired torque signals are fed to the estimation algorithm, a pre-processing step is needed to filter the high frequency noise in the raw signal. A curve-fitting algorithm is applied to fit a smooth curve to the measured data, Figure 10.

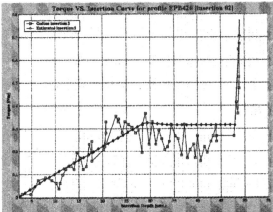

Figure 9: The torque signature signal and the fitted curve

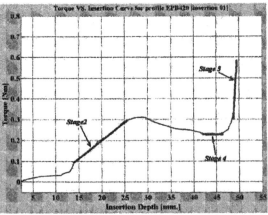

Figure 10: Individual screw insertion stages highlighted in the measured curve.

Using the NRM, the estimated unknown parameter converges rapidly to a stable value of $\mu = 0.25$, Figure11.

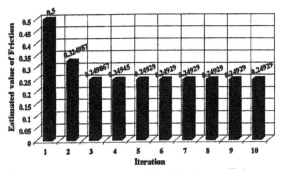

Figure 11: The convergence of the friction coefficient over 10 iterations.

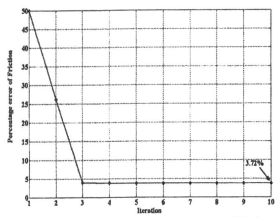

Figure 12: The percentage error of the estimated friction.

Figure 12 presents the percentage error between the estimated value and the actual value, 0.24. As can be seen from Figure 12, the percentage error after convergence is 3.72%.

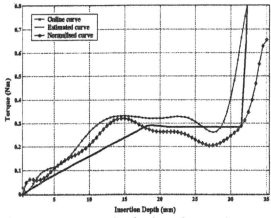

Figure 13: Experimental and estimated torque signature signals

Figure 13 shows the experimental torque signature curves for two separate insertions. The first curve is the torque signal used to estimate the unknown parameter, μ. In both cases the same screw type and plate materials and hole geometries were used (Table 1).

In order to obtain an indication of the quality of the estimation process, the root mean square (RMS) error between the acquired torque signal and the reconstructed ideal signal using the estimated value for μ is computed. The results for the RMS error of these experiments for insertions 1 and 2 are 0.228 and 0.211, respectively. These results show that the original and the estimate-based signals are in good agreement. This approach can be used to develop an automated on-line monitoring strategy based on the RMS error value.

8 CONCLUSIONS

This paper presents an integrated approach for the on-line estimation of friction properties during self-tapping threaded fastenings. The technique, based on the Newton Raphson method, estimates unknown friction properties of the screw insertion system, based on system dynamic equations. The estimated friction properties can be used to predict ideal insertion signals, providing a basis for developing automated monitoring algorithms. Test results are presented to validate the estimation methodology, and it is shown experimentally that the unknown parameter in question can be estimated within a relatively small error range.

9 ACKNOWLEDGEMENTS

Mongkorn Klingajay is funded by the Thai Government through the Rajamangala Institute of Technology (RIT), Bangkok, Thailand.

10 REFERENCES

1. Seneviratne, L.D., Ngemoh, F.A., Earles, S.W.E. (1992). Theoretical Modeling of Screw Tightening Operations, Proceedings of ASME European Joint Conference on Systems, Design and Analysis, 1:189-195.

2. Ngemoh, F.A. (1997). Modelling the Automated Screw Insertion Process, Ph.D. Thesis, University of London.

3. Visuwan, P. (1999). Monitoring Strategies for Self-Tapping Screw Insertion Systems, Ph.D. Thesis, University of London.

4. Seneviratne, L.D., Ngemoh, F.A., Earles, S.W.E., Althoefer K. (2001). Theoretical modelling of the self-tapping screw fastening process, Proceedings of IMechE, Part C, Journal of Mechanical Engineering Science 135-154.

5. Klingajay, M. Seneviratne, L.D. (2002). Friction Estimation in Automated Screw Insertions, Proceedings of ISRA International Symposium on Robotics&Automation 1:682-688.

6. Klingajay, M., Seneviratne, L.D., Althoefer, K. (2002). Parameter estimation during automated screw insertions, Proceedings of IEEE International Conference on Industrial Technology (ICIT'02) 2:019-1024.

7. Klingajay, M., Seneviratne, L. D., Althoefer, K. Zweiri, Y.H. (2003). Parameter Estimations for threaded assembly based on the Newton Raphson method, Proceedings of IEEE International Conference on Automation 1:864-869.

8. Klingajay, M., Seneviratne, L.D. Althoefer, K. (2003). Identification of threaded fastening parameters using the Newton Raphson Method, Proceedings of IEEE/RSJ International Conference on Intelligent Robots and Systems (IROS2003) 1:5055-2060.

9. Klingajay, M., Seneviratne, L.D. Althoefer, K. (2004). Model Based Monitoring Strategies for Threaded Assembly Systems, Proceedings of 11th IFAC Symposium on Information Control Problems in Manufacturing, Brasil, 2004.

10. Greig D.M. (1980). Optimisation, Longman London and New York, 1980.

11. Optimisation Toolbox for use with Matlab, The Math Works.

APPENDIX

The mathematical screw insertion model is included below. The torque signature curve is represented by five piece wise linear curves.

Stage 1 - Screw Engagement $\{ 0 \leq \phi \leq \alpha \}$

$$\tau_1 = 1/\alpha \, R_s \, A_c \, \sigma_{UTS} \cos\theta.\phi + \mu \, R_f \, K_f \, \sigma_f \cos\theta(\phi)$$

Stage 2 - Screw Breaks through $\{ \alpha \leq \phi \leq \phi_b \}$

$$\tau_2 = R_s \, A_c \, \sigma_{UTS} \cos\theta + 2\mu \, R_f \, K_f \, \sigma_f \cos\theta(\phi - \alpha)$$

Stage 3 - Screw Full-Breakthrough $\{ \phi_b \leq \phi \leq \phi_b + \alpha \}$

$$\tau_3 = 1/\alpha \, R_s \, A_c \sigma_{UTS} \cos\theta(\phi_b - \phi + \alpha) + \mu \, R_f \, K_f \, \sigma_f \cos\theta(\phi_b + \phi - \alpha)$$

Stage 4 - Screw Sliding $\{ \phi_b + \alpha \leq \phi \leq \phi_t \}$

$$\tau_4 = 2\mu \, R_f \, K_f \, \sigma_f \cos\theta(\phi_b)$$

Stage 5 - Screw Tightening $\{ \phi \geq \phi_t \}$

$$\tau_5 = \left[\left(\frac{\mu_h (D_{sh}^3 + D_s^3)}{3(D_{sh}^2 - D_s^2)} \right) + \left(\frac{D_s + D_h}{4} \right) \left(\frac{\pi\mu(D_s + D_h) + 2P}{\pi(D_s + D_h) - 2\mu P} \right) \right] \left[\frac{K_{th} K_{tb}}{K_{th} + K_{tb}} \right] \left[\frac{P(\phi - \phi_t)}{2\pi} \right]$$

www.elsevier.com/locate/ifac

TOOL WEAR MONITORING IN MACHINING PROCESSES: REVIEW AND NEW APPROACH

M. Correa[1], C.R. Peres, J.R. Alique

Industrial Automation Institute - CSIC, Campo Real km. 0.200, Arganda del Rey
C.P. 28500, Madrid, Spain. {macorrea, clode, jralique@iai.csic.es}

Abstract: Research field such as monitoring tool condition have taken great relevance, and its repercussion has been demonstrated on the machining quality improvement. This paper presents a review and the state-of-the-art of the tool wear monitoring. Key issues presented include sensorial techniques, control, signals processing methods, and decision algorithms. In addition, a preliminary draft of the general model for the tool wear estimation is presented. This estimation is incorporated as a worthiness variable of a pre-established objective function, for example, maximizing productivity. This model is part of the intelligent supervisory system based on multi-objective.

Keywords: Mechanization, Sensors; Signal processing; Model based recognition, Monitored control systems.

1. INTRODUCTION

Machine tool automation is an important aspect for manufacturing companies facing the growing demand of profitability and high quality products as a key for competitiveness. The purpose of supervising machining processes is to detect interferences that would have a negative effect on the process but mainly on the finished workpieces. As part of this global objective, it is necessary to maximize the Material Remove Rate (MRR) minimizing the tool wear rate, and maintaining superficial and dimensional quality of all the workpieces within pre-established specifications.

During the last years, a significant amount of time and research effort has been dedicated to tool monitoring, since tool breakage represents around 20% of all the machine stops. Tool wear also causes a negative effect on the workpieces quality, specifically on its dimensional precision, finishing, and superficial integrity.

Proposed techniques on tool monitoring include variables such as chatter, breakage, tool wear(especially fast wear) and changes on cutting conditions such as cutting speed, depth, feed and tool inputs/outputs.

Techniques of artificial vision have been used with relative success in laboratory tests in post-process, when the tool is not in contact with the workpiece. However, those techniques have not presented the same results under other than in an ideals conditions environment.

All of the above give us confidence to assure that to solve the problem of monitoring the tool wear in real time is a very important goal to achieve in the near future.

The present paper has been divided in seven sections. In the second section, a summary of how the tool wear is recognized and which are the utilized techniques is presented. In the two following sections, sensors used on signals acquisition for tool monitoring and the processing techniques of those signals are evaluated. In the fifth section, classification methods are shown, as well as a summary of some outstanding researches on the subject of the tool wear supervision for the milling processes. In the sixth section, a new approach for tool wear monitoring is proposed into a hierarchical frame of cooperation between complementary tasks, and finally the conclusions are presented in the seventh section.

[1]Scholarship holder of CSIC in the I3P program, finance for the European Social Fund.

2. TOOL MONITORING

Tool wear is inherent to the process of cutting metals. Usually the operators of machine tools focus their inspections on to detect values that exceeds the admitted limit value, verification of the operation sequences, and tool conditions. Such inspections are limited to comparing the state of the used tool with a new one and base on their expert knowledge to determine yet by how long the tool could be used.

Physically, to determine tool wear can be performed by using direct measurements such as optic exploration of the tool's tip, electric measurements of the resistance to contact between the tool and the workpiece and chip's radioactive analysis. Indirect measurements include cutting forces, vibrations, surface roughness, temperature, and thermoelectric effects or acoustic emissions.

Sensibility of several components of the cutting forces to the tool wear has been the theme of many researches. In most cases, the radial component of the cutting force is considered because it shows a high correlation with the tool edge wear. The wear parameter most used is the width of the worn side that is in better state than rest of the sides. However, many authors do not report if they utilized the average or the maximum of that parameter, or to measure the longitude of the wear on the best or worse side. Another utilized method is the Taylor's tool life equation, also called "theory of Taylor tool life" shown in the equation (1) whose calculations are explained thoroughly in (Elbestawi, et al., 1991).

$$VT^n = C \qquad (1)$$

Where V is the cutting speed, n it is the index of combination between tool and workpiece material which can be found in manuals of machining (Hicks, 1997). T is the life tool and C a constant value.

When proposing a reliable and flexible method for monitoring tool condition, it should be taken in consideration the following principal tasks:

− Fast detection of collisions, accidental contacts of the tool or the toolholder with the workpiece, or with components of the machine tool.

− Identification of tool breakage or fissure in cutting edges.

− Estimation or classification of tool wear caused by abrasion or other influences.

The most common method proposed for recognition of the tool state is the use of the cutting forces (Cook, 1980). This method has the great advantage of to be sufficiently documented and has been recognized as the most outstanding to detect tool state.

3. SENSORS

Traditionally sensors used for the supervision of tool conditions are divided in direct and indirect. The direct ones are not widely used in industrial applications of real time because of its high costs and difficulty of installation.

The indirect sensors are relatively economic and small to be used in monitoring tool state, including signals acquisition of cutting forces, temperature, machining, measurement of vibrations and acoustic emissions during the machining process. Given that these methods are indirects it is necessary to find the correlation between sensorial signals and the tool-wear states.

According to researches carried out in this field (Cho, et al., 1999; Kuo, and Cohen, 1999; Landers, and Galip, 2000), the signals with higher potential to be used in industrial applications are the cutting forces and the acoustic emissions.

Acoustic emission sensors have the advantage of to characterize with accuracy the tool wear during the cutting process. They are very sensitive signals and in high-frequency, however they can be contaminated easily by the noise. As the sampling frequency differs significantly, in some studies only ultrasonic sounds were kept for analysis. Although some authors suggest not to discard audible sounds because they can have important information.

Likewise other variables reflect the tool state indirectly, such as the power level of the spindle or displacement motor, where the power required is affected by the applied forces (Peres, 2000). Also the roughness of the piece is considered an approach for tool wear detection. The most usual manner of measuring the roughness is to use a surface tester or using an active sensor consistent of a luminous source and a camera for to measure the amount of light being reflected from the mechanized workpiece surface.

In the last years, the use of the sensorial fusion technique has increased. This technique measures different variables by means of several sensors for to find a manner of relating these measurements in the search of a common objective, in this case the tool wear detection. Systems with a single sensor could be used but with an appropriate number of statistically independent characteristics. Sensorial fusion of signals sources with bad quality would give worst results than the use of an unique sensor. Even more, if for this theme of research the cutting conditions have not been considered, because is possible that the sensors signals can be affected by the cutting parameters, for example the variations on feed rate (Silva, et al., 1998; Silva, et al., 2000).

4. SIGNALS PROCESSING

To reduce the influence of interferences such as noise and others not measurable on signals, have been used low pass filters, mainly. However some authors disagree of this method considering that important parts of information would get lost.

Another technique, although not very common, utilizes analytic models derived from physical models of the machining processes using the definition of cutting angles developed by Kienzle (Kienzle and Victor, 1957). With this technique the current values of cutting conditions are used as inputs to the model. These models calculate correction factors to describe the influence of cutting conditions on the magnitude of the force in their three directions and under the supposition that new tools are being utilized.

4.1. Characteristics extraction

The machining processes are very complex due to their non-linearity which makes difficult to divide the characteristics of these processes in categories. Characteristics describe the signals in a given time interval and they should be estimated in the elect domain. Next the most widespread division according to the available literature is presented.

- Signal's spectrum characteristics in the frequency domain can be estimated by using Fourier transform.

- Signal's characteristics in the time domain with the use of polynomials coefficients.

- Signal's statistical characteristics in the statistical domain, for example, with parameters of probability distribution.

- Using the SFS (Sequential Forward Search) algorithm. The variance of the characteristics is analized: intra-class variance within a characteristic and the inter-classes variance between different characteristics. In an optimum group of characteristic, the inter-classes variance must be maximized, while the intra-class variance must be minimize.

It is important to find which characteristics are independent of the cutting conditions since they contribute to the dynamics knowledge of the process. This will be useful at the moment of finding the tool wear dependence during the cutting process.

5. CLASSIFICATION METHODS

At the moment, sensors have been studied in depth as well as signal processing algorithms, and significant advances have been achieved within artificial learning. However, to be able to monitoring the tool state in an effective way, it is necessary to achieve artificial learning methods and decision make procedures with capacity of associating the selected variables with the tools conditions.

Nowadays, the classification methods for monitoring more used are divided in two categories: the weighty methods, that include patterns recognition, such as the Artificial Neural Networks (ANN), the Fuzzy Logic (FL) and the Genetic Algorithms (GA), and the decomposition methods such as Decision Tree (DT) and Systems Based on Knowledge (SBK).

As discussed above, the machining process is influenced by the cutting conditions, like example can be mentioned the workpiece and the tool geometry. ANN, FL, GA and a combination of ANN and FL have been used to model non-lineal dependences between the extracted characteristics from the measured signals and the cutting conditions for a determined tool (Achiche, et al., 2002; Balazinski, et al., 2002; Dimla, and Dimla, 1999; Lo, 2003; Zhou, et al., 2002). However the generalization capability achieved at the moment is not enough to get a supervisor adaptable to the necessities of the industry since its use is commonly restricted to a specific machine tool and for a small range of cutting conditions. This seems to be the main obstacle to achieve an efficient supervisor system.

Even if exist the consensus that ANN are indicated as the most appropriate method to obtain the expected results, to propose a system that uses ANN presents inconveniences to who develops the pattern such as to select inappropriate characteristics. This fact would increase the risk of network overtraining and somehow its capacity of generalization could be worse.

Usually an ANN only estimates a single output, there are few studies that propose several outputs, such as tool wear classification and chatter detection, estimation of width and length of the worn side, or estimation of width of the worn side and crater depth. At the moment, the best results have been achieved with Multilayer Perceptron networks being the most frequently used architectures, and specifically with the training algorithm backpropagation (BP).

Sick (Sick, 2002) made a meticulous revision of the researches were carried out along the last decade focused on monitoring of tool wear in turning processes. He points out the lack of validation of the results obtained from the experiments as well as the absence of tests with different patterns or statistics that allows the researchers to determine the used characteristics. He also highlights the lack of simulation experiments and independent sets of experiments for optimization tests. This author has a very particular point of view and stand out that the temporal information is only considered in very few publications. Bukkapatman & Kumara (Bukkapatman, et al., 2000) affirm that "The wear is continuous, and not a falling monotone phenomenon. Consequently, the exact abstraction of wear growth requires the past information, implying that the architecture of the neuronal network should have some internal memory, these requirements match with the Recurrent Neuronal Network (RNN)". Suggests that the experiments using dynamic neuronal networks applied to turning processes produces best results. The most promising methods within this context are the polynomial approach technique and networks with delay elements in feed forward direction.

Altintas & Yellowley (Altintas and Yellowley, 1987) proposed an algorithm that can identify the axial cutting depth and the radial width independently, based on two orthogonal cutting forces measurements in a perpendicular plane to the spindle axis of machine tool. This proposal is based on two components of the cutting forces (Fx and Fy) which constitute a lineal functions of the cutting depth. They compose a force index Q(t) that determines independent parameters of the cutting depth, equation (2).

$$Q(t) = Fx(t) / Fy(t) \qquad (2)$$

This research has been taken as a base for posterior researches to detect tool state in the milling processes.

In Elbestawi et al. (Elbestawi, et al., 1991) were proposed an approach for on-line monitoring flank wear, based on the magnitude of sensibility to individual cutting force harmonics, and how is this correlated with cutting edge flank wear. The forces acting on a dynamometer were synchronously gathered and averaged at time domain. The tool wear was determined averaging cutting edge wear on each tooth, measured with microscope. Later, the same team [12] modeled conventional milling including the effects of the cutting dynamics and tool wear.

Quan et al. (Quan, et al., 1998) presented a characterization of tool wear conditions, starting from obtained signals of an acoustic emission sensor and from a cutting force sensor utilizing multi-steps analysis, band frequency analysis, and machining parameters. These parameters are used as inputs in a neuronal network trained with an enhanced back propagation algorithm which returns the tool-wear estimation.

Wilkinson et al., (Wilkinson, et al., 1999) intends to predict tool wear by using acoustic emission (AE) and roughness. A clear correlation between the signals of the acoustic emission sensor and the tool wear flank, in continuous cutting processes, was found. In a previous study it was demonstrated that AE signal's intermediate frequency between the 0.1 - 1 MHz range decreases with increments of the flank wear for the end milling. The authors demonstrated the relationship between the space spectrum frequency measured from a machining surface and the tool wear during a end milling process.

6. NEW APPROACH FOR TOOL WEAR MONITORING

To develop a system for tool wear detection in real time is a complex project. To achieve this goal we propose to add from the cutting process some indirect variables which could be measured with sensors that are easy to install such as acoustic emission sensors. They could be embedded in the machine tool, in this way they would not interfere on its maneuverability, as can be appreciated in figure 1.

The tool wear estimation is proposed like one module of the multi-modular system. Several cooperative modules supply information to a Multi-objective Intelligent Supervisor, then it calculates new parameters and acts on the cutting process in order to obtain optimal cutting conditions according to objective functions, see figure 2.

Fig. 1. Acoustic emission sensors installed in the machine-tool.

The Tool State module carries out a estimation of the tool wear which is associated to the cutting parameters; due to their interrelation this module is directly connected to the Superficial Quality prediction module.

For the tool wear estimation module, the proposal is to integrate information acquired by acoustic emission sensors, analyzing the first 12 harmonic frequencies, once it has been proved that they contain the most outstanding information. Additionally, it will include some machining parameters such as feed rate (f), cutting depth (a), spindle speed (s) and some data from the Superficial Quality module, as the workpiece characteristics. For instance, we would incorporate characteristics such as hardness, piece geometry, cutting tool geometry, cutting tool diameter and material, and chip geometry. The Superficial Quality module would be able to predict the superficial roughness, which will be another input variable for the Tool State module. In fact, the Superficial Quality module would also account for the geometric modeling which contributes with dynamic information on the cutting process and therefore on the tool state at cutting time.

The novelty of this proposal is the intent of integrating different modules. The direct relationship between superficial quality and tool's life has been demonstrated by several authors. However, they have always been approached as independent problems. Currently, a focus on finding a common objective function valid for both modules is lacking in this field, regardless of the advantage of having well-known variables for each case.

The main advantage of this module is the use of frequency measurements as the main contributors of valid information. Since these measurements can be acquired in real time, and integrated into experimental methods and artificial intelligence techniques such as ANN, it is possible to think that they would render a valuable solution for the industry.

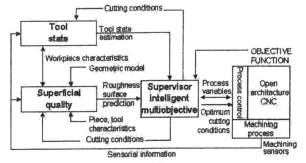

Fig. 2. Multi-modular supervisor system.

As was previously stated, the backpropagation algorithm is the one that has provided better results. In this case, the proposed module could incorporate new variants to the BP algorithm which, among other things, would add convergence speed such as the Quick BP and the Resilient BP. Likewise, an alternative to keep in mind to a backpropagated multilayered perceptron (recommended for real-time situations), is the CMAC neural network. In contrast, FL will serve as a base to create a neurofuzzy system that would allow classifying the tool wear in fuzzy sets: new, half-new, half-worn and worn. This classification was decided based on the fact that qualitative qualification is thoroughly accepted for most of the machine tool users.

Based on the information delivered by several modules, the Multi-objective Intelligent Supervisor, keeping in mind the objective functions, will decide about the optimal cutting conditions and it will adjust the appropriate parameters during the process.

During the first experimentation phase we have confirmed the clear incidence of the tool wear in the superficial finishing, this is reflected in the cutting force and in the own machining surface.

All cutting tests were performed on a HS1000 Kondia milling machine, equipped with a Sinumerik 840D open CNC. The workpiece material used for testing was the GGG40 quality steel with a profile of the dimensions 180 x 105 mm, with a maximum depth of cut of 2mm. A two-flutted Karnasch 30.6472 carbide end mill with 12 mm in diameter was used. These tests were performed at different cutting speeds and feed rates to determine the effect of new and worn tools in high speed roughing.

In the first test, the spindle speed was sp = 10.000 rpm and the cutting speed wasVc=800m/min. In the second test, the programmed spindle speed was sp=7000 rpm and the cutting speed was Vc=600 m/min. The results obtained are shown in Table 1, of these we can deduce that only with a medium worn tool the average roughness index (Ra) is affected largely, in some location critical increasing in more than 400%, this effect is evidenced observing the peak values where the increase was 800%.

Is evident the direct influence of the cutting speed in the quality surface with worn tool, if the cutting speed is smaller, the average roughness will be much lower.

Rq is standard deviation of Ra, and Rmax measure extreme values. Figures 3, 4.

Table 1. Comparison of sensorial signatures for different tool conditions.

	Vc=800 m/min		Vc= 600 m/min	
New tool	Ra(μ)	F(N)	Ra(μ)	F(N)
Mean Value	299.66	1.43	230.10	1.23
Peak	1560	1.7	927.6	1.5
Worn tool	Ra(μ)	F(N)	Ra(μ)	F(N)
Mean value	328.47	5.66	231.65	2.67
Peak	1003.4	12	927.6	5

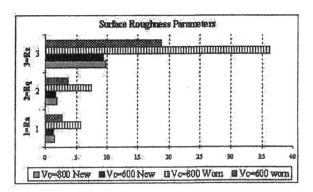

Fig. 3. Surface roughness parameters, incidence of cutting speed.

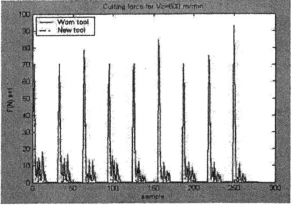

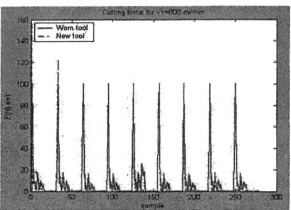

Fig. 4. Incidence of cutting speed in cutting force. (a) Vc= 600 m/min. (b) Vc=800 m/min.

7. CONCLUSIONS

Hot topics during the last years, which account for most of the research in milling processes, are sensors types used for supervision, and the development of techniques for modeling . These issues are decisive for designing a reliable supervision system. However, a reliable system that could be transferred to the industry has not been designed yet, although good results have been achieved during laboratory tests.

Even Fuzzy Logic is the Artificial Intelligence technique with better acceptance within the industrial field; it is the use of Artificial Neural Networks, together with Fuzzy Logic, which has been demonstrated to be the most effective method to achieve the best identification and classification of tool wear. Therefore, it allows us to be able to propose a new approach for monitoring the tool wear with suitable adaptations and with performance in real time.

The combination of the signal processing with soft-computing promises to offer additional information on the cutting process. Also, it allows achieving a better use of the measuring hardware (sensors, filters, converters, etc.).

It is a proven fact that the Backpropagation algorithm provides good results when modeling the milling processes (Correa, 2003), and in turn with models for tool wear prediction. This provides a good starting point to our proposal of monitoring the tool wear.

Based on previous studies, the proposal of a module to monitor tool state arises framed within a modular system, which is an innovative proposal to integrate aspects of tool state and superficial quality to a hierarchical control interacting with open architecture of Computer Numerical Control (CNC), the new generation of commercial numerical controls.

REFERENCES

Correa, M. (2003). Estudio de técnicas de inteligencia artificial aplicadas al modelado del proceso de fresado. *Trabajo Fin de Maestria, Universidad Politécnica de Madrid.*

Lo S.P. (2003). An adaptative-network based fuzzy inference system for prediction of workpiece surface roughness in end milling. *Journal of Materials Processing Technology.* 142, pp. 665-675.

Achiche S., Balazinki M., Baron L. and K. Jemielniak (2002). Tool wear monitoring using genetically-generated fuzzy knowledge bases. *Eng. Appl. of Artificial Intelligence.* 15, pp. 303-314.

Balazinski M., Czogala E., Jemielniak K. and J. Leski (2002). Tool condition monitoring using artificial intelligence methods. *Engineering Applications of Artificial Intelligence.* 15, pp. 73-80.

Sick B. (2002) On-line and Indirect tool wear monitoring in turning with artificial neural networks: a review of more than a decade of research. *Mechanical Systems and Signal Processing.* 16(4), pp. 487-546.

Zhou Y., Li S. and R. Jin (2002). A new fuzzy neural network with fast learning algorithm and guaranteed stability for manufacturing process control. *Fuzzy Sets and System.* 132, pp. 201-216.

Bukkapatman S.T.S., Kumara S.R.T. and Lakhtakia (2000). Fractal estimarion of flank wear in turning. *Journal of Dinamic Systems, Measurement, and Control, Transaction of the ASME.* 122, pp. 89-94.

Landers R. and A. Galip (2000). Model-based machining force control. *ASME Journal of Dynamics Systems, Measurement, and Control.* 122(3), pp. 521-527.

Peres C. (2000). Supervisión inteligente del proceso de mecanizado en una máquina-herramienta. *Tesis doctoral, facultad de Informática UPM.*

Silva R.G., Baker K.J. and S.J. Wilcox (2000). The adaptability of tool wear monitoring system under changing cutting conditions. *Mechanical Systems and Signal Processing.* 14(2), pp. 287-298.

Cho D.W., Lee S.J. and C.N. Chu (1999). The state of machining process monitoring research in Korea. *International Journal of Machine Tools & Manufacture.* 39, pp. 1697-1715.

Dimla E. and Snr. Dimla (1999). Application of perceptron neural networks to tool-state classification in a metal-turning operation. *Engineering Applications of Artificial Intelligence.* 12, pp. 471-477.

Kuo R.J. and P.H. Cohen (1999). Multi-sensor integration for on-line tool wear estimation through radial basis function networks and fuzzy neural network. *Neural Networks.* 12, pp. 355-370.

Wilkinson P., Reuben R.L, Carolan T.A. and J.D.C. Jones (1999). Tool wear prediction from acoustic emission and surface characteristics via an artificial neural network. *Mechanical Systems and Signal Processing.* 13(6), pp. 955-966.

Quan Y., Zhou M. and Z. Luo (1998). On-line robust identification of tool wear via multi-sensor neural network fusion. *Engineering Applications of Artificial Intelligence.* 11, pp. 717-722.

Silva R.G., Reuben R.L., Baker K.J., and S.J. Wilcox (1998). Tool wear monitoring of turning operation by neural network and expert system classification of a feature set generated from multiple sensors. *Mechanical Systems and Signal Processing.* 12(2), pp. 319 – 332.

Hicks, T.G. (1997). *Handbook of Mechanical Engineering Calculations.* McGraw-Hill Professional. 1 edition October 1, 1997.

Ismail F., Elbestawi M., Du R. and K. Urbasik (1993). Generation of milled surfaces including tool dynamics and wear. *Transaction of ASME, journal of engineering for industry.* 115, pp. 245-252.

Elbestawi M.A., Papazafiriou T.A. and R.X. Du (1991). In-process monitoring of tool wear in milling using cutting force signature. *Int. J. Mach. Tools Manufact.* 31(1), pp. 55-73.

Altintas Y. and I. Yellowley (1987). The identification of radial width and axial depth of cut in peripheral milling. *Int. J. Mach. Tools Manufact.* 27(3), pp. 367-381.

Cook, NH. (1980). Tool Wear Sensors. *Wear.* 62, pp. 49–57.

Kienzle, O. and H. Victor (1957). Spezifische schnitkrafte beider metallbearbeiting. *Werkstoff technik und maschinenbau.* 47(5), pp. 224-225.

ELSEVIER
IFAC
PUBLICATIONS
www.elsevier.com/locate/ifac

REDUCING POTENTIAL RISKS BY PREVENTING EVENTS: A CASE STUDY

Marlon Núñez, Raúl Fidalgo and Rafael Morales

Department of Languages and Computer Sciences
mnunez@lcc.uma.es, rfm@lcc.uma.es, morales@lcc.uma.es

Abstract: The performance of execution tasks may be improved by incorporating problem prediction capabilities to autonomous systems with the purpose of reducing potential risks. Learning from past failure experiences might help autonomous systems to predict adverse conditions or failures in advance and take preparatory actions. A new architecture level is emerging, the preventive level, which benefit from recent advances in temporal data mining methods for event prediction. This paper summarizes the main functions of this level and presents a testing procedure for an autonomous mobile robot in a simulated adverse environment.

Keywords: machine learning, autonomous mobile robotics, preventive behavior, simulators, navigation, chaos.

1. INTRODUCTION

The quality of execution tasks may be improved by providing autonomous systems with event prediction facilities. An autonomous system could avoid problems or prepare some preventive or preparatory actions to face inevitable problems.

An example of the need of such a preventive behavior, which also serves as a motivation, is the following reflection from the New Millennium Space Program (Bernard and Pell, 1997) with regards to the use of current planning architectures for space probes: "If the primary engine breaks, the system may only be able to switch to the backup engine if it has been warmed up. Future remote agents should have the capability to anticipate such possible failures, or even opportunities, and to then build plans that provide the necessary resources so the system is prepared for many possible futures."

The area of prediction for autonomous systems has received relatively little attention to date. The type of preventive level is missing from current-generation agent architectures, with the result that many agent systems behave in short-sighted way.

This paper is organized as follows: Section 2 presents the preventive level and its main functions. Section 3 explains the BPL event prediction method. Section 4 presents the testing procedure of this level architecture in the field of virtual robotics. Section 5 presents the conclusions.

2. PREVENTIVE LEVEL

In addition to the three-level architecture developed in many autonomous systems (Pell, *et al.*, 1998; Washington, *et. al.*, 1999) consisting on the deliberative, executive, and reactive levels, the new preventive level is needed to learn from past experiences trying to discover failure temporal patterns. This might help autonomous systems to predict failures in advance and take preparatory actions.

The preventive level might offer the following services:

1. Creation and updating of an Event Prediction Model.
2. Generation of short- and mid-term predictions, and,
3. Responding to model consultations given hypothetical situations.

---
[1] This work was partially supported by CICYT project MOISES TIC2002-04019-C03, Spain.

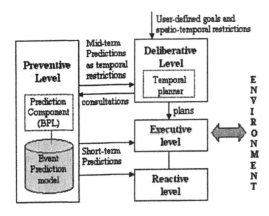

Fig. 1. Preventive four-level architecture.

Figure 1 illustrates the four architecture levels, including the proposed level. A preliminary view of this architecture was introduced in (Núñez, *et al.*, 2001). The prediction component uses the BPL method (Núñez, 2000; Núñez, *et al.*, 2002) to receive the current status of the system and its environment, translate it into events (i.e., faults events, warnings), and discover recurrent behavior patterns. Periodically, BPL generates temporal rules for the modeling of recurrent behaviors (if any).

Predictions consist of: an event type (e.g. Alarm= "Battery<5%"), the most probable future temporal interval of occurrence of the event, and the confidence of the prediction.

3. LEARNING FAILURE PATTERNS FROM PAST EXPERIENCES

Learning temporal patterns from past experiences is the key factor for problem prediction. As far as we know, the only method that predicts future temporal intervals of occurrence of problem events (alarms, failures) is the BPL learning method. The BPL method discovers temporal patterns taking into account observed events, static attributes of the observed systems and their environments. BPL considers a continuous time model. An event may happen at any arbitrary point of time. Temporal patterns are used to predict target events.

This method consists in three components (see figure 2). The first one, the summarizer, consists in analyzing input events from a situation for building summaries of that situation in specific times. Each summary, called a behavior summary, has a numeric class that is the temporal distance from the moment of the summary to the occurrence of a target event. The second component, the learner, takes the behavior summaries of all situations as training examples for the construction of a regression tree for each target event to be predicted and translates it into a set of prediction rules.

Every rule shows the system and environmental conditions and predicts the expected time interval of the next occurrence of a target event. A rule may also predict that a target event may not occur. For instance, a rule may express that if an event does not

occur, then a target event will probably not happen in an interval time.

The third component, the predictor, takes unlabeled behavior summaries and performs predictions.

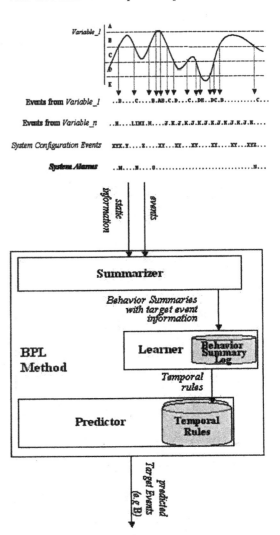

Fig. 2. Illustration of the processes involved in pattern discovery for event prediction.

3.1 Construction of behavior summaries

In order to construct behavior summaries the Summarizer has to analyze the possible consequences of every event (observed or abstract) associated with the problem, thus the following natural assumptions are made: an event could be a consequence of some temporal correlation of past events, static characteristics of the observed system, and static characteristics of the environment; on the other hand, an event could have an effect in the future for a limited period of time, called a Latency Window.

Because of the above, two systems in the same state situated in two different environmental conditions behave differently. The event history of a system is not enough to predict future events. For this reason, this process constructs summaries with static and dynamic attributes of the system and its environment (see figure 2), as well as temporal correlation predicates: repetitions of events, duration of current states, and oldness of past events.

Events are expressed as *[situationID, eventGenerator, attributeID, newValue, timeOfOccurrence]*. *SituationID* is the identifier of an observed situation. A situation is made up of a *system* and its *environment*. The *eventGenerator* is the source of the event, which may be the system or its environment. *AttributeID* is the identifier of a dynamic attribute, whose value has changed to *new Value* (observed or abstract) at the specified *time of occurrence*. An example of an event is: *[4, system, activity, SatDown, 12:20:05]*, which indicates that in situation 4, the *system*, say dog Rin-tin-tin, has a dynamic attribute called *activity* that changed to the value *SatDown* at time 12:20:05. Another example: *[4, environment, temperature, 12-15, 12:33:30]* which indicates that in the *environment* of situation 4, the temperature changed to the range 12-15 at time 12:33:30.

An event schedules the construction of several behavior summaries in the future during its latency window. Since our time model is continuous, if a behavior summary is intended to be scheduled very near in time to another existing behavior summary, then it is not scheduled. Figure 3 illustrates the way BPL monitors the consequences of *eventX*.

Note that *eventX* schedules four behavior summaries during its latency window. The latency window is the period of time during which *eventX* is supposed to affect the future. Figure 3 also shows that target *eventY* occurs within the latency window of *eventX*. The behavior summary BS_1, will register the temporal distance from its construction time, t_1, to the time, t_y, of the next target *eventY*, which is $t_y - t_1$. The temporal distances of BS_2 and BS_3 will be: $t_y - t_2$ and $t_y - t_3$, respectively. On the other hand, BS_4 will be labeled depending on the next occurrence of target *eventY*. If the next target *eventY* occurs very far in the future, BS_4 will have as a label a special symbol, called *itDidNotOccur*, registered in the behavior summary as a negative constant.

Once an event arrives several actions are performed on behavior summaries. BPL may update the values of some behavior summaries or set the value of the class on others. The numeric class is filled in later on when the specific target events arrive. Note that BPL does not have to construct a behavior summary periodically, but only when an event arrives.

A behavior summary summarizes a situation at a specific instant. Each behavior summary is constructed as a learning example to predict a specific target event. It has a numeric class, which is the temporal distance from the construction of the behavior summary to the next occurrence of that target event. Each behavior summary contains temporal information about past events and current values of the system and environmental attributes. This temporal information is described in terms of new temporal features for each type of event during its latency window such as: the number of *repetitions* of events of that type; the *duration*, which is the

temporal distance from the last event of that type whose value is valid at construction time; and *oldness* which is the temporal distance from the last event of that type. Other new features may be included, such as the duration of past distances, or distance from past repetitions, or binary information about a specific order of events; however, we consider *repetition*, *duration*, and *oldness* as fundamental.

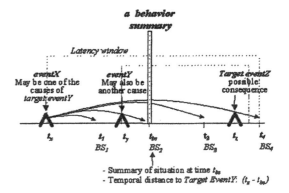

Fig. 3. Monitoring the consequences of an event

A behavior summary is characterized by the situation, the target event to learn about and the time of construction of the *bs*. Briefly speaking, the fields of a behavior summary are:
- Static Attributes, which are attributes inherent to the analyzed objects,
- Dynamic attributes, which indicate variable states at the time of construction of the behavior summary.
- Oldness of past events, which is the distance from past events.
- Repetition of events, which is the frequency of past events during a latency window,
- Duration of current value, which is the distance to the event that led to the current state,
- Numeric class of the BS, which is the temporal distance from construction time of the *bs* to the next occurrence of the target event.

3.2 Regression and temporal refinement.

The Learner takes the behavior summaries as examples for growing regression trees and then translates these into temporal patterns (see figure 2). A behavior tree is a set of temporal patterns used to predict time intervals in a continuous time form.

Before constructing the regression tree, an intermediate process is applied that deals with chaotic behaviors. Any kind of regression tree method may be applied, being CART (Breiman, *et al.*, 1984) the most known. We used a simplified version of the EGR regression algorithm (Núñez, 2000b) for this task. The complete version of this algorithm uses "*is-a*" taxonomies and *costs* associated with each attribute, grows regression trees that are shorter and more general than CART method (Núñez, 2000b).

EGR, as any regression tree algorithm, guarantees the mutual exclusion between its leaves (rules). That is, the same conditions will lead to, at most, only one prediction.

The last step is to refine regression trees. This consists in adding the latency window data used during the summarization process to the calculated attributes. The resulting trees, called behavior trees, are easily understandable by users. All the leaves of a behavior tree allow us to obtain the predictions for the same target event. We may translate behavior trees to rules for clarity reasons. As an example, a BPL rule may be:

IF *Repetition*(dog.activity=jump)= [9, 22) *inLast* 3
    minutes, and *Oldness* (dog.IngestedFood=water)
    = [15, 26) minutes,
THEN (dog.IngestedFood=water) *withinNext* [50,
    83) minutes. *Confidence* 78%

This temporal pattern says that: "if dog jumped [9..22) times in the last 3 minutes, and the time passed since the dog drank water was 15 to 26 minutes, then the dog will probably drink water within the next 50 to 83 minutes. Confidence of the pattern: 78%". Another example is the following:

IF Duration(dog.activity=walking)=[63, 86)
    minutes,
THEN (dog.IngestedFood=water) *inNext* [0, 25)
    minutes. Confidence 85%

This temporal pattern says "if the dog has been walking for 63 to 86 minutes, then the dog will probably drink water in the next 0 to 25 minutes. The confidence of the rule is 85%".

Figure 4 shows a behavior tree that predicts a service degradation event in computer.

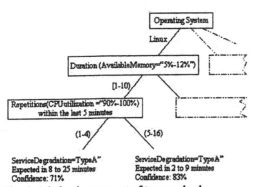

Fig. 4. A behavior tree: set of temporal rules.

3.3   Prediction of target events

Once the behavior trees are built, the unlabeled behavior summaries are sent to the Predictor to predict their corresponding class, that is, to predict target events.

If the values of the attributes of an unlabeled behavior summary meet the set of conditions of a behavior rule, then the prediction of the rule is fired. The previous prediction, if any, regarding that target event is no longer valid. If the attributes do not meet

any rule condition of a behavior tree, the previous prediction is simply deleted, and therefore no prediction regarding the corresponding target event is made. Because of the use of a regression tree method (Breiman *et al*, 1984; Núñez, 2000b), at most one rule within the same behavior tree may be fired.

Every behavior tree updates the prediction of its target event. That means that several predictions, associated to different target events, may be fired. If this happens, no prediction is omitted since every one of them is possible in continuous time.

BPL has been used for predicting communications events (Núñez, *et al*, 2002), and solar events (Núñez, *et al*. 2004).

4. TESTING PLATFORM

Testing a four-level architecture is a very difficult task. For that reason we have experimented on a simulated and reduced problem. We simulated an outdoor mobile robot and its environment that consists of a network of roads with slopes, obstacles and two terrain conditions (wet and dry).

The environment also has adverse conditions: a wet terrain, which is the consequence of a simulated weather environment. A wet terrain may appear according to a non-linear function using the Duffing differential equations system. The robot may go to a *shelter* to protect from rain. The autonomous robot has to discover patterns in the rainy behavior for predicting the specific rain conditions (target events), which will affect its future navigation.

4.1 Project objectives

The goal of the robot is to go to a certain place in the road network. The robot may fail if it consumes all its battery.

A muddy road segment increments the cost of the navigation (battery consumption, time to pass it). If the muddy segment has a high slope, the robot might not be able to climb it.

In order to improve its performance, the robot needs to learn:

1. The robot-environment dynamics.
2. To predict continuously when the terrain will become muddy (because of the rain).

The reactive level has to learn the robot-environment dynamics. The preventive level has to learn to predict the rain and therefore when the terrain becomes muddy.

Based on experimentation, the robot has to learn by itself strategic decisions, like going to a shelter or changing the path to the goal.

The preventive level is based on the BPL event prediction method. The reactive and executive levels are based on a specific reactive planner (Terrón, 2004) that is integrated to an incremental decision tree learner (Núñez, and Fidalgo, 2004). In order to face this problem, we have built the Koala simulator of the robot and its environment, and a training system. A specific training methodology has been specially conceived to make the system acquire all possible reactive and preventive navigation knowledge.

4.2. The Koala simulator

A simulator, called the Koala Simulator (Carrillo, 2003), was specifically developed for this project. This tool simulated a real robot, called KOALA from K-Team (Switzerland), in a virtual outdoor environment. We simulated each of its internal variables: sonars, infrared sensors, battery discharge, direct current through battery, and the engine and its electro-physical behavior. We have added two sensors, which are not present in the commercial robot: a rain detector and a detector of muddy terrain. Figure 5 shows the graphical user interface of the Koala simulator. This interface shows the physical effect (velocity, position, sonar and infrared observations) as well as the battery discharge when it navigates in the environment.

4.3. Training system

The training system (Martos, 2003) was developed specifically for this project with the purpose of forcing the robot Koala to experience all possible types of obstacle and navigations problems in the virtual environment presented by the simulator.

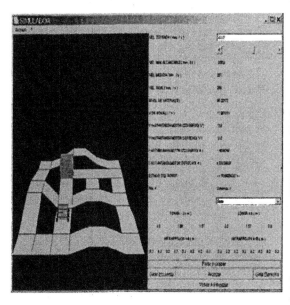

Fig. 5. Graphical user interface of the Simulator tool.

The training system forces the robot to experience a large number of possible situations by forcing the initial conditions of the robot and a road segment. In figure 6, the he Training System has created a road segment with mud and a slope of 30 degrees. In general, the training system creates specific characteristics and configures the conditions of each possible situation: terrain type (dry and muddy) slope (-90 to 90 degrees). For each condition, the training system also puts the initial battery level so that the robot could know the consequences of having no energy.

In the case of situations that require special maneuvers, the training system allows an operator to carry out this careful action, so every navigation action may be learnt by the reactive level, based on the OnlineTree learning method (Núñez et al, 2004). Another important purpose of the training system is to evaluate the control behavior of a robot that works in an autonomous way, counting cases of success, failure, time and consumed battery, and generating statistical data.

We have developed all applications in Java language. All developed components have been tested using 2.0 GHz PCs with 512 Mbytes of memory.

4.4 Results

Reactive behavior: The reactive control component of the virtual Koala robot learned incrementally: obstacle avoidance; knowing which slopes it may climb and in which conditions; and knowing speed conditions for turning right/left around muddy/dry corners. In (Terrón, 2004) specific details on reactive results are shown

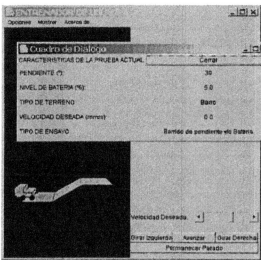

Fig. 6 Graphical user interface of the training system. Training case: slope of 30 degrees and muddy terrain.

Preventive behavior: This level has been developed to predict the mud in the terrain according to a non-linear function using the Duffing differential equations system.

As we have explained before, the wet terrain appears according to a non-linear function using the Duffing differential equations system (Moon, and Holmes, 1979). In this way we simulate the effects of rain. We empirically validated the BPL prediction method by learning to predict the target events: "Intense rain" and "moderate rain".

The non-linear behavior of the rain was taken from another and very different system: The Double Well Oscillator. This oscillator is a system in which there is a mass that vibrates on a filament. The displacement $x(t)$ of the mass is modeled mathematically by a differential equation, called the

Duffing equation. If its parameters are configured appropriately, including a force $F(t) = F_0 \cos(wt)$ exerted on the mass, we have the differential equation $x''+x'-x+x3 = F_0\cos(t)$. For some values of F_0 the system shows several kinds of behaviors, from periodic and nonlinear to chaotic behavior. In our experiments we configured the differential equation to produce a chaotic behavior (Hilborn 2000). In chaotic behavior only short-term predictions are possible.

Then, we assume that x was the rain intensity using the Duffing equations. The upper band of x was the "intense rain", and the mid band was the "moderate rain". We make BPL discover a model of dynamic behavior for predicting these events.

For measuring the predictions error we used Relative Mean Square Error (RMSE), a typical measurement used in regression. It is relative because it is the ratio between the error of the analyzed predictor and the error of a naive media predictor. The RMSE error of a predictor is measured by calculating the sum of square distances between the analyzed predictor and the real value. If the RMSE error is greater than 1, it means that the analyzed predictor is worse than the naive media predictor. Permitted RMSE errors obtained by regression algorithms for typical problems fall bellow 0.3.

BPL was run in batch mode every time to simulate an incremental fashion. The RMSE error of BPL for predicting upper band was 0.062, which is a satisfactory result. The RMSE error for predicting the mid band was 0.387, which is not a good result. That is, predicting "intense rain" was easier than predicting "moderate rain". The preventive level learned satisfactorily short-term predictions related to "intense rain" in its virtual environment.

An important goal, not tested yet, is the translation of BPL rules to TILCO temporal logics (Bellini, et al., 2000). When a problem is predicted BPL sends the translated rules to the planner, which executes TILCO rules. The planner will take into account all temporal restrictions thrown by BPL for generating new temporal plans.

5. CONCLUSIONS

A new architecture level called the preventive level is presented. The main roles of this level are: the creation and updating of an event prediction model; the generation of real-time predictions, and response to model consultations given a hypothetical situation. The construction of the event prediction model is based on the discovery or previous recurrent behaviors.

The BPL method of constructing event prediction models has proved to be a valuable tool for predicting short-term as well as mid-term target events when patterns exist. Testing this kind of architecture is a difficult task. We presented a simulation and testing platform for making an autonomous system learn reactive behavior. This platform is also used to learn to predict failure or adverse events, in our case intense rainy conditions.

The long-term challenge is that autonomous systems may discover intelligent behaviors by themselves. Otherwise, autonomous systems will not be fully operative in problems with frequent adverse conditions.

6. REFERENCES

Bellini P., Mattolini R. Nesi P. (2000) Temporal Logics for real-time system specification, *ACM Computing Surveys*, Vol.31.

Bernard D. and Pell B. (1997) Designed for Autonomy: Remote Agent for the New Millennium Program. In: *Proc. of the Fourth Intl. Symposium on Artificial Intelligence, Robotics, and Automation for Space.*

Breiman L., Friedman J., Olsen R., and Stone C (1984), *Classification and Regression Trees*, Wadsworth Group.

Carrillo M. (2003), *Simulator of a mobile outdoor robot situated in a 3D environment*, final dissertation. Industrial Eng. U. de Málaga.

Hilborn R. (2000). *Chaos and Nonlinear Dynamics*. Oxford University Press.

Martos J. (2003). *Trainer of a mobile outdoor robot situated in a 3D environment*, final dissertation. Industrial Engineering, Universidad de Málaga.

Moon, F. C., & Holmes, P. J. (1979). A magnetoelsastic strange attractor. *Journal of Sound Vibration*, 65, 285.

Núñez M. (2000a). Learning Patterns of Behaviour by Observing System Events, in: *Lecture Notes in Artificial Intelligence*, Vol. 1810, Machine Learning: ECML 2000, Springer-Verlag.

Núñez, M. (2000b). Generalised Regression Trees, In: *Proceedings of the 14th International Conference of Statistical Computing (Compstat)*, Physica-Verlag, 367-372, The Netherlands.

Núñez, M., Morales R., Triguero F., & Figuero. R. (2001). Event Prediction: Towards Preventive Planning." Proc. of On-Board Autonomy-2001, European Space Agency Press. The Netherlands.

Núñez M., Morales R. and Triguero F. (2002) Automatic Discovery of Rules for Predicting Network Management Events. In: *IEEE Journal on Selected Areas in Communications*, 20, 4.

Núñez M. and Fidalgo R. (2003), *Incremental construction of decision trees*, University of Malaga, Technical Paper 03-04, Málaga.

Núñez M., Fidalgo R. and Morales R. (2004) Learning patterns from events and other multivariate data. In *Book of Abstracts of Euro Electro Magnetics 2004*, Magdeburg (Germany).

Pell B., Bernard D., Chien S., Gat E., Muscettola N., Nayak P, and Williams B. (1998) An Autonomous Spacecraft Agent Prototype, *Autonomous Robotics*, 5, Kluwer Academic.

Terrón S. (2004). *Reactive planning architecture and its integration with learning systems*, final dissertation, Computer Sciences, U. de Málaga.

Washington R., Golden K., and Bresina J., Smith D., Anderson C., and Smith T. (1999) Autonomous Rovers for Mars Exploration, *IEEE Aerospace Conference*

ELSEVIER
IFAC
PUBLICATIONS
www.elsevier.com/locate/ifac

MODELING THE CONTROL OF A DISCRETE EVENT MANUFACTURING CELL

Eduardo A P Santos and Marco A B de Paula

*Pontifical Catholic University of Paraná, Graduate Program on Production and Systems Engineering. Curitiba,
80215 901, Brazil*
eduardo.portela, marco.busetti@pucpr.br

Abstract: The following paper presents a methodology for modelling the control of a manufacturing systems class. One uses the Supervisory Control Theory as a control system synthesis tool. One proposes then a set of models in subsystems automata and in operational specifications which one wants to impose, according to the manufacturing system configuration. These models will be used in the plant controllers synthesis process.

Keywords: automation, manufacturing systems, discrete event systems, automata, supervision.

1. INTRODUCTION

The manufacturing systems considered throughout the present work are used for the pieces manufacturing, requiring multiple operations for processing and/or assembling. Each operation is performed within a station, and the station is physically linked to the following station through a pieces transportation system (Groover, 2001) (Askin & Standridge, 1993).

Such systems main control goal is their own subsystems co-ordination as long as a series of individual tasks are accomplished, assuring the system correct performance. Various approaches have been used in the manufacturing systems control design, one points out the Supervisory Control Theory (SCT) (Ramadge & Wonham, 1989). The named approach allows the controllers automatic synthesis from the system dynamics modeling in open-loop (plant) and in the designed behavior specification by controllable languages and automata.

Considering the automated manufacturing systems design involves a large quantity of operations specifications related to the subsystems, the SCT presents a limitation factor - the exponential growth of the number of states relating to the number of the system constituting elements. In previous works, Queiroz and Cury (2000a,b) proposed dealing with the limitation factor by exploring, besides the specifications modularity (Wonham & Ramadge, 1988), the natural plant modularity in large-sized systems. Instead of building a sole monolithic controller for the whole plant, in the proposed modular approach one seeks to build, as long as possible, a local controller for each specification, modeling it only according to the subsystems affected by its action, one intends that the resultant supervisors are locally modular, that is, the

supervisors set action generates a behavior in the closed-loop identical to the monolithic supervisor's Queiroz and Cury (2000a,b). The modular control approach is quite worthwhile in the sense of promoting greater flexibility, greater computational efficiency and safety in the control's applicability.

The task of building proper automata representing each subsystem behavior (without control action) and each specification is not simple and requires some extent of experience regarding manufacturing systems modeling. Cassandras and Lafortune (1999) propose techniques in order to represent in automata the specifications in natural language common to the practice. Although to a certain extent the classification is generalizing the specifications classes, the problem of building or adapting them to the herein approached systems still exists. Thus, despite of the SCT advantage being placed on the controllers automatic synthesis, the building of the models for the plant and the specifications might depend on the experience and the designer's inspiration.

One will demonstrate throughout the paper that the subsystems and the operational specifications modeling can change according to the system configuration, causing an extra difficulty in the manufacturing systems task.

2. MODELING AND CONTROL OF DISCRETE EVENT MANUFACTURING SYSTEMS

Discrete event systems (DES) are dynamic systems that evolve in accordance with the abrupt occurrence of events. They are generally asynchronous (not clock driven) and non-deterministic (some events may occur spontaneously). For example, an event may be the arrival of a customer in a queue, completion of a task or failure of a machine in a manufacturing system, transmission of a message in a communication network, termination of a computer program, variation of a set point in a control system, etc. Such systems are encountered in a variety of fields, for example in manufacturing, robotics, computer and communication networks, traffic, and logistics (Kumar & Garg, 1995).

The Supervisory Control Theory introduced by Ramadge and Wonham (1989) is a general approach to the synthesis of control systems for DES. Given a DES describing the uncontrolled behavior, the plant, and a specification for the controlled behavior, a supervisor can be automatically synthesized to control the plant to stay within the specification. Ramadge and Wonham´s theory of supervisory control uses formal languages and automata to model both the uncontrolled DES and the specification for the controlled behavior. In their approach, a DES execution is modeled as a sequence of events. The set of such events forms a language and represents all the possible executions of the system. The basic problem of supervisory control is to modify the open-loop behavior of a DES by eliminating sequences of events from the system behavior. The objective is to restrict the behavior of the system so that is contained in a desired behavior, called specification. This is achieved by constraining the discrete event generator to execute events only in strict synchronization with another system, called supervisor.

2.1. Modeling of a discrete event system

In the framework of Ramadge and Wonham (1989), the system spontaneously generate discrete events $\sigma \in \Sigma$ classified as controllable ($\sigma \in \Sigma_c$), when the event can be disabled by the control system, or uncontrollable ($\sigma \in \Sigma_u$). Let Σ^* be the set of all finite chains of elements in Σ, including the empty chain ε. A language is a subset of Σ^*. The behaviour of a DES may be modelled by languages that, when regular, are represented by automata. An automaton is a quintuple $G = (\Sigma, Q, \delta, q_0, Q_m)$, where Σ is the event set, Q is a set of states, $q_0 \in Q$ is the initial state, $Q_m \subseteq Q$ is the subset of marked states and $\delta: \Sigma \times Q \to Q$, the transition function, is a partial function defined in each state of Q for a subset of Σ. Then, G is characterized by two languages: its closed behaviour $L(G)$ and its marked behaviour $L_m(G)$. The prefix closure of a language L is denoted by $\overline{L}$. Automata are illustrated by state transition diagrams that are directed graphs where nodes represent states and labeled arcs represent events. Marked state nodes are double lined and the initial state is pointed by an arrow. Intercepted arcs indicate controllable events.

For purpose of control logic design, in the modeling process internal procedures of DES can be abstracted to discrete events that allow for a consistent expression of specifications and of the open loop system behavior. Taking into account the decentralized structure of the system, the plant is then represented as a set of completely asynchronous subsystems, named product systems (Ramadge & Wonham, 1989).

Let a product system be composed of subsystems $G_i = (\Sigma_i, Q_i, \delta_i, q_{0i}, Q_{mi})$, $i \in I = \{1,...,n\}$, and, for $j=1,...,m$, let generic specifications $E_{gen,j}$ be languages respectively defined in $\Sigma_{gen,j} \subseteq \Sigma$. For $j=1,...,m$, the local plant $G_{loc,j}$ related to $E_{gen,j}$ is defined by

$$G_{loc,j} = \underset{i \in I_{loc,j}}{\|} G_i, \text{ with}$$

$$I_{loc,j} = \{k \in I \mid \Sigma_k \cap \Sigma_{gen,j} \neq \varnothing\},$$

where the operator $\|$ represents synchronous composition (Ramadge & Wonham, 1989) . Thus, the local plant $G_{loc,j}$ is composed only by original subsystems directly restricted by $E_{gen,j}$. For $j=1,...,m$, the local specification $E_{loc,j}$ is defined as $E_{loc,j} = E_{gen,j} \| L_m(G_{loc,j})$ (Queiroz & Cury, 2000a,b).

2.2. Local modular control approach

The work of Queiroz and Cury (2000a,b) extends the framework of Ramadge and Wonham (1989) in order to avoid state-space explosion by exploiting modularity of specifications and decentralization of plant. By this local modular approach, the plant is represented as a set of asynchronous automata and the specifications are expressed locally over the affected sub-plants. Optimal supervisors are computed from the local specifications and are finally reduced by a minimization algorithm.

In this paper, a supervisor is represented by an automaton S, whose state changes are dictated by the occurrence of events in the plant G. The control action of S, defined for each one of its states, is to disable in G events that may not occur is S after an observed chain of events. A task of the closed loop system is considered complete only when it is marked by the plant and by the supervisor. A supervisor S is non-blocking if $\overline{L_m(S/G)} = L(S/G)$. The necessary and sufficient condition for the existence of a non-blocking supervisor S that reaches a given specification $K \subseteq Lm(G)$ $(Lm(S/G) = K)$ is the controllability of K (Ramadge & Wonham, 1989). K is said to be controllable (with relation to G) if $\overline{K}\Sigma_u \cap L(G) \subseteq \overline{K}$. The class of controllable languages contained in K has a supreme element called $SupC$ (K, G). When the control problem comprises multiple specifications we can design a monolithic supervisor for the entire set of specifications or a modular supervisor for each specification. By the local modular approach, for $j=1,...,m$, non-blocking supervisors $S_{loc,j}$ are computed directly on the respective local specifications $E_{loc,j}$ so that $L_m(S_{loc,j}/G_{loc,j}) = SupC$ $(E_{loc,j}, G_{loc,j})$. The local modularity of the set of supervisors is verified if:

$$\underset{j=1}{\overset{m}{\|}} \overline{L_m(S_{loc,j}/G_{loc,j})} = \overline{\underset{j=1}{\overset{m}{\|}} L_m(S_{loc,j}/G_{loc,j})}$$

This condition, according to Queiroz and Cury (2000a,b), assures that the modular approach doesn't cause any loss of performance relating to the monolithic approach. The results presented by Queiroz and Cury (2000a,b) show that the locally modular synthesis induces a natural non-centralized structure for supervisors. In fact, each local supervisor only needs to exchange information with its corresponding local plant. In the case of changes in the plant or in the specifications, respected the non-blocking condition, the control modules can be redesigned, based only on local information. A distributed control system with more flexibility and higher computer simplicity is obtained.

3. MODELING THE MANUFACTURING SYSTEMS

The SCT appliance as well as the local modular synthesis assume the obtainment of proper models for the subsystems and for the specifications one is trying to impose. Regarding the manufacturing systems class approached by the present research, the cells basic operation is represented by figure 1. According to the figure, the pieces inputting the system pass through multiple processes until the designed final product. A specific working station fulfills each process, and each one is linked to the next through a manipulation or transportation system.

In modeling terms, one can initially consider each station relating to a plant's subsystem. Thus, one must associate to each station a specific model in automaton representing its behavior without control action. It is still necessary to define which states need to be observed within the model. The specifications are built from the global system configuration and from the various subsystems requirements one wants to impose. Regarding the approached systems, a fundamental specification is the existence of a correct piece flow in the system and that all the operations are performed in a coordinated manner. One will demonstrate such basic specification can be represented in various forms, depending on the adopted configuration (or the type of equipment used) in the manufacturing plant.

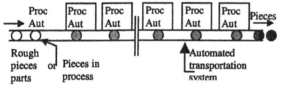

Fig. 1. Typical configuration of an automated transference line.

3.1. Modelling the subsystems

According to SCT, the plant modelling to be controlled can be obtained as follows:
1. Identify the subsystems or equipment set involved in the system to be controlled;
2. Build the basic model as a DFSA (Deterministic Finite State Automaton) Gi, for each equipment i involved in the system in the most synthetic manner;
3. Calculate the most refined Representation per Product System (RPS), making the composition of the subsystems synchronous;

4. Define the control structure Γ, through the identification of controllable and non-controllable sets of events Σ_c e Σ_u, affecting the system.

In many cases this phase consists of a relatively simple task, since the plant spatial configuration itself allows the modeller in selecting the various existing subsystems. For instance, in a transference line as shown by figure 1, it is possible to associate to each workstation an automaton Gi representing its basic behaviour. In other words, the system physical modularity to be controlled will be represented by the set of automata Gi correspondent to each workstation.

Attention must be paid to identify each state of equipments. One must seek the compatibility between the co-ordination functions expected by the control system and the correct states identification for the various equipment composing the plant. In general, one can identify the following states in equipment constituting manufacturing systems:
• Inactive state (generally initial state);
• Active states or operating (eventually an equipment can have different operational states);
• Breakage or failure state;
• Intermediate state (s).

Figure 2 illustrates four examples of models which can represent manufacturing systems. The events labelling the transitions indicate the system changes of state. The first case, figure 2a, presents an equipment basic representation given by a two states automaton (g0 and g1), being the initial state and an operational state (or a operating state). However, there is some equipment that works in distinctive operation modes. For example, a transportation system being shared by two workstations can be represented by a three states automaton, figure 2b, being the initial state (g0 - inactive), a second representing the operation of supplying one station (g1) and a third state representing the operation of supplying the second workstation (g2). Both active states (g1 and g2) are interconnected to the initial state by events correspondent to the beginning and the end of each activity's operation.

The possible intermediate states represent a refinement of the active state correspondent to the first case automaton. Whereas such states are included in the model, one gets a better observation of the system free behaviour. For instance, the automaton shown by figure 2c represents a subsystem with a sole intermediate state (g1). Consequently, one can build operational specifications that couldn't have been built using, for instance, a two states automaton. However, one must be careful in building a very synthetic model, due to the exponential growth of the number of states in the synthesis process.

The breakage and failure states are necessary when the designer wants to impose co-ordination specifications considering the subsystems abnormal operation. Thus, one can consider that all the equipment has failure liability, although the real issue is related to the control system level where it will be addressed. Typical cases found in the literature consider the breakage and failure states in specification terms, for instance, when one wants to impose priority for the equipment repair. Figure 2d illustrates the failure state inclusion (g2).

3.2. Modelling operational specifications

Continuing with the SCT methodological proposition, the next step for the locally modular supervisors is modeling each specification individually, considering only relevant events. A first specifications set is related to the transportation system, asynchronous or synchronous. The first one is characterized by the independence among stations transportation, or else, there are distinctive mechanisms transferring pieces from one station to another. On the contrary, the synchronous is characterized by a sole transporter, when moving a piece from a station to the other, the remaining pieces are simultaneously transferred to the subsequent stations (Groover, 2001). In both cases, one wishes to impose a correct pieces and parts flow inside the system. If using asynchronous transporters, one can reach that specification through constraints on the system buffers (position where the pieces and parts are visible). Such constraints are represented by imposing buffer underflow and overflow non-occurrence.

Figure 3 illustrates a small cell where the subsystem G1 is supposed to transport the piece from buffer 1 to buffer 2. When performing the operation, the workstation G2 processes and finally the subsystem G3 removes the piece from buffer 2 to buffer 3. In this case, one considers that buffer 2 have unitary capacity. The specification E1 assures the overflow and underflow non-occurrence of buffer 2; the specification E2 assures the piece is processed after the arrival at buffer 2; the specification E3 assures that the piece is removed from buffer 2 only after the end of the activity performance.

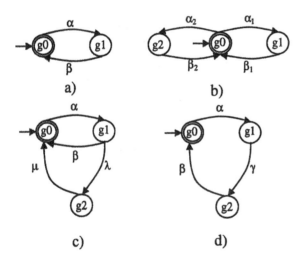

Fig.2. Models of subsystems by automata.

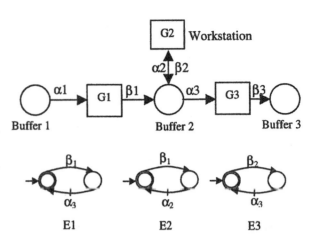

Fig. 3. Example of specifications model when using asynchronous transportation

According to the local modular appliance, the local plants are obtained through a composition of the subsystems sharing events with their correspondent specifications E_1, E_2 e E_3. The following generators are then obtained (considering model Gi as shown by figure 2a): $G_{loc,1} = G_1 \parallel G_3$, $G_{loc,2} = G_1 \parallel G_2$ and $G_{loc,3} = G_2 \parallel G_3$. The local specifications are obtained through the synchronous composition of generic specifications E_1, E_2 e E_3 with their correspondent local plants: $E_{loc,1} = E_1 \parallel L_m (G_{loc,1})$, $E_{loc,2} = E_2 \parallel L_m (G_{loc,2})$ and $E_{loc,3} = E_3 \parallel L_m (G_{loc,3})$. Then it is possible to calculate the maximum controllable languages enclosed in $E_{loc,j}$ ($j \in \{1,2,3\}$), that is, $SupC$ ($E_{loc,j}$, $G_{loc,j}$).

As previously mentioned, the synchronous transportation is characterized by the simultaneous movement of pieces among stations. This characteristic provides changes relating to the operational specifications of the system being designed. The two main synchronous transportation systems are the swivel tables and the conveyor belt. Figure 4 illustrates a four positions swivel table, the first related to the piece arrival (P1), the second related to the execution of the first operation (P2), the third related to the execution of the second operation (P3) and the fourth related to the removal of the piece from the table (P4).

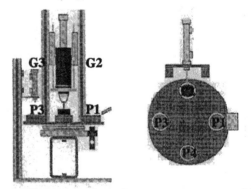

Fig. 4. Example of synchronous transporter: four positions swivel table with operations relating to drilling and testing.

The correct pieces flow throughout the swivel table shown by figure 4 is obtained by the specifications E_a, E_{bi} and E_{ci}. One considers the swivel table as a subsystem (G0) and the remaining stations that supply and remove pieces from the table (G1 and G4, respectively) and those executing processes in other table positions (G2 and G3). The subsystems G1 and G4 are not presented by figure 4.

The specifications E_a, E_{bi} and E_{ci} are shown by figure 5. The specification E_a alerts against the removal of pieces from the table's position P1 (without having a piece) or from the position P2 (without processing the piece by G2). The specifications E_{bi} prevent the synchronous transporter from working while the workstations (G2 e G3) and/or the supplying subsystem and the pieces removal (G1 and G4, respectively) are operating. The specification E_{c1} models the correct pieces supplying from station G1 to the table position P1. Thus, it avoids the pieces or parts overlay during positions, processing without material during position P2 and executing the transportation without the realization of the activity related to the station G2. The specification E_{c2} prevents the station G2 from performing two successive activities, putting station G3 in operation without the piece's processing by G2 and the beginning of the synchronous transport with the piece processed by G2 and not processed by G3. The specification E_{c3} prevents the station G3 from performing two successive activities, putting the transport G4 without the piece processed by G3 and setting the synchronous transportation with a piece in the table's fourth position P4.

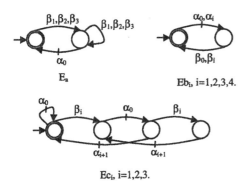

Ec$_i$, i=1,2,3.

Fig. 5. Specifications models when using synchronous transporters.

The local plants are obtained through the composition of the subsystems sharing events with their correspondent specifications E_a, E_{bi} (i=1,2,3,4) and E_{ci} (i=1,2,3). The following generators are then obtained: $G_{loc,a} = G_0 \parallel G_1 \parallel G_2 \parallel G_3$; $G_{loc,bi} = G_0 \parallel G_i$ (i=1,2,3,4) and $G_{loc,ci} = G_0 \parallel G_i \parallel G_{i+1}$ (i = 1,2,3). The local specifications are obtained through the synchronous composition of generic specifications E_a, E_{bi} e E_{ci} with their correspondent local plants: $E_{loc,a} = E_a \parallel L_m (G_{loc,a})$; $E_{loc,bi} = E_{bi} \parallel L_m (G_{loc,bi})$ and $E_{loc,ci} = E_{ci} \parallel L_m (G_{loc,ci})$. Then, one calculates the maximum controllable languages enclosed in $E_{loc,a}$, $E_{loc,bi}$ and $E_{loc,ci}$, that is, $SupC(E_{loc,a}, G_{loc,a})$, $SupC(E_{loc,bi}, G_{loc,bi})$ and $SupC(E_{loc,ci}, G_{loc,ci})$. It is possible then to calculate the maximum controllable languages enclosed in $E_{loc,j}$ ($j \in$ {a,b1,b2,b3,b4, c1,c2,c3}), that is, $SupC(E_{loc,j}, G_{loc,j})$.

Usually there are systems using synchronous transportation where intermediate positions without workstations are supposed to exist. This configuration becomes necessary due to two aspects: physical constraints or anticipation of delay time in transferring pieces from one station to the next (Groover, 2001). The pieces transference delay time anticipation is necessary when particular types of processes are scheduled. The piece, after such process (for instance, painting, glueing), needs to observe a specific lead-time before going to the next workstation. Thus, each intermediate position between two workstations is anticipated in order to counterbalance the time of delay.

The existence of positions without activity changes substantially the system operation specifications, when compared to those previously shown. The fundamental aspect to be observed is related to the non-existence of one event indicating the activities end of operations (event β) in the intermediate duct (see specifications E_{ci} - figure 5).

The proposed solution then is to associate to each position without activity a sole state automaton with one controllable event in self-loop. The automaton indicates the piece has reached the position and it is ready to be transferred to the next station or to the other position without activity. Thus, one solves the lack of indication issue related to end of an activity and also makes possible to use the same structure of the presented specification on figure 5 (E_{ci}).

Consider now a conveyor belt, where 5 positions are anticipated: first for receiving (P1), second for executing an operation (P2), third without activity (P3), fourth with another operation (P4) and a fifth for the removal of the piece from the conveyor belt (P5), as illustrated by figure 6.

The new specifications set shown by figure 6 is presented by figure 7, where the indicating generator model is correspondent to the non-activity position and represented by the automaton Pi. Regarding the specifications shown by figure 5, the following differences are pointed out: a new specification of mutual exclusion between indicating generator Pi and the synchronous transporter ($E_{b2'}$) is presented; the specifications $E_{c2'}$ e $E_{c2''}$ appear, besides those already present in the model (E_{c1} e E_{c3}). The specifications $E_{c2'}$ e $E_{c2''}$ guarantee the correct piece flow between positions P2 and P3 and between P3 and P4, respectively.

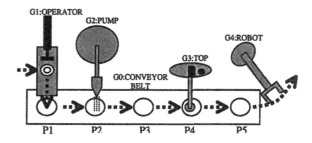

Fig. 6. System using synchronous transporter with position without activity (P3).

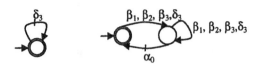

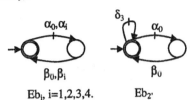

P₃, (number of position without activity).   E_a

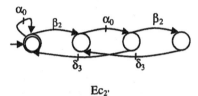

Eb_i, i=1,2,3,4.   Eb_2'

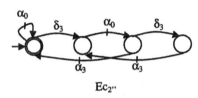

Ec_2'

Ec_2''

Fig. 7. Specifications models for the systems shown by figure 6.

The specification $E_{c2'}$ prevents the subsystem G2 from repeating the activity, the indication (δ_3) without processed material by G2 and the transport execution G0 before the indication (δ_3). The specification $E_{c2''}$ prevents P2 from indicating the same material successively, puts the subsystem G3 without material in position P4 and sets the transport G0 into motion with the G3 non-processed material in the position 4 (see figure 6).

The local plants are obtained through composition of the subsystems sharing events with their correspondent specifications E_a, E_{bi}, $E_{b2'}$, E_{ci}, $E_{c2'}$ e $E_{c2''}$. The following generators are then obtained: $G_{loc,a} = G_0 \| G_1 \| G_2 \| G_3 \| P_3$; $G_{loc,bi} = G_0 \| G_i$, $i = 1,2,3,4$; $G_{loc,b2'} = G_0 \| P_2$; $G_{loc,ci} = G_0 \| G_i \| G_{i+1}$, $i = 1,3$; $G_{loc,c2'} = G_0 \| G_2 \| P_3$ and $G_{loc,c2''} = G_0 \| P_3 \| G_3$. The local specifications are obtained through the synchronous composition of generic specifications E_a, E_{bi}, $E_{b2'}$, E_{ci}, $E_{c2'}$ e $E_{c2''}$ with their correspondent local plants: $E_{loc,a} = E_a \| Lm(G_{loc,a})$; $E_{loc,bi} = E_{bi} \| Lm(G_{loc,bi})$; $E_{loc,b2'} = E_{b2'} \| Lm(G_{loc,b2'})$; $E_{loc,ci} = E_{ci} \| Lm(G_{loc,ci})$; $E_{loc,c2'} = E_{c2'} \| Lm(G_{loc,c2'})$ and $E_{loc,c2''} = E_{c2''} \| Lm(G_{loc,c2''})$. The maximum controllable languages enclosed in $E_{loc,a}$, $E_{loc,bi}$, $E_{loc,b2'}$, $E_{loc,ci}$, $E_{loc,c2'}$ and $E_{loc,c2''}$, that is, $SupC(E_{loc,a}, G_{loc,a})$, $SupC(E_{loc,b2'}, G_{loc,b2'})$, $SupC(E_{loc,ci}, G_{loc,ci})$, $SupC(E_{loc,c2'}, G_{loc,c2'})$ and $SupC(E_{loc,c2''}, G_{loc,c2''})$ are then calculated.

5. CONCLUSION

On the complexity that involves the modelling through automata of the desired specifications for the system, the above set of proposed models allows the systemization of the controllers synthesis tasks. The models obtained from the local supervisors allow the reuse of such models in other project and make it easier to modify the control structure of a specific manufacturing system.

It is important to state also the reuse of formal modelling tools and SED controller's synthesis. For that reason, the control system reliability is increased, not only by systemizing its development but also by including the logic formalism from the SCT, guaranteeing the completion of established specification.

REFERENCES

Askin, R. G. and C. R. Standridge, 1993, *Modeling and analysis of manufacturing systems*, New York: John Wiley & Sons.

Cassandras, C. G. and S. Lafortune, 1999, *Introduction to discrete event systems*, Kluwer Academic Publishers.

Groover, M. P., 2001, *Automation, production systems and computer integrated manufacturing*, Upper Saddle River: Prentice Hall.

Kumar, R. and V. Garg, 1995, *Modeling and control of logical discrete event systems*, Kluwer Academic Publishers.

Queiroz, M. H. and J. E. R. Cury, 2000a, "Modular control of composed systems", Proceedings, American Control Conference, Chicago, USA.

Queiroz, M.H. de and J.E.R. Cury, 2000a, *Modular supervisory control of large scale discrete-event systems*, Discrete Event Systems: Analysis and Control. Kluwer Academic Publishers, pp. 103-110. (*Proc. WODES 2000*, Ghent, Belgium)

Ramadge, P.J. and W. M. Wonham, 1989, "The control of discrete event systems", Proc. of IEEE, Special Issue on Discret Event Dynamic Systems, **77**, pp. 81-98.

Wonham, W. M. and P. J. Ramadge, 1988, "Modular supervisory control of discrete event systems", Mathematicas of control, signals and systems, **1**, pp. 13-30.

PUBLICATIONS
www.elsevier.com/locate/ifac

Experimental 3-D Visual servoing for FMS applications

Gustavo Schleyer, Gastón Lefranc

Escuela de Ingeniería Eléctrica.
Universidad Católica de Valparaíso, Chile.
E-Mail: glefranc@ieee.org

Abstract: This paper presents a visual system that gives the information of the position, height and orientation of several objects presented in the working area of the robot manipulator. With this information, the robot manipulator's effector can pick and place objects to a specific position. A pair of stereo camera produces the feedback obtaining a particular position of the effector of robot. The servoing implemented is based on vision stereo lateral model. This servoing system is tested experimentally in real time on a 5 degrees-of-freedom Scara Manipulator, including the stereo cameras and image processing using Matlab.

Keywords: Servoing Systems, Robotic Manipulators. Stereo vision, Kinematics control,

1. INTRODUCTION

Flexible Manufacturing Systems (FMS) utilizes in its Cells robotics manipulators to perform different task as pick and place materials, parts or products, assembly a products, and quality control. These tasks require speed and precision, to have some economical advantages and good engineering. Conventional robot manipulators have limited accuracy in positions and need time to perform task.

A servoing system provides these requirements. A servoing system uses a visual system to control the position (and others one) of a robotics manipulator. This visual servoing does not need to know a priori the coordinates of the work piece and could not need a robot teaching, allowing having not repetitive tasks, especially in assembly. The vision feedback control loops have been introduced in order to increase the flexibility, the speed and the accuracy of robot system (Brady, 1989; Hutchinson et al, 1996).

Vision-based robot control is classified into two groups (Weiss et al, 1987): position-based and image-based control Systems. In a position-based control system, the input is computed in the three-dimensional Cartesian space (3-D visual servoing) (Wilson et al, 1996). The position of the target with respect to the camera is estimated from image features corresponding to the perspective projection of the target in the image. There exits several

methods to recover the pose of an object, all of them based on the knowledge of a perfect geometric model of the object and the calibration of the camera to obtain unbiased results. In an image-based control system, the input is computed in the 2-D image space (2-D visual servoing) (Espiau et al, 1996).

An image-based visual servoing is robust with respect to camera and to robot calibration errors. However, its convergence is theoretically ensured only in a region around the desired position. Except in very simple cases, the analysis of the stability with respect to calibration errors seems to be impossible, since the system is coupled and nonlinear.

A new approach is called 2-1/2-D visual servoing since the used input is expressed in part in the 3-D Cartesian space and in part in the 2-D image space (Malis et al, 1999).

There exist several techniques to extract 3D information. Some of them, called direct sense, estimates the distance to an object based in the measurement of the time of the transmission and reception of a wave known the propagation media. This can be done by laser, ultrasound and radar. The disadvantage is that can measure one point at a time. Other technique is to use the shadow to compute the depth of the object. (Irving et al, 1987). A method for determining depth from focus (Ens, et al, 1993) relates the distance from camera to objects out of focus, needing two images. The technique using encoded light pattern, the objects are illuminated from one point, in a plane, or a mesh of points through a projector with a position and orientation known respect to the camera (Vuylsteke and Oosterlinck, 1990).

The technique that uses two perpendicular cameras in specific positions obtains two images to compute the space information of the object.

Stereo Vision utilizes two cameras focusing the same object from different views and then to determine from the differences of the images, the distance of the objects by triangulation. There are several models for Stereo Vision as Lateral Model, Axial Model, and Generalized array of stereo cameras (Alvertos, Brzakovic, 1989).

A previous work of the authors, is related with a Servoing System using one camera applied to a pick and place task (Lefranc, Cano, 2002).

In this paper presents a visual system that gives the information of the position, height and orientation of several objects presented in the working area of the robot manipulator. With this information of the robot manipulator's effector can pick and place objects to a specific position. A pair of stereo camera produces the feedback obtaining a particular position of the effector of robot. The servoing implemented is based on vision stereo lateral model. This servoing system is tested experimentally in real time on a 5 degrees-of-freedom Scara Manipulator, including the stereo cameras and image processing using Matlab.

2. The 3-D Visual servoing

The system has a stereo vision with two web cameras, based on vision stereo lateral model, that is to say both cameras separated by a horizontal displacement and perpendicular to objects. The cameras are mounted as shown in Fig. 1, including illumination system (two 100 W of power).

Figure 1. Camera systems and the illumination.

A computer's program control the images capture, the image processing, computes the pose of the objects in the scene, and the control of the robotic manipulator. The robotics manipulator is an IBM 7547, Scara type. The complete system is in the Laboratory on de Robotics, Artificial Intelligence and Advanced Automation in the Escuela de Ingeniería Eléctrica de la Pontifical Universidad Católica de Valparaíso, Chile. See Figure 2.

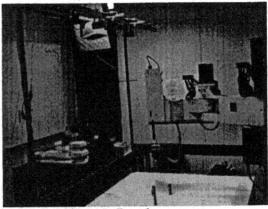

Figure 2. Complete system.

In the next figure 3, a lateral plane of the system can be observed. The work space is restricted to the common vision of the two cameras. The position and height of the objects can be computed

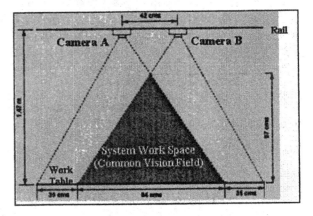

Figure 3. Work space

The restriction is considering only the objects in the common work space of the stereo vision. Pixels outside of that space are coloured to avoid them.

3. Images Processing.

Images captured by cameras are in RGB format, where each pixel is combination of blue, red and green colours. The RGB image is transformed to a grey scale image and the colour information is now bright information. With this kind of image the objects are contrasted with the background, obtaining two frequencies of gray level.

Then, it is possible image segmentation using a threshold chosen according to the illumination, the objects and the colours along the scene. The gray image is transformed to a binary image as shown in Fig. 4, where the black colours are objects and white pixels are background. This mean that the image is matrix with each cell has a '0' or '1' value depend on if a black or white pixel. The resolution used is 288*352 pixels, with a matrix of 288 rows and 352 columns.

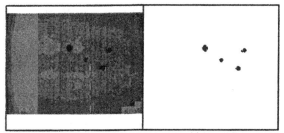

Figure 4. Gray scales Image and binary image.

The next step is to eliminate the noise using and appropriate threshold. The noise is due to errors in image capture and problems in the work area.

For objects identification, it is used the tag technique assigning a number to each one in the image, from left to right and up to down, searching cells with "1" value in the first set of cells, the next set tag is "2" and so on, until "n" that is the object number. The result is presented in the Fig. 5.

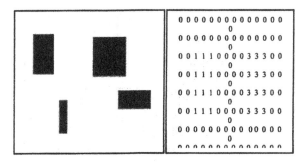

Figure 5. Object's tags.

An algorithm utilize Tag's object to compute the object centroid and its (x, y) position. To find centroids, the image matrix is taken, store extreme positions of all elements different from "0", obtaining the length and width, and then dividing them. The algorithm gets the centroids without considering the orientation of the objects. The result, for the objects in Fig. 4, is shown in the Fig. 6.

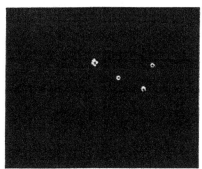

Figure 6. Object's Centroids.

The object orientation is necessary to compute, to know the rotate angle for the effectors of the robotics manipulator can pick the object. The algorithm uses the centroid of the object to determine the distance to every contour point of the object, choosing the minimal distance, that is to say the orientation of the object. The orientation angle θ is the argument of a complex number in the corresponding quadrant. Figure 7 presented how is the plane complex origin cantered in the centroid of the object and the θ angle.

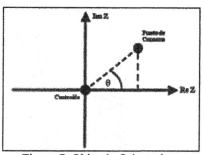

Figure 7. Object's Orientation.

Procedure is repeated for every object.

4. Coordinates of the robotics manipulator.

4.1 Computing (x, y) coordinates.

The coordinates of the centroids of the objects within the common work space, is expressed as a row and column of the pixel of the object in the image matrix. However, this information is not sufficient to have (x, y) coordinates for the manipulator can pick the object. It needs a relationship between image coordinates and the manipulator coordinates. As this relationship is not the same in the whole work space, this space is divided in segments with different relationships between the two coordinates. Each segment has 5 measurements of centroid points and the coordinates that the robotics manipulator has to have to pick objects. The relationship between the column of the centroid pixel and the X coordinate of the robotics manipulator, are in Fig. 8.

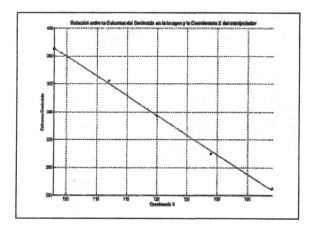

Figure 8. Relationship between centroid common en the image and X coordinate of the robotics manipulator

From Fig. 8, a linear relationship is found, and it is possible determine X from equation (1) using linear progression.

X = -2.8282904*(Column of Centroids) + 676.89183 (1)

Equation (1) transforms the information of column of centroids pixel of the object, in the X coordinate to send to robotics manipulator. Equation (1) is valid only in the segment that is jeans it has to have equation for each segment.

To obtain the Y coordinate and to send to robotics manipulator, a relationship between X and Y coordinates is put in a Figure 9.

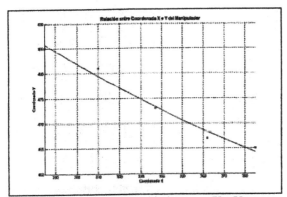

Figure 9. Relationship between X e Y Coordinates of the Manipulator.

The Relationship between X and Y coordinates are related using quadratic progression to obtain a second grade polynomial in equation (2).

$$Y = 5.203 * 10^{-4} * (X)^2 - 0.5636046 * (X) + 604.126309 \quad (2)$$

Equation 2 determines Y coordinate to send to robotics manipulator. The same procedure is used to obtain the equation in each segment with corresponding data. The row and column of the centroids permits to identify the segment in the work space where the objects are and then to assign the right equation, to compute (x, y) coordinates for the robotics manipulator.

To reduce errors due to perspective distortion, the coordinate measurements are done using the image nearest of the object. The second image is employs to determine the depth, the Z coordinate.

4.2 Compute the Z coordinate.

The equations (3), (4) and (5), from Stereo Vision Lateral Model, computes Z coordinate.

$$\begin{bmatrix} x' \\ y' \\ z' \end{bmatrix} = I \begin{bmatrix} x \\ y \\ z \end{bmatrix} + \begin{bmatrix} \Delta x \\ 0 \\ 0 \end{bmatrix} \quad (3)$$

$$\frac{x_0}{x_L} = \frac{y_0}{y_L} = \frac{\lambda - z_0}{\lambda} \quad (4)$$

$$\frac{x_0 + \Delta x}{x_R + \Delta x} = \frac{y_0}{y_R} = \frac{\lambda - z_0}{\lambda} \quad (5)$$

Find the value of z_0 from the preceding equations; it is the equation (6).

$$z_0 = \lambda \cdot \left(1 - \frac{\Delta x}{\Delta x + x_D - x_I}\right) \quad (6)$$

Where:

x_D and x_i: represent the column coordinate of the pixel of the centroid from image captured by right and left cameras

Δx: represents the distance between cameras along the x axis, equal to 103 pixels.

λ : is focal length of the cameras lens.

z_0 is proportional to height of the object, but it is not the z coordinate of the manipulator. To compute the real coordinates utilizing a scale factor 'F', transforming the image coordinate z_0 into z coordinate in the real space. In the coordinate system of the robotics manipulator, the z = 0 value is the highest of z axis, and the z = -249 value is the lowest one, however, the minimum value of z = -220, avoid that effectors crushes against the table. It is necessary to have a term that it puts "0" value of z_0 with the zero of the manipulator. This zero adjust term 'T_0' is in the scale factor is included in the equation (6) of the Stereo Vision Lateral Model, obtaining the equation (7).

$$Z = T_0 + F \cdot \left(1 - \frac{\Delta x}{\Delta x + x_D - x_I}\right) \quad (7)$$

λ (Focal length) is included in the scale factor.
Z value is the coordinate sent to robotics manipulator.

To calibrate the system, the term values of zero adjust term 'T_0' and scale factor 'F', it is utilized two objects with height known, placed in the same point in the common work space of the system. It computes z coordinate for robotics manipulator used to pick each object. The difference between the columns coordinates of the centroid pixel of the two images ($X_D - X_i$). This data are in the Table 1.

Table 1. Data for 'To' y 'F'

	Object A	Object B
z Coordinate	-219	-135
$X_D - X_i$	93	102

Putting these values in the equation (7) gives the system equation (8).

$$T_0 + \left(\frac{93}{103 + 93}\right) \cdot F = -219$$
$$T_0 + \left(\frac{102}{103 + 102}\right) \cdot F = -135 \quad (8)$$

The values obtaining are: $T_0 = -1946.6$ and $F = 3640.9$. This values is replace in the equation (7), to obtain equation (9) to compute z coordinate of the robotics manipulator in the segment of the work space, where the measurement are made

$$Z = -1946.6 + 3640.9 \cdot \left(1 - \frac{103}{103 + x_D - x_I}\right) \quad (9)$$

Equation (9) transforms directly the displacement the centroid pixel of the object between each image of the stereo vision system, in the z coordinate of the robotics manipulator. Due to perspective distortion of the object, because of the cameras are not perpendicular to the objects, the equation (9) is valid only in small segment of the work space of the manipulator.

4.3 Compute of the rotation coordinates.

The object orientation is determined by orientation algorithm and then it has to have a transformation to relate algorithm coordinates and manipulator coordinates. The algorithm coordinates considers a positive rotation in the unclockwise, and the positive rotation in the manipulator is clockwise. There exists a difference of 18° between both coordinate systems computed by equation (10).

$$\theta_{Manipulador} = (-1) \cdot \theta_{Algoritmo\ de\ Orientación} + 18° \quad (10)$$

Finally, the R coordinate is sent to robotics manipulator to have the turn angle of the effector, is equal to the value $\theta_{Manipulado\ r}$ obtained.

5. Evaluations

5.1 Errors in Coordinates (X, Y, Z).

Errors values are the difference between the right coordinate to send to robotics manipulator and the coordinate computed by 3D Servoing. If it has an inappropriate error, the manipulator picks bad the object, or fall down or placed in a wrong position. To determine error percentage it uses 15 cylinders test objects with 2.5 cm, in three groups of five objects with height of 3.5, 8.0 and 15.0 cm.

The different sizes permits evaluating the accuracy of the (X, Y, Z) coordinate computation, especial the Z coordinate.

The five test objects are put in a random way in each of the 4 quadrants and compute the error. The (X, Y, Z) theorical coordinates are the position where the effector of the robotics manipulator has to move to pick the object.

The percentage of error is computed comparing the theorical coordinate with the coordinate obtaining with the computer vision using the equation (10).

$$Error(\%) = \left| 100\% - \frac{(System\ coordinate) \cdot 100\%}{Theorical\ Coordinate} \right| \quad (10)$$

Table 2, 3 and 4 present the percentage of error for different kind of cylinders in the 4 quadrants. The coordinates are expressed in mm. For example, if the Z coordinate is -120 and the position computed by computer vision system is -110, implies that the manipulator will take the object 1 cm up the expected value.

Table 2. Percentage Error for h = 3.5 cm. Cylinder

		Theorical Coordinate			System Coordinate			Error (%)		
		X	Y	Z	X	Y	Z	X	Y	Z
I Quadrant	1	329	666	-230	324	668	-207	1,5	0,3	10,0
	2	272	749	-230	268	751	-230	1,5	0,3	0,0
	3	226	609	-230	221	615	-230	2,2	1,0	0,0
	4	172	738	-230	167	737	-219	2,9	0,1	4,8
	5	110	704	-230	104	709	-230	5,5	0,7	0,0
II Quadrant	6	253	842	-230	244	843	-230	3,6	0,1	0,0
	7	187	921	-230	183	923	-185	2,1	0,2	19,6
	8	119	993	-230	113	995	-196	5,0	0,2	14,8
	9	79	839	-230	73	841	-240	7,6	0,2	4,3
	10	-7	979	-230	-12	982	-240	71,4	0,3	4,3
III Quadrant	11	-95	792	-230	-101	790	-205	6,3	0,3	10,9
	12	-145	690	-230	-151	688	-170	4,1	0,3	26,1
	13	-229	736	-230	-233	737	-221	1,7	0,1	3,9
	14	-306	779	-230	-311	781	-230	1,6	0,3	0,0
	15	-366	676	-230	-369	678	-230	0,8	0,3	0,0
IV Quadrant	16	-76	855	-230	-78	850	-198	2,6	0,6	13,9
	17	-165	921	-230	-173	920	-230	4,8	0,1	0,0
	18	-242	979	-230	-246	977	-230	1,7	0,2	0,0
	19	-261	882	-230	-264	882	-207	1,1	0,0	10,0
	20	-360	935	-230	-366	935	-230	1,7	0,0	0,0

Table 3. Percentage Error for h = 8.0 cm. Cylinder

		Theorical Coordinate			System Coordinate			Error (%)		
		X	Y	Z	X	Y	Z	X	Y	Z
I Quadrant	1	295	745	-185	299	741	-196	1,4	0,5	5,9
	2	302	643	-185	293	652	-185	3,0	1,4	0,0
	3	209	637	-185	201	642	-185	3,8	0,8	0,0
	4	172	738	-185	167	737	-219	2,9	0,1	18,4
	5	83	709	-185	73	714	-119	12,0	0,7	35,7
II Quadrant	6	278	816	-185	274	821	-53	1,4	0,6	71,4
	7	210	941	-185	205	949	-196	2,4	0,9	5,9
	8	173	877	-185	167	883	-185	3,5	0,7	0,0
	9	71	996	-185	65	999	-141	8,5	0,3	23,8
	10	41	932	-185	42	939	-221	2,4	0,8	19,5
III Quadrant	11	-66	703	-185	-58	700	-185	12,1	0,4	0,0
	12	-146	753	-185	-150	747	-113	2,7	0,8	38,9
	13	-254	799	-185	-255	791	-185	0,4	1,0	0,0
	14	-282	716	-185	-278	711	-170	1,4	0,7	8,1
	15	-338	676	-185	-346	667	-185	2,4	1,3	0,0
IV Quadrant	16	-76	855	-185	-71	852	-173	6,6	0,4	6,5
	17	-144	941	-185	-144	940	-185	0,0	0,1	0,0
	18	-253	994	-185	-252	994	-174	0,4	0,0	5,9
	19	-268	901	-185	-260	898	-170	3,0	0,3	8,1
	20	-367	879	-185	-375	877	-185	2,2	0,2	0,0

Table 4. Percentage Error for h = 15 cm. Cylinder

		Theorical Coordinate			System Coordinate			Error (%)		
		X	Y	Z	X	Y	Z	X	Y	Z
Quadrant I	1	302	643	-120	290	647	-120	4,0	0,6	0,0
	2	272	749	-120	270	741	0	0,7	1,1	100,0
	3	192	783	-120	191	787	-174	0,5	0,5	45,0
	4	121	770	-120	134	718	-129	10,7	6,8	7,5
	5	44	824	-120	35	825	-129	20,5	0,1	7,5
Quadrant II	6	253	842	-120	241	843	-119	4,7	0,1	0,8
	7	195	966	-120	198	979	-120	1,5	1,3	0,0
	8	171	903	-120	159	908	-73	7,0	0,0	39,2
	9	107	860	-120	95	861	-120	11,2	0,1	0,0
	10	18	957	-120	14	959	-141	22,2	0,2	17,5
Quadrant III	11	-63	682	-120	-45	653	-110	28,6	4,3	8,3
	12	-122	815	-120	-104	797	-111	14,8	2,2	7,5
	13	-200	756	-120	-194	745	-69	3,0	1,5	42,5
	14	-278	779	-120	-282	775	-120	1,4	0,5	0,0
	15	-361	799	-120	-375	782	-120	3,9	2,1	0,0
Quadrant IV	16	-81	980	-120	-63	981	-85	22,2	0,1	29,2
	17	-101	857	-120	-84	845	-68	16,8	1,4	43,3
	18	-223	941	-120	-212	937	-120	4,9	0,4	0,0
	19	-296	901	-120	-298	900	-120	0,7	0,1	0,0
	20	-389	932	-120	-404	938	-120	3,9	0,6	0,0

5.2 Errors in Orientation Coordinate (R).

The 3D Servoing developed, the cylinder is interpreted as a circle and it can pick with any orientation angle. The test objects are a parallelepiped of 3.7 cm. heights and square base of 3.2 cm for side. To evaluate the orientation coordinate, in the workspace is marked theorical angles every 10 grades from 0° to 180°. Then, these angles are compared with the angles obtained from stereo vision system, using equation (10) and the percentages errors are in Table 5.

Table 5. Percentage Error for orientation.

θ Theorical	θ System	Error (%)	θ Theorical	θ System	Error (%)
0°	0°	0	100°	90°	10,0
10°	0°	100,0	110°	135°	22,7
20°	0°	100,0	120°	135°	12,5
30°	27°	10,0	130°	135°	3,8
40°	27°	32,5	140°	135°	3,6
50°	27°	46,0	150°	135°	10,0
60°	45°	25,0	160°	180°	12,5
70°	90°	28,6	170°	180°	5,9
80°	90°	12,5	180°	180°	0,0
90°	90°	0,0			

Each segment of the workspace has different equations doing difficult the evaluation. The values in the segments are not equal. The system work well: it can determine successful the position in (X,Y) plane, the height (Z axis) and the orientation of the test objects, The accuracy is better in small, objects and focus perpendicular to the cameras. The system is able to work with several objects in the workspace, computing the position in 3D, to transport them to other specific position, previously defined. The errors are produce mainly for the distortion due to perspective of the object and it is not correctly focus. This distortion can be reduce doing the segmentation of the workspace of the computer vision system and relating the equations of stereo vision with height of the objects and the centroid point from the captured images.

6. Conclusions

It has presented a visual system that gives the information of the position, height and orientation of several objects presented in the working area of the robot manipulator. This 3D servoing system gives this information, the robot manipulator's effector can pick and place objects to a specific position. A pair of stereo camera produces the feedback obtaining a particular position of the effector of Scara 7475 robotics manipulator. The servoing implemented is based on vision stereo lateral model.

The system has the capacity of identifying the spatial position and the orientation of several objects in steady state, presented in the common work space of stereo vision cameras, then that information is sent to the manipulator to pick and place objects. The (x, y) robotics manipulator coordinates are obtained applying equations based on centroids of the objects. To compute z coordinate is utilized equations of the Stereo Vision Lateral Model, with a zero adjust term and scale factor to coincide the original coordinate system with the manipulator coordinate system. To compute the orientation angle, an algorithm is used. Depend on what place of quadrants, there exist errors, choosing the low one. The stereo vision is evaluated in a Flexible Assembly Cells.

7. References

L. E. Weiss, A. C. Sanderson, and C. P. Neuman, "Dynamic sensor based control of robots with visual feedback," *IEEE J. Robot. Automat.*, vol. 3, pp. 404–417, Oct. 1987.

Alvertos, N., Brzakovic, D. "Camera Geometries for Image Matching in 3-D Machine Vision", IEEE Transactions on Pattern Analysis and Machine Intelligence, Vol. 11, N°9, September 1989, pp. 897-914.

Irvin, R.B. and Mckeown, D.M., "Methods for Exploiting the Relationship Between Buildings and Their Shadows in Aerial Imagery", IEEE Transactions on Systems, Man, and Cybernetics, Vol. 19, N°6, December 1989, pp. 1564-1575.

Vuylsteke, P. and Oosterlinck A., "Range Image Acquisition with a Single Binary - Encoded Light Pattern", IEEE Transactions on Pattern Analysis and Machine Intelligence, Vol. 12, N°2, February 1990, pp. 148-164.

Espiau, B.; Chaumette, F.; Rives, P., A new approach to visual servoing in robotics, Robotics and Automation, IEEE Transactions on , Volume: 8 Issue: 3 , June 1992, Page(s): 313 –326.

Ens, J. and Lawrence, P., "An Investigation of Methods for Determining Depht from Focus", IEEE Transactions on Pattern Analysis and Machine Intelligence, Vol. 15, N°2, February 1993, pp. 97-107.

Hutchinson, S., Hager, G.D. and Corke, P.I., (1996). A tutorial on visual servo control. IEEE Trans. On Robotics and Automation 12, 651-670

Wilson, W.J.; Williams Hulls, C.C.; Bell, G.S., Relative end-effector control using Cartesian position based visual servoing, Robotics and Automation, IEEE Transactions on , Volume: 12 Issue: 5 , Oct. 1996, Page(s): 684 -696.

Nakamura, Shoichiro, Análisis Numérico y Visualización Gráfica con MATLAB, Prentice – Hall Hispanoamérica, S.A., México 1997.

Malis E., Chaumette F., and Boudet S., 2-1/2-D Visual Servoing, Robotics and Automation, IEEE Transactions on, vol. 15, no. 2, pp. 238-250, April 1999.

Umbaugh, S., Computer Vision and Image Processing, Prentice Hall PTR, NJ, 1999.

Lefranc, G., "Servoing Systems: A Tutorial", IEEE International Symposium on Robotics and Automation, 2002, Toluca, México.

Lefranc G., Cano F., "Sistema Servoing Pick and Place", Congreso Latinoamericano de Control Automático, 2002, Guadalajara, México.

Leigthon F., Lefranc G., "Flexible Assembly Cell using Scara Manipulator". MCPL2004, Chile.

This work was possible thanks to the Research Project DGIP 2004, grant by the Pontifical Catholic University of Valparaiso, Chile

PUBLICATIONS
www.elsevier.com/locate/ifac

TOWARDS AN OPTIMISED APPROACH TO CONTROL AMHS IN WAFER MANUFACTURING PLANTS

Jairo R. Montoya-Torres[1,2] and Jean-Pierre Campagne[3]

Centre Microélectronique de Provence and Centre Génie Industriel et Informatique
École Nationale Supérieure des Mines, F-42028 Saint-Étienne, France
E-mail: montoya@emse.fr

[2] *STMicroelectronics, 300mm Manufacturing Programs*
Z.I. de Rousset, F-13106 Rousset cedex, France

[3] *Laboratoire de Productique et Informatique de Systèmes Manufacturiers*
Institut National de Sciences Appliquées de Lyon, F-69621 Villeurbanne, France

Abstract: The management of modern wafer factories is a very hard task. Earlier research studies in this industry have been primarily focused on the optimal dispatching of lots on machines. Nowadays, productivity and ergonomic reasons impose the use of Automated Material Handling Systems (AMHS) to transfer wafer-lots between work centres. Hence, in addition to the traditional problem of lot dispatching, it is now necessary to manage optimally the AMHS to attain high throughput rates and short cycle times. Traditional methodologies on the study of AMHS in IC wafer manufacturing use simulation-based models. These models, however, do not provide analytical methods which are preferable to develop an optimal AMHS control system. This paper proposes an optimisation approach based on discrete-event simulation, where the objective is to improve the system productivity by using an efficient AMHS control policy. A mathematical analysis is presented, as well as a framework for a global optimisation of the production system.

Keywords: automation, material handling, mathematical programming, simulation, VLSI manufacturing.

1. INTRODUCTION

Managing short cycle times, maintaining high utilisation rates of equipment and resources along with guaranteeing on time delivery and high throughput rates are well-known targets in high-tech manufacturing industries, such as the semiconductor industry. For instance, in new generation semiconductor fabrication facilities (fabs), the seamless and reliable operations of Automated Material Handling Systems (AMHS) used for the transportation and the storage of material is a prerequisite for an efficient fab logistic. This paper presents our work in progress on the optimisation of AHMS control in semiconductor manufacturing plants.

Nowadays, with the advent of the new wafer generation, mainly because of productivity and ergonomic reasons, wafer fabrication needs a high level of process automation. In addition to the traditional tool dispatching problem, the IC industry has to deal with two other problems: (i) the selection of the right AMHS, and (ii) its optimisation. The first problem implies to both select the best facility configuration and decide the type of technology that should be used for the handling system. Solving the second problem helps to increase the system productivity by minimising cycle times and maximising

throughput at a given service rate. This optimisation may be possible by controlling traffic parameters and vehicle scheduling and routing rules.

The automated material handling literature for semiconductor manufacturing has mainly been focused on the design of the AMHS layout (Montoya Torres *et al.*, 2004). Concerning the operational issues, wafer transport management is often called a "non-value-added" activity (but without it, no value can be added), and not much attention has been given to date in the literature. Since having the AMHS as the bottleneck in the factory is unacceptable (Goff and Wilkey, 2003), appropriate analysis, techniques and tools are needed to carefully explore and plan the operations of the AMHS. Specific applications in semiconductor manufacturing done to date in the literature have mainly focused on simulation studies of inter-bay wafer transportation, i.e. transfer of lots between manufacturing departments (Lin *et al.*, 2001; Jimenez *et al.*, 2002). Although those studies provide some insights on the behaviour of AMHS, the drawback is that they consider the AMHS as an isolated system, thereby neglecting the influence of variability originated in the rest of the wafer fab. It is important to note that the optimal design of individual unit processes does not guarantee an efficient manufacturing facility, because the economic viability of semiconductor fabs is governed by both throughput and cycle times. In addition, those approaches rely on *ad-hoc* discrete-event simulation programs in order to identify the conditions in which certain vehicle dispatching rules perform well. This is because the IC wafer industry has considered that the concept of complete, cost-efficient and flexible automation of transport, storage and handling procedures will inevitably lead to automated material handling systems much more complex to operate. Under this assumption, the IC manufacturers have considered, without formal analytical proof, that it is impossible to test these systems using analytical methods such as mathematical programming.

The main advantage of using discrete-event simulation-based methodologies is that simulation allows to analyse operations of production and logistic systems at a great level of details and to consider dynamic variations (i.e. stochastic events, uncertainty). Although simulation supports the comparative evaluation of different AMHS logistic control policies, classical simulation models do not provide analytical methods which are required to develop feasible or optimal AMHS control policies. On the other hand, many analytical methodologies are available to support decision-making for the management and control of production and logistics systems. Among these, mathematical programming has received widespread acceptance in industry. Optimisation models however often require simplifications of the production system and aggregation of the production equipment. Moreover, optimisation models applied in industry are often deterministic. Hence, it seems appealing to combine simulation and optimisation in order to benefit from the advantages of both worlds.

This paper aims to propose an optimisation methodology based on a discrete-event simulation approach for the AMHS control in wafer fabrication. Since total cycle times of wafers are greatly increased by transportation times, the objective function will consider the problem of the optimal assignment of vehicles to lot transport requests to minimise transport time. In addition to a basic integer linear programming model, a relaxation of the problem is presented and lower bounds on the performance metrics are derived. In the last part of the paper, we present a framework for a global optimisation approach using simulation.

2. MATHEMATICAL ANALYSIS

2.1. Assumptions and Notations.

The development of the mathematical model is based on the following assumptions:
- Machine layout and material flow paths are completely specified, including fixed travel distances, machine locations, and guided path directions. Consequently, they are considered as constraints for the optimisation of operations management.
- The inter-resource material flow rates in terms of load per time period are known. They are calculated from the production routings (sequence of operations) of the wafers to be manufactured and the product demands over the manufacturing horizon.

- Whenever a transporter visits a machine or the output station of a resource group, there always exist materials to be transferred.
- Only horizontal material movements are considered (i.e. AGVs, overhead transporters or automated rail carts, commonly used in wafer factories).
- Transporters are assumed to be identical (i.e. they have identical travel speeds) and with unit load capacity.
- For the traffic management problem, the control at intersection points of the unidirectional guided path is sufficient to avoid collisions.
- Transporters and machines are supposed to be reliable.
- Preemption is not permitted, neither for operations nor vehicle trips.
- The number of wafers started in a given period is known and constant over the whole manufacturing horizon. This assumption is realistic since lot release policies are defined at higher decision levels. This also allows us to formulate the objective function in order to minimise the total transportation time of lots. Thus, as stated by Little's Law (Little, 1961), it is easy to compute the throughput rate for the production period under study.

There exist three types of transporter operations between a pair of production equipment. These definitions are adopted from (Ioannou, 1995) and are the following:
- A loaded move is the transporter operation from the pick-up point $p(i)$ to the delivery point $d(i)$. The set of loaded moves is denoted by L, and its cardinality $|L|$ will be referred to as n throughout the text and also corresponds to the total number of lot production operations.
- An unloaded move is the transporter operation from one drop-off point to the next pick-up point, during which no work-in-process is carried. The set of possible unloaded moves is denoted by U.
- A complete move is the concatenation of a loaded move and an unloaded move. The set of complete moves is denoted by C. Each element of C will be noted as (i,j), where i is the loaded move included in the complete move, and j is the loaded move that follows i in the sequence of a transporter.

Assuming that there exists a path between each pick-up/delivery pair, we can verify that after the completion of a loaded move, a transporter can perform an unloaded move to the origin of any other loaded move in L. If the pick-up and delivery points of a station coincide, this unloaded move may be of zero distance. As a result, either three or only two production resources may be included in a complete move. In the latter case, either $d(i)=p(j)$ or $p(i)=p(j)$ and $d(i)=d(j)$, where $i,j \in L$ are consecutive moves performed by the same transporter.

Let N and V be, respectively, the set of lots to be transported during the time horizon T, and the set of available transporters to transfer lots in the facility. The total cycle time of a lot is composed of actual processing time, total transport time and total waiting time (for both processing and transport). Assuming intelligent production scheduling rules are used in the shop floor to minimise total waiting time, and if the objective is to minimise the total cycle time of lots, one strategy is to minimise the total transport time. Therefore, assuming that times θ_i,

θ_i^p, and θ_i^d reflect the time needed to perform loaded move i, to pick-up the lot from $p(i)$, and to deliver the lot to $d(i)$, respectively, then the total transport time t_{ij} associated to move $(i,j) \in C$ is defined as:

$$t_{ij} = \begin{cases} \theta_i + \theta_i^p + \theta_i^d & \text{if } i \neq j \\ \infty & \text{if } i = j \end{cases} \quad (1)$$

2.2. Mathematical Formulation.

To formulate the problem of minimising total transportation times in a wafer fabrication facility, we define a binary variable $x_{ij}^k = 1$ if move $j \in L$ is performed following move $i \in L$ by transporter $k \in V$, and 0 otherwise.

Based on the assumptions and notation described above, the formulation of the wafer transportation problem (WTP) is as follows:

Problem WTP

$$\text{Minimise } Z = \sum_{(i,j) \in C} \sum_{k \in V} t_{ij} x_{ij}^k \quad (2)$$

Subject to:

$$\sum_{(i,j) \in C} \sum_{k \in V} x_{ij}^k = 1 \qquad \forall\, i \in L \quad (3)$$

$$\sum_{(i,j) \in C} \sum_{k \in V} x_{ij}^k = 1 \qquad \forall\, j \in L \quad (4)$$

$$\sum_{(i,j) \in C} x_{ij}^k = \sum_{(j,i) \in C} x_{ji}^k \quad \forall\, i \in L, \forall\, k \in V \quad (5)$$

$$\sum_{i \in W} \sum_{j \in W} x_{ij}^k \leq \sum_{i \in W} \sum_{j \in L} x_{ij}^k - 1$$

$$\forall\, W \subset L : 2 \leq |W| < \sum_{(i,j) \in C} x_{ij}^k \quad \forall\, k \in V \quad (6)$$

$$\sum_{(i,j) \in C} t_{ij} x_{ij}^k \leq T \qquad \forall\, k \in V \quad (7)$$

$$\sum_{k \in V} x_{ij}^k \leq |V| \qquad \forall\, (i,j) \in C \quad (8)$$

$$x_{ij}^k \in \{0,1\} \qquad \forall\, (i,j) \in C, \forall\, k \in V \quad (9)$$

The objective function (2) considers the total transport time. Constraints (3) and (4) ensure that each loaded move in L is performed by exactly one transporter. Since transporters perform closed sequences of moves, both constraints are required to guarantee a feasible assignment of complete moves to transporters. The set of constraints (5) forces each transporter to follow continuous paths in terms of loaded moves. This set of constraints can also be seen as flow conservation constraints. That is, consider each transporter as a distinct flow of unit intensity. The moves assigned to this transporter represent the nodes of an auxiliary graph. Flow conservation is imposed on each node of this graph, to guarantee continuous movement of the transporter. Since vehicles perform closed sequences of moves, no flow sources or sinks are present, and the equality holds for each node of the graph (i.e. for each move in L). The exponential set of constraints (6) enforces sub-tour elimination, guaranteeing the existence of a single tour for each transporter. Note that a tour is a sequence of moves in the form $(i_1, i_2, ..., i_l)$. This type of

constraint is encountered in the formulation of the well-known Travelling Salesman Problem (TSP), where the salesman has to visit a set of cities, before returning to his starting point. Constraints (7) limit the time that each transporter operates to the studied period T. The set of constraints (8) assures that number of actually used vehicles is at most equal to the total number of available vehicles in the system during the production period. Finally, constraint (9) forces the variable x_{ij}^k to assume binary values 0 or 1.

Note that our formulation is closely related to the well-known vehicle routing problem (VRP), which is proven to be NP-complete (Garey and Johnson, 1979). This means that it is unlikely to have efficient (polynomial time) algorithms to solve it. In our problem, no central depot is present and the demand at each node, which represent a loaded move, is one unit.

In order to guarantee a feasible solution to problem WTP, each complete move should satisfy $t_{ij} < T/2$ for all $(i,j) \in L$. The case in which there exist some completes moves with $t_{ij} > T/2$ may lead to an empty feasible solution space, if these moves have to be performed by the same transporter.

2.3. A Relaxation and Lower Bounds.

The problem that results by considering only the transportation times in the objective function of problem WTP, by removing the capacity constraints (7), and by ignoring the transporter indices, is the well known assignment problem (or minimal weight matching problem (Korte and Vygen, 2002)) which is presented below.

Problem A

$$\text{Minimise } Z_a = \sum_{(i,j) \in C} t_{ij} x_{ij} \quad (10)$$

Subject to:

$$\sum_{(i,j) \in C} x_{ij} = 1 \qquad \forall\, i \in L \quad (11)$$

$$\sum_{(i,j) \in L} x_{ij} = 1 \qquad \forall\, j \in L \quad (12)$$

$$x_{ij} \in \{0,1\} \qquad \forall\, (i,j) \in C \quad (13)$$

The optimal solution Z_a^* of A provides for every loaded move $i \in L$ the loaded move $\phi(i)$ that should follow i, such that the total cost of transportation (total transportation time) is minimised. Z_a^* and the associated $\phi(i)$ are derived in polynomial time by the Hungarian algorithm (Korte and Vygen, 2002). Note that Z_a^* bounds from below the component of the transportation time of lots in the objective function of problem WTP, i.e. $Z_a^* \leq Z_{opt}^* = \sum_{(i,j) \in C} \sum_{k \in V} t_{ij} x_{ij}^{k*}$, where x_{ij}^{k*} is the optimal number of vehicles actually used for transporting lots in the facility. As suggested in (Ioannou, 1995), this lower bound on the transportation time can be used to obtain a lower bound on the number of transporters. The expression is given in equation (14):

$$R_a^* = \left\lceil \frac{Z_a^*}{T} \right\rceil \qquad (14)$$

2.4. Discussion and Remarks.

In the integer linear programming model for wafer transportation problem developed above, production routes where modelled as sequences of operations through the machines. Thus, it is potentially equally applicable to the more classical flow-shop or job-shop systems. In fact, the re-entrant nature of semiconductor manufacturing is highly related to the physical location of machines. From the conceptual viewpoint, however, it allows the modelling of the process either as a flow-shop or a job-shop depending on the products manufactured.

On the other hand, this model contains a large number of binary variables and constraints. To quantify the computational complexity of the model, it is useful to compute the number of decision variables and constraints as functions of the problem parameters, which basically include the cardinality $|V|$ of the set of available transporters and the number of loaded moves per time period, $|L|$. Given the number of binary variables and constraints, it is easy to determine whether a certain problem instance can be solved by explicit or implicit enumeration methods. Unfortunately, it might not be possible to solve practical instances of **WTP** due to its size. The total number of binary variables x_{ij}^k, that link the loaded moves to the network path and to the transporters, is given by $\Psi = |V| \times |L|$. The number of constraints in the transportation problem **WTP** is examined by evaluating the sets over which each constraint is expressed. Thus, the total number of constraints is given by $\Gamma_{total} = (|L|+1) \times (|L|+|V|) + e^{|L|}$.

In addition to the computational complexity of the mathematical model, wafer semiconductor manufacturing is a very complex manufacturing process. The fabrication process is mainly characterised by: a large number of process steps requiring both simple and batch machines, some with sequence-dependent set-up times and some with time constraints between process steps, re-entrant flows, unpredictable rework operations after inspections, R&D lots inducing priorities for on-line production dispatching, shared tools across tool groups, unreliable equipment, and cycle time dependent yield loss. These characteristics are not easy to model and contain nonlinearities, combinatorial relationships and uncertainties that cannot be modelled effectively by simply listing a linear objective and a set of linear constraints. Therefore, recent practices in modelling wafer manufacturing have largely embraced simulation-based methodologies. Those reasons justify the use of discrete-event simulation with optimisation models in order to analyse the impact of wafer transportation strategies on the behaviour of modern real-life wafer factories.

3. TOWARDS A GLOBAL OPTIMISATION APPROACH USING SIMULATION

Mathematical models combined with simulation is a very important tool for optimising material flows and logistics in wafer fabs. In the search for flexible and reasonably priced production, material flow and transport concepts, simulation-based optimisation may become an increasingly invaluable aid in factory management and control. By using simulation, alternative production concepts can be analysed and weak points be spotted in time as a result of the dynamic behaviour of a complex material flow and production systems.

The global simulation-based optimisation approach presented in this section is motivated by the general configuration of a Decision Support System (DSS). The three major components of a DSS are input data, analytical tools, and graphical interface. The input data are normally in the form of a database which contains the information needed to make decisions. This database can also include certain parameters and rules, such as the vehicle speed, product mix, and lot dispatching rules. The analytical tool can use operations research algorithms, simulation modelling and other rule-based procedures. The graphical interface (i.e. charts, spreadsheets) are used to display the results of the DSS analysis and to help users to grasp the large quantity of output data.

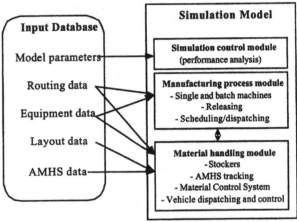

Fig. 1. Integrated simulation-optimisation framework

Figure 1 shows the framework, which consists of an input database and a simulation model. The input database is used to model input data including model parameters, routing data, tool data, factory layout data, and AMHS data. The simulation model is a simulation-based analytical tool that models a virtual automated fab by importing model input from the input database. This simulation model must represent the actual fab in order to make accurate decisions. The third component of the DSS, the graphical interface, is included in the control module (performance analysis) of the simulation model. The hypothesis and assumptions needed to implement this global simulation-based optimisation approach are the same than the ones presented in Section 2.1 for the mathematical analysis of the wafer transportation problem. Let us now detail the components of this simulation-optimisation framework.

The manufacturing process model is composed of a tool model (equipment, machines), a scheduling/dispatching system, and a lot release system. The set of machines to process the lots of wafers are classified into sequential (i.e. single wafer-based) machines and batch machines. Each tool model performs the operations of pre-processing (i.e. reading bar code, opening wafer-carrier, and docking), processing (i.e. loading wafers into the equipment, processing wafers, and unloading), and post-processing (i.e. closing the wafer-carrier, reading the bar code, undocking, etc.). The routing table determines the processing time of the equipment at each step. Batch processing machines, such as furnace and clean wet tools, process a batch of a given number of lots at a time. Batches are formed at the stocker, and the tool does not call the lots until a batch is formed or a maximum waiting time has been reached (WNLTT rule). Once a batch is formed, the lots belonging to that batch are transported individually by vehicles of the material handling system. When the entire batch has reached the equipment and entered the buffer, the batch enters the pre-processing, processing, and post-processing operations. Upon completing the post-processing operation, the lots of the batch are individually transported back either to the stocker or to their next destination. The scheduling/dispatching and releasing systems determine the equipment dispatching rule and lot releasing policy, respectively. Various dispatching rules exist in the literature, but most of the wafer fabs uses common-sense dispatching rules, such as, among others, First-In-First-Out (FIFO), Shortest Time to Bottleneck Resource (STBR), Shortest Time to Next Visit (STNV). The FIFO rule selects lots in the same order of arrivals (i.e. the first lot to be executed is the one with the earliest time of arrival at the equipment queue). In practice, however, a wafer fab usually classifies products into regular lots, hot lots, and super hot lots by assigning different priority values. In some fabs, it is also possible that regular lots have different priorities, which may be determined by using other dispatching rules, such as Critical Ratio (CR) or Earliest Due Date (EDD). When lots have priority, the dispatching is done by ranking them in a list according to their priority. Then, lots are executed based on the First-in-the-rank dispatching rule (i.e. the one with highest priority is treated first). Wafer release policies are also modelled in our framework. Usually, start dates are defined at higher decision levels, and production control is constrained to respect the schedule. For simulation purposes, however, two lot release policies can be modelled: random and deterministic. The former is based on an exponential distribution of lot arrivals and is used to capture the randomness on the lot starting process. In the second policy, lots are released according to a predetermined schedule.

The material handling module consists on the AMHS tracking (including the network flow and the fleet of vehicles), stockers, a material control system (MCS), and a vehicle dispatching and control module. Inter-bay and intra-bay are treated as an unified system in which it is possible to transport a lot carrier from one processing tool to another without passing by intermediate storage. Loops in the AMHS network are considered to be unidirectional, and each loop has a series of stopping points at which vehicles can stop to load/unload a lot carrier. Each bay has a stocker where lots wait to be processed on one of the corresponding machines. The network path also has a

series of control points. When vehicles arrive at these control points, they can ask for and receive information from the MCS concerning the operations they have to perform next (i.e. pick up another carrier, travel to the park area, etc.). The vehicle dispatching module consists of vehicle-initiated dispatching rules and load-initiated dispatching rules (Egbelu and Tanchoco, 1984). A vehicle-initiated dispatching rule deals with the situation that a vehicle has the choice of a transport task when multiple jobs (lot carriers) are waiting for pick-up simultaneously at different locations in the fab. A load-initiated rule deals with the problem of matching a task to a vehicle when multiple empty vehicles are idle and waiting for a transportation task assignment. In the literature, various authors have proposed several heuristic rules (Egbelu and Tanchoco, 1984), which can be combined with the mathematical formulation presented in previous sections.

The simulation control module is the main control panel, while the manufacturing process and material handling modules are core components of the simulation model. Users may modify simulation parameters, lot dispatching and release policies, AMHS control strategies, etc. This module also serves as a reporting module for performance analysis and provides real-time decision-making information. It may be possible to define several on-line reports, including WIP information, wafer movement and throughput reports, utilisation reports, and cycle time reports. Analysts should be able to export data to perform off-line statistical analysis.

4. INDUSTRIAL APPLICATION

The semiconductor fabrication model considered in this study is based on a real-life plant located in the South of Italy. The factory layout is divided into a certain number of bays (aisles) that contain the equipment to process wafers. This bay configuration creates a large amount of material flow between bays, especially since the wafer fabrication process is highly re-entrant. The bay configuration, however, has the advantage of allowing maintenance during equipment operation without disruption production (Cardarelli and Pelagagge, 1995). The equipment used in wafer manufacturing is very sensitive due to the tight tolerance that is required, and it is down for preventive maintenance and repairing a large portion of time. Having the processing equipment of a particular type collocated allows maintenance passages to be constructed between the bays for monitoring and repairing the equipment and thus ensures that the production can continue during these maintenance periods. Hence, the maintenance areas are separated from the clean room space and allow access to some parts of the equipment without disrupting production. In the fab, there are a total of 241 processing machines grouped in 71 single-server and multi-server station families. Multi-server stations consist of several identical parallel machines. We assume that all visits by all lots to a specific station have the same processing time distribution. The process flow of wafer lots consists of a total of 694 processing steps, and exhibits the cyclic re-entrant flow that characterises wafer manufacturing where each lot visits the same workstation many times. Since each lot flows 39 times through the photolithography station, the process flow we consider is referred to as a 39-mask process in semiconductor terminology. The transport operations in the fabrication plant modelled here consist of both inter-bay and intra-bay transfers, and are performed by an unified automated material handling system, which allows direct tool-to-tool delivery. The AMHS layout consists of an unidirectional path network with vehicles for the transfer of lots between bays and within bays. The order in which lots are executed at each workstation depends on the scheduling policy.

In semiconductor manufacturing firms, the cost of capital is a key successful factor and the competition is time-based. Hence, the most important performance indexes are assumed to be flow time related measures, with a constraint on the average utilisation rate of the whole fab. During simulation runs, we have collected statistics on cycle time, transport times, equipment utilisation rates and throughput. Results are presented in figures 2 and 3. As shown in figure 3, the lower bound on the number of vehicles can be used as a start point to perform simulation experiments, so that the optimal number may be computed using the results of the simulation study. For the factory under study, the "optimal" number of vehicles required in the system is 22 vehicles. Notice that because of the model complexity, as explained in section 2.4, the maximum CPU time required to run the model is 33,4 hours.

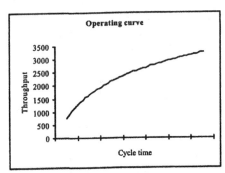

Fig. 2. Factory operating curve

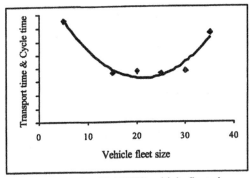

Fig. 3. "Optimal" AMHS vehicle fleet size.

5. CONCLUSIONS AND FURTHER WORK

In this paper, we presented a framework for optimisation of AMHS control in wafer semiconductor factories. The main objective is to combine the advantages of mathematical-based optimisation models and discrete-event simulation. An integer linear programming model for the wafer transportation problem was proposed. Its analytical study drove us to develop a relaxation in order to compute lower bounds on both the objective function

and the vehicle fleet size. In addition, a modelling simulation framework was presented to support the comparative evaluation of different AMHS logistic control policies and to capture particular characteristics and the stochasticity of the wafer production process.

In future work, we want to implement the proposed simulation-optimisation methodology at the wafer fab described in the last section of this paper. We intend to test the actual computational complexity using real data. The relaxation should allow us to obtain comparative results with an acceptable computational effort. This information about the performance indexes can then be used to improve the performance of the simulation.

ACKNOWLEDGEMENTS

The work presented in this paper was partially supported by the French Foundation of Technological Research under grant CIFRE-621/2002. Acknowledgements are due to L. Vermariën and Prof. S. Dauzère-Pérès for fruitful comments and insights.

REFERENCES

Cardarelli G., Pelagagge P.M. (1995) Simulation tool for design and management optimisation of automated interbay material handling and storage systems for large wafer fab. *IEEE Transactions on Semiconductor Manufacturing*, **8**, 44-49.

Egbelu, P.J. and M.A. Tanchoco (1984). Characterization of automatic-guided vehicle dispatching rules. *International Journal of Production Research*, **22**, 359-374.

Garey M.R., Johnson D.S. (1979) Computers and intractability: A guide to the theory of NP-completeness. W.H. Freeman and Company.

Goff G.L., Wilkey A. (2003) Hybrid 300mm wafer handling. *European Semiconductor*, April, 19-23.

Ioannou G. (1995) Integrated manufacturing facility design. PhD thesis, University of Maryland, USA.

Jimenez J., Kim B., Fowler J., Mackulak G., Chung Y.I., Kim D.J. (2002) Operational modeling and simulation of an inter-bay AMHS in semiconductor wafer fabrication. *Proceedings of the 2002 Winter Simulation Conference*, p. 1377-1382.

Korte B., Vygen J. (2002) Combinatorial Optimization: Theory and Algorithms. Springer-Verlag.

Lin J.T., Wang F.K., Yen P.Y. (2001) Simulation analysis of dispatching rules for an automated interbay material handling system in wafer fab. *International Journal of Production Research*, **39**, 1221-1238.

Little J.D. (1961) A proof for the queuing formula $L=\lambda w$. *Operations Research*, **16**, 651-665.

Montoya-Torres, J.R., Vermarien, L., Campagne, J.P., Marian, H. (2004) Design and operation of automated material handling systems for IC wafer semiconductor manufacturing. *Proceedings of the Industrial Simulation Conference*, p. 175-179.

PUBLICATIONS
www.elsevier.com/locate/ifac

ONTOLOGY-BASED APPROACH
FOR THE FUTURE PRODUCTION OF EUROPEAN INDUSTRY

Raffaello LEPRATTI
Ulrich BERGER

Brandenburg Technical University at Cottbus
Chair of Automation Technology
P.O. Box 10 13 44, 03013 Cottbus – Germany
{lepratti;berger.u}@aut.tu-cottbus.de

Abstract: In order to support European industry in its transition process towards the knowledge-based enterprise a set of novel information-based tools for enabling knowledge, skill and data transfer is needed. Their design depends on the organic and functional enterprise infrastructure features and relations between the heterogeneous agents involved across the whole value added chain. This paper presents two ontology-based approaches aiming at overcoming interoperability barriers arising in communication process among humans and machines. The first one focuses on computer-supported human collaboration and human-machine interaction by means of natural languages, enabling semantic independence of shared knowledge and data contents. The second one proposes an approach for machine data exchange and sharing, applying standards as highly extruded common knowledge.

Keywords: Knowledge-based Systems, Man-Machine-Interaction, Manufacturing System, Natural Languages, Ontologies and Machine Data Exchange.

1. THE LOT-SIZE ONE PARADIGM

European industry is in transition process from a mass production industry towards a knowledge-based customer- and service-oriented one, which aims at a production model on demand, mass customization, rapid reaction to market changes and quick time-to-market of new innovative products.

In this transition, it faces the challenge to produce according to a *lot-size one paradigm* at low cost and high quality. A customizing in final products leads to a strong individualization of product features, which influences the normal course of the product life cycle making risky investments and resource plans in production.

Following this vision, networked, knowledge-driven and agile manufacturing systems stand out as necessary key elements towards this future production scenario, which shall allow European industry long-term competitiveness improvements above all by added values in product-services.

The context of collaborative engineering and manufacturing has witnessed a striking expansion in all fields of the value added chain. In spite of a successful employment of a set of information-base tools, knowledge, skill as well as data transfer it

shows many inefficiencies and hurdles (Goossenaerts et al., 2002) and still represents the major problem toward the achievement of a suitable and efficient infrastructure ensuring in future the establishment of the knowledge-based enterprise.

Therefore, research efforts and technology development on information infrastructures are ongoing, addressing a. o. information architecture, methodologies, ontologies, advanced scenarios, standard machining tools and services. These should contribute in providing a holistic solution for a knowledge-based engineering and manufacturing architecture, which must feature a systemic dynamic learning behaviour, where innovation emerges from new complex interaction forms between integrated technologies, human resources, management and organizations in all phases of the valued-added chain, i.e. (i) production preparation, (ii) planning and programming as well as (iii) process execution. Hence, new solution approaches shall allow above all operational knowledge acquisition and knowledge feedback in computer-based collaborative engineering and manufacturing.

Both, already existing and arising industrial know-how should be gathered together either manually from human experiences (usually by experts) and by use of intelligent cognitive sensing systems, or automatically derived from human interventions (e. g. short process corrections at shop-floor level) or responses of other machine equipments. Through knowledge retrieval mechanisms, also machines acquire intelligence reaching the necessary challenging level of efficiency and robustness. Such visionary workflow architecture is shown in Fig. 1.

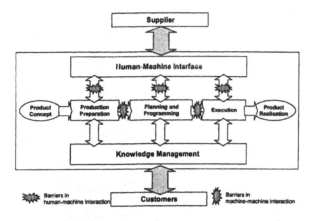

Fig. 1. Basic buildings block structure of the knowledge-based enterprise

2. INTEROPERABILITY ISSUES

If, on the one hand, the enterprise structure proposed in Figure 1 could represent an adequate solution to meet the growing market challenges, on the other hand, it shows to be also ambitious in connection with its functional requirements.

A smooth global information flow between all actors involved in this process is the most important aspect for ensuring the correct process behaviour. However, as shown in Figure 1, still too many complications evolve when trying to find standard criteria for interoperability across the entire heterogeneous human qualifications as well as cultures and machine programming languages setting.

On the one hand, possible understanding problems arise, while two persons try to communicate with each other as consequence of discrepancies in their cultural and/or professional backgrounds. They are incline to cognitive perceive and mentally process same situations of the real world in subjective ways, referring these to different models (so called mental models). Thus, also two interaction partners, even though speaking the same language and using identical terminologies, can misunderstand each other, since vocabulary terms represent merely etiquettes of cognitive categories.

On the other hand, in machine-to-machine communication, incompatibilities in data structure or code languages (different syntax and semantic rules) are major reasons of impediments in transferring information from a software system to another one, which uses distinct technology solutions.

2.1 State-of-the-Art

Some standards enabling interoperability in engineering and manufacturing have been already

successfully employed. Some relevant examples are here described:

The KIF (Knowledge Interchange Format) (Genesereth and Fikes, 1992) as well as the KQML[1] (Knowledge Query and Manipulation Language) allow interchange of information and knowledge among disparate software programs - either for the interaction of an application program with an intelligent system or for two or more intelligent systems - to share knowledge in support of co-operative problem solving with the possibility to structurally represent knowledge at a meta-level.

The STEP ISO 10303 (STandard for the Exchange of Product Data) (Fowler, 1995) addresses to the exchange and sharing of information required for a product during its life cycle (such as parametric data like design rationale, functional specification and design intent). STEP is nowadays a well-known standard for real world product information modelling, communication and interpretation. Some examples are STEP AP-203 (Application Protocol) (Configuration Control for 3D Design of Mechanical Parts and Assemblies), STEP AP-214 (Core Data for automotive Mechanical Design Processes), STEP AP-224 (Mechanical Parts Definition for Process Planning Using Machining Features), STEP-240 (Machining Process Planning) and STEP-NC (ISO 14649 Industrial automation systems and integration Physical device control) (Richard et. al., 2004).

Finally, CORBA (Common Object Request Broker Architecture) (Object Management Group, 1995) and COM/DCOM[2] (Distributed Component Object Model) provide neutral - both platforms and languages independent - communication between remote applications based on object oriented distributed technology, allowing different clients and/or servers connected within a network to live as individual entities able to access to the information they need in a seamless and transparent way. Both solutions aren't standards but are widely used e. g. for the development of agent-based systems.

2.2 Requirements for an Innovative Architecture

While shown standard and further solutions have been successfully proved and employed, they are often too strong task-oriented in their applications or remain just one-off solutions. At present, a generic knowledge management concept for architecture as shown in Figure 1 has not been developed. The extent of knowledge and skill transfer in engineering and manufacturing is often strong limited.

New requirements for innovative and generic holistic knowledge-based enterprise architectures such capability in upgrading different heterogeneous systems, transparent data exchange among them, distributed open environments and improved information sharing go therefore beyond the actual state-of-the-art. To foster the transition process of European enterprises, big efforts in the research of further suitable concepts are still needed.

[1] http://www.cs.umbc.edu/kqml/

[2] http://www.microsoft.com/com/default.asp

In the next section two ontology-based approaches, which aim at improving interoperability, will be presented. However, while the first one bases on the use of ontologies and addresses mostly the semantic standardization of computer-supported human-human communication as well as human-machine interaction by the use of natural languages, the second one focuses on overcoming complications in data exchange among heterogeneous software applications of machines and equipments.

3. THE ONTOLOGICAL APPROACH

3.1 The Role of Ontology

In today's production systems the development of communication and production technologies becomes not only more efficient but also more complex. The employment of heterogeneous technologies represents challenges facing professional and cultural staff requirements, which has to work with. This stresses the importance of a novel knowledge management solution able to archive semantic standardization of knowledge and data contents and provide task-oriented as well as user-based redistribution of stored information. A corresponding building block knowledge management architecture concept is illustrated in Figure 2.

However, it is difficult to identify a unified knowledge and data form, when considering the different nature of tasks needed across the whole value added chain. According to Figure 1, three different forms of knowledge and data are identified: (i) The so called 1-D interaction form, i. e. textual, is for instance still the most common way used for information exchange in scheduling tasks during both product preparation and planning phase. (ii) 3-D technologies of the Digital Factory have gained importance in the last years above all with regard to process & planning activities and represent the most profitable way to design production environments (e. g. planning of human and machine activities and machine programming). Finally, (iii) graphical (2-D) technologies such as interactive platform systems support user-friendly on-line process corrections at shop-floor level.

Under these circumstances the need of standard procedures for an efficient processing of knowledge contents, which are able to acquire, filter and retrieve data of different *multi-dimensional* sources, is assuming more and more an essential role.

In Figure 2 the core of the architecture is represented by the Ontology Filtering System (OFS). It plays this important role enabling semantic autonomy of different information contents independently of their nature of being. All *multi-dimensional* data sources mentioned above could be processed in an equivalent manner, i. e. knowledge contents of different forms are stored in the OFS knowledge data base in a standard data structure according to a pre-defined set of semantic definitions and relation rules. This offers a number of advantages: On the one hand, it supports knowledge retrieval and representation in a task-oriented manner according to the specific user and

machine requirements and, on the other hand, it facilitates the computer-supported knowledge exchange among humans or machines and between human and machine avoiding possible semantic ambiguities of knowledge contents.

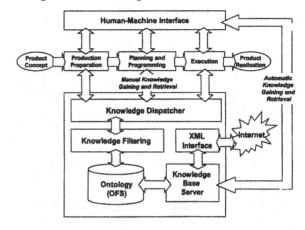

Fig. 2. Ontology-based knowledge management architecture

In the next section a mathematical description of the applied ontology is presented. It will be described on the basis of the natural language use. A related prototype has been developed as shown in (Berger et al., 2004) and already experimentally validated within a research initiative focused on application within the automation technology domain - compare with (Lepratti and Berger, 2004).

3.2 The Ontologies for Stanardisation of Knowledge and Data Contents

Although the use of natural languages still represents an hazard solution approach due to possible misinterpretations, which could arise during the interaction process as consequence of syntactical, lexical and extensional ambiguities connected to their domain of use, they represent the most familiar and understandable communication form for human beings. Following Winograd's theory (Winograd, 1980), assuming that there is no difference between a formal and a natural language, one finds also proper reasons for all the efforts to formalize knowledge expressed by natural languages.

The so called Ontological Filtering System (OFS) removes possible ambiguities in natural languages by means of a semantic network, in which words are chained together hierarchically per semantic relations. This network consists, on the one hand, of a set of words selected for a specific domain of application and used as key words, in order to standardize information contents for the machine data processing. On the other hand, it encloses a set of additional words, which could be used from different persons in their natural communication, since there are more ways to express the same knowledge meaning. These words could have different abstraction degrees in their meaning (so called *granularity*). Thus, some words are more general in their expression than others, while others can go very deep with their meaning. A simplified example of this semantic network is given in Fig. 3. According to their specification level all words – key words and additional words - are linked together by means of semantic relations such as *hypernymy, hyponymy, synonymy* or *antonymy*. A parser within the OFS processes knowledge contents and leads back words meanings to these ones belonging to the set of pre-defined key words. In this way, one can say, the OFS provides a semantic filtering function. A mathematical description could better explain how it works.

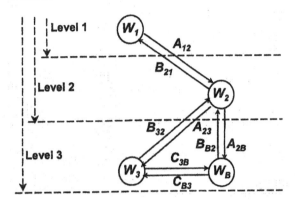

Fig. 3. Example of Ontological Network

Considering W as set of chosen words belonging to the natural language and chosen for a specific domain of use:

$$W_{NL}=\{w_1, w_2, ..., w_n\} \qquad (1)$$

and a set of key words W_B, which represents the key terminology selected to formalize knowledge contents:

$$W_B=\{w_{1B}, w_{2B}, ..., w_{nB}\} \qquad (2)$$

using following set R of semantic relations of natural languages such as: hypernymy (A), hyponymy (B), synonymy (C) and antonymy (D):

$$R=\{A, B, C, D\} \qquad (3)$$

one can define the ontological network as the following ordered triple:

$$ON=<W, R, S> \qquad (4)$$

where W represents the addition set of terms $W_{NL} \cup W_B$ and S takes the specification level of each element of W into account. According to Figure 3, relations between the elements of W can be included in a relation matrix $\Re$:

$$\Re = \begin{bmatrix} 0 & A_{12} & 0 & 0 \\ B_{21} & 0 & A_{23} & A_{2B} \\ 0 & B_{32} & 0 & C_{3B} \\ 0 & B_{B2} & C_{B3} & 0 \end{bmatrix} \qquad (5)$$

Multiplying $\Re$ by the transposed vector W^T.

$$\Im = W^T \cdot \Re = \begin{bmatrix} W_1 \\ W_2 \\ W_3 \\ W_4 \end{bmatrix} \begin{bmatrix} 0 & A_{12} & 0 & 0 \\ B_{21} & 0 & A_{23} & A_{2B} \\ 0 & B_{32} & 0 & C_{3B} \\ 0 & B_{B2} & C_{B3} & 0 \end{bmatrix} \qquad (6)$$

one attains the system of equations $\Im$, which reflexes the structure of the ontological net in turn:

$$\Im = \begin{cases} W_1 = A_{12} \cdot W_2 \\ W_2 = B_{21} \cdot W_1 + A_{23} \cdot W_3 + A_{2B} \cdot W_B \\ W_3 = B_{32} \cdot W_2 + C_{3B} \cdot W_B \\ W_B = B_{B2} \cdot W_2 + C_{B3} \cdot W_3 \end{cases} \qquad (7)$$

Furthermore, on the basis of the following figure 4

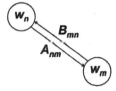

Fig. 4: Simple relation in the Ontological Network

one deduces the simple semantic relation (8)

$$\begin{cases} W_n = A_{nm} \cdot W_m \\ W_m = B_{mn} \cdot W_n \end{cases} \Rightarrow A_{nm} \cdot B_{mn} = \gamma \qquad (8)$$

where γ represents an empty element, since paths from W_n to W_m and vice versa are equivalent over A_{nm} and B_{mn}. Similarly, this counts also for:

$$C_{nm} \cdot C_{mn} = \gamma \qquad (9)$$

Resolving (7) as functions of W_B using (8) and (9), one obtains following results:

$$\Im = \begin{cases} W_1 = A_{12} \cdot A_{23} \cdot C_{3B} \cdot W_B + A_{12} \cdot A_{2B} \cdot W_B \\ W_2 = A_{23} \cdot C_{3B} \cdot W_B + A_{2B} \cdot W_B \\ W_3 = C_{3B} \cdot W_B + B_{32} \cdot A_{2B} \cdot W_B \\ W_B = W_B \end{cases} \qquad (10)$$

In this way, every equation of (10) gives the number of different semantic paths, which lead a specific given element w_i of W to the corresponding key word W_B.

The structure of the Ontology Filtering System presented in this section should be easily extended to any other languages or data structure. When considering the natural language L with a specific vocabulary of symbols W, one can rearrange the definition used above assigning elements of a further specific application domain L' to symbols of W.

In the following section it will be shown, how to extend or modify the OFS Knowledge Base and in particular the ontological net.

3.3 The Engineering Data Exchange Approach

In section 4, the ontological approach for knowledge management shows how knowledge contents expressed in specific language can be computer-processed, i. e. standardized in their meaning. Also data exchange among heterogeneous software applications of machines and equipment in engineering represents an important issue towards the development of the holistic architecture of Figure 1. In the next section an overview of possible problems and a conceptual approach solution will be discussed.

3.3.1 Standard for Engineering Data Interoperability

Nowadays in engineering domains, the data communication process represents a crucial aspect within the digital product creation process. On the one hand, an heterogeneous set of software tools is applied. Thus, a variety of data formats and structures describing the same engineering object lead to incompatibilities. Furthermore, with development of new information technologies, the more digital simulation tools are employed for complicated scenarios, the more complex data become. It ranges from plain text to 2-D, 3-D geometries with semantic information. On the other hand, data communication within the extended enterprise of figure 1 makes the exchange of data with customers, partners or suppliers for a specific engineering object more complex. Therefore, data compatibility of various engineering tools in the extended enterprise represents the essential requirement in exchanging data between different applications.

A second approach bases on the use of knowledge-derived engineering standards, with which the encompassing architecture of Figure 1 should be composed.

3.3.2 Knowledge-derived Standard-based Data Exchange Architecture

The High Level Architecture (HLA, IEEE 1516) for enterprise-wide and in external supply cooperation respectively, describes the test platform for performing distributed simulations. As to the data exchange requirements, i. e. engineering data compatibility, engineering knowledge retrieval and application, a corresponding architecture should be built up. The architecture of engineering data exchange using knowledge-based standard is depicted in Figure 5. Its components are:

1) Engineering tool set: A typical HMI, which is an aggregation of engineering IT tools and Computer Integrated Manufacturing (CIM) machine for engineering i.e. specification, design, analysis, planning, manufacturing, inspection, services, etc. This interface for computer application must support like: (i) access to data, (ii) exchange of information and (iii) Multiple views of product data.

2) Converter: The interface between the software (machine) and the knowledge-based engineering standard backbone. The standard should fit for the entire digital product creation process, i.e. product concept, planning and programming, execution, and product realization. For each milestone in this process chain, a universal converter should be available.

3) Engineering data management: It synchronizes i.e. saves/provides engineering data for corresponding tools and realizes standard-specified data management, using data bank functions, i.e. data configuration and interface accessibility.

4) Connection mechanism for external integration: It is based on the net interface to the engineering data management system. This connection includes the engineering portal and engineering XML interface for the Internet application. For building an extended enterprise, this connection is an important element.

5) Knowledge management: (described in section 4) Its connection to the data standard for engineering is described in section 4.3.3.

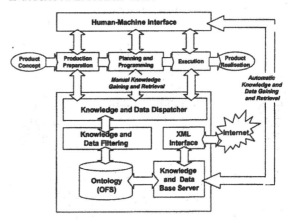

Fig. 5: Standard based knowledge management architecture in product creation

3.3.3 Standard for Knowledge Representation

A standard is a consistent definition or behaviour established by custom, authority, or consensus. In other words, standard is a higher knowledge level as greatest common denominator syndrome. Standards must be taken into account regarding to many variants or options to maximize performance. The data standardisation is a process of knowledge accumulation, sharing, description, application guiding. Each entity in standard can represent the knowledge in all relevant engineering fields and be directly used in engineering tool sets for the intelligent application.

The connection between engineering data standard with knowledge management i.e. ontology-based knowledge management architecture, can be described as follows: i) the Knowledge Management System retrieves knowledge and data from the engineering domains, divides these into different application areas through the knowledge dispatching system and filters their semantic contents. At this point, knowledge and data could be archived into the ontology data management system. ii) The standard structures and data formats are defined only after the standardization of knowledge syntax and semantics. Finally, the conversion of standardized ontology into specific standard data formats is to be done.

3.3.4 A Possible Application Scenario

In engineering domain, rapid system functional development of Computer-Aided Design (CAD) and Computer-Aided Manufacturing (CAM) are progressed. Computer Aided Process Planning (CAPP) system working with 3D-Geometry is far from the engineering interest, because: (i) A CAPP system has the upstream (CAD) and downstream (CAM) application which are at dynamic unbalanced technology levels, this brings the complexity to be integrated. (ii) CAPP itself is very complex. Process planning uses not only planer know-how but also a variety of heuristic rules and logical decision. Conventional computer algorithmic programs can do little with regard to the logic inference. (iii) The most important reason is that standard description of manufacturing process is still missing there. The standard for CAx domain is usually the STEP format.

The STEP 3D product data exchange has been achieved in an industry-practical way (ProSTEP, 2004). However, STEP is still a development activity and has its limited consideration fields (Michael, 2001). For CAPP system such as AP 240 process planning and STEP standard definition for machine tool is in development. (iv) Due to the application of heterogeneous CAx systems in supplier chain, the process data exchange is blocked (Zhang and Alting, 1994). Thus, knowledge and standard for process planning should be acknowledged through a knowledge-derived standard-based data exchange architecture.

This visual scenario for process planning is described in (Cai et. al, 2004). The OEM planning engineering cooperates with the machine supplier using own planning systems for process definition and the supplier integrates the entire process as well as orders the equipment and cutting tool from its own sub-suppliers. The Internet supporting collaborative data exchange module (ISCDE) is applied. This connection permits data exchange with the engineering data management system through an engineering portal server and engineering XML interface. In this way, suppliers are embedded and the so-called extended enterprise is built up.

This approach is still in progress but promises interoperability enhancements in the collaborative process planning within the extended enterprise. Also the knowledge-based standard makes the entire CAx-engineering in up / downstream application continuously. The next step foresees the connection of the knowledge system, in order to retrieve data standard for the specification of the engineering data management systems for the process planning.

CONCLUSION

Within the entire product life cycle knowledge acquisition, retrieval and application can be found in every manufacturing and engineering phase. A knowledge management system for the future production of European industry should embrace the entire enterprise structure, giving the necessary structure flexibility to need the growing market challenges such as rapid product creation with higher quality and short time-to-market strategy. As shown in section 4, ontologies help, on the one hand, in standardizing knowledge and data semantic contents in the communication among humans as well as in the human-machine interaction and, on the other hand, they support the knowledge-derived standard based data exchange. Corresponding architectures have been described.

ACKNOWLEDGMENT

We are indebted to Mr. Jing Cai for his contribution.

REFERENCES

Berger, U.; Lepratti, R.; Erbe, H.: "Human-Robot Collaboration in Automated Manufacturing". In: Proc. IFAC-MIM 2004, October, Athen, 2004..

Cai, J; Wayrich, M.; Berger, U.: „Digital Factory Trigged Virtual Machining Process Planning for Powertrain Production in Extended Enterprise": In: Proceedings of: Mechatronics and Robotics (MechRob04), P. Drews (ed.), Aachen, Germany, 2004, pp. 1072-1077.

Genesereth, M. R. and Fikes, R. E.: "Knowledge interchange format, version 3.0", reference manual. Technical report, Logic-92-1, Computer Science, Stanford University, 1992.

Goossenaerts, J. B. M.; Arai, E.; Shirase, K.; Mills, J. J.; Kimura, F. (2002): "Enhancing Knowledge and Skill Chains in Manufacturing and Engineering". In: Proc. of DIISM Working Conference, 2002.

Lepratti, R.; Berger, U.: "Enhancing Interoperability through the Ontological Filtering System". In Proceeding of the 5th IFIP Working Conference in Virtual Enterprises, Kluwert, Boston-London, 2004, pp. 183-190.

Michael J. P.: "Introduction to ISO 10303 - the STEP Standard for Product Data Exchange," Journal of Computer Information Science Engineering, 1(1), pp.102-103.

Object Management Group: "The Common Object Request Broker Architecture and Specification (CORBA)", Revision 2.0, 1995.

Richard, J.; Nguyen, V. K., Stroud, I. (2004): "Standardisation of the Manufacturing Process: IMS/EU STEP-NC Project on the Wire EDM Process". In: Proceedings of the IMS International Forum 2004, May 17-19, Cernobbio-Italy, 2004, pp. 1197-1204.

Winograd, T.: "What does it mean to understand language?". Cognitive Science 4, 1980, pp. 209-241.

Zhang H.C., Alting L.: "Computerized manufacturing process planning systems" Chapman & Hall, 1994, pp. 188- 198.

ELSEVIER
IFAC
PUBLICATIONS
www.elsevier.com/locate/ifac

Heuristics for efficient inventory management in a serial production-distribution system with constant and continous demand

J. BOISSIÈRE, Y.FREIN et C. RAPINE

Laboratoire GILCO - Gestion Industrielle, Logistique et Conception

46 Av. Félix Viallet

38031 GRENOBLE – France

julien.boissiere@gilco.inpg.fr

SUMMARY : *In this paper, we present new heuristics for finding efficient policies in a serial distribution system. We consider a retailer that faces a continuous and constant demand at the end of the chain and a production facility with limited capacity at the start of the chain. Intermediate stages link both facilities. The heuristics we propose are based on two local-system optimisation. We use these optimisations to build up the policies stage by stage in a Myopic way. An experiment comparing the performances of our heuristics and some litterature results (power-of-two policies) is related at the end of the paper.*

KEY WORDS : *logistic chain, distribution, production, heuristics, inventory, stationary policies*

1 Introduction

For many years, inventory management has received great interest from industrials and researchers. Industrial performance became a major concern for firms about half a century ago, mainly because of the rise of competition. It became even more important in the seventies because of the recession. In the 90ies, the rise of the global world market dramatically increased competition among major companies, and today, the performance of communication medias and transport networks has put almost all firms in a world wide competition. In such a context, inventory management is of main interest for maintaining low costs in the distribution channel. In this paper, we are interested in finding management policies for a serial production-distribution system, with continuous and constant demand. On the production end we consider a capacity constrained production with limited capacity. Both sides are linked with intermediate storage facilities. We focus on stationary policies, because they are easy to implement and to manage. A stationary policy is a policy in which each facility orders a constant batch at regular intervals. We propose heuristics for obtaining efficient stationary policies in any serial production-distribution network with continuous demand. In section 2, we present a tour of the litterature related to inventory management, and we explain our contribution in this context. In section 3 we set up our model and give some basic results and procedures we use in the heuristics described in section 4. Experimentals results of the heuristics and a comparison with some classical results is also done in section 4. Section 5 concludes the paper.

2 Literature review

The major contributions in distribution inventory management has been published by Roundy in 1986 [10]. He studied a network with any physical structure, taking into account transportation costs and holding costs. Assuming constant and continuous demand, non-decreasing holding costs along the chain and infinite production capacity at the beginning of the chain, he re-used the famous power-of-two policies he introduced [9] for the one-warehouse-multiple-retailer problem. Again, he proves that these policies are very efficient (98% efficient) and gives a quick algorithm to find them. This work has been used and extended by many authors such as Atkin & Sun (1995) [1] who allowed backlogging or Federgruen & Zheng [6] who include time constraints for each period. Also Federgruen *et al* [5] studied the same model with general transportation costs (or set-up costs). Some other contributions in this field can be found in Graves *et al* 1993 [7]. In 1981, Williams [12] proposed heuristic techniques in such models. He restricted his study to stationary nested policies. Some researchs are also developped in discrete time environement (see e.g. van Hoesel *et al* [8] or Axsäter [2] for heuristic techniques). Recall that all these paper consider increasing holding costs along the chain and unlimited production capacity.

In this paper, we focus on serial systems. We do not make any assumption on the holding costs : the holding cost at a downstream stage can be lower than the one at an upstream stage. This situation happens in distribution systems, because no "real-value" is added to the product. Thus, one can imagine some cheap holding cost at a downstream stage thanks to more space, efficient handling facilities and so on.

We also consider some production capacity constraints at the start of the chain. The need for this add-on comes from the need of coordination in todays logistic-chains. The lack of this type of coordination will become apparent in section 4.3 when we compare our results to some literature results that ignore production capacity. This production capacity leads to continuous in-system flow. Together with unconstrained holding costs along the chain, this capacity constraint shows up the need for new approaches.

3 Model description, Basic results and procedures

3.1 The model

We study a N-stages serial system (see figure 1). Stages will be denoted by indice $i \in [1; N]$. The first facility (stage 1) is a production facility. There is a production capacity c per unit of time. The last facility (stage N) is a retailer that faces a constant and continuous demand with demand rate d. Without loss of generality, we assume that the production capacity c is equal to the demand rate d. Between stage i and its successor $i + 1$, there is a transportation route. We consider a fixed transportation cost C_i for route $\{i-1, i\}$: each shipment costs the same amount, whatever quantity is shipped. Because the demand is constant and known, we do not consider any lead time. Each stage can hold any amount of inventory. For stage i, there is an inventory holding cost h_i per unit per unit of time. We aim at optimizing the global cost of the logistic chain, restricting our search space to stationary policies. This means that we must determine the best period-set $\mathbf{T} = \{T_2, ..., T_i, ..., T_N\}$, where T_i is the period of replenishment between stage $i - 1$ (supplier) and stage i (customer).

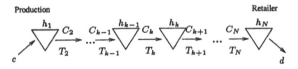

FIG. 1 – The N stages serial model

3.2 Multiple policies

Very interesting and efficient policies in such a system are multiple policies. A multiple policiy is a policy in which the ratio between two successive periods $\frac{T_i}{T_{i+1}}$ is an integer or the inverse of an integer. These policies are very natural ones as they simulate the use of a basic lot-size and some multiple of it. We can also expect them to perform quite well because of the saw-tooth pattern of the evolution of inventory levels. Furthermore they make the inventory cost of facility i easy to evaluate. Though we will not build up only multiple policies in our heuristics, they represent the major part of the policies we have generated. Moreover, there exist some theoritical results on this kind of policies that are presented in section 3.3.1.

3.3 Preliminary results

3.3.1 Performance of multiple policies

In Boissière et al [4] we study the same serial system restricted to 3 stages (see figure 2). This model is composed by a capacited production facility, that provides products to a retailer, that faces a constant and continuous demand. The flow goes through an intermediate facility. The objective is to find a stationary policy that minimizes the long-run average cost of a product, considering holding costs at all facilities, as well as transportation costs. In this context, we show that multiple policies are very closed to optimal (less than 1.5% more expensive). We use this preliminary result as a local-system optimisation in the N-stages model. This optimisation will be denoted MULTI along the paper.

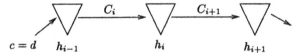

FIG. 2 – Model used for MULTI procedure

3.3.2 Constrained lot sizing

In Boissière et al [3], we provide an algorithm for finding an optimal lot-size in a constrained environement. The model we study in that paper is the following : there is a production facility, that provides products to an intermediate warehouse, which provides the retailer. The shipments period between the warehouse and the retailer is given (imposed by the retailer), and the objective is to find the policy that minimizes the production facility cost, plus the transportation cost between the production facility and the warehouse, plus the holding cost at the warehouse (see figure 3). We propose an algorithm that gives ε-optimal policy for any ε. This algorithm is based on increasing an integer value, and calculating the best policy for each step. We show in that paper that the algorithm converges, we observed that in practice, the computation remains very short. Again, we will use this result as a basic procedure in our heuristics, named CLS procedure (Constrained Lot-Sizing).

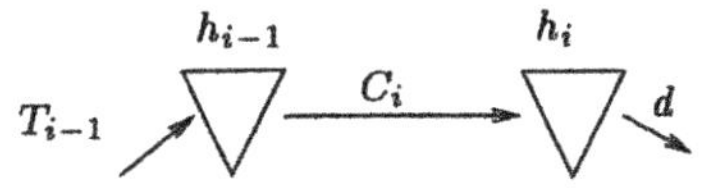

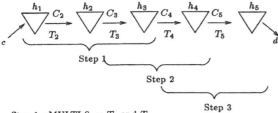

FIG. 3 – Model used for CLS procedure

3.4 Use of local procedure

Our heuristics for N-stages problems are based on system-MYOPIC-heuristics. Schwarz & Schrage [11] introduced this concept meaning that the global system optimisation is done by several local systems optimisation, which do not take the whole system into account.

In our approach we determine the periods one by one (or two by two). For determining new periods, we use the preliminary results related in 3.3.2 and 3.3.1.

We start by determining the 2 first periods at one end of the chain (e.g. T_2 and T_3), using MULTI procedure and considering that the 3rd facility faces constant and continuous output (or input, depending on which end of the chain we start, see section 4.1). These periods are then fixed, and we determine the 3rd period, 4th and so on. For example when T_3 is fixed we can find the optimal value for T_4 considering that stage 4 is the last stage and faces demand rate d, using CLS procedure. And we can go on each stage until determining T_N. An example is shown on figure 4.

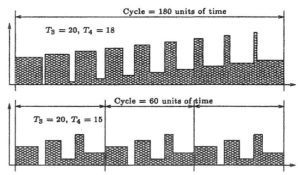

Step 1 : MULTI fixes T_2 and T_3
Step 2 : CLS fixes T_4
Step 3 : CLS fixes T_5

FIG. 4 – CLS based heuristic applied to a 5-stages system

We can also use MULTI procedure to build up the policy. Indeed, after determining T_2 and T_3, we can determine T_4 and T_5, assuming that facility 3 faces a constant input rate of $c = d$, and facility 5 a constant output rate d (see figure 5 for a 5-stages model).

Unfortunatelly, unlike CLS procedure, this action does not take into account the value of T_3. So, we may obtain a period T_4 that does not fit

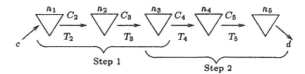

Step 1 : MULTI fixes T_2 and T_3
Step 2 : MULTI + Adaptation procedure fixes T_4 and T_5

FIG. 5 – MULTI based heuristic applied to a 5-stages system

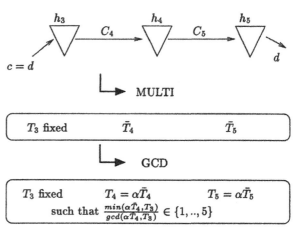

FIG. 6 – Inventory evolution at stage 3 for different values of T_3 and T_4

with T_3's value leading to very high holding costs at stage 3. See for example the top evolution on figure 6 for e.g. $T_3 = 20$ and $T_4 = 18$, $T_5 = 9$. Hence, there is a need for adapting the new period T_4 to T_3's value conserving the good coordination between T_4 and T_5. In the last example, we can expect $T_3 = 20$, $T_4 = 20$ and $T_5 = 10$ or $T_3 = 20$, $T_4 = 15$ and $T_5 = 7.5$ to perform much better (see figure 6)! This is why we have developed some adaptative procedures.

The first adaptation procedure is based on the fact that the holding cost at a stage is a decreasing function of the *gcd* of in and out periods (figure 7). In most cases moving from a small gcd (e.g. $gcd(18, 20) = 2$) to a bigger gcd (e.g. $gcd(20, 20) = 20$ or $gcd(15, 20) = 5$) does not need big changes in the value of the periods, so does not dramaticaly deteriorate costs in other systems

FIG. 7 – The GCD procedure

while improving holding costs. So, as an adaptation procedure, we chose to recalculate the two new periods while making this gcd greater than some given value. In practise we first compute $\hat{T}_4$ and $\hat{T}_5$ together (using MULTI), and we take the best value of α such that $\frac{min(\alpha\hat{T}_4, T_3)}{gcd(\alpha\hat{T}_4, T_3)} \in \{1..5\}$. Th criteria we evaluate is the cost of the system from stage 3 (holding cost function of T_3 (fixed) and T_4), to stage 5 (holding cost function of T_5).

The second adaptation heuristic combines both MULTI and CLS approach (called COMBINE procedure). A couple of multiple periods is computed using MULTI procedure. The ratio between both periods is conserved. The real value of the first period ("closed to constraint period") is then computed with CLS procedure.

4 Heuristics and results

4.1 Heuristic description

The heuristics we developped are characterized by 4 major attributes :

1. the starting point option. This attribute identifies if the construction of the policy starts from the production facility (denoted P-start), or from the retailer's facility (denoted R-start).

2. the go-and-back option. This attribute allows us to choose to do go-and-back procedures, meaning that after having built a first policy starting from one side (P or R-start), the heuristic goes back to the start along the chain, taking into account the last period found. For example, in the previous examples, after T_5 has been calculated, we recompute T_4, then T_3 and again T_2. We can allow either no back to start calculation (Go option), a complete go-and-back movement (GoBack option), or doing so up to convergence (Conv option). Notice that because we have no theoritical result on the convergence we decided to stop after a maximum of 100 go-and-back movements. However it appeared that very few instances computation went up to that value (about 100 out of 11,000). Because convergence is not detected by using the global cost parameter, but by identifying cycling moves, the heuristic takes the best policy ever generated until the cycles arise.

3. the optimisation procedure option. The attribute represents the optimisation procedure used at each step of the construction of a policy. A reduced local system including either 2-stages with constrained output period and constant input (for CLS, see figure 3) or 3 stages with constant output and input (for MULTI, see figure 2) is studied using one

the optimisation procedure (see section 3.4). These procedures give 1 or 2 periods to add to the in-construction policy. Notice that all heuristics start with a MULTI procedure for the first two periods.

4. the adaptation option. The adaptation attribute is only available with MULTI optimisation procedure. It represents some adaptative behavior on the periods that will be added to the policy. They are of 2 types : the first is the gcd procedure (called GCD adaptation, see section 3.4) and the second (COMBINE option) combines both MULTI and CLS procedures (section 3.4).

We have tested all possible heuristics according to this classification as well as some variants that were not convincing. We will present only a few of them, the others being either less good, or equivalent to the one presented.

The most natural heuristics are the P or R-start/GO/CLS procedures. These procedures build the policy stage by stage, using an optimal local procedure each step, starting from one side or the other, and do not reconsider their choices. Naturally, we extended these procedures with the GoBack and Conv options. Clearly, both evolutions improve the performance, but we will only study the Conv option here. Moreover, it arises that the R-start was a bad option for this construction style. Again, we only give P/Conv/CLS's results in section 4.3.

A second class of heuristic contains the ones that use the MULTI procedure for building up the policy. The most natural one here is the P or R-start/GO/MULTI heuristic. However, we do not talk about this one as it performs really poorly because of the lack of adaptative procedure. Indeed, in this case, two successive periods may be determined independently, thus leading to very expensive holding costs at some stages. The first heuristic we tested was the P-start/GO/MULTI/GCD, that forces a GCD adaption. Again we do not speak here of the similar R-start because it performs far less good. From that point, we developped the GoBack and Conv version of it, leading to P/GB/MULTI/GCD (not detailed in this paper) and P/Conv/MULTI/GCD. At last, we managed to use CLS as an adaptation procedure. For this last class of heuritics we tried both P-start and R-start (respectively P/Conv/MULTI/COMBINE and R/Conv/MULTI/COMBINE), using the COMBINE adaptative procedure. Finally we mixed both approach, to obtain the PR/Conv/MULTI/COMBINE procedure that takes the best policy from the two preceeding heuristics.

4.2 The experiment

We have tested several type of instances, from 3 stages up to 10 stages. There are 3 class of instances : in the first class (class 1), the non-decreasing holding cost assumption holds, in the second (class 2), this assumption holds as well, but the holding cost at the production facility can be greater than the next one, and in the third class (class 3, the balanced class), no assumption is made on holding costs.

We compare the results of the two first class with Roundy's power-of-two policies. In this context, Roundy's result (98% guaranteed performance) does not hold anymore. However, when the production holding cost is low compared to the others or the capacity is much larger than the demand, power-of-two policies may lead to very efficient chain. This is also true when the holding cost at the production facility becomes small, compared to the global cost (increasing number of stages).

Each result represents the average value of 1,000 random instances. No running time is given because it is too small to be relevant (11,000 instances within a few minutes).

4.3 The results

Table 1 compares the three heuriStics based on CLS optimisation procedure. We recall here that for 3 stages instances, all heuristics are equivalent to a unique MULTI procedure.

Class & Stages	R/GO/	P/GO/	P/Conv/
Class 1			
3	236	236	236
5	463	442	439
7	736	668	661
9	1046	909	900
Class 2			
5	469	479	474
7	721	702	690
10	1172	1062	1045
Class 3			
5	67	67	66
7	92	94	91
10	131	132	128

TAB. 1 – Average costs for 1000 instances CLS based heuristics

From these results, we see that R/GO/ is generally far less good than P/GO/. This is the major reason why we did not report any R/Conv/ heuristic. Anyway, R/GO/ is slightly better when the system is balanced (third class instances).

We see that waiting for convergence does not dramatically improves performance. As expected P/Conv/ is slightly better than P/GO/ (about 1%

for Class 1 instances), but becomes more interesting for balanced instances (up to 3% less costly). However, P/Conv/ is clearly the best heuristic among those, being beaten only once by R/GO/ by 1%.

In table 2, we compare the MULTI based heuristics. Here again, we do not report any R-start

Class & Stages	P/GO//GCD	P/Conv//GCD	P/Conv//COMBINE	PR/Conv//COMBINE
Class 1				
3	236	236	236	236
5	444	439	431	427
7	665	658	624	614
9	924	913	827	808
Class 2				
5	476	467	457	453
7	691	676	649	642
10	1061	1046	964	923
Class 3				
5	67	64.7	62	61
7	90	85	79	77
10	127	120	109	100

TAB. 2 – Average costs for 1000 instances for MULTI based heuristics

result because they perform pretty badly in the first class instances. If we compare P/GO//GCD and its extension P/Conv//GCD we see that the second one is of course better but the gains are not tremendous (about 1%) for class 1 instances. However, convergence becomes very interesting again when the instances become balanced, reaching some 5% improvement for class 3 instances.

Table 2 also reports the results of some combined heuristics. Cleary the COMBINE adaptative procedure performs much better than the GCD adaption.

We also see that taking the best solution with a P-start and R-start, even when reaching convergence greatly improves the general result, including balanced and unbalanced systems.

Generally speaking, we must admit that MULTI procedure with GCD adaptation rarely improves the basic CLS heuristic for unbalanced instances. Still, it is much better for balanced systems (see table 2 and 1). The COMBINE approach together with both P-start and R-start runs, clearly outperform any single procedure heuristics (up to 9% cheaper than any other).

Anyhow, these results do not allow us to have any idea about how our heuristics perform really. In order to have some basic idea of it, we compared these results with Roundy's power-of-two policies for class 1 and class 2 instances. To do so, we compute the optimal power-of-two policy on the N-1 stages system obtained by removing the production facility from our original system. The computed power-of-two policies is applied to the global system (re-including the production facility) to obtain the global cost. Results of this computation together with global results are reported in table 3. Recall that we could not compute power-of-two policies for class 3 instances.

Class & Stages	Power-of-two	P/ Conv/ CLS	P/ Conv/ GCD	PR/ Conv/ COMBINE
Class 1	Value			
3	240	-1.7%	-1.7%	-1.7%
5	421	4.3%	4.3%	1.4%
7	597	10.8%	10.2%	2.9%
9	776	16.4%	18.1%	4.5%
Class 2				
5	527	-10.0%	-11.4%	-14.0%
7	701	-1.6%	-3.6%	-8.4%
10	960	8.9%	9.0%	-3.9%

TAB. 3 – Overcost of the heuristics compared to power-of-two policies

From these results we can expect our last heuristic to work very well for any kind of network. Power-of-two policies are very efficient for non-decreasing-holding-costs-models even when adding some low holding cost at the production facility (class 1 instances. When the number of stages grows (class 1 instances), this last facility loses its importance and Roundy's result becomes very efficient, while our best heuristic stays up to 9 stages within less than 5% extra-cost.

When the holding cost at the production facility becomes greater (class 2 instances), it takes more importance, then our heuristics are all efficient, even for large number of stages.

5 Conclusion

In this paper, we have presented some heuristics for finding good inventory management policies for serial distribution system with constant and continuous demand, limited production capacited at the start of the chain, fixed transportation costs and arbitrary holding costs. The heuristics construct the policy stage after stage in a Myopic way. The basic procedures are based on 2 local-systems optimisation procedures. The best heuristic we developped combine both of these procedures to obtain very good results. Some perspectives for this work are to bound the loss induced by the use of these heuristics, and to try to adapt the global scheme we have developped to other types of networks.

References

[1] D. Atkins and D. Sun. 98%-effective lot sizing for series inventory systems with backlogging. *Operations Research*, 43(2) :335–345, 1995.

[2] S. Axsäter. Performance bounds for lot sizing heuristics. *Management Science*, 31(5) :634–640, 1985.

[3] J. Boissière, Y. Frein, and C. Rapine. Distribution de produits : comparaison de performance avec et sans coordination des acteurs. In *MOdélisation et SIMulation 2004*, 2004.

[4] J. Boissière, Y. Frein, and C. Rapine. Near optimal policies in a distribution chain with production capacity. In *Project Manufacturing and Scheduling 2004*, 2004.

[5] A. Federgruen, M. Queyranne, and Y-S. Zheng. Simple power-of-two policies are close to optimal in a general class of production /distribution networks with general joint setup costs. *Mathematics of Operations Research*, 17(4) :950–963, 1992.

[6] A. Federgruen and Y-S. Zheng. Optimal power-of-two replenishment strategies in capacited general production/distribution networks. *Management Science*, 39(6) :710–727, 1993.

[7] S. Graves, A. Rinnooy Kan, and P. Zipkin. *Handbook in operations research and management science*, volume Vol 4 : Logistics of production and inventory. North Holland, Amsterdam, 1993.

[8] S. Van Hoesel, H.E. Romeijn, D. Romero Morales, and A.P.M. Wagelmans. Polynomial algorithms for some multi-level lot-sizing problems with production capacities. Technical Report 24, Econometric Institute, Erasmus University Rotterdam, June 2002.

[9] R. Roundy. 98%-effective integer-ratio lot-sizing for one-warehouse multi-retailer systems. *Management Science*, 31(11) :1416–1430, 1985.

[10] R. Roundy. A 98%-effective lot-sizing rule for a multi-product, multi-stage production/inventory system. *Mathematics of Operations Research*, 11(4) :699–727, Novembre 1986.

[11] L.B. Schwarz and L. Schrage. Optimal and sytem myopic policies for multi-echelon production/inventory assembly systems. *Management Science*, 21(11) :1285–1294, 1975.

[12] J.F. Williams. Heuristic techniques for simultaneous scheduling of production and distribution in multi-echelon structures ; theory and empirical comparisons. *Management Science*, 27(3) :336–352, 1981.

PUBLICATIONS
www.elsevier.com/locate/ifac

CONTROL LAW SYNTHESIS ALGORITHM FOR DISCRETE-EVENT SYSTEMS

HENRY, Sébastien – DESCHAMPS, Eric - ZAMAI, Eric – JACOMINO, Mireille

Laboratoire d'Automatique de Grenoble (LAG)
ENSIEG - Rue de la houille blanche, Domaine Universitaire, BP 46
38402 Saint Martin d'Hères Cedex, France
Tel: (33)4 76 82 64 05, E-mail: henrys, deschamps, zamai, jacomino (@lag.ensieg.inpg.fr)

Abstract: Usually, in an industrial context, the design of discrete control law to drive manufacturing system is assumed off line by several experts. This is mainly due to the lack of a generic method to model the controlled system abilities. Consequently, not only the design of all the control laws mobilizes many PLC program developers, but also, in case of unexpected resource failures, the reconfiguration process can be only considered from a manual point of view. So, to bring a solution in this field, from a specific controlled system model, an automated control law synthesis is proposed to provide a law compatible with one of the CEI 61131-3 languages.

Keywords: Discrete control law, Synthesis, Reconfiguration, Manufacturing systems, Programmable Logic Controllers, IEC 61131-3.

1. INTRODUCTION

This paper deals with the dependability of automated systems and more particularly manufacturing systems. These systems are made of a supervision, monitoring and control (SM&C) system and a controlled system (see Fig. 1) defined by the set of resources. In this paper, a resource will only be a manufacturing or handling machine. From a hierarchical and modular point of view, a manufacturing system can be a plant, a workshop, a flexible manufacturing cell or a machine-tool. A product flow goes through all these systems. Depending on services offered by the controlled system and the customer's request, the control system applies control laws to act on the product flow.

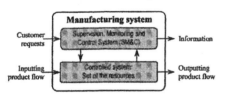

Fig. 1. A manufacturing system

This paper is mainly directed towards the automated control law synthesis. So, in the context of manufacturing system design, the objective is to reduce the time required to create a control law in one of the CEI 61131-3 languages and to minimize the debug time by creating correct code initially. Secondly, in the context of the dynamic reconfiguration, the resource failure reactivity and the customer's request variation will be definitely improved. So, this paper is organized as follow: in section 2, the functions of a manufacturing system are discussed. Then the required initial data for the

automated control law synthesis are defined. And they are compared with data considered by the classical synthesis approaches. So, section 3 presents the aim of the paper. In section 4, the controlled system model is submitted and section 5 presents other initial data. Thus, a discrete control law is synthesized with a specific algorithm, in section 6. The paper then concludes and gives several research directions with section 7.

2. SCOPE

The basic function defined by the norm NF X50-100, and in (see Web site Value methodology) of a manufacturing system is to increase the value of the product flow. It achieves its function by having an effect on the product flow, as in Fig. 1. This effect is defined for a product by the difference between its input state and its output state which are data for the control law synthesis problem. The secondary function of a manufacturing system is to guarantee the performances. Indeed to assure a profit to the company, the added product flow value must be superior to the manufacturing cost. From a control law point of view, the manufacturing cost is the total of the costs generated by the infringement of the security and environmental constraints and the costs resulting from the values of the productivity and quality criteria. In other words, the control law must respect the security and environmental constraints and optimize the productivity and quality criteria.

To achieve these functions, the manufacturing system must control the product flow evolutions. These are linked to the resource evolutions controlled by the control system according to interpreted or compiled control laws compatible with the CEI 61131-3 norm. Thus, the control laws of all the levels must respect the controlled system constraints (security and environmental) and the objectives (the product specifications, the values of the productivity and quality criteria), see (Mendez, et al., 2002). When an expert synthesizes manually a control law, he has in mind a picture of the resources evolutions and their possible effects on the product flow. In addition to the evolutions, this picture takes the controlled system constraints into account. As the expert, the control law synthesis method must have access to a controlled system model, to the initial state of the product flow and the resources (PF&R) and to a model of the objectives. For discrete event systems, the three synthesis approaches presented below are compared on the basis of the system model, the initial system state and the objectives.

To drive the controlled system, the supervised control presented in (Charbonnier, et al., 1999) is based on the Ramadge and Wonham theory. In this approach, the supervised system is a set of resources with their local control law given by an expert. From the automata system model, which is not obtained with a modeling methodology, the supervisor limits the system behavior to the most permissive, which respects the objective specifications. They must also be expressed with the automata. Thus they cannot express the optimization criteria. This method then does not guarantee performance in productivity and quality criteria.

The automated control code synthesis proposed by (Holloway et al., 2000) is based on the condition systems that are closely related to Interpreted Petri Nets. A component model representing a physical component is described with this formalism as well as the associated taskblocks. In a condition system, the output conditions result from marked places. For each output condition of the component model and from any marking, a taskblock defines all the control laws to reach the marking for which the output condition is true. Except for the low level, the set of control laws cannot be built because of its size. At the upper level for the coordination of the taskblocks, the authors propose to use the supervisory control approach.

According to (French, 1982), the general scheduling problem is to find a sequence of operations, carrying out one or several tasks, which is a feasible schedule, and optimal with respect to some performance criterion. An operation represents the use of a resource to produce a service. Contrary to the two previously described synthesis approaches, the scheduling is not a method with one algorithm and a controlled system model that is exactly defined with the modeled constraints point of view. Each scheduling problem defined by controlled system model and objectives (task(s) and criterion) has a particular solving algorithm. Thus in the context of

the reconfiguration, if a disruption modifies the scheduling problem, then a new algorithm must be applied. This is deeply penalizing for the expected reactivity. However, the controlled system model is always formalized by a set of operations with the constraints on or between the operations like precedence, preemption, splitting, and very rarely condition(s) on the resource states, etc, and with the resource properties like disjunctive, etc.

3. AIM OF PAPER

In comparison to other synthesis methods, scheduling methods are interesting from the point of view of the optimization criteria and the controlled system modeling with operations. In spite of the amount of works in the scheduling field, none of them has a very low-level controlled system point of view with so many operation constraints. First, to use the scheduling method to synthesize control laws, the controlled system model must integrate all the constraints. Then, from data (controlled system model, initial PF&R state, the objectives), the automated control law synthesis is developed. As in the scheduling methods, the first synthesis stage generates a precedence graph. The second one translates this graph to one of the CEI 61131-3 languages (see Fig. 2). The paper focuses on the first synthesis stage considering only duration criterion.

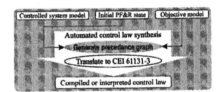

Fig. 2. Aim of paper

4. CONTROLLED SYSTEM MODEL

To achieve its basic function, the manufacturing system must control the product flow evolutions. But its control system does not drive directly the product flow evolutions, it drives the resource evolutions. Thus, we propose to represent the controlled system model by the set of the product flow evolutions, the resource evolutions and the links between the product flow and resource evolutions. With these links and from product flow evolutions respecting the objectives, we can deduce the required resource evolutions, and in this way we can automatically synthesize a control law. If the product flow and resource evolutions are modeled in a same operation, its model defines this link.

4.1. SAPHIR.

To perform the understanding of the proposed approach, this section presents an application example based on the loading system (see Fig. 3) of the research platform Saphir of the Laboratoire d'Automatique de Grenoble, in France. However, it is important to note that this approach can be generalized for all Discrete Events Systems (DES). This platform is dedicated to the assembly of camshafts. A rotating storage been made up of a tray with four places is used to receive until six different kinds of products. These products are identified by a weight identification system. Once a product has been identified, a central conveyor drives it to a sorting device. So, a robot takes the different products to assembly them. A shopworker is charged to fill the rotating storage and to empty the assembly station.

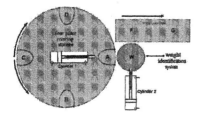

Fig. 3. Loading system of SAPHIR

4.2. Operation Model.

To choose the operations to reach the final product flow state given by the objectives, it is required to have information on the operation effect on the product flow (Henry, et al., 2004). To have this effect, the operation modifies the state of the resource which carries out the operation. Thus a resource evolution defined by initial, intermediate and final resource states is characteristic of an operation. Depending on the product flow state before the operation running, an operation can have no or several different effects on the product flow. For instance, the cylinder 1 (C1) state from the retracted position to the extend position is characteristic of the "Extend Cylinder 1" (EC1) operation. If there is a product in A, this operation will then have an effect on a product. But also if there is a product between A and W. Thus, an operation model is made up of a basic behavior describing the effect on the resource, no or several extra behaviors describing the possible effect on the product flow, as in Fig. 4.

Extend Cylinder 1 (EC1) Duration: 3 s		
Effect on the resource	**Associated constraints**	
C1 retracted position	C2 retracted position	No P between A and B No P between A and D
C1 intermediate position	C2 retracted position	No P between A and B No P between A and D
C1 extended position		
Effect on the product	**Associated constraints**	
P in A	Tray indexed position Tray null speed	No P between A and W No P in W
P between A and W	Tray indexed position Tray null speed	No P in W
P in W		
Effect on the product	**Associated constraints**	

Fig. 4. EC1 operation

Rotating the Tray a quarter Turn Clockwise(TTC) Duration: 4 s		
Effect on the resource	**Associated constraints**	
Tray indexed position Tray null speed		No P between A and W
Tray non-indexed pos. Tray non-null speed		No P between A and W
Tray indexed position Tray null speed		
Effect on the product	**Associated constraints**	
P in D	C1 retracted position	
P between D and A	C1 retracted position	
P in A		
Effect on the product	**Associated constraints**	
P in C		
P between C and D		
P in D		
Effect on the product	**Associated constraints**	
Effect on the product	**Associated constraints**	

Fig. 5. TTC operation

To carry out an operation, at least the basic behavior must be carried out. For that, the condition on the initial state of the resource must be true. For this reason, the initial resource state is called the condition of the basic behavior. And to carry out an extra behavior, the condition on the initial product state of this extra behavior must be true. The initial product state is called the condition of the extra behavior. In the effect model, the intermediate and final states are the result of the operation running. But the effect model does not take the security and environmental constraints into account. For instance, this model does not guarantee collision between the cylinders will be avoided. Thus, the EC1 running ends accurately only if constraints on the other resources and the product flow are respected before (pre-constraints) and during (constraints) the EC1 running, as in the associated constraints in Fig. 4. For a behavior, the set of pre-constraints and constraints is noted (pre-)constraints. If the condition of an extra behavior is false, the associated (pre-)constraints of this extra behavior must not be respected. Finally to optimize the control law, the quantifiable criteria like the time cycle are required. To assess these criteria, the operation modeling must give the operation features like the duration. The partial model of the rotating the Tray a quarter Turn Clockwise (TTC) operation presented in Fig. 5 shows an operation that can simultaneously have several extra behaviors contrary to the EC1 operation. Indeed, "P between A and W" state is the second-extra-behavior condition but its negation is a first-extra-behavior pre-constraint.

State variables of			Operation 1	Operation 2	Operation 3
	product flow	∈ product specifications	are modified	are not modified	are not modified
		∉ product specifications	can be modified	are modified	are not modified
	resources		can be modified	can be modified	are modified

Fig. 6. Three kinds of operations

For the product flow state, we can distinguish two kinds of product-flow state variables. Some state variables result from the product specifications like geometric forms, etc. And other state variables of the product flow result from the controlled system like their position, etc. According to the state variables modified by an operation running, we have identified three kinds of operations, presented in Fig. 6.

4.3. Information on the specific states.

When a product is located in F on the central conveyor, no operation offered by the loading system can evacuate this product. Nevertheless, the "No product in F" state is a pre-constraint and a constraint of the "extend C2" operation with an effect on the product. So an operation sequence to reach this state does not exist. To avoid searching for such sequence, information on this specific state must be represented in the controlled system model. The environment (sensors, other control systems, human operators) of the loading system provides this information. Thus, the controlled system model is the set of operations with a specific state list like "No product in F".

5. INITIAL PF&R STATE AND OBJECTIVES

Before defining the initial state and the objectives, two details must be clarified. Firstly, the loading system does not have any operation 1. So in the example presented, there is not any State Variable of the Product flow belonging to the product Specifications, called in this paper SVP∈S. Secondly, the operator can add or remove the products in C. He does not inform the control system about his actions. And the product presence (or absence) in the tray is known when the product arrives in A. So in C, the presence of a product is unknown. Now the initial state can be defined. For the state variables of the product flow not belonging to the product specifications (SVP∉S), the values are: no product in A, W, F and between all the points; unknown in B, C and D; non-identified for a product in B, C and D. For the state variables of the resources (SVR), the values are: C1 in the retracted position, C2 in the retracted position, the tray in the indexed position with a null speed. As soon as possible, an identified product must be put in F on the central conveyor. The final state of this objective defines only the values of some SVP∉S: product in F, identified for the product in F. The optimized criterion is the time cycle.

6. AUTOMATED CONTROL LAW SYNTHESIS

An operation is an elementary brick. The assembly of two bricks creates a link between these bricks. This link is not modeled in the controlled system model, but it must be found by a synthesis algorithm according to the objectives. Finally, a control law is a particular assembly of the elementary bricks. The proposed controlled system model does not represent all the possible control laws. Then, from the initial PF&R state and the objectives, the global synthesis method builds the acceptable control laws. Afterwards, it searches for the optimal control law according to the criteria. But the set of the acceptable control laws is too large to be built. So, we propose an automated control law synthesis step by step with local optimizations for the first synthesis stage that generates a precedence graph. Depending on the objectives, this graph will be acyclic or cyclic, as in Fig. 7. For instance before the preventive maintenance, some resources must be stopped in a specific state with a control law. And after the resources must be started up with an other control law. To synthesize these control laws, only a acyclic graph will be generated. On the contrary to manufacture several products, a same control law is carried out several time in succession provided that the input stock is supplied and the output stock is emptied. For the loading system, the operator must put products on the tray, and the central conveyor must evacuate the products in F. In this case, a acyclic graph and a cyclic graph are generated.

Depending on the three kinds of operations defined in Fig. 6, the acyclic graph generation is split into three algorithm steps (see Fig. 7) that generate: a first sequence with only operations 1, a second sequence with operations 1 and operations 2, and finally a acyclic graph with the three kinds of operations. The first sequence is defined to only modify the SVP∈S from their initial value to their final value with respect to the product constraints (assembly, etc). To respect the (pre-)constraints on SVP∉S to run the operations 1 of the first sequence, sequences of operations 2 are added between the operations 1. And finally to respect the conditions and the (pre-)constraints on SVR to run operations 1 and 2, sequences of operations 3 are added at the second sequence to generate a acyclic graph. The three kinds of operation sequences are generated with the same technique. This one is applied on the loading system for a sequence of operations 2.

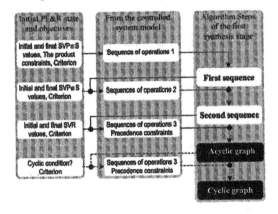

Fig. 7. Algorithm steps of the first synthesis stage

6.1. Second sequence.

Where sequences must be added? To obtain the second sequence, sequences of operations 2 are added to the first sequence. Depending on the initial and final SVP∉S values and the first sequence, we must know where sequences of operations 2 must be integrated. The i-1 and i operations are two consecutive operations of the first sequence. To run the i operation, the (pre-)constraints on the SVP∉S must be satisfied. If the SVP∉S values after the i-1 operation satisfy them, then no operation 2 must be added. Otherwise a sequence of operations 2 must be added between the i-1 and i operations. For the same reasons to run the first operation 1, a sequence of operations 2 can be added before this operation according to the initial SVP∉S values. Similarly, a sequence of operations 2 can be added after the last operation 1 to satisfy the expected final values of SVP∉S. The loading system doest not have any operation 1. Thus, only one sequence of operations 2 is generated. This sequence modifies the SVP∉S from their initial values to their final values.

PF&R states defined completely or partially? When the control law will be applied, the PF&R will have to be in a state defined by the initial state-variable values. If all these values are not defined (the product presence in B, C and D is unknown, for instance), then several PF&R states can respect these values. Thus, the initial PF&R state is partially defined. Even though from an initial PF&R state that is completely defined, the synthesized control law is generally more optimized. Indeed when the initial value of a variable is not known, a (pre-)constraint on this variable is considered non-satisfied. For instance, if the product presence in W is unknown, then the first extra behavior of the EC1 operation cannot be carried out (see Fig. 4). And inversely an extra behavior conditioned by this variable is likely to be carried out. Thus, the associated (pre-)constraints must also be satisfied. For instance, if the product presence in D is unknown, then the constraints associated with the first extra behavior of the TTC must be respected: C1 in a retracted position (see Fig. 5). Finally, an initial PF&R state that is partially defined can increase the number of constraints having to be satisfied and can reduce the number of satisfied constraints. Therefore from such a state, the number of carrying out operations can be reduced. And finally if a graph is found, it is generally less optimum than a graph found with a initial PF&R state that is completely defined. This is the consequence of the problem that is over-forced with a partial initial PF&R state. On the contrary when the final PF&R state is partially defined by the objectives, the problem is then less forced because more PF&R states respect the final state-variable values. For the loading system and the objectives defined in §5, the initial state and the final state are partially defined.

Build an automaton to find a sequence of operations 2. A state of this automaton is characterized by the SVP∉S values. An event is an operation 2. Depending on if the sequence sought is integrated between two operations 1 or before the first operation 1, the initial automaton state is defined by the SVP∉S values after the i-1 operation 1 or by the initial SVP∉S values. Similarly depending on if the sequence sought is integrated between two operations 1 or after the last operation 1, a state is a marker state when the (pre-)constraints on SVP∉S of the i operation 1 or the final SVP∉S values are satisfied with the SVP∉S values of this state. Finally to find a sequence of operations 2, the best way from the initial state to a marker state is sought according to the optimized criterion.

How build this automaton? From a A state, all the operations 2 able to run are sought. These operations are the executable operations 2 from the A state. Then for each executable operation 2, a state is calculated from the A state according to the operation model describing the effect on the SVP∉S values. The calculated state can be an existing state or a new state. When the state already exists, a transition is added to the transition function of the automaton. The event associated is operation 2 found. And if the state is new, the transition function is also updated and the new state is then added to the set of states. The previous technique is first applied at the initial automaton state and after at each new state, except if the new state is a marker state.

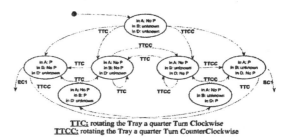

TTC: rotating the Tray a quarter Turn Clockwise
TTCC: rotating the Tray a quarter Turn CounterClockwise

Fig. 8. Partial automata of operations 2

Seek executable operations 2 from an automaton state to add a sequence of operations 2 between the i-1 and i operations 1. For a state defined by the SVP∉S values, an operation is considered as an executable operation 2, if it has at least an executable extra behavior that has only an effect on the SVP∉S. So, the extra-behavior condition defined by SVP∉S values must be true and the associated constraints on the PF must also be true by considering the SVP∈S values after the i-1 operation 1. The basic behavior and the (pre-)constraints on the SVR are not considered for the second-sequence generation. From the initial state of the loading system, two operations have one or several executable extra behaviors with an effect on the SVP∉S values: TTC and TTCC (see Fig. 8). For other operations, the extra-behavior conditions on the SVP∉S values are false like the extra behaviors of the EC1 operation.

Finally, to find here the sequence of operations 2 to reach the objectives defined in §5, the automaton has fifteen states, twenty-five arcs and two marker states. Part of it is presented in Fig. 8 with only the SVP∉S modified. From this automaton, the TTC, EC1, IP and EC2 sequence is found with the sequence length as the criterion to minimize (see Fig. 9).

6.2. End of the synthesis

Acyclic Graph (AG). The j-1 and j operations are two consecutive operations of the second sequence. To run the j operation, the conditions and (pre-) constraints on the SVR must be satisfied. If the SVR values after the j-1 operation do not satisfy them, then a sequence of operations 3 is added before the j operation, but it is not necessarily added after the j-1 operation. Indeed operations 3 do not have an effect on the product; they can be run at the same time as operations 1 or 2. To minimize the time cycle, we search for the earliest time when each operation 3 can be run. A precedence constraint is added before operation 3, as for the RC1 operation 3 in Fig. 9. At the end of the third algorithm step, a AG is built, like the grey area in Fig. 9. This one puts one identified product on the central conveyor.

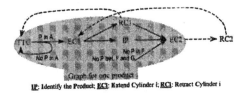

IP: Identify the Product; ECi: Extend Cylinder i; RCi: Retract Cylinder i

Fig. 9. The precedence graphs

Cyclic graph. To build the cyclic graph, the fourth algorithm step considers the possibility of applying again the previously defined AG to put a second identified product. To run the operations of the second AG at the earliest, operations 3 and the precedence constraints are added between the operations of both AGs. For instance, the TTC operation of the second AG can be run at the earliest after the RC1 operation of the first AG. But the EC1 operation of the second AG cannot be run after the last operation (EC2) of the first AG because the EC1 pre-constraints are not respected. The RC2 operation 3 must be added between EC2 of the first AG and EC1 of the second AG. To end the fourth algorithm step, both AG are merged. For instance the TTC operations for the first and the second AG are merged and so on (see Fig. 9).

Second synthesis stage: translate to SFC. In Fig. 10, we show that the translation of the cyclic graph to SFC is feasible. Each operation is associated with an action of one stage. And the SFC control structure is defined by the precedence constraints.

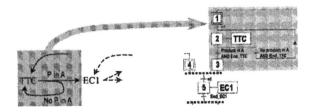

Fig. 10. Translate to CEI 61131-3

7. CONCLUSION AND FUTURE WORKS

In this paper, problem of controlled system modeling and the automated control law synthesis is dealt with. First, each resource is described by all the offered operations. The structure of a proposed operation model is generic and can be applied for all DES. This model is composed of the basic behavior and of no or several extra behaviors. A behavior is defined by an effect and the associated (pre-)constraints. All the operations with information on the specific states define the model of the controlled system. Then, from the initial state of the resources and the product flow, the first synthesis stage defined a precedence graph according to the objectives. Finally, the second synthesis stage translates the precedence graph to one of the IEC 61131-3 languages.

Future works will first focus on the validation of our synthesis approach in an industrial context inside a PLC programmers department. Second, based on such a validation, the synthesis algorithm will be integrated in PLC to test the dynamic reconfiguration abilities.

REFERENCES

Charbonnier, F., H. Alla, and R. David, "The supervised control of discrete-event dynamic systems", *IEEE Transactions on Control Systems technology*, Vol 7, No 2, March 1999.

French, S., *Sequencing and Scheduling, New York:* Halsted Press, 1982.

Henry, S., E. Zamaï, and M. Jacomino, "Real Time Reconfiguration of Manufacturing Systems", *accepted for IEEE International Conference on Systems, Man and Cybernetics*, Netherlands, October 2004.

Holloway, L.E., X. Guan, and al, "Automated Synthesis and Composition of Taskbloks for Control of Manufacturing Systems", *IEEE Trans. On Systems, Man and Cybernetics, Part B*, Vol 30, No 5, October 2000.

H.Mendez, E.Zamaï, B.Descotes-Genon, « Synthesis of a Reference Model for the Design of Monitoring Strategies », *Japan-USA Symposium on Flexible Automation*. July 15-17, 2002. Hiroshima, Japan.

Value Methodologie
www.value-eng.org/catalog_monographs.php.

PUBLICATIONS
www.elsevier.com/locate/ifac

Ambient Services for Smart Objects in the Supply Chain Based on RFID and UPnP Technology

Aldo Cea (*), Eddy Bajic

Research Center for Automatic Control - CRAN - CNRS UMR 7039
University Henri Poincaré, Nancy, BP 239, 54506 Vandoeuvre-les-Nancy, France
aldo.cea@cran.uhp-nancy.fr ; eddy.bajic@cran.uhp-nancy.fr

Abstract: As management has developed throughout the years, the production and logistic processes have changed dramatically toward a product centric approach. Consequently, new capacities are expected for the products in their informational, sensory and decisional interactions with processes, information systems, operators and users. Nowadays, it is technologically feasible to assign an active role to a physical product along its lifecycle. In this case, the product acts as a services provider or a services requester based on the interactions between the product and the supply chain processes. This article presents an approach that aims to transform a product into intelligent product or smart object, featuring communication capabilities, information processing using local memory and networks to collect and process data. This object can perform perception and action capabilities within its environment with built-in sensor or actuator to newly interact in the supply chain.

In the proposed approach, product and data are upgraded to intelligent object and services, providing a high level of functional interactions in networked and ambient services architecture. The paper proposes a methodological solution in implementing the concept of intelligent product and services based on the RFID and UPnP technologies. Finally, a case study is presented to implement the proposed approach.

Keywords: Intelligent Product, Industrial Logistics, Automatic Identification, RFID, UPnP, Ambient Services, Ambient Network.

1. INTRODUCTION

In later years, the market requirements have evolved toward high standards of product quality, increasingly short response time and product customization (Helander and Jiao, 2002). The interactions between the processes, the operators and the product since manufacture level until its use require more information and automated or intelligent exchanges between partners (Karkkainen, et al., 2003), in a sure and relatively quick way. The requirements in the supply chain are as follows:

- To find out the exact status of production lines and product achievements;

- To customize the product as soon as possible during production according to client demand;

- To trace all the interventions carried out on the product;

- To ensure that the product reaches the customer under the best conditions of perenniality and traceability;

- To allow accessibility of extended information about product usage or characteristics;

- To develop interactions between the products and their environment;

- To offer services on product knowledge, usage and practice, maintenance, dismantling, recycling, destruction;

By means of Automatic Identification technologies by radio frequency (RFID), we will detail the mechanisms and the current works, so as to transform a physical object into intelligent actor who can interact with other objects in a services environment, during its lifecycle. On the other hand, UPnP (Universal Plug and Play) technology allows to conceptualize and to define the desirable characteristics of a distributed system integrated by product intelligent in ambient services architecture which will be explained later. Finally, a methodological proposition is presented incorporating our conceptual frame and RFID & UPnP technologies. A case study is proposed to show an industrial application of our proposition in the supply chain.

2. PRODUCT AND SERVICES

The guide line of this work is based on the following proposition: "the product is an actor who manages his evolution in cooperation with the different actors from the supply chain (supplier, producer, distributor, and consumer)" (Bajic and Chaxel 2002). We represented the product, (an actor in its lifecycle), as an intelligent product (Wong, et al. 2002).

This one can be defined as a physical and informational object with the following capacities:

1. it possesses a unique identification;
2. it is capable of communicating effectively with its environment (Kintzig, et al., 2002);
3. it can retain or store data about itself;
4. it deploys a language to display its features,

(*) PhD. Student who is being sponsored by the Universidad de Viña del Mar (UVM) – Chile and working as a research at the Research Center for Automatic Control (CRAN) - France.

production requirements etc...

5. it is capable of participating in or making decisions relevant to its own destiny (to what happens to it).

The physical and informational nature of a product is characterized conceptually as physical product and as virtual product. The **physical product** is a material entity identified and characterized by the intrinsic variables that describe it (such as forms, weight, size, etc.). The **virtual product** in addition is represented like information system, associated to some mechanisms of action and decision providing to the physical product new extrinsic capacities and possibilities of interaction capabilities with its environment.

In our approach, a **process** is considered as a system (material, software and human) that realizes information, decision and action activities, connected to a product. We break down the product lifecycle into a number of different processes associated with each supply chain phase. Figure 1 illustrates the dichotomy of both Physical / Virtual Entities of an intelligent product, the interactions between the physical products. The processes of the supply chain are realized by means of a set of services provided or requested by a product.

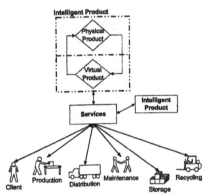

Figure 1. Interactions between intelligent products and processes by means of specialized services

These services provide essential help and support to the 'actors'. This in turn helps the user to get the most out of the product during different phases of its lifecycle. In addition, we define an **interaction** as a mechanism of exchange between a product and a process, allowing the deployment of an action initiated either by the product or by a process. The **service** concept definition is represented by a material and/or an information resource that offers a specific functionality, which can be locally available, or remotely through a network. Services are supported by the interaction concept.

An **actor** is thus a **service producer / service provider** or a **consumer / client** that asks for services offered by other actors.

In a classical approach, the product has usually a provider role and the processes are clients; the product is passive to its environment, and it responds to the process requests. Let's say for example:

product identification, product information research, ... In a reactive approach, the product becomes a client asking for services and receiving benefits while the processes are service providers. The product invokes demands and actions - called services – towards the host process; for example during product use, or product maintenance, storage, recycling ..., thus performing real product-driven interactions.

From the point of view of the technological implementation, RFID technology (Radio Frequency Identification) allows communication between the physical product and the virtual product by means of electronic tags mounted on the product, radio frequency protocols, and networks.

Finally, to answer to the implicit requirements of this proposition, as services offering and requesting in an automated supply chain environment, an ambient services architecture is necessary to host intelligent interactions. Further on in this essay we will analyze the features of such architecture and put the emphasis on UPnP (Universal Plug and Play) technology.

3. RFID TECHNOLOGY AND SMART OBJECTS

The automatic identification tech (Auto-Id) plays an important role in the identification method. In the near future, radio frequency identification (RFID) will be a real option for supply chain projects. RFID represent a reliable method for data transfer in object identification. Among all the Auto-Id technologies, barcode scanning allows the reading of a code printed on an object. This technology does not allow the reading simultaneously of a set of products with barcodes, and still need a direct line of sight to a product. On the other hand, electronic labels technology or "tags" are used to store information of a product. The stored information in the tag can be read or modified. It is possible to read simultaneously a set of tags. These tags can store from 64 bits up to some kilobytes and more in reading/writing modality. (Paret, 2001). A traditional RFID system contains RFID tags, tag readers that can read and write data, and a controller (computer) that controls the system. Additionally, the reading and writing process between the tag reader and the RFID tag can be realized from centimeters up to meters depending on the system characteristics.

The real innovation in RFID does not lies in the technology itself, but in its application in real-world situations (Brewer, et al., 1999). By attaching an electronic tag to a physical product, it can be automatically identified and located into the vicinity of a tag detection system. The identification of an object allows the union between the physical world (physical product) and the virtual world (virtual product). This union maximizes the use of the information and knowledge along the supply chain.

Conceptually, the terms of intelligent product and smart product (Kintzig, et al., 2002) are synonyms indicating the capability of an object to communicate with its active environment, and interact with its users or the other objects. A smart object is able to acquire, to receive and to distribute information in a near or distant environment, and is able to carry out diverse actions on its own initiative, or request help from. According to Upadhyaya, *et al.* (2002), a smart object is aware of its environment and is able to perceive its surroundings through sensors, work with peers using short-range wireless communication technologies, and provide context-aware services to users in smart environments. In addition, smart objects can provide increasingly sophisticated services to users in smart environments when they are able to take advantage of nearby handheld devices in an "ad-hoc" way.

Nowadays, several projects are looking for to develop the capacities of a smart product. We presented three of them, who seem particularly interesting:

• **EPC Global Initiative** (2003) - a joint venture between EAN International and the Uniform Code Council Inc. - is to attach an unique electronic product code (EPC) of 96 bits to the product (Brock, 2001). This integrated numerical code firstly allows the uniqueness identification of an object, yet also allows access to dynamic associated product information across an Internet network. This approach provides an extended information architecture based on Internet by means of a classical product requesting procedure.

• **The Disappearing Computer (DC)** is an EU-funded proactive initiative of the Future and Emerging Technologies (FET) activity of the Information Society Technologies (IST) research program. The aim of the initiative is to see how information technology can be diffused into everyday objects and settings, and to see how this can lead to new ways of supporting and enhancing people's lives that go above and beyond what is possible with the computer today (The Disappearing Computer Initiative, 2003).

• **The SIRENA Project** (2004) intends to create a service-oriented framework for specifying and developing distributed applications in diverse real-time embedded computing environments, including industrial automation, automotive electronics, home automation and telecommunications systems, where objects and products are real active components in industrial or day-to-day environment.

We think that the new paradigm of smart object provides an ability to embed new capabilities into object allowing extended access information up to complex services invocation, and interactions from virtually anywhere at any time, potentially transforming the way we live and work in a society of objects.

4. AMBIENT SERVICES ARCHITECTURE

In order to develop interactions between an intelligent product and the other actors of the supply chain, we must define and characterize the ambient services concept. These ambient services, where the product is a **service provider** or **service requester**, demand the transparency of all the offered services in an environment and associated control mechanisms. **Ambient service** is an abstract view of a system that provides information management capabilities, processing capabilities and event messages in an ambient network, representing a working environment. A product inserted in a local area network or a wireless network can appear or disappear in a **domain of work**. The access to the intelligent product in domain of work demands a spontaneous configuration and identification of nodes and their associated services in the ambient network.

Ambient architecture based on Internet technologies have been developed integrating concepts such as ad-hoc network, mobility management, and service discovery (Duda, 2003). A generic Ambient Services Architecture for intelligent products interaction management needs the following standard services:

▪ **Identification:** To know the product unique identity in the ambient network. In our approach, the identification process is made by means of electronics tags.

▪ **Localization:** To know where the product is.

▪ **Information brokering:** To identify the information sources allowing the execution of a given service.

▪ **Service Discovery and Registry:** To automatically discover and advertise for services in the ambient networks using communication mechanisms: to localize service, to retrieve service information and call parameters, to announce service offering. Each new service provider is registered at the time of its connection to the network.

▪ **Service Invocation:** Process by which a service is requested by a client. A service has a name, an access mechanism and characteristics that describe it.

▪ **Event notification:** When a change happens in the parameters or variables of a service, the interested users receive a notification of this fact.

Currently most known services architectures are Jini, UPnP, OSGi, CORBA. The services protocols such as Salutation, SDP, SLP, WSDA allow to carry out the Process Service Discovery. After analysis, we consider UPnP architecture well suited to developing the smart objects interactions concept related to the wide open and distributed aspects of the domains of works and offering pereniality according to standards TCP/IP, XML and Web technologies (Jeronimo and Weast, 2003).

5. UPNP TECHNOLOGY

The UPnP™ Forum is an industry initiative designed to enable simple and robust connectivity among stand-alone devices and PCs from many different vendors, leading the way to an interconnected lifestyle (UPnP Forum, 2004). The Forum consists of more than 700 vendors, including industry leaders in consumer electronics, computing, home automation, home security, appliances, printing, photography, computer networking, and mobile products. The UPnP is a distributed, open networking architecture that leverages Internet and Web technologies, such as Hypertext Transport Protocol (HTTP), Simple Object Access Protocol (SOAP), Generic Event Notification Architecture (GENA), Simple Service Discovery Protocol (SSDP) and eXtended Mark-up Language (XML) (Intel UPnP, 2004). The generic architecture UPnP includes the two following entities: Devices (or controlled devices) and Control Points, as shown on figure 2.

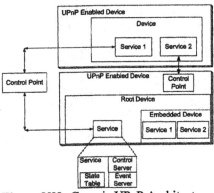

Figure N°2: Generic UPnP Architecture.

The term **Device** is used to define a logical container of others devices and services. The **Services** are logical entities providing a specific service to UPnP device network. Services are controlled by Control Points. A service exposes actions and models its state with state variables. A service in an UPnP device consists of a state table, a control server and an event server. A state table models the state of the service through state variables at run time and updates them when the state changes. A control server receives actions request, executes them, updates the state table and returns responses. An event server publishes events to interested subscribers anytime the state of the service changes. On the other hand, the **Control Point** is the logical entity that can control specific services. A control point in an UPnP network is a controller capable of discovering and controlling other devices. After discovery, a control point could: retrieve the device description and get a list of associated services; retrieve service descriptions for interesting services; invoke actions to control the service; subscribe to the service's event source. Any time the state of the service changes, the event server will send an event to the control point. It is expected that devices can incorporate control point functionality, and vice-versa to enable true peer-to-peer networking.

5.1 UpnP functionalities

We can distinguish six basic functionalities in UPnP:
- **Network Addressing:** Devices obtain an IP address through Auto-IP or Dynamic Host

Configuration Protocol (DHCP) mechanism.
- **Discovery:** Control Points search for devices and services, and devices advertise their services.
- **Description:** Once a control point finds a device or service of interest, it requests a description document. Devices and services respond by sending XML description document that define the actions and attributes they support.
- **Control:** The control point invokes the actions described in the XML description documents associated with the services they control. These actions are executed by the services and typically cause changes in the service states and attributes.
- **Eventing:** The Control Point subscribes to event servers hosted by the services, which allows them to receive events from a specific service they are interested in.
- **Presentation:** If a device has a URL for presentation, then the control point can retrieve a page from URL's device, allowing a user to control the device and/or view device status.

Figure N° 3 incorporates the two points of view on the functionalities of the UPnP architecture: Device Side (Device A) and Control Point Side (Device B).

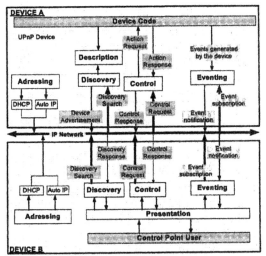

Figure N° 3: UPnP Device and Control Point

The upside part, shows the device functionalities when joined to the network: identified by IP addressing, discovered by discovery search, and controlled by requests issued from a Control Point. A device informs of its description and generates events to the domain of work. In the lower part of figure 3, a Control Point is identified in the network, and constantly generates searches of new devices, gets device description, gets service descriptions, invokes actions towards the devices and generates subscriptions to events generated by devices.

6. METHODOLOGICAL PROPOSITION

Nowadays, the dynamic industrial and logistic context, demands high standards and interoperability for processes performance. In that respect, we propose a methodological offer so as to create an effective integration between a physical product and the interactive world of the information technologies. Our offer is applicable in multiple domains of work, along product lifecycle. To present our approach we will use three generic cases:

• **Case One - UPnP assisted passive object:** Firstly, we can say that the automatic identification technologies by radio frequency, allow the transfer of information between a physical product and an identification system. An electronic tag can carry the product identifier (code) and optionally the product description, for example: manufacturer information, production information, etc. In addition, the product identifier in the tag can act as pointer or link towards an information system. Case one on the Figure 4 represents a product carrying a tag and plugged in UPnP architecture. The merger between the object and an UPnP device represents the intelligent product entity. The intelligent behavior begins with the automatic identification of the product by an UPnP device using a RFID interface. When the device recognizes an object entering into the ambient network, it automatically installs the services associated with the identified product type. Thus the services now available in the UPnP device represent a virtual image of the product parameterized by the information stored in the tag or in a remote information system. At that time, all the services associated to the product are known and can be remotely called by all UPnP control points in the ambient environment. In terms of industrial application, in storage and warehouse processes, when a physical object is received at entrance, it is automatically identified by an interface RFID attached to an UPnP device in charge of inputs management. On the basis of a permanent product identification activity, the logical part of the device, depending on the nature of the product, sets up the corresponding services in the UPnP ambient architecture : product description (Service1), storage conditions needed for the product (Service2), extended product information, ... (Cea, et al. 2004). Also, the product is a passive entity managed by the device, called UPnP assisted passive object.

• **Case Two - UPnP assisted active object:** In this case, additionally to the characteristics described in the previous case, the UPnP device has Control Point capabilities to act as an active entity. Intelligent behavior is thus supported by the product by means of a dual role of a server (as a service producer) and a consumer (as a service requester). A software layer in the control point is parameterized by identified product information to manage product decision making and corresponding services calls in the UPnP ambient architecture. Continuing with the previous example, the intelligent product can demands services among the following: "Looking for a storage place for x times with y, z dimensions?", "What is my current GPS position?", ... The intelligent product makes its decision process according to the answers generated by the available server entities in

the UPnP architecture. A PDA (Personal Data Assistant) with RFID Interface and WiFI can figure such an UPnP interface (Cea, et al., 2004).

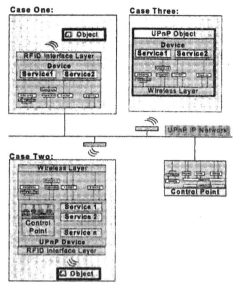

Figure N° 4: Methodological proposition integrating UPnP architecture with a RFID smart object.

▪ **Case Three - UPnP integrated object:** In this situation, the UPnP device is embedded into the physical product, with no more need of RFID communication. For example, a temperature sensor in the storage process, can act as an object offering temperature measurement service in the environment answering to the demands of an UPnP Control Point. Wireless technology is an alternative contribution to mobility and to automatic identification of a product, for the use of intelligent services in an UPnP network (Zahariadis, et al., 2004). Embedded computer power and energy storage or energy supplying actually impose limitations on such industrial development.

In the following section, we will present a case study showing a test case application of our proposition.

7. CASE STUDY

Our case study conceptually analyzes the application of the methodological proposition in a **Warehouse**. A warehouse is a traffic place of products, in which they are located and stored. The basic system is composed for pallets and products identifiable due to the tags, RFID readers, PDA's, UPnP Control Points and Temperature Sensors. We describe the behavior of an intelligent product in this context.

7.1 UPnP Architecture Components

Figure N° 5 shows the components of the test case UPnP Architecture. **RFID reader systems** are situated at the entrance, at the exit of the warehouse and in some places fixed inside the warehouse to determine the product position. The **tags** allow the pallet / product identification. In addition, the tag contains Product Storage Conditions: storage temperature, product's volume, weight, and product's expiration date.

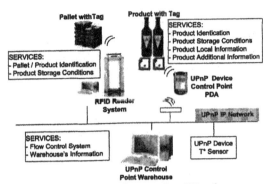

Figure N° 5: Architecture in a Warehouse.

Each PDA device contains the following basic services associated to the product: **Product Identification; Product Storage Conditions, Product Location** in the warehouse; and **Product Additional information** obtained from pointer to product database. The PDA allows the user to define message subscriptions to show the product's status depending on its expiration date and its temperature condition. The aims of Control Point Warehouse are to manage the **Flow Control System** and the **Warehouse's Information**. For this, the RFID reader system records the action that item x passed point y at time z. Thus, out-of-stocks situations can be controlled. It is possible at any moment to recall the entire history of products from their production to the current state. The **vector** (item description, position, time and temperature) allow the product representation in a database. The **Database** can confirm that items have been stored in the appropriate place and can show more **Product Additional Information**, such as the traceability and recycling information. Additionally, temperature can be monitored by **Device Temperature Sensor.** If limit values are exceeded, this can be notified to an UPnP Control Point.

In this case, the role of an intelligent product consists in asking the warehouse system if conditions are favorable for its storage. This internal and automatic process reflects the behavior of a product in an ambient network environment demanding quality standards for its warehousing. In addition, an intelligent product can offer product information (identification and description) to the UPnP Control Points figuring actors in the warehouse layout.

8. CONCLUSION

Nowadays, industry can benefit from the new results reached in the RFID technology and in the information technologies field. We presented how a physical product can be transformed into an intelligent product, functionally integrated in an ambient network for decision making and actions invocation by means of services. A product is no longer just a physical part carrying data or information, but tends to an active entity that can act as a service provider or as a service requester, i.e. a smart object. Ambient services architecture allows the basic functionalities standard to deploy. We demonstrate the potentialities of UPnP and RFID as open and flexible standards to develop product services modeling and components design to support intelligent product interactions in an ambient services network architecture based on internet standards. The case studied shows the feasibility for implementing our conceptual proposition. The methodological approach is continued to define classes of services and demonstrators corresponding to product lifecycle phases in the supply chain.

REFERENCES

Bajic E., Chaxel F. (2002). Holonic Manufacturing with Intelligent Objects. *IFIP. BASYS.* Mexico.

Brewer A., Sloan N., and Landers T. L (1999). Intelligent tracking in manufacturing. *Journal of Intelligent Manufacturing.* 10, pp 245-250,

Brock D. L (2001). The electronic product code. *Auto-Id Center,* MIT, Cambridge.

Cea A., Bajic E. and De Matteis Ph. (2004). Une approche de modélisation des interactions produits-processus par les objets communicants, *MOSIM.* France.

Duda A. (2003). Ambient Networking. *Smart Objects Conference,* Grenoble, France.

EpcGlobal (2003). http://www.epcglobalinc.org/

Helander M. G. and Jiao J. (2002) Research on e-product development (ePD) for mass customization. *Technovation* 22, pp 717-724.

Intel UPnP Technology (2004) http://www.intel.com/technology/upnp/

Karkkainen M., Holmstrom J., Framling K. and Artto K. (2003). Intelligent products – a step towards a more effective project delivery chain. *Computer in industry* 50, pp 141-151.

Kintzig G., Poulain G., Privat G. and Favennec P. (2002) *Objets Communicants.* Hermès. France

Jeronimo M. and Weast J. (2003). *UPnP Design by Example.* Intel Press. USA.

Paret D. (2001). *Identification radiofréquence et cartes à puce sans contact,* Ed. Dunod. France

The Disappearing Computer Initiative (2003) http://www.disappearing-computer.net/

The SIRENA Project (2004). http://www.sirena-itea.org/Sirena/Home.htm

Upadhyaya S., Chaudhury A. , Kwiat K. and Weiser M. (2002). *Mobile Computing. Implementing Pervasive Information and Communications Technologies.* Kluwer Acad. Publishers. USA.

UPnP Forum (2004), http://www.upnp.org

Wong C.Y., McFarlane D., Ahmad Zaharudin A. and Agarwal V. (2002). The intelligent product driven supply chain. *IEEE SMC,* Hammamet.

Zahariadis Th. and Pramataris K. (2004). Broadband consumer electronics networking and automation. *Int. J. Commun. Syst;* 17:27–53.

ELSEVIER
IFAC
PUBLICATIONS
www.elsevier.com/locate/ifac

A CONTROL-MONITORING-MAINTENANCE ARCHITECTURE. APPLICATION TO A ROBOT-DRIVEN FLEXIBLE CELL

E. Rocha Loures[1][2], J. C. Pascal[1]

[1] *Laboratory for Analysis and Architecture of Systems – LAAS/CNRS, Toulouse-France*
[2] *Graduate Program in Production System Engineering – PUCPR, Curitiba-Brazil*
e-mail addresses: loures@laas.fr, jcp@laas.fr

Abstract: This paper proposes a control-monitoring-maintenance architecture (CMM) for flexible manufacturing system (FMS). This framework is based on a hierarchic and modular model structured in CMM modules. The maintenance and its integration into the CMM system results in improvement of the system functionality in terms of availability, efficiency, productivity and quality. These efforts acts as a basis for the control architecture of a robot-driven flexible cell, connected to the Ethernet-TCP/IP.

Keywords: Discrete Event Systems, Maintenance, Supervision, Petri Nets.

1. INTRODUCTION

The flexible manufacturing system (FMS) can be considered as a system able to produce a wide variety of products under variable production conditions. Classically, this type of system is constituted by a set of elements and devices such as workstations, transport systems, programmable logic controllers, robots, vision systems, connected to an industrial communication network. According to the increase of the manufacturing cell dimensions, the control and supervision of its structure becomes more complex. Thus, to keep flexibility regarding its decision organization, a hierarchic and modular structure is proposed (Zamai et al., 1997). Hierarchic architectures have been a subject of many works the last decades (Gershwin et al., 1988), (Giua and DiCesare, 1994), (Srinivasan and Jafari, 1993). It represents the framework basis of recent approaches concerning the control and monitoring of the FMS (W. Jeng and Liang, 1998), (Pascal, 2000).

Maintenance and its integration with the control and monitoring system enable to improve system functioning regarding availability, efficiency, productivity and quality. Thus, it is possible to implement corrective and preventive actions, making easier the repairs and servicing over the process elements as well as a better control provision of tools and repair parts. Different methods do not take maintenance into account as an isolated issue, but making it part of a supervision, control and monitoring system. Some works can be mentioned as the proposal of a maintenance model integrated into a supervision-control architecture performed by (Ly et al., 2000), (Berruet, 1998), the maintenance and the reconfiguration aspects in (Tang and Zhou, 2001), and the management of maintenance policies

proposed by (Abazi and Sassine, 2001). Most of these approaches focus the maintenance integrated into the monitoring systems in order to support the planning and resources checking, where the real time conditions are not pointed out. In (W. Jeng and Liang, 1998), a SMT (supervisor-monitor-troubleshooter) framework presents an integrated design method for supervisory control, monitoring and troubleshooting where real time aspects are considered. Our approach looks into this context through a CMM framework.

One of its advantages is to consider the control-monitoring-maintenance functions simultaneously. The maintenance must be considered as one of the main functions of the system, linked to the control and monitoring functions. The hierarchical and modular decomposition enables to implement maintenance policies, limiting its effects over the global system, allowing it to preserve its availability. Another advantage is the fact that, during maintenance action, the monitoring is kept active and guarantees a coherent intervention in the process.

In this paper, section 2 describes the control-monitoring architecture as a basic ground for the works. Then the integration of maintenance aspects is described leading to the proposition of the CMM architecture. The section 3 presents an application of this architecture, which is under implementation in a real robot-driven flexible cell.

2. CONTROL-MONITORING-MAINTENANCE ARCHITECTURE

The real time control of the complex Discrete Events Systems (DES) is classically solved by structuring the control system, both hierarchic and modular, on a four-level decision organization: planning, real time scheduling/supervision, co-ordination and local control (Subias et al., 1997). The control functioning principle in a hierarchic context requires that only the module, which has emitted a request for the lower level, is able to modify it (Zamai et al., 1997). Therefore, for modifying a request that has been emitted, i.e., to reconsider the current control, the monitoring system must take the "request/report" sequence into consideration. Thus, it is desirable a monitoring system following the control hierarchic levels.

Maintenance, according to (Ly et al., 2000), is defined as a group of technical and administrative actions, including supervision and control operations, to maintain (*preventive maintenance*) or re-establish (*corrective maintenance*) an entity in a specific state or in specific operating safety conditions (RAMS – reliability, maintainability, availability and safety) (Ly et al., 1999). The maintenance policies can be classified in three great categories (Abazi and Sassine, 2001): corrective, preventive and mixed. In all three cases, the triggering of a maintenance operation is a decision process based on the state (measured or estimated) of considered resources (Ly et al., 2000).

The triggering of maintenance tasks through a continuous decision process inspired us to enrich the CM module (Zamai et al., 1997) with a maintenance module and then to define the *control-monitoring-maintenance (CMM) model*, as shown in figure 1.

the process, e.g.: starting an operation, a request is simultaneously transmitted to the reference model that estimates a temporal window of a process event occurrence (end of operation). If the report issued from the process is received by the control model within the allocated temporal window, the control and the process model evolve simultaneously. Otherwise, the reference and control models can no longer evolve. At this moment it is possible to conclude that a process fault just occurred and the diagnosis module is thus required.
- *Diagnosis*: it proceeds the identification of the incident to which symptoms have been acknowledged. This task is simple when it deals with a not ambiguous symptom, even though it may also need a more sophisticated treatment using, for instance, an information system which completes the information reporting absence of process report (Subias et al., 1997).
- *Recovery decision*: its role is, according to diagnosis information, to modify the control models, activate urgent procedures, trigger recovery procedures and, finally, decide about propagation of failure treatment to the upper level.
- *Emergency*: provides the possibility, for the decision block, to immediately take action on the process, carrying out pre-defined and priority sequences (re-start, setting idle position, etc).
- *Interface operator*: it is the user interface that allows dialogue with human operator each time its intervention is necessary. The operator presents a central decision role.

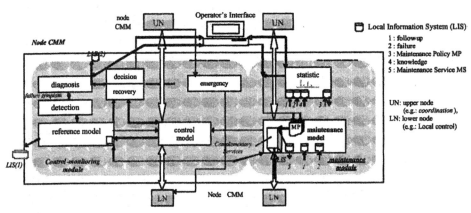

Fig. 1. Control-Monitoring-Maintenance (CMM)

2.1 Control-monitoring module

First, let us consider the control-monitoring module. Each node is made up of the following elements:
- *Control Model*: comprises the operation sequence to be executed, which corresponds to a specific handling of the process.
- *Reference Model*: it is a model of the process which describes its possible states. The state of this module is verified at the moment of a control request and at the moment of process signal (report) knowledge. The essential role of this block is to maintain, continuously, the most perfect image of the real process state.
- *Detection*: the reference model is connected to the control model – it evolves in parallel and in real time during current operations of the process. At the moment which the control model emits a request to

The monitoring process follows a detection-diagnosis-recovery sequence. A confinement and propagation mechanism from the decision block decides if the fault is treated at the proper level (confinement) or if a request must be emitted to the upper level node (propagation).

2.2 The maintenance module – the CMM node

Figure 1 shows the architecture of a *control-monitoring-maintenance model* (CMM) at a given level. The insertion of the maintenance module did not modify the functionality of the described CM module elements.

First it is necessary to point out the purpose of the *decision* and *recovery* block concerning the CMM structure. As soon as this block is required by the

200

diagnosis block, it triggers different actions. If a risk exists for the operator or the process, the decision block triggers an emergency procedure and informs the upper level. Otherwise, it establishes a recovery point and the specific sequence making possible to access it. If the expected state is not coherent with the upper level vision, it informs about it. If no recovery process can be established, it sets the system into a maintenance state, informs the upper level and requires the operator. This is the *corrective maintenance* scenario. In this case it is necessary to "repair the defective material", i.e., eliminate fault effects in a way to reach the system regular operation. This requires human intervention. The operator can access a maintenance module that offers a group of maintenance operations which are not necessarily the same as those of the control. The control module remains in the blocked state during all intervention period. At the end of the maintenance activity, the system is redirected to a coherent state concerning the control model, by the maintenance module which authorizes its evolution.

The maintenance module consists of two blocks: the *statistic block* and the *maintenance model*. The purpose of the *statistics block* is to use all information issue from the process and control (process data registering) to calculate, estimate and establish a preventive maintenance plan. Its role is to inform the operator that a preventive maintenance operation becomes necessary. This block depends on a *Local Information System* (LIS). The LIS is updated by different blocks of the CMM module and by external information (operator, equipment information, maintenance policies, etc). Concerning the maintenance, it presents the following structure:
- *LIS (policy)*: updates the maintenance intervention database. This registering makes future interventions easier (Abazi and Sassine, 2001);
- *LIS (knowledge)*: contains information supplied by operators (experience return), information and indicators about the production system (MDT, MTTF, MTTR, MUT);
- *LIS (follow-up)*: memorizes all control activities, their start and end dates; it enables to keep a control and a process image when a failure occurs;
- *LIS (failure)*: memorizes information about detection, diagnosis and the decision regarding a failure.

In addition, it can be used to update the temporal windows of the reference models in order to estimate the average operation time, so as to have a more precise representation of the process.

The *maintenance model* suggests a group of services and/or sequences of specialized operations that are linked to a maintenance policy and that are not necessarily foreseen by the control under regular operation. A similar context may be fond in (Ghoshal et al., 1999), where some concepts were taken into account and adapted into our approach.

Maintenance is triggered by the operator after an unresolved fault case – it is the *corrective maintenance* policy, or triggered by the statistic block – it is the *preventive maintenance*. When maintenance is required (either corrective or preventive), the maintenance model of the module inhibits all pre-set operation sequences (regular operating conditions) at the control model level. At this point the maintenance module controls the process. The maintenance mode is, therefore, synchronized with the evolution of the reference model. In fact, taking account of the process restrictions enables to provide a safety mechanism, making impossible for the operator to trigger operations that are not in conformity with the process state. It is necessary to mention that, before any maintenance procedure, an image of the control state is captured by the local information system (LIS) to replace, after intervention, the process and control in a coherent state.

The communication with the upper level consists in informing the maintenance state and, eventually, the estimated duration for implementing the maintenance policy. This information is received by the upper maintenance module (node) which informs the control-monitoring module. The consideration of this information by the upper CMM module may lead to a reconfiguration process (Berruet, 1998). The reconfiguration process consists of the updating of FMS architecture flexibility to compensate the lack of services caused by the faulty resource. This can result in a simple displacement of the operations sequence or the use of available resources, enabling the expected services.

3. Application to a robot-driven flexible cell

3.1 The flexible cell

The manufacturing cell is composed by four *working stations* served by a *conveyor system* and its *pallets* with parts to be treated. Each working station has a *robot* and a *vision system,* two parts stocks and a working desk. This cell is represented in figure 2. This working station can perform assembly and disassembly, control and calibration operations.

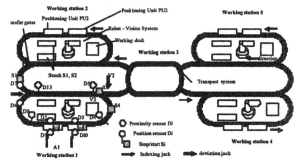

Fig. 2.   Robot-driven flexible cell

The *functional organization* of the cell is hierarchic and modular, structured in 5 levels:

- *local control* of robots, vision systems and zones of the conveyor system at the lowest level.

- the *coordination level* organized into a *local coordination* of the working stations (robot, vision system), a *coordination of the conveyor system* (organized and controlled by 5 modules) and a *general coordination*.

- at the highest levels the *supervision, scheduling and planning (production management)*.

3.2 The CMM module

The *functional* structure of a local coordination CMM node is shown in figure 3. It proposes a more flexible structure allowing the choice and the configuration of a service issue from the upper level (taking a set of pre-defined services into account). Thus, the node functionality and flexibility increases through the offering of this set of services, layer and modular organized leading to a local 'virtual hierarchy'. It results a better interface with the upper level (*general coordination*) and lower level (*local control of robots and vision systems*). All this functionality is extended to the maintenance module

Let us start with the *assembly* of a spare part issue from the stock, on top of another part located above a working desk. The *control, reference and maintenance* models concerning this coordination level are described in figure 4, making up the CMM coordination module.

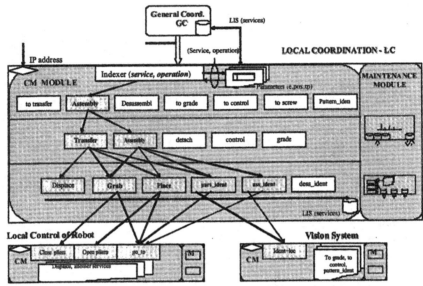

Fig. 3. The CMM module

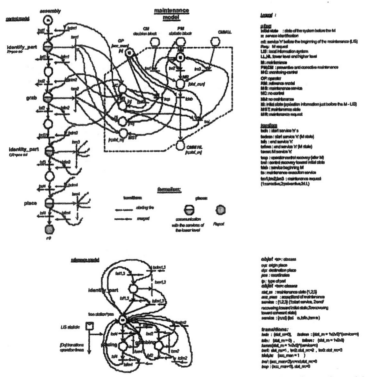

Fig. 4. CMM reference-control-maintenance models for coordination of working stations

202

The representation of these models is based on *Petri Nets with Objects (RdPO)* where the tokens hold the information (token attributes). The interpretation associated to transitions based upon these attributes leads to the tests and actions execution (Zamai et al., 1997).

Let us consider the control model. The *assembly* service starts a sequence of low level services: *'identify_ part'* service, which identifies the part location regarding its recovery (offered by the local control of the vision system), *'grabbing'* service which grabs and moves the part in a placing position, *'placing'* service which corresponds to the assembly part and *'grabbing'* service for the evacuation of the assembly part (offered by the robot local control (figure 5)).

The associated reference model represents the process restrictions (mutual exclusion of operations and estimated operational times).

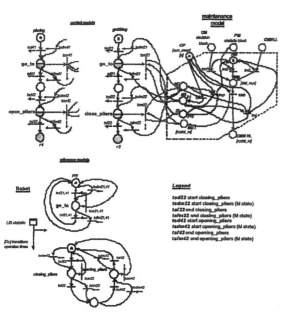

Fig. 5. Local command CMM models for robots

The communication between these models is implemented by means of a transition fusion method: as soon as the service control model *'part_identification'* is demanded, which corresponds to the transition firing *'tsd1'*, the *'tsd13'* transition associated to it at the reference model, is simultaneously fired (under the condition that its input place *'free_station'* is marked, meaning that the requested service is coherent with the process state). Therefore, the reference model evolves toward the *'part_identification'* state. The request is transmitted to the lower level, in this case, the local control of the vision system. If the execution report of the lower level is not received within a temporal window defined by the reference model, the reference and control models can no longer evolve. A failure is detected and the monitoring process is triggered (detection, diagnosis, recovery). If the execution report is received within the temporal window, the control and reference models can evolve (simultaneous *'tsf1'* and *'tsf13'* firing). The reference model evolves toward the *'free_station'* state and the control model can request of the *'grabbing'* service.

Therefore, simultaneous *'tsd2'* firing occurs in the control and reference models and the changing of the latter to the *'taken'* state also occurs. The service request is transmitted to the lower level, in this case, to the robot's local control

At this level, the control model carries out a sequence of elementary controls over the process: *'go_to'*, then *'close_pliers'* (figure 5). Two reference models represent the process behavior associated with the pliers and the arm movement. The evolution of control and reference models is made as described previously.

When this last sequence operation (*'close_pliers'*) ends, an execution report of the *'grabbing'* service (*'cr2'*) is transmitted to the upper level CMM module. Afterwards, the *'place'* and *'grab'* services are executed.

Let us suppose that a failure is detected at the end of the final assembly part recovering: the end-of-execution report of the *'close_pliers'* is not received within temporal window. Thus, the decision block requires the operator for a corrective maintenance action (*CM-tm1-MR* sequence at the maintenance model of the local control CMM module). The operator acknowledges the module in a maintenance state and the upper level is informed about it (*OP-tdeb-MST/CMM NS* sequence). At the upper level, this information will be taken into account by the decision block and the operator, who will decide, or not, about placing the module also in the maintenance state (*CMM LL-tm3-MR-tdeb-MST*). The maintenance state of the module means the suspension of the on-going operation and its registering due to its recovery (*MST-tsm22-NC/IS* sequence).

The operator at this moment is able to require services or operations available in the control model or trigger operation sequences that are different of the regular control (*OP/NC-ts-MS-tsdmn*). The detection mechanism still takes into account the restrictions of the process through the reference model (*'tsdmn'* and *'tsfmn'* transitions). The sequences, tests or maintenance operations depend on the operator's experience. He always has the power to decide about the execution of operation over the process. It is possible to use the local information system (LIS) data as a reference: historic data of failures, interventions, adopted maintenance policies, MTBF, MTT indexes, etc. Considering our case, the detection of a failure at the *'close_pliers'* service, the operator is able to require:
- the opening of the pliers (*OP/HS-ts-MS-tsdm42-open_pliers* sequence) and if the end-of-execution report (opening/closure signal value is followed and strength sensor is in 0) is received within temporal window (leading to *tsfm42* firing and *NC* marking), it is possible to conclude that the opening/closure pliers device is working
- the pliers repositioning (*OP/NC-ts-MS-tsdm21-go_to-tsfm21-NC* sequence) to make a closing test of the pliers with maximum safety;
- the closing of the empty pliers (*MS-tsdm22-*

203

close_pliers sequence). If the end-of-execution report (value of the closing/opening signal is followed and the strength sensor has an empty closing value) is not received within the time window, it is possible to conclude that the closing device is not working (rupture in air supply, for example).

Once repaired, the operator is able to perform tests again, requiring a specific part in the stock ('*identify_part*' service of the CMM coordination module, after '*taking*' and finally '*placing*' it in the stock, for example). And finally, the operator must replace control in a coherent state – to the breakdown initial state (registered by the stick of the *IS* location), i.e., at the beginning of the '*taking*' service (*OP/NC/IS-trei-MS-tsdm2 sequence*) or proceeds the finalization of the suspended operation before leaving it to the general control (*OP/NC/IS-trop-MS-tsdm2 sequence*). In both cases the reference models are equally replaced in coherent states. All interventions are registered in the Local Information Systems (LIS policy) to make future interventions easier.

6. Conclusion

This study presented a hierarchic and modular control-monitoring-maintenance architecture, organized in CMM modules. In each module the interaction of a control module, reference models of the process (allowing the detection of failures) and a maintenance module (consisting of a maintenance model and a statistics module) can be observed. This module enables to carry out different maintenance policies, more specifically curative and preventive ones. The statistics block, coupled to a local information system, enables to register all information issue from the process and control, estimates and establish a preventive maintenance plan and helps the operator during a maintenance intervention. The operator can access, via maintenance model, the services and operations offered by the control and can sequence them differently from the expected process. In addition, it is possible to, eventually, access services available, but not used in the current process. The evolution of reference models during control stages and during maintenance stage results in the assurance of a coherent use of the process. This integration process, the maintenance tasks integrated into the control and monitoring structure, are the main advantage of our approach.

The proposed architecture is actually being implemented in a robot-driven flexible cell, connected to the Ethernet TCP/IP. Therefore, it is necessary to study the problems presented in the remote maintenance context e.g.: the remote configuration of the CMM modules, network and devices, the reliability, safety and real time conditions. Moreover, the planning and scheduling levels will be considered leading to the definition of interfaces with the CMM nodes.

References

Berruet, P. (1998). Toward an Implementation of Recovery Procedures for FMS Supervision. *Proc. of Information Control Problems in Manufacturing*, pp. 371-376, Nancy, France.

Chaillet-Subias, A., Zamai, E., Combacau, M. (1997). Information flow in a control and monitoring architecture. *Proc. of the IEEE International Symposium on Industrial Electronics*, , pp.53-58, Guimaraes, Portugal.

Gershwin, S.B., Caramanis, M., Murray, P. (1988). Simulation experience with a hierarchical scheduling policy for a simple manufacturing system. *Proceedings of the 27th IEEE Conference on Decision and Control*, vol. 3, pp. 1841 – 1849.

Ghoshal, S. et al. (1999). An Integrated Process for System Maintenance, Fault Diagnosis and Support. *Invited paper in Proc. IEEE Aerospace Conference*, Aspen, Colorado, USA.

Giua, A., DiCesare, F. (1994). Petri net structural analysis for supervisory control. *IEEE Transactions on Robotics and Automation*, vol. 10, pp. 185-195, Issue: 2.

Jeng, W.-H, Liang, G.R. (1998). Reliable automated manufacturing system design based on SMT framework. *Computers in Industry 35*, pp.121-147.

Ly, F., Toguyeni, A. K. A. and Craye, E. (2000). Indirect predictive monitoring in flexible manufacturing systems. *Robotics and Computer Integrated Manufacturing*, Elsevier Science Ltd., n. 16, pp. 321-338.

Ly, F., Simeu-Abazi, Z. and Leger, J-B. (1999). Maintenance terminology review. *Report of the Research Group of Safety Functioning Production*, France.

Pascal, J-C. (2000). A modular and hierarchical approach for supervisory control of batch processes. *4th International Conference on Automation of Mixed Processes: Hybrid Dynamic Systems (ADPM'2000)*, pp.369-374, Dortmund (Allemagne).

Simeu-Abazi, Z. and Sassine, C. (2001). Maitenance Integration in Manufacturing System Evaluation and Decision. *International Journal of Manufacturing Systems*, vol. 13, n. 3.

Srinivasan, V.S., Jafari, M.A. (1993). Fault detection/monitoring using time Petri nets. *IEEE Transactions on Systems, Man and Cybernetics*, vol.. 23, pp. 1155-1162.

Tang, Y., Zhou, M.C. (2001). Design of reconfigurable semiconductor manufacturing systems with maintenance and failure. *Proc. of the IEEE International Conference on Robotics and Automation – ICRA*, vol.1, pp. 559 – 564.

Varvatsoulakis, M.N., Saridis, G.N., Paraskevopulos, P.N. (2000). Intelligent organization for flexible manufacturing. *IEEE Transactions on Robotics and Automation*, vol. 16, pp. 180 – 189, Issue 2.

Zamai, E., Chaillet-Subias, A., Combacau, M., Bonneval , A.De (1997). A hierarchical structure for control of discrete events systems and monitoring of process failures. *Studies in Informatics and Control*, vol. 6, n. 1, pp. 7-15.

ELSEVIER
IFAC
PUBLICATIONS
www.elsevier.com/locate/ifac

FLEXIBLE ASSEMBLY CELL USING SCARA MANIPULATOR

Felipe Leighton   **Gastón Lefranc**

Escuela de Ingeniería Eléctrica.
Universidad Católica de Valparaíso, Chile.
E-Mail: glefranc@ieee.org

Abstract: In this paper is presented an assembly flexible cell, the product design for assembly of switches and the conditions to perform. The assembly flexible cell has a robotic manipulator, a computer vision system and a conveyor. The cell model is done by Petri Nets. The system is part of a flexible manufacturing system (FMS) designed previously.

Keywords: Flexible Manufacturing Cells, Assembly, CIM, Petri Nets.

1. INTRODUCTION

Within the area of the manufacturing, CIM model (Computer Integrated Manufacturing) is used to automate all the activities in a company: the organization, storage (ASRS) manufacturing (CAM) engineering (CAE) and Quality Control. These activities are supported by local area computer networks and related database. In CAM includes fabrication, assembly and quality control of products, utilizing robotics manipulators, computer vision, etc., to facilitate automation (Lefranc, 1993). These industrial systems, allow flexible their production and to automate their processes, without human intervention.

Assembly systems are widely used in several industries. These systems are composed of many automated parts, robotics manipulators and sensors to complete the production task accurately. However since these systems are composed of many components, it is difficult to deal with unexpected situations. For example, an undetected error may propagate and end up as a detectable failure which may cause the whole line to stop its operation. In this case, it may take considerable time to diagnose the system and identify the main reasons for the failure. There exist approaches to detect and to predict failures, but the usage of these approaches is limited (T. C. Cao, A.C., Sanderson, 1992), (Q. Jing et al, and 1996).

The ability to produce a variety of products through the combination of modular components is a meaningful benefit of product modularity. Reference (K. Ulrich and K. Tung 91) described five different ways that modular products are developed in industry. Component-swapping modularity can be achieved when two or more alternative types of components are paired with the same basic product body to create different product variants. This modularity is often associated with the creation of product variety as perceived by the customer.

In this paper, the assembly cell is modelled by Petri nets.

The Assembled Flexible Cell is part of a developed Flexible System in the Laboratory of Robotics, Artificial intelligence and Advanced Automation Lab, in the School of Electrical Engineering of the Pontifical Catholic University of Valparaiso. FMS includes: an ASRS cell (Automatic Storage Retrieval System) (Andrada, Lefranc, 2002) with an executive system and database (Maizares, Lefranc, 2003), and others cells (manufacturing and quality control).

In this work is presented an assembly flexible cell, the product design for assembly of switches and the conditions to perform. The assembly flexible cell has a robotic manipulator, a computer vision system and a conveyor. The cell model is done by Petri Nets. The system is part of a flexible manufacturing system.

2. ASSEMBLY CELL STRUCTURE.

The assembled flexible cell of is constituted by a robotics manipulator, a computer vision system and an automated transport system. The manipulator is IBM 7547 type Scara, as shown in Fig. 1.

Figure 1. Robotics manipulator and conveyor.

The automated system of transport is conformed by four conveyor belts that form a "circuit" around the robotics manipulator and the assembly area (Fig. 2). Its function is to relate the different cells of the FMS system, ASRS cell flexible cell, assembly flexible cell, manufacturing flexible cell and quality control flexible cell. Through this system of transport, pallets are transferred to one cell to another.

Figure 2. Pallets in the conveyor.

The computer vision system, fig 3, is utilize for parts recognition, to determine heights, centroids and orientation angle of objects to be pick for the robotics manipulator, to take pallets, to identify what pieces are in pallets, etc. For assembly, an exact position is needed in the work space to make the assembly job. In order to prove the cell a wood figure is assembled, and then switches with simple, double or triple plug, according to the programmed configuration.

Figure 3. The Computer Vision System

3. PLANNING OF THE ASSEMBLY

In order to increase the ease of reconfigurability, it is beneficial to automate as much of the work from design to implementation as possible. This approach reduces the number of errors in the control code. The implementation should be modular in order to avoid the state space explosion problem.

The steps of a system design procedure for the manufacturing system are based on a model of the system and its specification of behaviour. These models consist of several interacting entities modelling the resources of the system and their intended behaviour.

The following consideration about the product to assembled are: composition, union of part, number of pallets, numbers of product to assembly, parts to assembly.

Composition: Two kind of assembly: A simple system is assembled, conformed by three pieces, as figure 4, two of cubical form is seen, the first with dimensions of side 5.5 cm and height of 2.8 cm, the second with dimensions of side of 3.5 cm and height of 3.7 cm and one of cylindrical form, with a diameter of 2.5 cm and one height of 4 cm. Next, switches with simple, double or triple plug, according to the programmed configuration.

Union of each subpart: For the geometric figure a simple method of insertion will be used. For the assembled of switches will be used an automatic insertion.

Devices to make the assembled one: The devices used are robotics manipulator, a computer vision system and an automated transport system. Pallets of 15*15 cm transport different Kits with the pieces to assembly, through the conveyor belts. Figure 6.

Figure 4. Wood piece Figure 5. Switch

The number of pallets required: They transfer the parts and products. Four pallets are used: one for the geometric system, and three for switches.

The number of products that is desired to obtain: As objective one looks for to assemble a number between three and seven products.

The planning of the assembled one of a product can be divided in the following parts (Lefranc, 1996):
    Receiving Module
    Programmer Module
    Generating Module of Sequences of Assembled.

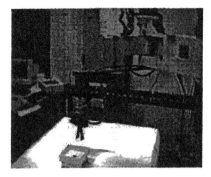

Figure 6. Pallet

The information (composition, union of part, number of pallets, numbers of product to assembly, parts to assembly) is put in the Receiving Module of information and it stores in a data base by lots or product groups. Previously to the reception of information of the lots of products, the cell requires to be described and to be formed.

The Programmer Module of the flexible assembly cell takes the information from the data base, amount and type of product to assemble, and must generate the plans of work of the cell, that is to say, a chronogram of lots of products to assemble, and another chronogram for each one of those lots of production. The module consists of an algorithm that determines the assembled repetitive set of, which is defined with the intention of using a limited number of pallets of fixed capacity. Therefore, for assembling one of a lot of products is divided in equal parts, being each one of them a repetitive set. Each one of these sets begins with the positioning of pallets with parts within the work area. The diagram of figure 7, show a simple view of the operation of the iterations of assembled.

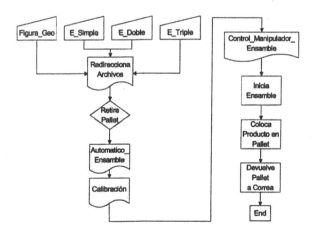

Figure 7. Assembly Iterations

The generating module of sequences of assembly, for this project was not considered, since the possibility of varying the design of the components does not exist, for the case of the switch is already delimited.

4. Design of the effector ("Gripper").

In the assembly of a product, several factors are combined: the election of "gripper", the election of acting to take an object using the manipulator. A generated plan activates the sensors (systems of vision, sensors ultrasonic, etc.) that they cause that the manipulator effector be located on the object, takes it and begins the assembly task. To choose "the gripper" is an important step in the process of pick and place parts (Cano, Lefranc, 2002). In the present industry "grippers" is used like passive fingers. In fact a good design allows a simple pair of passive fingers to take different forms of objects and to make the taken pieces more flexible. For the particular design of "grippers" for the manipulator scara, the following steps were followed:

4.1. Study of the parts to pick and place

They are going away to assemble products that contain pieces of different form. For the wood object, three pieces, two in form of bucket and one of cylindrical form exist. For the switch, the pieces are the covers, the bases and the plugs. Each one these parts are in the pallet and it must be retired by the manipulator and to place it in the work area.

4.2. Additional requirements.

Within the additional requirements to consider, they are due to the maxim height that can reach the manipulator and the maxim opening of the effector. These two points go related to the dimensions of the effector (to gripper). On the other hand one is due to consider as they pay attention the effector to the manipulator, for it is due to have present the adjustment options, easy assembly and disassembling. Another important point is how it is going away to retire pallet of the conveyor belt. This represents difficulties for effectors and it has to consider in the future design.

4.3. Solutions for steps 1 and 2.

As they are going to use different geometric forms, is necessary giving a small curve in its inner face of "gripper". It is used a porous material that it facilitates the holding. Referring to the material of construction is used a plastic material denominated duroplax.

4.4. Development of the designs that combine the preceding modular solutions.

Taking the last conditions in the previous points, the designs are seen in fig 8.

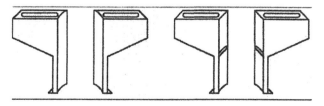

Figure. 8 Effectors Forms.

The final result of the effectors is observed in Fig. 9.

207

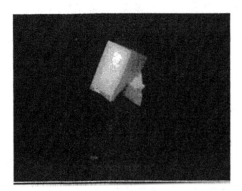

Figure 9. Final Effectors.

4.1. Design of the "Kits"

For the kits design, similar steps are followed for the design of the "grippers". Nevertheless, it is restricted to the fact on pallets of dimensions already established. Kit is constituted by all those pieces that are grouped to be assembled and form an end product.

For the design of kit, a surface or material that allows the distribution of the pieces within pallets, without increasing of considerable form its weight is due to look for. Different materials were evaluated, such as plastic, puffs up and plumabic. The last one has better performance, as much by its weight as by its easy handling and adaptability to the different characteristics from the pieces. Once established the material of fixation within pallets, the position of the different pieces within this one settles down, considering its simple assembly and favouring the easy one taken from them by the manipulator.

For the geometric figure, there is no disadvantage in fixing the position, nevertheless for the plug, that counts on a greater number of pieces; it exists but difficulties, since a position is needed that the taking of each one facilitates its components, favouring its priority within the joint. Kits must have the necessary space for the taking of pallets from the conveyor belt to the work area. Kits, can be observed in figures 10 and 11.

Figure 10. Kit of wood. Figure 11. Kit of a plug.

5.   Computer Vision System.

The vision system must determine the space position of objects in stationary state, within the space of work of the robotics manipulator. It has to be able to recognize objects of different size, pick them, to transport and to assemble them according to the plan.

5.1. Digital image processing

The images are captured with web cam cameras. These images pass through a process of digitalization where it obtain the space coordinates of the image (x,y), and the amplitude of the intensity of light in each point or pixel of the image, which denominates quantification of the gray level. Figures 12.

Figure 12. Digital image

In fig 13 one is to the image segmentation separating the different objects in the scene.

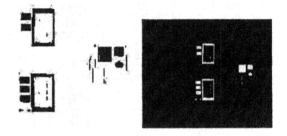

Figure 13. Segmentation   14. Thereshold

The value of this threshold depends on several factors, such as the amount of present objects in the image, the colours and the luminosity of the atmosphere. For this reason, it is necessary to develop a procedure that being able to fix the threshold most suitable. It use techniques based on the histogram are used. Fig.14.

The obtained binary image by means of the previous techniques can contain "noise". The noise corresponds to all those pixels of the image that do not represent the real scene, but that constitutes a distortion of the information contained in the image. This can happen if the surface of support of the objects or bottom of the image is not perfectly white due to the dust existence, any other type of small material or impurities, as also due to errors in the capture of the image.

It is necessary to use a technique allows to compute the centroid of an object of a binary image. Each row of the matrix of the image, from the row superior to the inferior one, crosses from left to right in search of pixels of value `1', if one of these pixels is found stores the position of the column and an accountant is increased. Once a complete row has been crossed, the positions of all their columns are added that have value `1' and it is divided by the value of the accountant, which has registered the total amount of columns with value `1' who contains the row, this gives like result the position column of pixel central of the segment of the object contained in the row. See figure 15.

208

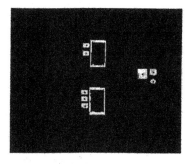

Figure 15. Centroid of the objects of the scene.

The direction of an object is determined by the minimum distance between its centre of mass and an element of his contour. Therefore, the first step to follow to calculate the direction of an object is to determine which pixels belongs to his contour. For it the gradient of the image is used.

This can be observed in following Figure 16 where the point corresponds to the centre of mass of the rectangle and point B was determined finding the minimum range between the centroid and the contour of the object, the union of these points forms a perpendicular straight line to the direction that must have the effector of the manipulator to take the object.

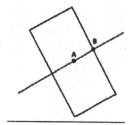

Fig 16, Pick an object

Table 1: Errors to Assembly product								
	Wood bucket and cylinder		Simple SW		Double SW		Triple SW	
	With retiring	Without retiring	With retiring	Without retiring	With retiring	Without retiring	With retiring	Without retiring
1	0%	0%	0%	0%	50%	0%	40%	20%
2	33%	0%	33%	0%	0%	25%	0%	20%
3	0%	33%	33%	0%	25%	0%	40%	0%
4	33%	0%	33%	0%	0%	25%	0%	0%
5	0%	0%	0%	0%	25%	0%	60%	0%
6	33%	0%	33%	33%	25%	0%	0%	40%
7	0%	0%	0%	33%	0%	0%	20%	40%
8	33%	33%	33%	0%	25%	50%	0%	60%
9	33%	0%	0%	33%	25%	50%	20%	20%
10	0%	0%	0%	0%	0%	0%	60%	20%

Table 2: Time of Assembly of a product.							
Geometrical Figure		Simple SW		Double SW		Triple SW	
With retiring	Without retiring	With retiring	Without retiring	With retiring	Without retiring	With retiring	Without retiring
1'4"	1'37"	1'12"	1'34"	1'25"	1'59"	1'45"	2'22"

6. Evaluations

For the evaluations, different tests were done. One test is to see if pallet is retired from the"circuit", and without retired. Each one of these tests was made for different products to assembly the geometric figure, and each type of plug. The results given by Table 1, reflect the percentage of error, by amount of products, in other words, if the product to assemble account of 3 pieces (geometric figure, simple plug) and fail in one, the result is 33% of error, or, if it counts of 5 pieces, and it fails in two, have an error of the 40%, (triple plug). Table 2 shows the times of assembly, like for the evaluations, the iterations, with and without retiring pallet from the conveyor.

To reduce the errors due to the bad position of the pallets in workspace, it uses a small mobile robot to place in right place. In Fig. 17 it can see in the workspace the pallet and the mobile robot. The robot receives the order from computer cell. Two wall permits to put the pallet in the position desired. This position is the best one for Stereo Vision System.

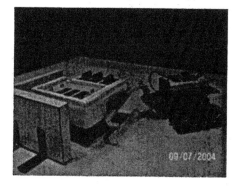

Figure 17. Mobile robot to improve perform

7. Petri Nets Model
8.

Assembly systems are composed of many components; it is difficult to deal with unexpected situations. An undetected error may propagate and end up as a detectable failure which may cause the whole line to stop its operation. In this case, it may take considerable time to diagnose the system and identify the main reasons for the failure.

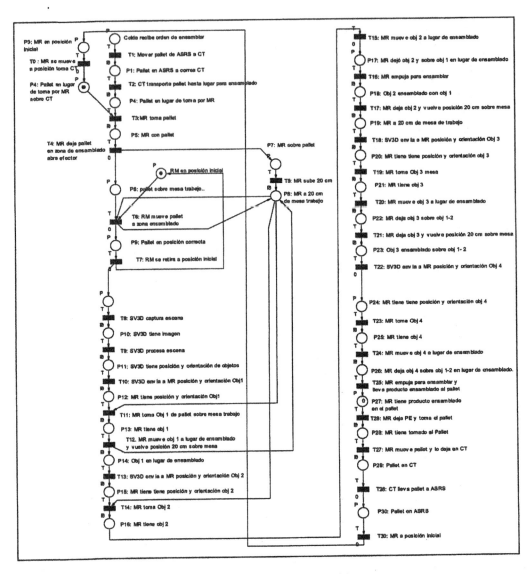

Fig. 18 Petri nets of SW assembly

Places	Table 3: Place definition
	Arrive assembly order to cell
P_1	Pallet in ASRS to CT
P_2	Pallet in conveyor CT
P_3	MR in initial position
P_4	Pallet in place to be pick by MR over CT
P_5	MR with pallet
P_6	Pallet in workspace
P_7	MR over pallet
P_8	MR at 20 cm from workspace
P_9	Pallet in right position
P_{10}	SV3D has digital image
P_{11}	SV3D has position and orientation of objects
P_{12}	MR has position and orientation of Obj1
P_{13}	MR has obj 1
P_{14}	Obj 1 in workspace
P15	MR has position and orientation obj 2
P16	MR has obj 2
P17	MR placed obj 2 over obj 1 in workspace
P18	Obj 2 assembled with obj 1
P19	MR at 20 cm from workspace
P20	MR has position and orientation obj 3
P21	MR has obj 3
P22	MR placed obj 3 over obj 1-2 in workspace
P23	Obj 3 assembled over obj 1- 2
P24	MR has position and orientation obj 4
P25	MR has obj 4
P26	MR placed obj 4 over obj 1-2 in workspace
P27	MR has PE in the pallet
P28	MR has the Pallet
P29	Pallet in CT
P30	Pallet in ASRS

Tr	Table 4: Transitions definition
T0	MR moves to CT
T1	Move pallet from ASRS to CT
T2	CT transports pallet to assembly place
T3	MR pick pallet
T4	MR place pallet in workspace and open effector
T5	MR up 20 cm
T6	RM moves pallet to assembly zone
T7	RM moves out
T8	SV3D captures scene
T9	SV3D processes scene
T10	SV3D send to MR, position and orientation of Obj1
T11	MR pick Obj 1 from pallet that is in workspace
T12	MR moves obj 1 to assembly zone and return to 20 cm
T13	SV3D send to MR, position and orientation of Obj 2
T14	MR pick Obj 2
T15	MR moves obj 2 to assembly zone
T16	MR pushes for assembly
T17	MR places obj2 and return to 20 cm from workspace
T18	SV3D send to MR, position and orientation of Obj 3
T19	MR pick Obj 3
T20	MR moves obj 3 to assembly zone
T21	MR places obj 3 and return to 20 cm from workspace
T22	SV3D send to MR, position and orientation of Obj 4
T23	MR pick Obj 4
T24	MR moves obj 4 to assembly zone
T25	MR pushes for assembly and pick PE to pallet
T26	MR place PE and pick pallet
T27	MR moves pallet and place in CT
T28	CT transport pallet to ASRS
T29	Pallet is place in ASRS
T30	MR moves to initial position

There exist approaches to detect and to predict failures, but the usage of these approaches is limited.

The Flexible Assembly Cell is modelled using Petri Nets that it permits to have a model, to simulate the behaviour of the cell and to evaluate its performance.

In Fig 18 It shows the Petri Model for the assembly of Switch with 2 plugs.

Definitions:

MR	Robotic Manipulator
ASRS	Automatic Storage Retrieval System
CT	Conveyor
RM	Mobile Robot
SV3D	Stereo Vision System
Obj 1	Switch cover
Obj 2	Switch internal cover
Obj 3	Plug 1
Obj 4	Plug 2
PE	Final assembly product

9. Conclusions

It has presented an assembly flexible cell, the product design for assembly of products and the conditions to perform. The assembly flexible cell has a robotic manipulator, a computer vision system and a conveyor. The cell model is done by Petri Nets. The system is part of a flexible manufacturing system (FMS) designed previously

The vision system recognizes the objects within the work area, establishing the position of each one of them. The illumination problem exists that affects the quantification of the objects, like in the determination of the centroid and the direction of each one of the subparts. This problem it solves with the installation of a system of illumination superior to the existing one and using a mobile robot.

A Petri Net model is used to simulate and evaluate detecting and predicting some situations

References.

Lefranc, G., *"La Manufactura Integrada por Computador: un Tutorial"*, Magazine Automática e Innovación de la Asociación Chilena de Control Automático, vol 1, 2, 1993, pg 45.

T. C. Cao, A.C., Sanderson, 1992, "Sensor-based error recovery for robotic task sequences using fuzzy petri-nets", Proceedings of the 1992 IEEE International Conference on Robotics and Automation, Vol.2, pp.1063-1069.

Q. Jing , W. Xisen, P. Zhihua, X. Youngcheng, 1996, "A Research on Fault Diagnostic Expert System Basedon Fuzzy Petri Nets for FMS Machining Cell", Proceedings of IEEE International Conference on Industrial Technology, pp. 122-125.

Andrada I., Lefranc G. "Sistema Experimental de un sistema de almacenamiento automatizado. I Parte. XV Congreso de la Asociación Chilena de Control Automático 2002.

Maizares M., Lefranc G., "Base de datos y programa ejecutivo para un sistema AS/SR". IEEE Latin-American Conference on Robotics and Automation. Chile, 2003.

Lefranc, Gastón, "Simulación de la generación de secuencias de ensamblado en una celda flexible de producción". Automática e Innovación de la Asociación Chilena de Control Automatico. Año 3. Volumen 2. N°7, 1996.

Lefranc G., Cano F., "Sistema Servoing Pick and Place", Congreso Latinoamericano de Control Automático, 2002, Guadalajara, México.

Schleyer G., Lefranc G., Experimental 3-D Visual servoing for FMS applications: MCPL'2004.

This work was possible thanks to the Research Project DGIP 2004, grant by the Pontifical Catholic University of Valparaiso, Chile

PUBLICATIONS
www.elsevier.com/locate/ifac

AN INTEGRATED DESIGN METHODOLOGY FOR AUTOMATED MANUFACTURING SYSTEMS

Eduardo A. P. Santos and Marco A. B de Paula

Pontifical Catholic University of Parana, Graduate Program on Production and Systems Engineering, Curitiba,
80215 901,Brazil
eduardo.portela,marco.busetti@pucpr.br.

Abstract: The current paper contributes to the development cycle of the automated manufacturing systems by proposing a more efficient procedural model. The developed approach is based on the concurrent performance of the physical and control sections, achieving an integrated approach providing time and resources economy. The purpose of this approach is to decrease the time cycle of the designed system, producing a reduction in implementation, integration, testing and maintenance efforts.

Keywords: manufacturing, automation, design, methodology, discrete event systems.

I. INTRODUCTION

The engineering efforts required to design automated manufacturing systems include different types of technological skills. In fact, a manufacturing plant design is made of complex tasks integrating mechanical, electrical/electronic engineering, control systems design and also computer science domain. The entire system is composed by mechanical parts for product handling, sensors and actuators for motion, control and supervision, and digital controllers, adequately programmed to perform the control tasks needed (Bonfe *et al.*, 2002).

An efficient design methodology for the development of the final system should take care of all the aspects concerning the design, specially those related to the integration between mechanical and their control software. Unfortunately, this is not always true in the manufacturing industry, where software for supervision and control is developed during the final phase of the plant design, usually with limited time schedule, according to the specifications of mechanical engineers and customers.

Moreover, specifications on the behavioral features of the system are usually expressed informally, with textual descriptions or even verbally. This fact leads to greater inefficiencies in the control software development, particularly relating to testing and maintenance, which are not supported by rigorous and unambiguous documentation.

This scenario leads to a very inefficient development cycle for manufacturing plants, and often few components can be reused (e.g. mechanical, electronic and software parts) in a subsequent design, increasing global costs. The design of a complex system may require the co-operation of several teams belonging to different cultures and using different languages. New specification and design methods are needed to handle these cases where different languages and methods need to be used within the same design.

The automated system design has been traditionally carried according to the computers' hardware and applications manufactures, assuming the physical process is already designed or implemented. It is possible to justify such procedure realizing that the design teaching on Mechanical Engineering and on Electrical, Computers, Mechatronics Engineering takes place in a fragmented way.

In fact, when a manufacturing system complexity increases and different technologies are required, the design team will need specialists from many domains, as well as Computing, Hydraulics and Pneumatics, Electronics, etc. Integrating their information can also be complex if each one uses its own concepts, diagrams and terminology. In this case, communication and organization issues will often appear.

Statistics from Brazilian industries present that historically up to 90% of development efforts are dedicated to solving post-installation problems; failure corrections or functionality errors on the automated plant (Moraes & Castrucci, 2001). It is extremely important to introduce new methods, techniques and tools on the methodologies of existing designs, in order to reduce the time and resource waste caused by maintenance and optimization.

This paper describes an integrated design methodology for the manufacturing systems, where physical and control design are performed simultaneously. The aim of the methodology is to achieve a better integration between mechanical and control design, in order to obtain a low cost design and reducing the general costs preventing post-installation problems.

II. MECHANICAL DESIGN FOR MANUFACTURING SYSTEMS

The systematization of mechanical design allows improvements in the development and the models comprehension. The systematic also satisfy the learning organization and the design automation needs. Contributions on the subject include Hubka & Eder (1988) and Pahl & Beitz (1989).

According to the authors' approaches, such models merge into a consensual model composed by different phases: clarification, conceptualization, initialization and finally detailed design execution. Such phases lead to intermediate results: specification, functional structure, preliminary layout and detailed documentation.

On the task's clarification phase, the problem is analyzed and information regarding the issue is pointed out. Moreover, a set of design specifications is produced. These specifications define the function and the properties required by the system as well as the constraints, such as standards and deadlines.

The conceptual design starts after the specifications, moves to identify one or more solution principles that attend the specified objectives. According to Pahl & Beitz (1989), the conceptual design is "the part of the design process on which by identifying the essential problems through abstraction, by establishing the functional structure and by the search of solution principles and its combination, the basic solution path is exposed through elaboration of a solution concept".

2.1 Functional description

There is a consensus among researchers of artificial intelligence, software development, technical systems designers, and mechatronic systems designers about functional description, they present it as fundamental in the design process, being the link between the requirements imposed by the customer and the design process itself.

The functional description is used by the designers to model the desired product or system into an abstract level that will lead to the development of the control model or the physical system capable to produce it. Different forms of functional description are proposed and analyzed in the literature. The German school of design builds a functional description based on the functions converting or driving matter, energy or information (Hubka & Eder, 1988; Pahl & Beitz, 1989; Pulm & Lindemann, 2001).

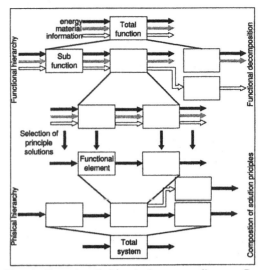

Fig. 1. Functional description according to German School (Umeda & Tomiyama, 1997).

According to Umeda & Tomiyama (1997) and Santos, *et al.* (2001), there are some difficulties in using the functional description in accordance with the German School. However, the article's authors point out that despite of the functional structure being highly abstract, it is the first model to incorporate information on the system structure to be built. According to the present work, the designer must use functional models, which progressively and objectively conduct the physical structure of the system being developed, and the action must be taken in the design conceptual phase.

Thus, at a certain point of the conceptual design, the functional description must express the physical interconnection of the elements, devices or machines. The central idea is to use a non-ambiguous model of functional description, aiming to introduce at this point (phase) the control system design. In this sense, the use of the Channel/Instance net (presented bellow) is proposed.

2.2 The Channel/Instance Net

The Channel/Instance (C/I) net consists of a diagrammatic representation using two basic elements: the active functional units (instances), represented by rectangles, and the passive functional units (channels), represented by circles, both linked by oriented arcs, according to figure 2 (Santos *et al.*, 2001). The channels indicate system components that support the flow of information, energy or material without causing changes on their states. For example, it can be mentioned pipes, wires, conveyor belts, buffers, memories, etc. The instances correspond to the place where the tasks are executed, such as machine components, workstations, chemical reactors, objects (softwares), etc.

In this paper, the use of C/I net is proposed as a central model of functional description for automated manufacturing systems (the main flow is material). The following aspects justify this choice: it's totally free from implementation or production solutions; it details the physical interconnection between machines or devices (the channels are the place where the matter flows through).

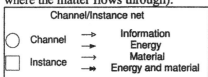

Fig. 2. Channel/Instance net.

III. CONTROL DESIGN FOR MANUFACTURING SYSTEMS

Since a large portion of the establishing of a manufacturing system on the shop floor is consumed by its control system, significant research has been carried out to develop conceptual design of such systems. Existing formalisms on modeling manufacturing systems include Markov chains, queuing theory, Petri nets and Supervisory Control Theory (Cassandras & Lafortune, 1999).

3.1 The Supervisory Control Theory

The Supervisory Control Theory (SCT), initiated by Ramadge & Wonham (1989), is a powerful framework for the synthesis of control for Discrete Event Systems (DES). This theory provides a systematic algorithm for computing the minimally restrictive supervisor that confines the open-loop system behavior (plant) in a specification.

From a certain DES describing the uncontrolled behavior, the plant, and a specification for the controlled behavior, a supervisor can be automatically synthesized in order to pilot the plant to perform within the specification. Ramadge and Wonham's theory of supervisory control uses formal languages and automata to model both the uncontrolled DES and the specification for the controlled behavior.

In the framework of Ramadge & Wonham (1989), the system spontaneously generates discrete events $\sigma \in \Sigma$ classified as controllable ($\sigma \in \Sigma_c$), if the event can be disabled by the control system, else uncontrollable ($\sigma \in \Sigma_u$). Let Σ^* be the set of all finite chains of elements in Σ, including the empty chain ε. A language is a subset of Σ^*. The behavior of a DES may be modeled by languages that, when regular, are represented by automata. An automaton is a quintuple $G = (\Sigma, Q, \delta, q_0, Q_m)$, where Σ is the event set, Q is a set of states, $q_0 \in Q$ is the initial state, $Q_m \subseteq Q$ is the subset of marked states and $\delta: \Sigma \times Q \to Q$, the transition function, is a partial function defined in each state of Q for a subset of Σ.

In their approach, a DES execution is modeled as a sequence of events. The set of such events forms a language and represents all the possible executions of the system. The objective is to restrict the behavior of the system in order to contain it in a desired behavior, called specification. This is achieved by constraining the discrete event automaton to execute events only in strict synchronization with another system, called supervisor.

Although supervisory control theory provides a systematic method to model the control of a DES, the control-space explosion problem still limits its shop floor applications. Due to the specifications synchronous composition that generates an exponential growth in the model states number and, in consequence, increases the computational complexity to be addressed.

3.2 Local modular control approach

The approach of Queiroz & Cury (2000a,b) extends the framework of Ramadge and Wonham (1989) to avoid state-space explosion by exploiting modularity of specifications and decentralization of plant. By this local modular approach, the plant is represented as a set of asynchronous automata and the specifications are expressed locally over the affected sub-plants.

According to Queiroz & Cury (2000a), a modular supervisor should be more promptly modified, updated and maintained. For example, if one subtask is changed, then it should only be necessary to redesign the corresponding local supervisor. In other words, the overall local supervisor should exhibit greater flexibility than its monolithic counterpart.

IV. INTEGRATED APPROACH

As seen in the previous section, the functional description is a fundamental phase in the conceptual design. A system conception model is obtained from a function structure model. Equally, the conception of the control system is made after modeling the physical plant behavior and the required specifications. Thus, this paper proposes an integrated functional description, where the conceptual model of the control system comes from the conceptual model of the physical system. The systematic procedures proposed herein, explore the functional description as the initial phase of the control system design.

4.1 Procedural model for integrated design

A functional description is proposed in order to integrate the physical and control design into conceptual design. It is done considering both models, the C/I net (that provides the system design with the functional perspective), languages and automata (that provide the behavior perspective). The procedural model proposed is presented in figure 3.

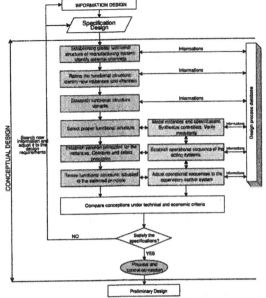

Fig. 3. Procedural model proposed for integrated design.

The first design process task is the specifications analysis according to the required functions in order to design the constraints. The specifications must be grouped by categories, for example, product description: dimension, weight, material; the physical environment: entering and exiting places, feed and retrieve (manual, automated, batch, etc); the objectives: reduce operations, reduce de processing time, etc; the control system: centralized control installation, need of monitoring functions etc.

From the specification list, it generates an abstract functional structure of the system to be built, named total function by Pahl & Beitz (1989). This stage consists basically on identifying the parts and the final products and from that, determining the entering and exiting channels where the raw parts or intermediate parts will be stored. Thus, the abstract functional structure must determine the main instance of a system through the relationship between the initial and final products.

Next, the designer details the functional structure required for the system. It is suitable to have a draft of the functional structure, it can be obtained by describing (exploding) the process (normally, several ones) that is necessary to the main system flow (entering and exiting of matters). In a general scheme, the first refinement takes into account the execution of the main tasks, where a channel must be previously selected (during the design) for each of these systems' tasks.

Gradually, the functional structure is developed through the identification of new channels and instances in order to satisfy all the operational requirements. The functional structure must be developed and detailed more to obtain a description that allows designers to verify the operations that must be executed to achieve the product, it means, all the processes and/or necessary assembling were performed. Figure 4 shows a refinement process.

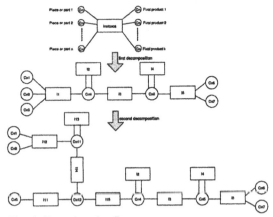

Fig. 4. Functional refinement.

The functional refinement must not impose a unique functional structure. Nevertheless, the functional description force is exactly on the possibility of creating and comparing in an abstract level, alternatives to functional structures. Obtaining functional variants through different configurations in C/I net is a process based essentially on the designer creativity, observation and experience. Thus, the designer decides the tasks order (processes and assembling), the existence of intermediate channels, ramifications or junctions on structure, the number of carrier instances, etc.

Having the variants of functional structures, the designer can select those more adapted to the design specifications. Such functional structure represented by the C/I net, describes the system under a functional perspective. A behavior perspective of the system can be also considered in designing to complete the functional description proposed in this paper.

The inclusion of the behavior perspective on the functional description is based on the matter behavior specification on the channels. Initially the models that represent the behavior of the instances are carried out. Thus, models describing the desired specifications of channels are determined. Those specifications impose instances' behavior restrictions to obtain the system global function. Finally, one can determine the controller imposing the correct flow of matter by limiting the instance operations.

The theory of supervisory control presented in the previous session is used within this stage to support the system's behavior description through the automata and languages representation. With the selected functional structures, the premises of the local modular control approach are used.

The representation is illustrated by Figure 5, where two automata (Instance 1 and 3) are used to describe the matter transport from channel A to channel C and another one (Instance 2) performs specific task on channel B. Such instances are modeled by a two states automaton: the initial state of instance 1 is inactive and it can change to active state. The occurrence of the events α (and β) results on the removal (and deposit) of matter from the entering (and exiting) channel of instances 1 and 3 and the occurrence of the events β results on the deposit of matter on the exiting channel. E1 specification guarantees that overflow or underflow in channel B will not occur. E2 specification guarantees that when the matter arrives on channel B, it is processed by instance 2. E3 specification guarantees that the matter will not be removed from channel B before the end of the instance 2 operation.

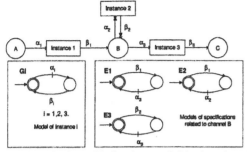

Fig. 5. Inclusion of behavior model in functional description.

Using specific algorithms, for example TCT (Wonham, 1999), one can synthesize local supervisors for the system and check the modularity property between supervisors. It assures that concurrent operations of the local supervisor are non-blocking. Applying these algorithms on the example presented by figure 5, one can obtain three supervisors, as showed by figure 6. One can notice that the supervisors are designed before selecting the solution's principles phase. Thus, the designer can have different options and choose the one matching the functional structure and the designed local supervisors.

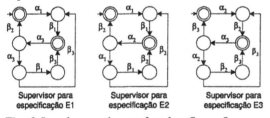

Fig. 6. Local supervisors related to figure 5.

According to the chosen option, the designer will select the solution principles for each instance or for instances' groups. For example, Instance 2 should carry the matter from channel A to channel B according to a pre-defined specification. Technologies used to execute the operation are selected by the designer based on specific criteria (i.e., industrial policies and customers preferences are important when selecting these technologies).

Pahl & Beitz (1989) recommend the use of a morphologic matrix to help the designer in organizing several solution principles for each instance of the functional structure. Another method of selection is the use of project catalogs or solutions database, as shown by figure 7.

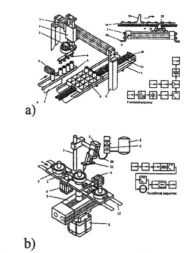

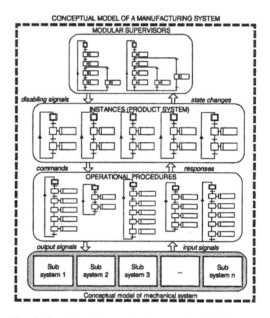

Fig. 7. Examples of solution principles: a)packing ; b)gluing (Festo, 2000).

Consequently, as showed by figure 8, the conceptual design is composed of: A functional description of the system in C/I net; Instance models and specifications in automata; Control structure (local supervisors) presented according to IEC 848 (1988); Conceptual models of the physical system.

Fig. 8. Result of integrated conceptual design proposed.

4.2 Operational specification set

The methodology proposed by Ramadge & Wonham (1989) for control's synthesis expects an adequate model for the system behavior specification. These specifications are defined by a set of restrictions on the channels, which by themselves determine the co-ordination operation of the instances.

According to Groover (2001), there are different applicable configurations to automated manufacturing system. Such configurations may involve junctions, branches (Y), ramifications or feedback on the matter flow line. The type of transportation system used also represents an important aspect to be considered. In this sense, the present paper associates operational specifications to manufacturing systems according to the configuration selected.

However, using automata to represent adequately each system's specification requires experience on the use of this formalism. Therefore, the paper proposes that a specification set is built in order to make the design task more efficient and systematic. Other than that, allows the reuse of control modules (local supervisors) in distinct projects.

V. CONCLUSION

The present research explores the functional description as a fundamental tool for the optimization of automated systems design, extending some concepts for a better fitting to the systems under study. Opposing the sequential engineering, where plant conceptions and control systems are considered non-integrated, this approach aims to take into account aspects related to different technologies or areas. In this sense, design process becomes more efficient and reliable, reducing time and resource. The general contributions of the paper associated with the low cost design are: approaching the design issue in an integrated method (physical and control system); structuring the conceptual design in order to reach a global conception; proposing a new functional description model; inserting in the design formal tools for modeling and synthesis; creating an operational specifications set.

A hierarchical and decentralized control structure can be naturally obtained, providing advantages such as good maintenance, error detection and easier modifications on the physical structure of the system. For example, the designer will be able to replace a specific mechanism in operation by another one without having to modify the whole control structure of the system. In addition, functional structure parts can be used in subsequent designs using similar specifications.

This work presents many others aspects must be studied in order to reduce the production costs. The continuity of the research looks for a better efficiency in the modularity verification process. Because, even knowing that the local modular approach brings computational economy, the high complexity of this modularity can result in a costly control synthesis, which can be an obstacle for a larger scale of production.

Another topic is related to the expansion of the specification structure. It is necessary to include complex configuration of manufacturing systems, allowing the reuse of control programs at a larger scale.

REFERENCES

Bonfe, M., C. Donati and C. Fantuzzi (2002). An application of software design methods to manufacturing systems supervision and control. In: *Proceedings of the IEEE International Conference on Control Applications*, Glasgow, UK.

Cassandras, C. G., S. Lafortune (1999). *Introduction to discrete event systems*, Kluwer Academic Publishers.

Festo (2000). *Handling pneumatics: 99 examples of pneumatic applications*, Festo AG & Co.

Groover, M. P. (2001). *Automation, production systems and computer integrated manufacturing*, Upper Saddle River: Prentice Hall.

Hubka, V., and W. E. Eder (1988). *Theory of technical systems*, Springer-Verlag, Germany.

International Eletrotechnical Comission, *IEC 848*, Preparation of function charts for control systems, Switzerland, 1988.

Moraes, C. C., and P. L. Castrucci (2001). Industrial Automation Engineering, LTC, Brasil (in portuguese).

Pahl, G., and W. Beitz (1989). *Engineering design – a systematic approach*, Springer-Verlag.

Pulm, U., and U. Lindemann (2001). Enhanced systematics for functional product structuring. In: International Conference on Engineering Design (ICED'01), Glasgow, UK.

Queiroz, M. H. and J. E. R. Cury (2000a). Modular control of composed systems. In: *Proc. Of the American Control Conference*, Chicago, USA.

Queiroz, M. H and J. E. R. Cury (2000b). *Modular supervisory control of large scale discrete-event systems*. Discrete event systems: analysis and control, Kluwer Academic Publishers (proc. WODES, Ghent, Belgium).

Ramadge, P. J., and W. M. Wonham (1989). The control of discrete event systems. In: Proc. Of the IEEE, **77(1)**:81-98.

Santos, E. A. P., V. J. De Negri and J. E. R. Cury (2001). A computational model for supporting conceptual design of automatic systems. In: Proc. of International Conference on Engineering Design (ICED'01), Glasgow, UK.

Umeda, Y., and Tomiyama (1997). Functional reasoning in design. IEEE Expert, intelligent systems & their applications. vol. 12, n.2.

Wonham, W. M. (1999). Notes on control of discrete event systems. Course notes for ECE 1636F/1637S. Dept. of Electrical and Computer Engineering, University of Toronto, Canada.

PUBLICATIONS
www.elsevier.com/locate/ifac

SOFTWARE ARCHITECTURE OF AUTONOMOUS AGENTS

Grzegorz Dobrowolski and Edward Nawarecki

*Institute of Computer Science, AGH University of Science and
Technology, Kraków, Poland*
e-mail: grzela@agh.edu.pl

StreszczenieA software architecture of the autonomous agents based on the idea of abstract
state machines — ASM is proposed. It generalizes often used in software solutions the model
of a finite state machine. The proposed architecture establishes links between well founded
notions and recognized mechanisms of multi-agent systems and procedures of systematic
design and implementation of them as computer-network applications.

Keywords: multi-agent systems, software engineering, abstract state machines

1. INTRODUCTION

Abstract state machines — ASM (Börger and Stärk, 2003) become a very popular and important research direction in the field of design and specification of software systems nowadays. ASM is proposed as a general model of software that suits both simple applications and complex concurrent distributed systems as well. Due to the generalization, relations between ASM (ASM-based methods) and other commonly used techniques of software specification and verification like: UML, Petri nets etc. are preserved. It allows to foresee that ASM will augment the software engineer's tool kit soon.

Architecture is meant here as a bunch of means necessary for realization of autonomous software agents (consequently — multi-agent systems) selected according to a particular model of agency. It can be seen as a skeleton that supports implementation of agent-based software dedicated to various fields of applications.

Although study on multi-agent systems (Weiss, 1999) have being carried out for several years and there exist a proposal of standardization as well as implementation platforms prepared according to it, a lack of the universal architecture of the agents effective also in a process of production of software is still observed.

Not further developed the idea of agent-oriented UML (Odell *et al.*, 2000) gave mainly the well worked out inter-agent communication layer of the implementation platforms (Bellifemine *et al.*, 1999; Poslad *et al.*, 2000). There are also known various concrete constructions of agents but only in detailed descriptions (e.g. architecture taking advantage of rule-based systems (Dobrowolski *et al.*, 2003)).

A software architecture of the autonomous agents based on the idea of ASM is proposed in the article. It generalizes often used in software solutions the model of a finite state machine. One of the essential differences here is that state transitions are not caused by external signals (input of an automaton) but are the effects of autonomously taken decisions that may be only loose consequences of stimuli from the agent's environment.

The article consists of: premises of modeling an agent, a sketch of the corresponding model, some ideas of software engineering with respect to a multi-agent system that justifies structuring of design and programming at levels of an agent and a module of an agent. Finally, it describes the proposed architecture and gives indications how constitutive features of an agent can be reflected in the ASM framework.

2. AN AGENT AS A BLACK BOX

Let us start with analysis of possible interactions that can arise between an agent closed in a *black box* and its neighborhood.

What is obvious, the interactions are divided into two categories depending on direction: the agent's influence on the neighborhood (the environment or other agents of a system) and *vice versa*.

Each of the possible influences can be considered in two aspects. The information one that is of the most interest for the scope and, so-called here, physical one that embraces the rest of features of the influence. Differentiation of the aspects comes from observation that majority of real influences and, in consequence, these that are faced in multi-agent systems do not follow patterns: sender – information channel – receiver, causer – physical impact – object, concurrently. The difference becomes deeper when changes of information and physical states of both sides are taken into account.

Let us analyzed how the information and physical aspects look like in the cases of elementary agent–neighborhood interactions.

Perception. An agent gets information from neighborhood and changes his information state (e.g. memorizes the information). In the physical aspect he influences himself (powers of his sensors).

Receiving a message. It is the same as the perception in both discussed aspects.

Sending a message. An agent sends information and simultaneously affects neighborhood in the physical sense (e.g. uses a communication channel). The case can be also regarded as the information influence on the neighborhood.

Agent's influence on the neighborhood. As an example we can have here getting resources from the neighborhood. The information aspect of that arises via **perception**.

Physical impact on an agent. It may be, for example, collision with an unrecognized physical obstacle on the course of an agent-robot. In our analysis the agent is an object of the physical influence that has no direct information characteristic. This case is not commonly understood this way. Probably, it comes from the behind-the-scene assumption about existence of yet another unspecified type of influence that can be called *auto-perception*.

Although auto-perception is in both discussed aspects an internal (reflexive) influence, it should be mentioned here as the element that completes logically the above presented schema. **Auto-perception** produces

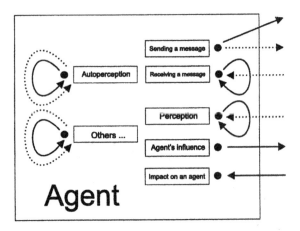

Rysunek 1. Interaction between an agent and neighborhood (broken arrows — information aspect, continuous — physical)

information about the agent's internal state, his actions, effects of external influences. There are other internal influences also.

Figure 1 presents an agent as a black box together with possible interactions with neighborhood. Arrows depict their information and physical aspects. The internal influences are also shown under labels: auto-perception and **others...**

Assuming that the schema is complete, i.e. there is no real influence that cannot be classified as belonging to one of the types discussed above or reflected as a combination of some of them, the following conclusions can be drawn:

(1) Excluding physical impact on an agent, the rest of agent–neighborhood interface is fully controlled by the agent himself.
(2) Physical influences of the neighborhood, completely independent of an agent, are registered by him (if agent can do that) via auto-perception actions.

The next assumption is that an agent operates in a discrete time regime. Then the influences can be regarded as finite sequences of realized actions (elementary influences). Moreover, excluding the case of physical impact of the neighbourhood ones more, the agent autonomously build these sequences in the sense of choice of both actions and respective moments of their executions.

According to the above considerations, *receiving a message* can be structured without loss of generality as follows. An agent performs periodically a sequence of equal actions — let us call them *listen for* — which check whether a message has come (auto-perception). If yes — the actual action of receiving is done.

Because a similar analysis can be carried out with respect to the rest of influences types, an elementary action of an agent falls into one of two general groups:

(1) **external actions** that create the agent's interface (actual receiving a message in the example above),
(2) **internal actions** that constitute indispensable support for the external ones (e.g. listen for).

External actions are performed by an agent in quasi-continuous manner (periodically, step-by-step). From the outer perspective (based on external actions performed) he can be seen as a discrete event-driven system instead. It must be stressed here that even if it is possible to show a relation between an external action and event in the neighbourhood, this action is undertaken autonomously.

From the assumption about synthesis of action sequences stems that a basic (decision) mechanisms of the agent's model is **sequential initiation of appropriate actions**. Taking decision about initiation opens possibility of concurrent realization of actions. If concurrency is unnecessary or can not be implemented, the mechanism becomes of sequential realization of actions.

Concluding consideration about the agent model, it is necessary to mention how the sequences of actions start and end. Let us assume that an agent start with some impulse from outside, which activate the sequential initiation of actions, and gets its end with the specific terminal action also decided by the mechanism.

3. ELEMENTS OF THE AGENT'S MODEL

A sketch of a formal model of an agent which follows considerations of the previous section is shown beneath. This way the model reflects mostly the agent's ability to co-exist with other agents in the common environment — features that are decisive in establishing a system. Obviously, the model defines also some conditions for abstract internal architecture of an agent.

Definicja 1. Agent Λ is a three-tuple of the form:

$$\Lambda = \{ A, S, F \subset S \times A \times S \} \quad (1)$$

gdzie: A finite set of actions (elementary) of agent Λ;

S finite set of internal states of agent Λ;

F three-element relation describing permitted succession of states and actions of agent Λ — in the given state the agent can perform an action (the second element) that leads him to a new state (the third element).

Set A represents all of the 'a's actions. It is possible to interpret the definition as describing the 'a's behaviour *from outside*. Then A describes a subset of external actions only.

Relation F reflects the possible combinations of states and actions. Each state implies a subset of actions allowed. This is a domain of decision for the mechanism of initiation of actions. Each elementary action can be chosen several times. In consequence the chosen actions are executed.

Questions with respect to action realisation as well as many others issues of the model can be left beyond the scope of the paper without loss of readability but are available in (Dobrowolski, 2002; Dobrowolski, 1998).

4. AGENT AS AN ABSTRACT STATE MACHINE

Abstract state machine has been chosen as the base for a definition of the agent's software architecture for two important purposes. ASM has good theoretical foundation. Both the automaton and its state are defined in terms of abstract algebra. So it can be awaited that the formulated architecture will be justified and easy for theoretical development.

From other hand, applied in description of algebras operations (static function in general sense) are augmented with — so called — dynamic functions that introduce into them algorithm elements traditionally implemented in programming languages. This way ASM acquire expression power of the languages and can be easily approved by practicing engineers. At the same time, such augmented algebraic structure gains interpretation of (computer) memory and changes of its contents — switching from state to state.

Due to practical orientation of the paper, only indispensable for the agent's architecture elements of ASM theory will be reported beneath in rather general manner. A reader is referred to (Börger and Stärk, 2003) for details.

4.1 *Elements of ASM*

ASM can be defined as a finite set of formulas of the shape

$$\{ \text{if } condition_i \text{ then } updates_i \}_{i=1,...,n} \quad (2)$$

where $condition_i$ is a predicate of a given, assumed for the particular ASM logic, and $updates_i$ is a set of assignment

$$f(t_1, \ldots, t_n) := t \quad (3)$$

each of which is understood as a change of value (or definition) of function f for the values of parameters equal to $t_1, \ldots, t_n$ to the new value t (for the sake of readability indexes which identify a concrete function and, next, ASM are neglected).

Functioning of ASM is cyclical performing of the following operations. All $\{condition_i\}_{i=1,\ldots,n}$ are evaluated concurrently and those $updates_i$, for which conditions are true, are marked. Next — also concurrently — the marked updates are realized through appropriate calculations of functions and assignments. End of the longest assignment completes a cycle.

Assignments (3) is performed as corresponding operators of programming languages in general. Firstly, the parameters are calculated according to their list (here: $v_1, \ldots, v_n$), next — value of the function ($v = f(v_1, \ldots, v_n)$). To make the value accessible in the next cycle, its identification — called location — is introduced via the function name and list of its calculated parameters. Location l can be interpreted as a piece of memory (with the address), which stores value v of the function, and the whole operation — update of the location (memory) to the value of (l, v).

Coming from cycle to cycle, we have in general to kinds of locations: those which stay untouched or have been just modified.

If for a given ASM we put together all locations possible during its life time, they create memory of the automaton. Departing from some state (starting) of the memory and performing appropriate locations updates in each cycle, we obtain evolution of the automaton state. For unambiguous determination of a next state it is necessary the bunch of updates in each cycle to be not conflicting i.e. the following formula to hold for each pair in a cycle

$$(l, v) \wedge (l, v') \Rightarrow v = v' \qquad (4)$$

It means that although the functions can use whatever data describing the state, their results must be located in distinct places of the memory.

Functions (updates and locations) as well as conditions can be differentiated according to their features into some types. So we have *static* functions — not affected during ASM run and *dynamic* ones, which algorithm depend on the state of an automaton. In turn, dynamic functions fall into three categories: *controlled*, *in* or *monitored*, *out*, and *shared*. *Controlled* functions can be modified only by other updates of the same automaton while *in* functions are modified only by the neighbourhood (reflect its influence) and, in consequence, can occur at right-hand side of assignment (3) only. On the contrary, *out* functions (modelling influence of an automaton) can stay only at left-hand side of (3)

but are accessible for the neighbourhood. *Shared* functions combine features of *controlled* and *in* ones. Static and *controlled* functions are identified as *internal* ones while the rest groups *external* functions.

The assumption is that in functions (updates, locations) are determined in each state as well as synchronization of *shared* functions realization is established.

It is obvious that functions have different numbers and types of arguments — *signatures*. Signature of all functions define the signature of ASM.

4.2 Architecture of an agent

Although methodology based on abstract state machines serves with possibility of working with modules (here: sub-ASM) or procedures, we start our discussion with the agent's modular structure (see fig. 2), which is generated by specificity of his applications and implementation conditions. Next, these modules (especially one of them) will be described in terms of ASM to introduce the proposed architecture.

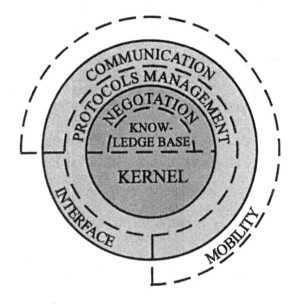

Rysunek 2. Modules of a software agent

From the point of view of establishing a multi-agent system the most important is a sphere which embraces modules of communication and interface. The former is dedicated to interactions with other software agents, the latter — a human agent. These two modules can be present and utilized in extent depending on tasks allocated to a particular type of agents in the system.

If an agent is mobile, the sphere is embraced by yet another module (sphere) that realized mobility of his code and calculations. It can be arranged in this way because mobility influences in fact just the structure of information flows not their contents.

All the above supports multi-direction and multi-aspect interchange of information that, in turn, needs management also with respect to its semantics. An unifying idea here is to use protocols (communication, interaction), which occurs to be powerful enough to stimulate and realize even very complicated cooperation patterns among agents. So is a need for a module (sub-sphere) of protocols management.

The rest of the semantic analysis is carried out in a central module of an agent — *kernel*. Internal structure of a kernel strongly depends on the function of an agent in the system as well as applied algorithms and data structures. As examples we can propose here: negotiation management module, knowledge base etc. It is obvious that a kernel is the most important and, usually, the most complicated agent's module. It is also the most specific for agent-based techniques.

Therefore, let us focus our attention on the kernel remembering that the rest of modules can be also represented by ASM. Moreover, some of them do not need any deeper analysis, e.g. protocols management module is exploiting an idea of finite state machine, interface module is being built using one of the popular graphic libraries. The point of departure towards a kernel architecture are conclusions of sections 2 and 3. They also dictates the way of description of the agent's features and mechanisms in terms of ASM.

Action ith of an agent is represented in ASM by expression $updates_i$ in a formula of the shape (2), no matter the action is internal or external. Updates of an action are of either *controlled* or *in* or *out* type depending on an information aspect of it with respect to a kernel (e.g. auto-perception, perception, sending a message, respectively [see fig. 1]).

Expressions $condition_i$ of the all specified agent's actions create in common the mechanisms of initiation of appropriate actions. On assumption that in a single step of an automaton a single action can be chosen at most, the mechanism works sequentially as it was postulated. If more than one condition is true than cause-effect relations are also preserved. There is due to the main assumption of ASM that a set of updates is not contradictory in each step (see eq. (4)). Then all triggered actions can be regarded as a single but compound action.

Now it is very easy to define the **agent's state** as an union of his updates (locations), i.e. the state of ASM that describes the agent.

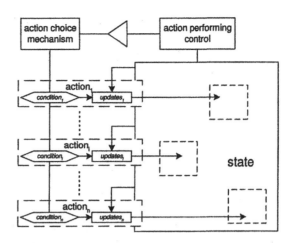

Rysunek 3. Software agent architecture based on ASM

Having the main elements of the agent's (a kernel) architecture defined, we can illustrate the idea in figure 3. Some details are discussed further.

Controlled updates model internal actions of a kernel and thus an agent. *In* and *out* updates of a kernel create its interface with outer modules. Some of them can be, in fact, a main part of execution of some external actions of an agent. Let us consider *receiving a message* as an example. The main part of it is *in* update that modifies kernel state in the appropriate location. The location stores the message contents and some additional information about communication like progress of the interaction protocol used. Analysis of the contents that can but has not to follow (in any period of time) is modelled as independent *controlled* updates with its own location. Of course, the analysis has access to the location of the true receiving action.

In this way the proposed architecture realizes yet another postulate about the (limited) **agent's autonomy**. In the above example the agent must receive a message (it is originally forced by the neighbourhood) but analysis of it and possible consequences are decided by him (analysis is an autonomous action).

The necessary condition of *receiving a message* is completion of the message by the communication module. It can be done in such a way that the appropriate $condition_i$ is dependent on a flag in location of the receiving action of the kernel.

Actions that are directed outside in their information aspect are realized analogously. A kernel executes *out* updates, which location is available to dedicated module. In general, execution of some (*external*) actions involves a few modules that cooperate as it is shown above.

Thus application of ASM supports design of **modular architecture**. It is easy to consider also hardware elements then some locations can be realized as device registers that work for an agent as a part of interface with real neighbourhood. Cooperation of modules can be designed in both synchronous and asynchronous modes.

5. SUMMARY

Architecture of a software agent based on the idea of abstract state machine is presented in the paper. The proposed architecture establishes links between well founded notions and recognized mechanisms of multi-agent systems and procedures of systematic design and implementation of them as computer-network applications. The proposed architecture in particular allows for:

- an effective definition of the agent's state, which is a basic notion in theoretical consideration of an agent and systems of them;
- an independent implementation of the agent's actions, which opens possibility to design learning agents (acquiring abilities to perform new actions);
- an independent design and implementation of the agent's modules;
- specification and verification of agent-based software done according to the proposed architecture in the range of formal methods as well as practical procedures (light) applied directly in a process of software production.

Summarizing, steady development and growing scope of application of ASM idea must be underlined, which positively sketch perspectives of the proposed architecture.

REFERENCES

Bellifemine, F., A. Poggi and G. Rimassa (1999). JADE — A FIPA-compliant agent framework. In: *Proc. of the 4th Int. Conf. on the Practical Applications of Agents and Multi-Agent Systems (PAAM-99)*. London, UK. pp. 97–108.

Börger, Egon and Robert Stärk (2003). *Abstract State Machines*. Springer-Verlag.

Dobrowolski, G., R. Marcjan, E. Nawarecki, S. Kluska-Nawarecka and J. Dziaduś (2003). Development of INFOCAST: Information system for foundry industry. *TASK Quarterly* 7(2), 283–289.

Dobrowolski, G. (1998). Network operating agents as a mean for decentralized decision support systems. In: *Management and Control of Production and Logistics MCPL'97* (Z. Binder, B.E. Hirsch and L.M. Aguilera, (Eds.). Vol. 2. IFAC/PERGAMON. pp. 393–398.

Dobrowolski, G. (2002). *Technologie agentowe w zdecentralizowanych systemach informacyjno-decyzyjnych*. Vol. 107 of *Rozprawy Monografie*. Uczelniane Wydawnictwa Naukowo-Dydaktyczne Akademii Górniczo-Hutniczej im. S. Staszica. Kraków.

Odell, J., H. Van Dyke Parunak and B. Bauer (2000). Extending UML for agents. (opublikowane w WWW).

Poslad, S., P. Buckle and R. Hadingham (2000). FIPA-OS: the FIPA agent Platform available as Open Source. In: *Proc. of the 5th Int. Conf. on the Practical Application of Intelligent Agents and Multi-Agent Technology (PAAM 2000)* (J. Bradshaw and G. Arnold, (Eds.). The Practical Application Company Ltd.. Manchester, UK. pp. 355–368.

Weiss, G. (Ed.) (1999). *Multiagent Systems: A Modern Approach to Distributed Artificial Intelligence*. The MIT Press.

ELSEVIER
IFAC
PUBLICATIONS
www.elsevier.com/locate/ifac

NICHING TECHNIQUES BASED ON SEXUAL CONFLICT IN CO-EVOLUTIONARY MULTI-AGENT SYSTEM

Rafał Dreżewski* Krzysztof Cetnarowicz*

*Department of Computer Science,
AGH University of Science and Technology,
Kraków, Poland
drezew@agh.edu.pl

Abstract: Evolutionary algorithms often suffer from premature loss of population diversity what limits their adaptive capacities in dynamic environments and leads to location of single solution in case of multi-modal fitness landscapes. Niching techniques for evolutionary algorithms are aimed at locating more than one optima of multi-modal functions. Sexual selection resulting from sexual conflict and co-evolution of female mate choice and male display trait is considered to be one of the ecological interactions responsible for speciation. This paper introduces the co-evolutionary multi-agent system with speciation by sexual conflict and its formal model. Such system is applied to multi-modal function optimization and the results from runs against commonly used test functions are presented.

Keywords: evolutionary algorithms, multi-agent systems, co-evolution, niching

1. INTRODUCTION

Evolutionary algorithms (EAs) have demonstrated in practice efficiency and robustness as global optimization techniques. However, they often suffer from premature loss of population diversity what results in premature convergence and may lead to locating local optima instead of a global one. What is more, both the experiments and analysis show that for multi-modal problem landscapes a simple EA will inevitably locate a single solution (Mahfoud, 1995). If we are interested in finding multiple solutions of comparable fitness, some multi-modal function optimization techniques (*niching methods*) should be used. Niching techniques (Mahfoud, 1995) are aimed at forming and stably maintaining niches (species) throughout the search process, thereby allowing to identify most of desired peaks of multi-modal landscape.

The understanding of speciation still remains a greatest challenge for evolutionary biology. The biological models of speciation include *allopatric models* (which require geographical separation of subpopulations) and *sympatric models* (where speciation takes place within one population without physical barriers) (Gavrilets, 2003). Sympatric speciation may be caused by different kinds of co-evolutionary interactions including *sexual selection*.

Sexual selection results from co-evolution of female mate choice and male display trait where females evolve to reduce direct costs associated with mating and keep them on optimal level and males evolve to attract females to mating (*sexual conflict*) (Gavrilets, 2003). The proportion of two sexes (females and males) in population is almost always 1 : 1. This fact combined with higher females' reproduction costs causes, that in the majority of cases, females choose males in the reproduction process according to some males' features. In fact, different variants of sexual conflict are possible. For example there can be higher females' reproduction costs, equal reproduction costs (no sexual conflict), equal number of females and males in population, higher number of males in population (when the costs of producing female are higher than producing male), higher number of females in population (when the costs of producing male are higher than producing female) (Krebs and Davies, 1993).

In the following sections the previous work on sexual selection as a population diversity and speciation mechanism for evolutionary algorithms is presented. Next, the formal model of co-evolutionary multi-agent system based on the sexual conflict, in which females' reproduction costs are higher than males', is presented. In such a system two sexes co-evolve: females and males. Female mate choice is based on values of some important features of selected individuals. Also the operator of grouping individuals into reproducing pairs is introduced. Such system is applied to multi-modal function optimization and compared to other techniques.

2. PREVIOUS RESEARCH ON SEXUAL SELECTION AS A SPECIATION MECHANISM

Sexual selection is considered to be one of the ecological mechanisms responsible for sympatric speciation (Gavrilets, 2003). Gavrilets (2003) presented a model, which exhibits three general dynamic regimes. In the first one there is endless co-evolutionary chase between the sexes where females evolve to decrease the mating rate and males evolve to increase it. In the second regime, female alleles split into two clusters both at the optimum distance from the male allele and males get trapped between the two female clusters with relatively low mating success. In the third one males answer the diversification of females by splitting into two clusters that evolve toward the corresponding female clusters. As a result the initial population splits into two species that are reproductively isolated.

Todd and Miller (1997) showed that natural selection and sexual selection play complementary roles and both processes together are capable of generating evolutionary innovations and biodiversity much more efficiently. Sexual selection allows species to create its own peaks in fitness landscapes. This aspect of sexual selection can result in rapidly shifting adaptive niches what allows the population exploring different regions of phenotype space and escaping from local optima. The authors also presented the model of sympatric speciation via sexual selection.

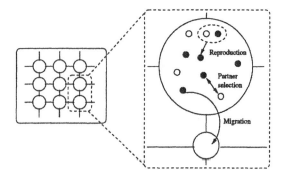

Fig. 1. Co-evolutionary multi-agent system with sexual selection used in experiments

Sánchez-Velazco and Bullinaria (2003) proposed gendered selection strategies for genetic algorithms. They introduced sexual selection mechanism, where males are selected on the basis of their fitness value and females on the basis of the so called indirect fitness. Indirect fitness is the weighted average of the individual's fitness value, age, and potential to produce fit offspring compared to her partner's direct fitness. Mutation rates are different for each gender. The authors applied their algorithm to Traveling Salesman Problem (TSP) and function optimization.

Sexual selection as a mechanism for multi-modal function optimization was studied by Ratford, Tuson and Thompson (1997). In their technique sexual selection is based on the seduction function value. This function give a low measure when two individuals are very similar or dissimilar and high measure for individuals fairly similar. The Hamming distance in genotype space was used as a distance metric for two individuals. The authors applied their mechanism alone and in combination with crowding and spatial population model. Although in most cases their technique was successful in locating multiple peaks in multi-modal domain the strong tendency to lose all the peaks except one after several hundreds simulation steps was observed.

As it was presented here, sexual selection is the biological mechanism responsible for biodiversity and sympatric speciation. However it was not widely used as maintaining genetic diversity, speciation and multi-modal function optimization mechanism for evolutionary algorithms. It seems that sexual selection should introduce open-ended evolution, improve adaptive capacities of EA (especially in dynamic environments) and allow speciation (location of different optima in multimodal domain) but this is still an open issue and the subject of ongoing research.

3. CO-EVOLUTIONARY MULTI-AGENT SYSTEM WITH SEXUAL SELECTION

The main idea of *evolutionary multi-agent system (EMAS)* is the modeling of evolution process in multi-agent system (MAS) (Cetnarowicz et al., 1996). The basic EMAS model allows the evolution of only one species. The co-evolutionary multi-agent system (CoEMAS) model allows modeling of biological speciation mechanisms based on co-evolutionary interactions (including sexual selection), competition for limited resources, and geographical isolation (Dreżewski, 2003). CoEMAS can be applied, for example, to multi-modal function optimization (in stationary and dynamic environments) and multi-objective optimization.

The system presented in this paper is the CoEMAS with sexual conflict (SCoEMAS). The mechanisms used in such system include: sexual conflict and co-evolution of sexes (higher female reproduction costs), sexual selection based on the location of agents in fitness landscape (females choose males), and forming reproducing pairs.

3.1 SCoEMAS

The *SCoEMAS* may be described as 4-tuple:

$$SCoEMAS = \langle E, S, \Gamma, \Lambda \rangle \tag{1}$$

where E is the environment of the $SCoEMAS$, S is the set of species ($s \in S$) that co-evolve in $SCoEMAS$, Γ is the set of resource types that exist in system, the amount of type γ resource will be denoted by r^γ, Λ is the set of information types that exist in system, the information of type λ will be denoted by i^λ. There are one resource type ($\Gamma = \{\gamma\}$) and four information types in $SCoEMAS$ ($\Lambda = \{\lambda_1, \lambda_2, \lambda_3, \lambda_4\}$).

3.2 Environment

The environment of $SCoEMAS$ may be described as 3-tuple:

$$E = \langle T^E, \Gamma^E = \Gamma, \Lambda^E = \{\lambda_1, \lambda_2\} \rangle \tag{2}$$

where T^E is the topography of environment E, Γ^E is the set of resource types that exist in environment, Λ^E is the set of information types that exist in environment. The topography of the environment is given by:

$$T^E = \langle D, l \rangle \tag{3}$$

where D is directed graph with the cost function c defined: $D = \langle V, F, c \rangle$, V is the set of vertices, F is the set of arches. The distance between two nodes is defined as the length of the shortest path between these nodes.

The l function makes it possible to locate particular agent in the environment space:

$$l: \quad A \to V \tag{4}$$

where A is the set of agents, that exist in $SCoEMAS$.

Vertice v is given by:

$$v = \langle A^v, \Gamma^v = \Gamma^E, \Lambda^v = \Lambda^E \rangle \tag{5}$$

A^v is the set of agents that are located in the vertice v. Agents can collect two types of informations from the vertice. The first one includes all vertices that are connected with the vertice v and the second one includes all female agents (of sex fem) that are located in the vertice v:

$$i^{\lambda_1} = \{u: u \in V \wedge \langle u, v \rangle \in F\} \tag{6}$$

$$i^{\lambda_2} = \{a^{fem}: a^{fem} \in A^{ind_i, fem} \cap A^v\} \tag{7}$$

where $A^{ind_i, fem}$ is the set of agents of sex fem and species ind_i ($ind_i \in S$).

3.3 Species

The set of species is given by:

$$S(t) = \{ind_1(t), \ldots, ind_{ns}(t)\} \tag{8}$$

where ns is the number of species, that exist in the system in time t. The changes in the number of species are connected with their mutual location in the fitness landscape.

Each of the species of individuals is defined as follows:

$$ind = \langle A^{ind}, SX^{ind}, Z^{ind}, C^{ind} \rangle \tag{9}$$

where A^{ind} is the set of agents that belongs to species ind. There are two sexes within each species: females and males ($SX^{ind} = \{fem, mal\}$). The set of actions for species ind is defined as follows:

$$Z^{ind} = \{die, get, unlink, seekfem, accept, \\ clone, rec, mut, givef, givem, migr\} \tag{10}$$

The set of relations with species that exist in the $SCoEMAS$ for the species ind_i is given by:

$$C^{ind_i} = \left\{ \xrightarrow{ind_i, get-} \right\} \tag{11}$$

The $\xrightarrow{ind_i, get-}$ relation models the intra- and inter-species competition for limited resources:

$$\xrightarrow{ind_i, get-} = \{\langle ind_i, ind_j \rangle: ind_i, ind_j \in S, \\ j = 1, \ldots, ns\} \tag{12}$$

where get is the action of taking resource from the environment and the "$-$" sign indicates that action get performed by individuals of species ind_i has the negative effect on the fitness of individuals that belongs to the same or other species.

3.4 Female sex

The *fem* sex of species *ind* is defined as follows:

$$fem = \langle A^{fem}, Z^{fem}, C^{fem} \rangle \quad (13)$$

where A^{fem} is the set of agents of sex *fem* ($A^{fem} \subseteq A^{ind}$). The set of actions that agent a^{fem} can perform is defined as follows:

$$Z^{fem} = \{die, get, unlink, accept, clone, \\ rec, mut, givef, migr\} \quad (14)$$

where *die* is the action of removing agent from the system (when it runs out of resource), *get* action allows agent getting some resource from the environment (the resource γ is given to the individuals proportionally to their fitness values), *unlink* is the action of quitting from the reproducing pair formed with the individual of sex *mal*, *accept* is the action of accepting the agent of sex *mal* as a partner for reproduction (agent a^{mal} is accepted when it is located in the same minima as the agent a^{fem}, here the modified version of *hill-valley* function is used (Ursem, 1999), and there is greater probability of accepting agents closer in genotypic space to the a^{fem} agent, according to Euclidean metric). *clone*, *rec*, and *mut* actions are responsible for, respectively, child creation, mutation with self-adaptation (Bäck *et al.*, 1997) and intermediate recombination (Booker *et al.*, 1997) of its genotype. *givef* action gives some resource of type γ to the child. *migr* action allows the migration within the environment.

The set of relations with sex *mal* is defined as follows:

$$C^{fem} = \left\{ \xrightarrow[givef-,givem-]{fem,accept+} \right\} \quad (15)$$

$$\xrightarrow[givef-,givem-]{fem,accept+} = \{\langle fem, mal \rangle\} \quad (16)$$

where *accept* is the action of choosing individual a^{mal} for reproduction (which has the positive effect on its fitness) by agent a^{fem}. The action *accept* results in performing action *givef* and *givem* by, respectively, agent a^{fem} and a^{mal}. These actions transfer some amount of resource γ to the child, what results in decreasing the fitness of agents a^{fem} and a^{mal}. The relation $\xrightarrow[givef-,givem-]{fem,accept+}$ models the sexual conflict over the rate of reproduction because *givef* action results in much stronger decrease of fitness than *givem* action.

3.5 Male sex

The *male* sex is defined analogically to the *fem* sex, see equation (13). The set of actions that agent a^{mal} can perform is defined as follows:

$$Z^{mal} = \{die, get, unlink, seekfem, \\ givem, migr\} \quad (17)$$

where *die*, *get*, *unlink*, and *migr* actions are the same as in case of *fem* sex. *seekfem* is the action that sends messages to females located in the same place, when agent a^{mal} is ready for reproduction (the amount of resource is above the given level). *givem* action is analogical to *givef* action of *fem* sex, and the only difference is that males give four times less resource to child than females. There is no relations with *fem* sex ($C^{mal} = \emptyset$).

3.6 Female agent

Agent *a* of sex *fem*, that belongs to some species $ind \in S$ ($a \equiv a^{ind,fem}$) is defined as follows:

$$a = \langle GN^a, Z^a, \Gamma^a = \Gamma, \Lambda^a, PR^a \rangle \quad (18)$$

where GN^a is the genotype (consisted of real-valued vector of objective variables and vector of standard deviations used in mutation with self-adaptation). The set of agent's actions $Z^a = Z^{fem}$, see equation (14). The set of information used by agent a $\Lambda^a = \{\lambda_1, \lambda_3, \lambda_4\}$. Information of type λ_3 is defined as follows:

$$i^{\lambda_3} = \{a_i^{mal} : \text{agent } a \text{ is paired with } a_i^{mal} \text{ agent}\} \quad (19)$$

Information of type λ_4 includes the time t_{pair} of forming pair with agent a_i^{mal}:

$$i^{\lambda_4} = \{t_{pair}\}, \quad (20)$$

PR is a set of agent's profiles with the order relation $\trianglelefteq$ defined:

$$PR^a = \{pr^{res}, pr^{rep}, pr^{mig}\} \quad (21a)$$
$$pr^{res} \trianglelefteq pr^{rep} \trianglelefteq pr^{mig} \quad (21b)$$

where pr^{res} is the resource profile (this is the basic profile, which goal has the higher priority), pr^{rep} is the reproductive profile, and pr^{mig} is the migration profile. Within pr^{res} profile all strategies connected with the amount of resource are realized ($\langle die \rangle$, $\langle get \rangle$). Within pr^{rep} profile all strategies connected with the reproduction process ($\langle unlink \rangle$, $\langle accept, clone, rec, mut, givef \rangle$) are realized. These strategies use informations i^{λ_3} and i^{λ_4}. Within pr^{mig} profile the migration strategy ($\langle migr \rangle$), which uses information i^{λ_1}, is realized.

3.7 Male agent

Agent *a* of sex *mal*, that belongs to some species $ind \in S$ ($a \equiv a^{ind,mal}$) is defined analogically to $a^{ind,fem}$, see (18). Genotype GN^a is defined identically as in the case of $a^{ind,fem}$ agent. The set of agent's actions $Z^a = Z^{mal}$, see equation (17). The set of information used by agent a $\Lambda^a = \{\lambda_1, \lambda_2, \lambda_3, \lambda_4\}$.

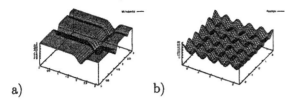

a)   b)

Fig. 2. Michalewicz (a) and Rastrigin (b) test functions

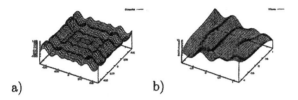

a)   b)

Fig. 3. Schwefel (a) and Waves (b) test functions

Information of type λ_3 includes agent a_i^{fem} with which the agent a is connected, analogically as in case of female agent — see (19). Information of type λ_4 includes the time of forming reproducing pair with agent a_i^{fem}, see (20).

The set of profiles is defined analogically as in the case of female agent, see (21).

pr^{res} and pr^{mig} profiles are identical as in the case of female agent. Within pr^{rep} profile all strategies connected with the reproduction process ($\langle unlink \rangle$, $\langle seekfem, givem \rangle$) are realized. These strategies use informations i^{λ_2}, i^{λ_3} and i^{λ_4}.

4. SIMULATION EXPERIMENTS

First simulation experiments were aimed at testing if *SCoEMAS* described in previous section is able to form and stably maintain species located in the minima of multi-modal fitness landscape. Also, the comparison to deterministic crowding (DC) niching technique (Mahfoud, 1995) and EMAS without any niching mechanisms was made.

Four widely used multi-modal test functions: Michalewicz, Rastrigin, Schwefel and Waves (see fig. 2 and 3) were used as the fitness landscapes in the experiments (Potter, 1997; Ursem, 1999).

Figures 4 and 5 show the location of SCoEMAS individuals in fitness landscape (Rastrigin function) during the typical simulation. At the beginning there are 50 females (represented with triangles) and 50 males (represented with squares). It can be seen that as the simulation goes on the individuals reproduce and locate themselves near the minima in multi-modal domain. What is more the subpopulations are stable, and do not disappear throughout the simulation.

Figures 6 and 7 show the average number of located minima from 20 simulations. The minima

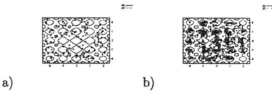

a)   b)

Fig. 4. The location of individuals in SCoEMAS during the 0th (a) and 50th (b) simulation step (Rastrigin function)

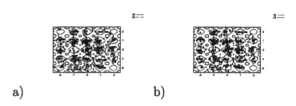

a)   b)

Fig. 5. The location of individuals in SCoEMAS during the 500th (a) and 5000th (b) simulation step (Rastrigin function)

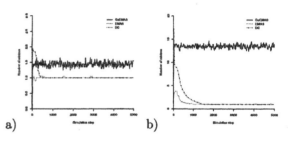

a)   b)

Fig. 6. The number of located minima of Michalewicz (a) and Rastrigin (b) function

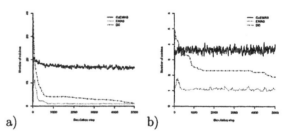

a)   b)

Fig. 7. The number of located minima of Schwefel (a) and Waves (b) function

was classified as located when there was at least one individual closer than 0.03 for Michalewicz function, 0.05 for Rastrigin function, 10.0 for Schwefel function and 0.025 for Waves function. The experiments was made for three techniques: SCoEMAS, EMAS, and DC.

The SCoEMAS stood relatively well when compared to other techniques. It stably maintained minima during almost the whole simulation. Although DC quickly located greater number of minima but there was quite strong tendency to lose almost all of them during the rest part of simulation. Simple EMAS, without any niching mechanisms was not able to stably populate more than one minima. It turned out that in case of multi-modal landscape it works just like simple EA.

The results indicate that simple EMAS can not be applied to multi-modal function optimization without introducing special mechanisms such as co-evolution. DC technique has some limitations — it has the strong tendency to lose minima during the simulation (this fact was also observed in (Watson, 1999)). CoEMAS with sexual selection is able to form and stably maintain species but still more research is needed.

5. CONCLUDING REMARKS

The idea of *co-evolutionary multi-agent system (CoEMAS)* allows us to model many ecological interactions between species, such as predator-prey and host-parasite co-evolution, mutualism, sexual conflict and co-evolution of sexes, etc.

In this paper sample CoEMAS with sexual conflict and resulting co-evolution of two sexes was presented. This system was applied to multi-modal function optimization. It properly formed and stably maintained species of individuals located in the minima of multi-modal fitness landscapes. SCoEMAS was able to detect and stably maintain more minima than EMAS without niching mechanism and deterministic crowding niching technique.

Future research will include comparison of other variants of sexual conflict (different costs of reproduction for sexes and production of female and male individual resulting in different proportions of individuals of each sex in population). Also, more detailed comparison to other niching techniques and the parallel implementation of CoEMAS using MPI is included in future research plans.

REFERENCES

Bäck, Th., D.B. Fogel, D. Whitley and P.J. Angeline (1997). Mutation. In: *Handbook of Evolutionary Computation* (Th. Bäck, D. Fogel and Z. Michalewicz, Eds.). IOP Publishing and Oxford University Press.

Booker, L.B., D.B. Fogel, D. Whitley and P.J. Angeline (1997). Recombination. In: *Handbook of Evolutionary Computation* (Th. Bäck, D. Fogel and Z. Michalewicz, Eds.). IOP Publishing and Oxford University Press.

Cetnarowicz, K., M. Kisiel-Dorohinicki and E. Nawarecki (1996). The application of evolution process in multi-agent world to the prediction system. In: *Proc. of the 2nd Int. Conf. on Multi-Agent Systems — ICMAS'96*. AAAI Press. Osaka, Japan.

Dreżewski, R. (2003). A model of co-evolution in multi-agent system. In: *Multi-Agent Systems and Applications III* (V. Mařík, J. Müller and M. Pěchouček, Eds.). number 2691 In: *LNAI*. Springer-Verlag. Berlin, Heidelberg.

Gavrilets, S. (2003). Models of speciation: what have we learned in 40 years?. *Evolution*.

Krebs, J.R. and N.B. Davies (1993). *An Introduction to Behavioural Ecology*. Blackwell Science Ltd.

Mahfoud, S. W. (1995). Niching methods for genetic algorithms. PhD thesis. University of Illinois at Urbana-Champaign. Urbana, IL, USA.

Potter, M. A. (1997). The Design and Analysis of a Computational Model of Cooperative Co-evolution. PhD thesis. George Mason University. Fairfax, Virginia.

Ratford, M., A.L. Tuson and H. Thompson (1997). An investigation of sexual selection as a mechanism for obtaining multiple distinct solutions. Technical Report 879. DAI Research Report.

Sánchez-Velazco, J. and J.A. Bullinaria (2003). Gendered selection strategies in genetic algorithms for optimization. In: *Proceedings of the UK Workshop on Computational Intelligence — UKCI 2003* (J.M. Rossiter and T.P. Martin, Eds.). University of Bristol. Bristol, UK.

Todd, P.M. and G.F. Miller (1997). Biodiversity through sexual selection. In: *Artificial Life V: Proc. of the Fifth Int. Workshop on the Synthesis and Simulation of Living Systems* (C.G. Langton and K. Shimohara, Eds.). The MIT Press. Cambridge, MA.

Ursem, R.K. (1999). Multinational evolutionary algorithms. In: *Proceedings of the Congress on Evolutionary Computation* (P.J. Angeline, Z. Michalewicz, M. Schoenauer, X. Yao and A. Zalzala, Eds.). Vol. 3. IEEE Press. Mayflower Hotel, Washington D.C., USA.

Watson, J.-P. (1999). A performance assessment of modern niching methods for parameter optimization problems. In: *Proceedings of the Genetic and Evolutionary Computation Conference* (W. Banzhaf, J. Daida, A.E. Eiben, M.H. Garzon, V. Honavar, M. Jakiela and R.E. Smith, Eds.). Vol. 1. Morgan Kaufmann. Orlando, Florida, USA.

PUBLICATIONS
www.elsevier.com/locate/ifac

EVOLUTIONARY OPTIMISATION OF FEM SIMULATION PARAMETERS OF COOLING AND SOLIDIFICATION PROCESS

Aleksander Byrski* Marek Kisiel-Dorohinicki*
Stanisława Kluska-Nawarecka**

* Department of Computer Science
AGH University of Science and Technology
Kraków, Poland
e-mail: {olekb,doroh}@agh.edu.pl
** Department of Computer Science in Industry
AGH University of Science and Technology
Foundry Research Institute, Kraków, Poland
e-mail: nawar@iod.krakow.pl

Abstract: The paper presents a general idea of parametric optimisation of FEM simulation model. Evolutionary algorithms are condsidered as a particular optimisation technique that may be used in this case. The considerations are illustrated with a simple optimisation experiment of a cast cooling process.

Keywords: finite element method, evolutionary algorithms, termophysical processes

INTRODUCTION

Physical properties of the metal resulting from its crystalline structure (including durability and flexibility), strongly depend on the course of the crystallization process. At the same time, the process of forming crystalline microstructure depends on the kinetics of the spatial distribution of temperature in the cast being cooled. Experimental study of the temperature in the whole volume of the cast is practically impossible, and even the temperature measurements, carried out simultanously in selected points of the cast are both hard to perform and very expensive. As a consequence, the only possible way to study the thermophysical processes in the cooled casts is the construction of mathematical models, that are able (at least approximately) to represent these phenomena.

Mathematical models of physical phenomena accompanying cast cooling and solidification (mainly heat propagation and phase transitions) have been the subject of many publications (Carslaw and Jaeger 1959, Yang *et al.* 1998). Yet because of their complexity, their effective use became possible only when numerical methods (particularly the finite differences method and finite elements method – FEM) allowing to get approximate results of differential equations describing thermophysical processes were developed. The simulation realized using these methods (particularly three-dimensional ones), enables the study of the temperature distribution in the whole volume of both the cast and the form, with additional possibilities of changing the border conditions (e.g. introducing thermical insulation or cooling).

* This work was partially supported by State Committee for Scientific Research (KBN) grant no. 4 T08A 026 25 and 11.11.110.493

Considering the simulation of cooling and solidification in casting processes, the main difficulty is the lack of knowledge of the form parameteres, which cannot be measured directly. In this case, one of possible solutions is a two-stage simulation process. In the first stage the form parameters are identified, by the application of the simulated temperature function to the temperature function obtained from the actual cast. The second stage is search for the optimal cooling conditions, with already determined parameters of the cast-form structure (Górny et al. 2001, Górny et al. 2002).

It turns out that finite element method (FEM) may be very effective for solving engineering problems in general, yet it is often difficult to use, because of many possibilities of modelling a specific task. Thus the paper first presents the general idea of "tuning" (parametrical identification) of a FEM simulation model (being the first stage of the cooling and solidification simulation process), based on the results obtained from physical experiments. The possibilities of applying evolutionary algorithms as a solving (optimisation) technique are demonstrated using the problem of assigning values of heat transfer coefficient of the moulding mass, based on the temperature measurements, obtained from the real cast cooling process.

FEM MODEL PARAMETERS OPTIMIZATION

Considering a typical case of using FEM method, complete knowledge about the simulated phenomenon and object parameters (substance features) is assumed. In such case (fig. 1) lack of compatibility between the results obtained from physical experiment and these obtained from simulation may be the result of improper method's parameters applied – e.g. improperly chosen shape of the finite elements, or the geometry of the mesh. However, contemporarily used, complex

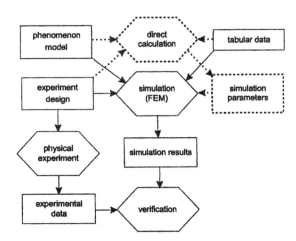

Fig. 2. The course of FEM simulation experiment with direct computation of parameter values

FEM tools, usually take over the responsibility of choosing proper parameters of the method, and thus these issues are outside the scope of this paper.

In many cases it turns out, that the knowledge about simulated phenomenon is not sufficient to point out an exact simulation model. The problem may concern both the properties of the simulated objects, as well as the phenomena accompanying simulated processes. It is possible to include them as the parameters (decision variables) describing simulation model, which usually become the function of the simulated objects temperature. Doubtless, the most effective approach is to calculate the values (dependencies) of these parameters directly from the experimental data (fig. 2), as it was shown in (Kobayashi et al. 1989). However, this approach requires construction of simplified analytical model of phenomenon, which will introduce significant inaccuracies to the results obtained (of order even equal to 100%).

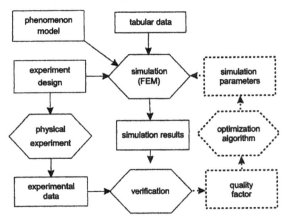

Fig. 3. The course of experiment with optimisation of FEM simulation parameters

Alternatively one may look for the values of simulation parameters in algorithmic way (fig. 3), by "matching" the simulation results to the obtained experimental results, with the assumption, that

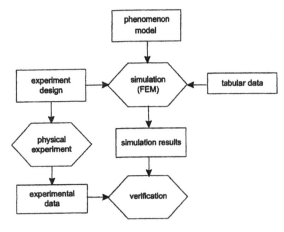

Fig. 1. The course of a typical FEM simulation experiment

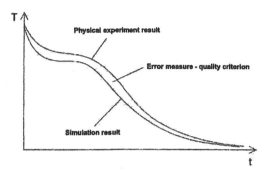

Fig. 4. Temperature measurements in physical experiment and simulation being the basis of criterion formulation

the results of conducted physical experiment are representative enough for simulated phenomenon – cf. inverse solving method (Drezet *et al.* 2000). Comparison between the physical and simulated experimental results, may be the basis for calculation of the quality factor (fig. 4) describing currently assumed set of simulation parameters. It makes it possible to use a plethora of iterative optimization methods, which do not require the knowledge of problem analytical model. An example of these methods are evolutionary algorithms.

EVOLUTIONARY ALGORITHMS

Evolutionary algorithms are based on iterative transformation of the *population of individuals* representing the set of potential solutions of the given problem (Fogel 1995). Evolution consists on generating consecutive generations, using so called *genetic operators* (or *variation operators*) and the process of *selection*. The main tasks of the genetic operators are random modification (*mutation*) and the exchange (*recombination*) of the individuals' genetic material. Selection of the best individuals is performed basing on the *evaluation function* – the measure of the fitness of each individual, being the equivalent of the quality of the solutions represented by them. The process of evolution should tend to generate better individuals and to find the needed (usually approximate) problem solution. General outline of classical evolutionary algorithm (Bäck 1996):

```
begin
    t := 0
    initialize P[0]
    evaluate P[0]
    while not terminate do
        begin
            P'[t] := variation P[t]
            evaluate P'[t]
            P[t + 1] := select P'[t] ∪ Q
            t := t + 1
        end
end
```

The values of the variable $t = 0, 1, 2, \ldots$ point out consecutive iterations of the algorithm. The initial population $P[0]$ is usually generated randomly. In each step t, λ individuals are generated using variation operators, that make up an intermediate population $P'[t]$. Selection is based on choosing μ individuals, which create subsequent population $P[t + 1]$, from the intermediate population $P'[t]$ and additional set of individuals Q (usually $Q = P[t]$ or $Q = \emptyset$ in so called generational evolutionary algorithms). The stopping condition is usually related to the number of iterations t_{max} and/or observing the stagnation of the process of evolution (generating the solutions of similar quality).

EVOLUTION OF CASTING COOLING PROCESS PARAMETERS

In the case of modelling of casting cooling processes, the simulation parameters (decision variables) are usually thermophysical features of the substance, and the interactions between particular components (parts of the model) or external influences. They are usually functionally related to temperature of the (point of the) object or the temperature difference among chosen object points (or objects' points). The functional relationship may be found by calculation the values of the parameter for the chosen values of temperature, that may be the basis for its interpolation in the whole domain. Then, for one needed parameter in the function of temperature $p(T)$ the vector of the individual may be described as:

$$x = [x_1, \ldots, x_n] \qquad (1)$$

and

$$x_i = p(T_i) \qquad (2)$$

where $T_1, \ldots, T_n$ are chosen temperature values defining specific interpolation points.

At the same time, the temperature is the easiest to obtain parameter describing the casting cooling process (macro scale). The comparison beween the temperature measurements carried out in specific points of examined physical object and simulation ones may be very convenient for the calculation of fitness (optimization criterion) for the evolutionary algorithm (fig. 4).

Let us consider a simple (below 1000 finite elements), two dimensional cooling process of the mould (fig. 5). Let us consider, that there is given the course of the changeability of the temperature values in one point of the cast (e.g. being the result of the measurements gathered during the physical experiment). Let the aim of the calculations be

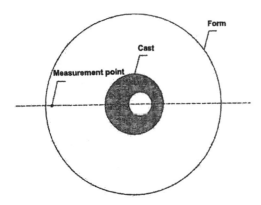

Fig. 5. Geometry of simulation model with measurement point marked

the finding of the *heat transfer coefficient* of the form, being the function of the temperature:

$$x = [x_1 = \lambda(T_1), \ldots, x_n = \lambda(T_n)] \quad (3)$$

where $\lambda(T_i)$ represents the values of the heat transfer coefficient for chosen temperature values T_i, pointing out the interpolation points.

The function to minimize is the difference between the obtained for given individual (exactly interpolated temperature function of the heat transfer coefficient value) and the target temperature function:

$$fitness(x) = \sqrt{\sum_{i=1}^{N}(T(t_i) - T^*(t_i))^2} \quad (4)$$

where:

t_i – consecutive time points representing the steps of simulation,

$T(t_i)$ – temperature values in the measurement point obtained in the simulation for the given individual,

$T^*(t_i)$ – target temperature values.

EXPERIMENTAL RESULTS

In the considered computational experiment, discrete values of the heat transfer coefficient were assumed. They belonged to the interval $(0, 1\rangle$ with the step of 0.05 and five interpolation points, of its temperature function. The target function for the temperature in the cooling time of the cast were obtained from the simulation for the given heat transfer coefficient function of the temperature:

i	$T_i[K]$	$\lambda(T_i)[\frac{W}{m}]$
1	300	0,5
2	310	0,6
3	330	0,7
4	345	0,8
5	400	0,9

For the remaining simulation parameters, tabular values were assigned, and in order to simplify, assumption was made, that the mould was made of technically pure copper, and the form – of silica (sand).

Classical genetic algorithm was used (binary representation, proportional selection) with following parameters:

- mutation probability: 0.05
- crossover probability: 0.85
- count of individuals: 16

In fig. 6 the temperature target function is shown, as well as the temperature function of random values of the heat transfer coefficient for selected values of temperature. Although all the values were picked randomly from the interval $(0, 1)$, so the differences relatively to the target values could not be significant, the obtained function varies from the target. It proves the importance of the heat transfer coefficient value of the form for the course of considered casting cooling process.

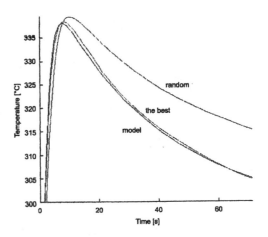

Fig. 6. Target course of temperature as well as the ones obtained for random and the best individual in the population

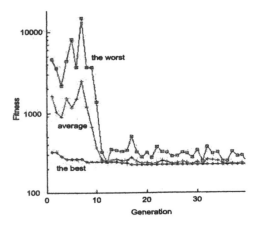

Fig. 7. Fitness in consecutive generations of the algorithm

Although not too large population was used, the implemented algorithm produced satisfying result – the function of temperature in the measurement's point for the best individual that can be seen infig. 6 is very similar to target. In fig. 7 the values of individuals' fitness are shown (the best, the worst and average) in consecutive epochs of the algorithm (warning: on the y-axis logarithmical scale is used). It can be seen, that the best solution is getting better, at least as long, as the diversity of the population is maintained.

CONCLUSION

The presented results of the simple computational experiment prove the possibilities of the application of evolutionary algorithms to "tuning" (parametric optimization) the FEM simulation model. Of course, the propriety of the obtained model depends on the number of degrees of freedom (optimized parameters) and the number of criterion components (accesible experimental results). In the future, the implementation of the optimization platform in the distributed environment is planned, which will allow further increasing of the numebr of degrees of freedom of considered models, and the number of criterion components, owing to this, increase of computational accuracy may be expected.

REFERENCES

Bäck, Th. (1996). *Evolutionary Algorithms in Theory and Practice.* Oxford University Press.

Carslaw, H.S. and J.C. Jaeger (1959). *Conduction of Heat in Solids.* Oxford University Press.

Drezet, J.M., M. Rappaz, G. Grn and M. Gremaud (2000). Determination of thermophysical properties and baundry conditions of direct chill-cast aluminium alloys using inverse methods. *Metall. and Mater. Trans.* **31A**, 1627–1634.

Fogel, David B. (1995). *Evolutionary Computation: Toward a New Philosophy of Machine Intelligence.* IEEE Press.

Górny, Z., S. Kluska-Nawarecka and H. Połcik (2001). Pilot casting and test performed to evaluate temperature field and solidification kinetics in castings made from HC copper. *Acta Metallurgica Slovaca* **7**(3), 221–225.

Górny, Z., S. Kluska-Nawarecka, H. Połcik and S. Bieniasz (2002). Contributions to numerical simulation of solidification of the industrial castings. *Archives of Foundry* **2**(2), 114–119.

Kobayashi, S., S. Oh and T. Altan (1989). *Metal forming and the finite-element method.* Oxford University Press.

Yang, H., L. Zhao, X. Zhang, K. Deng, W. Li and Y. Gan (1998). Mathematical simulation on coupled flow, heat, and solute transport in slab continuous casting process. *Metallurgical and Materials Transactions B.*

ELSEVIER
IFAC
PUBLICATIONS
www.elsevier.com/locate/ifac

BALANCING OF PRODUCTION LINES — EVOLUTIONARY AGENT-BASED APPROACH

Leszek Siwik, Marek Kisiel-Dorohinicki

Institute of Computer Science
AGH University of Science and Technology
Kraków, Poland
e-mail: {siwik, doroh}@agh.edu.pl

Abstract: This work introduces a new evolutionary approach to solving the problem of production line balancing, formulated as a multiobjective optimization task. Novelty of the proposed method consists in application of an evolutionary multi-agent system (EMAS) instead of classical evolutionary algorithms. Decentralization of the evolution process in EMAS allows for intensive exploration of the search space and effective approximation of the whole Pareto frontier. In the paper the application of the proposed technique as a core of a decision support system for balancing of production lines is described.

Keywords: evolutionary computation, multi-agent systems, multiobjective optimization, balancing of production lines

1. INTRODUCTION

Efficient and flexible production systems are the crucial part of contemporary enterprises. Prices of offered goods and also the level of profit, the company competitive position etc. depend, to a large degree, on the level of its costs. The low cost level directly increases the unit margin and simultaneously allows the company to establish the competitive price which, in turn, may increase demand for the product. As the operational profit depends directly on the turnover and unit profit, a production subsystem seems to be one of the most important parts of the contemporary production companies. Unfortunately, designing optimal production lines belongs to the class of the NP-hard problems. Moreover, during the production line designing process a lot of – often contradictory – factors have to be taken into consideration (so it is so-called a multiobjective problem). Therefore solving such problems based on some analytic model and using *classic* solving methods occurs to be fruitless due to a large number of dimen-

sions, different types of variables (continuous, discrete, binary), possible non-linearity, discontinuities of formulas (also performance functions) of the model.

In the above context, a need for a computational method arises that deals with multiobjective models in order to obtain some approximation of the Pareto frontier, and moreover:

- is to a large degree independent on analytic shape of the model,
- allows for approximation of non-coherent Pareto frontiers,
- allows for small changes of the model during computation.

Evolutionary algorithms possess at least some of this characteristics, yet they must be equipped with selection mechanisms effective for multiple criteria. To make them operational for production line balancing problems it is also necessary to define a specific encoding scheme and possibly dedicated variation operators. In the paper a new

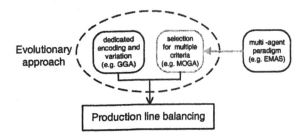

Fig. 1. The core of the approach: agent-based evolutionary solver for multiobjective balancing of production lines

technique is proposed, which combines elements of evolutionary computation with selected features of multi-agent systems, and may constitute a core of multiobjective optimization solver (instead of classical approach, e.g. MOGA – as illustrated in fig. 1).

The paper is organized as follows. Section 2 introduces basic concepts related to assembly line balancing. In this section a mathematical model considered in the course of this paper is also presented. Section 3 describes some issues connected with evolutionary approach to solving the ALBP. In section 4 a short overview of evolutionary techniques applied to solving multiobjective problems and the basics of agent-based approach are presented. Some experimental results showing advantages and disadvantages of the proposed approach are described in section 5.

2. ASSLEMBY LINE BALANCING PROBLEM

Generally, assembly line balancing problem consists in assigning technological tasks needed to manufacturing product(s) to workstations with respect to some objective(s). In the literature a significant amount of different variants of this problem can be found. Some of these models for the sake of significant simplification were designed rather for theoretical considerations. However, research on these models laid the foundations of building more realistic models, and, which is more important, of developing procedures finding optimal solutions in case of real-world insatnces of the ALBP. The most general classification of the ALBP distinguishes a SALBP (Simple Assembly Line Balancing Problem) and a GALBP (Generalized Assembly Line Balancing Problem). Of course models which potenially could be applied to real-world production systems are embraced by the GALBP, and thus research on these models have been recently intensified.

Relatively limited application area of the SALBP results to a large degree from non-realistic assumptions on which it is based (e.g. manufacturing one homogeneous product, no assignment restrictions besides the precedence constraints, serial line layout with one-sided stations, all stations equally equipped etc.). Additionaly, it has turned out that because of huge importance of production sub-system, as well as because of complexity and costs of such systems, it is impossible to build model that could be applied to real-world problems if it is based on optimizing only one objective. That is why a generalization of the ALBP consists in giving some of the SALBP limitations up and in more realistic approach to the optimization, i.e. optimization according to more than one objective. More realistic models result also from more economical than mathematical approach to solving the problem. Therefore in the literature among the GALBP can be found among other things cost-oriented models, multiobjective cost- and capacity-oriented models or profit-oriented models (Becker and Scholl 2003a, Becker and Scholl 2003b).

In the model proposed we pass over some limitations of the SALBP and, furthermore, it is based on two objectives which are the subject of optimization. Yet it is not so complicated as typical GALBP.

Let us consider a production system that may be characterized as follows:

- The goal is to optimize a production line manufacturing a group (family) of similar products (variants).
- The production line consists of N "universal" working stations, i.e. any technological task can be performed at any station.
- Each product of the family is characterized by the same technological route. In other words, exactly the same technological operations in exactly the same order have to be done for each variant.
- From the technological point of view, the only difference between products (variants) is the time of execution of individual tasks.

According to the above assumptions a production line balancing problem can be formulated as follows:

- $G = (O, A)$ - directed acyclic graph of a product family (the nodes of O represent technological operations, whereas the arcs of A represent a precedence relation),
- S - number of stations,
- V - number of product variants,
- N - number of operations ($\#O$),
- t_v^n - duration of n-th operation on variant v [1],
- N_s - number of operations assigned to station s,

---

[1] Quantities N, S, V, $t_v^n(i, j)$ and graph G are the input data

- $t_{v,s} = \sum_{n_s=1}^{N_s} t_v^{n_s}$ - duration of all operations assigned to station s - on variant v,

- $\bar{t}_s = \frac{1}{V} \sum_{v=1}^{V} t_{v,s}$ - average duration of station s,

- $\bar{t} = \frac{1}{S} \sum_{s=1}^{S} \bar{t}_s$ - average duration of all stations,

- σ_s - standard deviation of station s. Where
$$\sigma_s^2 = \frac{1}{V} \sum_{v=1}^{V} (t_{v,s} - \bar{t}_s)^2$$

- $\sigma = \frac{1}{S} \sum_{s=1}^{S} \sigma_s$ - standard deviation of production line. This quantity expresses the difference between the workload of product variants on each station,

- $\Psi = \frac{1}{\bar{t}} \sum_{s=1}^{S} (\bar{t}_s - \bar{t})^2$ - represents imbalance of production line.

The goal is to optimize the assignment of technological operations to working stations. Thus, assuming that $x_{s,n} = \{0,1\}$ denotes assignment between task n and station s, decision variables form a vector:

$$x = [x_{1,1}, \ldots, x_{S,N}]^T$$

which is the subject of minimization with respect to two criteria:

$$F = [\Psi, \sigma]^T$$

where Ψ and σ are described as above.

A production line balancing problem in its general form belongs to the class of NP-hard problems (polynomial solving algorithm for such problems does not exist). Furthermore, even simplified variants of the production line balancing (where we pass over the precedence constraint, or issues connected with grouping of similar products) are equivalent to so-called bin-packing problem, known as a typical NP-hard problem.

3. EVOLUTIONARY APPROACH

The ALBP has been a subject of frequent considerations so far and numerous solutions to this problem have already been proposed (Rekiek 2001). This is because of the influence of production line quality on the company's economical efficiency as well as because of the complexity of the problem itself. The review of different approaches applied to this problem includes both – dedicated algorithms searching exact solutions (such as dynamic programming, branch and bound methods or graph-based methods), and approximated methods searching suboptimal solutions, utilizing various (meta)heuristics as well. From the last group evolutionary computation techniques are of

our interest here so these very methods and problems connected with such approach are described here more widely.

One of the most important problem that appears in case of using of the EA to solving the ALBP is a representation scheme. During our research GGA-based representation has been used – however it has to be mentioned that this encoding scheme is not the only possible way to adapt the genetic algorithm to solving the ALBP. At least three other encoding schemes can be used during solving this problem (Becker and Scholl 2003a, Becker and Scholl 2003b):

- standard encoding - in this approach (which to some degree is opposite to the GGA-based representation) number of genes in each chromosome is equal to number of tasks, and each of them contains number of the station to which this task is assigned,
- order encoding - in this approach each chromosome is defined as a precedence feasible sequences of tasks,
- indirect encoding - there exists genetic algorithm solving the ALBP in which solutions are represented in an indirect manner. In this group solutions consisting in coding priority values of tasks or in coding sequence of priority rules and corresponding construction schemes to be applied for generating (decoding) the solutions have to be mentioned.

Besides the representation scheme, the very important problem in case of the GA is a appropriate fitness function. Because a lot of variants of the ALBP have been proposed - also a lot of fitness functions can be found in the literature. In case of the SALBP the most characteristic functions are undoubtedly the following:

- measuring the squared average relative deviation from a full station load,
- reducing the imbalance and minimizing the number of stations,
- considering the degree of imbalance and the absolute deviation of the number of stations used to the lower bound LM1,
- summing up the maximal station time and a penalty term for precedence violations,
- taking as fitness value the line efficiency reduced by a penalty term for infeasible surplus stations.

In the previous sections it was already emphasized that solving production line balancing problem consists de-facto in forming groups of technological tasks and assigning obtained groups to working stations so that the profit of the company assuming a given production level is maximized. An example of adapting the genetic algorithm to grouping problems is GGA (Grouping Genetic Algorithm) (Falkenauer 1996). The most interesting parts, that make the GGA different from classical genetic algorithms, are the encoding scheme and the form of genetic operators. In the GGA the number of genes within each chromosome is compatible with the number of produced groups and each gene encodes objects which are to be included in the particular group (so each gene represents de facto one of the decision variables forming vector x described by the equation 2). A chromosome that groups eight objects into four groups is presented in fig. 2 as an illustration.

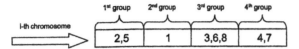

Fig. 2. The GGA - chromosome with objects grouped into four groups

Due to specific encoding scheme in the GGA a specific *crossover* operator is proposed as well. Firstly, two crossing points are determined at random for each parent and thus two crossing sections (one within each parental chromosome) are obtained. Then, the crossing section coming from the first parent is injected at the beginning of the crossing section of the second parent. As a result some objects can occur twice within the second parental chromosome. Consequently, duplicated objects have to be removed from genes coming from the second parental chromosome.

Mutation operator in the GGA can consist in: creating a new group from randomly selected objects (coming from other groups), deleting a randomly selected group (objects which form this group are assigned to other groups) or shuffling objects among groups.

4. AGENT-BASED EVOLUTIONARY MULTIOBJECTIVE OPTIMIZATION

It is worth to mention in this place that solving multiobjective optimization problems consists - de facto - of two stages. Firstly, relying on the so-called relation of domination, the set of possible solutions (i.e. the Pareto frontier) is calculated. Then, Decision Maker (e.g. a human being) has to choose only one solution from the whole set of the equivalent non-dominated alternatives. During this process personal preferences of the Decision Maker are of great importance, however, in the course of the paper just a part of the software enabling creation of the Pareto frontier is discussed.

For the last 20 years a variety of evolutionary multicriteria optimization techniques have been proposed (Coello Coello *et al.* 2002, Deb 2001,

Osyczka 2002). Among the most important and interesting methods there are:

- elitist EMOAs – i.e. multiobjective evolutionary algorithms based on so-called elite-preserving operator. These operators allow the elites of a population to be directly carried over to the next generation (Rudolph's algorithm, distance-based Pareto GA, strength Pareto EA, thermodynamical GA, Pareto-archived evolution strategy, multi-objective messy GA, multi-objective micro GA, elitist multi-objective EA with coevolutionary sharing etc.),
- non-elitist EMOAs – i.e. multiobjective evolutionary algorithms which are focused on emphasizing all non-dominated solutions in a population equally and simultaneously on preserving a diverse set of multiple non-dominated solutions without exploiting mechanism of elitism (vector evaluated GA, vector-optimized evolution strategy, weight-based GA, random weighted GA, niched-pareto GA, predator-prey evolution strategy, distributed sharing GA, modified NESSY algorithm, nash algorithm etc.),
- constrained EMOAs – i.e. algorithms and techniques that enable to handle constraints connected with solving problem. In other words - these are algorithms which ensure that found Pareto-optimal solutions are simultaneously feasible (Jimenez-Verdegay-Gomez-Skarmeta's method, constrained tournament method, Ray-Tai-Seow's method etc.).

The research on evolutionary algorithms and, simultaneously, on agent systems enabled us to formulate a new computational paradigm of *evolutionary multi-agent systems* (EMAS). In consequence, a novel evolutionary, but alternative to the above-mentioned, approach to solving multiobjective optimization problems, may be proposed (Kisiel-Dorohinicki *et al.* 2001).

EMAS is a kind of *multi-agent system*, in which basic agent interaction mechanisms are designed so that evolutionary phenomena emerge at a population level (Kisiel-Dorohinicki 2002, Byrski *et al.* 2003). This means that agents are able to *reproduce* (generate new agents) and may *die* (be eliminated from the system) realizing the phenomena of *inheritance* and *selection*. Inheritance is to be accomplished by an appropriate definition of reproduction, which is similar to classical evolutionary algorithms. A set of parameters describing basic behaviour of an agent is encoded in its genotype, and is inherited from its parent(s) – with the use of mutation and recombination.

The principle of selection in EMAS corresponds to its natural prototype and is based on the existence of non-renewable resource called *life energy*, which is gained and lost when agents execute actions. Increase in energy may be considered as a reward for 'good' behaviour of an agent, decrease – as a penalty for 'bad' behaviour (of course which behaviour is considered 'good' or 'bad' depends on the particular problem to be solved). At the same time the level of energy determines actions an agent is able to execute. In particular, low energy level should increase possibility of death and high energy level should increase possibility of reproduction.

In order to find the approximation of the Pareto frontier for a given multicriteria optimization problem, agents of EMAS must act according to the specific energetic reward/punishment mechanism, which prefers nondominated agents (Kisiel-Dorohinicki *et al.* 2001, Socha and Kisiel-Dorohinicki 2002). This is done via so-called *domination principle*, forcing dominated agents to give a fixed amount of their energy to the encountered dominants. This may happen, when two agents communicate with each other and obtain information about their quality with respect to each objective function. The flow of energy connected with the domination principle causes that dominating agents are more likely to reproduce whereas dominated ones are more likely to die. This way, in successive generations, nondominated agents should make up better approximations of the Pareto frontier.

5. EXPERIMENTAL STUDIES

The proposed idea of an evolutionary multi-agent system for multiobjective optimization applied to the assembly line balancing problem proved working in a number of tests. Preliminary experimental studies of the approach were based on the GGA representation (as described in section 3). The implementation was realized with the use of AgWorld platform (Byrski *et al.* 2003) – a software framework based on PVM, facilitating agent-based implementations of distributed evolutionary computation systems (for further reference see *http://agworld.sf.net/*).

Let us consider ALBP as defined in section 2, with 7 stations, 8 operations and 8 product variants. Let us assume further that times needed to perform individual technological operations define the space of possible solutions as it is shown in fig. 3. Of course since the solutions are shown in the space of criteria all possible combinations of standard deviation and imbalance of production line are given, rather than all possible assignments of technological tasks to stations. The Pareto frontier is marked by a solid line - of course only for visualization purposes because the Pareto frontier in this case is a discrete set.

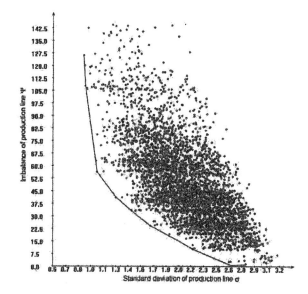

Fig. 3. All possible combinations of standard deviation and imbalance values for the considered production line

One of the most interesting issues regarding behaviour of the considered EMAS can be the information whether – and if so – how fast and how precisely the system is able to approximate the desired set of nondominated solutions. Initial population of 100 evolving agents and the first (generated at random) approximation of the Pareto frontier is shown in fig. 5a. It is of course extremely non-satisfying. Starting from this point – the population of agents acting according to the description presented in section 4 gradually improves the solution (the approximation of the Pareto frontier). In fig. 5b the Pareto set obtained by the system after 5000 generations is presented. This solution already seems to be acceptable comparing with the model Pareto frontier presented in fig. 3.

6. CONCLUDING REMARKS

Assembly line balancing problem serves us as an illustration of application possibilities of evolutionary multi-agent systems to difficult (NP-hard) optimization problems. The results obtained show that the approach may be successfully used for problems close to the those from engineering practice.

Of course up till now it is still too early to compare this method with various other heuristics supporting the ALBP known from the literature. Thus further research should concern surely the effectiveness of the proposed approach, especially in the case of more difficult (possibly constrained) optimization problems.

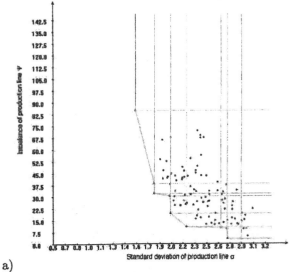

a)

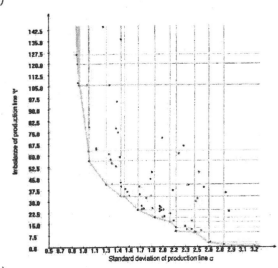

b)

Fig. 4. Initial population (a) and after 5000 generations (b) – approximation of the Pareto frontier (nondominated solutions) is marked by a solid line

REFERENCES

Becker, Christian and Armin Scholl (2003a). State-of-the-art exact and heuristic solution procedures for simple assembly line balancing. *Jenaer Schriften zur Wirtschaftswissenschaft.*

Becker, Christian and Armin Scholl (2003b). A survey on problems and methods in generalized assembly line balancing. *Jenaer Schriften zur Wirtschaftswissenschaft.*

Byrski, Aleksander, Leszek Siwik and Marek Kisiel-Dorohinicki (2003). Designing population-structured evolutionary computation systems. In: *Methods of Artificial Intelligence (AI-METH 2003)* (T. Burczyński, W. Cholewa and W. Moczulski, Eds.). Silesian University of Technology, Gliwice, Poland.

Coello Coello, Carlos Artemio, David A. Van Veldhuizen and Gary B. Lamont (2002). *Evolutionary Algorithms for Solving Multi-Objective Problems.* Kluwer Academic Publishers.

Deb, Kalyanmoy (2001). *Multi-Objective Optimization using Evolutionary Algorithms.* John Wiley & Sons.

Falkenauer, Emanuel (1996). A hybrid grouping genetic algorithm for bin packing. *Journal of Heuristics* 2, 5–30.

Kisiel-Dorohinicki, Marek (2002). Agent-oriented model of simulated evolution. In: *SofSem 2002: Theory and Practice of Informatics* (William I. Grosky and Frantisek Plasil, Eds.). Vol. 2540 of *Lecture Notes in Computer Science.* Springer-Verlag.

Kisiel-Dorohinicki, Marek, Grzegorz Dobrowolski and Edward Nawarecki (2001). Evolutionary multi-agent system in multiobjective optimisation. In: *Proc. of the IASTED Int. Symp.: Applied Informatics* (M.H. Hamza, Ed.). IASTED/ACTA Press.

Osyczka, Andrzej (2002). *Evolutionary Algorithms for Single and Multicriteria Design Optimization.* Physica Verlag.

Rekiek, B. (2001). Assembly Line Design. PhD thesis. Universite Libre de Bruxelles.

Socha, Krzysztof and Marek Kisiel-Dorohinicki (2002). Agent-based evolutionary multi-objective optimisation. In: *Proc. of the 2002 Congress on Evolutionary Computation.* IEEE.

ELSEVIER

IFAC

PUBLICATIONS
www.elsevier.com/locate/ifac

BEHAVIOR BASED DETECTION OF UNFAVORABLE ACTIVITIES IN MULTI–AGENT SYSTEMS

Krzysztof Cetnarowicz* Renata Cieciwa**
Gabriel Rojek***

Institute of Computer Science,
AGH University of Science and Technology,
Al. Mickiewicza, 30 30-059 Kraków, Poland
cetnar@agh.edu.pl
Department of Computer Networks,
Nowy Sacz School of Business – National-Louis University,
Ul. Zielona 27, 33-300 Nowy Sacz, Poland
rcieciwa@wsb-nlu.edu.pl
***Department of Computer Science in Industry,**
AGH University of Science and Technology,
Al. Mickiewicza 30, 30-059 Kraków, Poland
rojek@agh.edu.pl

Abstract: This article focuses on a problem of security in multi–agent systems. An approach to the security problem is discussed which refers to evaluating an agent on the basis of his behavior (actions which this agent undertakes). Presented mechanisms may enable to recognize and dismiss agents which are considered by other as undesirable or harmful in multi–agent system. Proposed mechanisms were implemented and tested in simulations. The results of tests are presented and discussed.

Keywords: multi–agent system, security, behavioral detection, ethically–social mechanisms

1. INTRODUCTION

Rapidly developed multi–agent technology makes possible the full flow of resources among open computer systems. Autonomic agents can yet freely migrate in the net without the knowledge of the owner or an administrator. Agents can also execute their tasks without anybody's knowledge. These tasks could be useful or destructive for the system on which an agent operate. An agent which migrate in an open system could be desirable or undesirable in a computer system. This ambiguousness attracts our interest to the problem of security in multi–agent systems. The main goal of our security approach presented in [2] is the division between:

- **"good"** (desirable) agents,
- **"bad"** (undesirable) agents, named also intruders.

2. ETHICALLY-SOCIAL INSPIRED SECURITY MECHANISMS

Taking into consideration the specific of functioning of an autonomous agent in multi–agent systems it could be stated, that:

- security system should be decentralized,
- security mechanisms should based on observation and evaluation of behavior of agent functioning in multi–agents system.

Above–mentioned paradigms are inspired by some of ethically–social mechanisms that function in human societies. Area of interest are mechanisms that prevent from misuses in societies in the way to enable secure functioning of an individual in the environment of others (possibly prejudicial for others) individuals. An individual in a society seems trustworthy if behavior of this individual could be observed by other individuals in a society and this behavior is evaluated by majority as good and secure [7,8]. The decision about trustworthy of an individual are made in society in the decentralized way — all individuals in a society make own decisions which form one decision of the society.

2.1 Decentralization of security system

Decentralization paradigm could be realized in multi–agent algorithms by means of equipping all agents (agents that exist in the protected system) with some additional goals, tasks and mechanisms. Those goals, tasks and mechanism are named division profile and should be designed to assure security for agents and the multi–agent system, which those agents assemble. So the agents will execute tasks that they have been created for and simultaneously will execute tasks connected with security of all agents in the system and / or the computer system. The name "division profile" is inspired by M–agent architecture which could be used to describe an agent (M–agent architecture was introduced among others in [1]).

2.2 Observation and evaluation of behavior

Undertaking actions by an acting agent should be seen as objects. Those objects create a sequence which could be registered by agents observing that acting agent. Registered objects–actions could be processed in order to qualify whether it is a "good" or "bad" acting agent. It should be mentioned, that the quoted notions of "good" and "bad" do not have absolute meaning. "Good" agent is a desirable agent for a definite multi–agent system in which evaluation takes place. "Bad" agent is an undesirable agent for a given system, although it can happen that it would be greatly desirable in a different multi–agent system.

3. DIVISION PROFILE

Division profile is a class of agent activity whose goal is to observe others agents in society in order to distinguish individuals whose behavior is unfavorable or incorrect ("bad") for the observer. Such distinguished "bad" individuals should be adequately treated (e.g. convicted, avoided, liquidated) what should also be formed by a division profile. As it was already stated, it is possible to equip every agent in the system with division profile mechanisms, so the security is assured by all agents existed in the system. Division profile is defined as:

$$a_d = (M_d, Q_d, S_d) \qquad (1)$$

where M_d is a set of division states m_d of agent a, Q_d is a configuration of goals q_d of agent's a division profile, S_d is a configuration of strategies s_d of agent's a division profile.

3.1 Division state

Division state m_d of agent a is represented as a vector:

$$m_d = (m_d^1, m_d^2, ..., m_d^{j-1}, m_d^j) \qquad (2)$$

where j is the number of neighboring agents; neighboring agents are agents which are visible for agent a (including itself; if all agents in the system are visible for agent a, j is equal to the number of all existing agents in the system); m_d^k is the factor subordinated to neighboring agent number k; this factor can be a number from any range and it indicates whether the agent number k is "good" or "bad" (low number indicates "good", high number indicates "bad").

3.2 How to fix the division state

Fixing of division state is inspired by immunological mechanisms — generation of T cells in the immune system. Immunological mechanisms should operate on actions made by observed agents. This approach is opposite to the one proposed in e.g. [5,6] in which immunological mechanisms operate on the resource's structure (fragments of code of programs).

Immunological inspired behavior evaluation has to be done on the basis of chains of actions performed by an observed agent. These chains are of the settled length l, so one chain contains l objects, which present undertaken actions by observed agent (one object represents one action). We should define the way how agent a will recognize (notice, but not estimate) actions undertaken by neighbors.

It is possible to store all actions undertaken by agents in the environment of multi–agent system. The action stored should be accompanied by the notion by whom a particular action has been undertaken. This method presumes the mediation of the environment and / or resources of the environment in the process of recognizing undertaken actions.

3.2.1. Detectors

The method of fixing the division state refers to the mechanism of immune system. Once detector's set is generated, this detector's set is used to find "bad" among presented sequences of action–objects. Taking into consideration the immunological inspiration of fixing of division state, functioning of division profile mechanisms can be split into 3 stages as it is shown onto Fig. 1.

In order to generate a set of detectors R, own collection W should be specified. This collection includes correct, "good" sequences of action–objects. This collection W should consist of action–object sequences of length l, which is undertaken by the agent–observer. This is correct, because of the assumption that actions which the agent undertakes are evaluated as "good" by him. Presuming there are stored h last actions undertaken by every agent, own collection W will contain $h - l + 1$ elements.

3.2.2. Algorithm of detectors generation

The algorithm of detectors generation refers to the negative selection — the method in which T–lymphocytes are generated. From set R_0 of generated sequences of length l those reacting with any sequence from collection W are rejected. Set R_0 contains every possible sequence (but it is also possible to use a set of random generated sequences). Sequence reaction means that elements of those sequences are the same. Sequences from set R_0 which will pass such a negative selection create a set of detectors R.

3.2.3. Behavior estimation of neighboring agents

First stage is a neighbor observation during which actions (and order of those actions) executed by neighboring agents are remembered. Those remembered actions create sequence N of presumed length h. After the next stage of detectors generation, generated detectors are used to find "bad", unfavorable agents. Every subsequence n of length l of sequence N is compared with every detector r from set R. If sequence n and r match, it means finding "bad", unfavorable actions. Sequence matching means that the elements of the sequences compared are the same. The number of matches for every observed agent is counted. On this basis behav-

ior estimation is made — division state $m_d = (m_d^1, m_d^2, ..., m_d^{j-1}, m_d^j)$ of agent-observer is modified to the $m_d' = (m_d^{1'}, m_d^{2'}, ..., m_d^{j-1'}, m_d^{j'})$, where j is the number of agents in the environment, $m_d^{k'}$ is assigned to the number of counted matches for agent number k.

3.3 Configuration of goals of agent's division profile

The way neighboring agents are treated is described by Q_d — configuration of goals q_d of agent's division profile. Configuration of goals of an agent is constant (however it is possible to design such a system, which in is possible the goal's adaptation). In the system described the configuration of goals consists only from one goal — liquidation neighboring agent (or agents) number k, if $m_d^k = \max(m_d^1, m_d^2, ..., m_d^{j-1}, m_d^j)$.

3.4 Configuration of strategies of agent's division profile

Actions, which should be undertaken by agent a in order to treat agent number k in the way described by the configuration of goal, are specified by S_d — the configuration of strategies s_d of agent's division profile. The configuration of strategies of the agent is constant and in the system described the configuration of strategies consists only of one goal: if the goal is to liquidate agent number k, a demand of deleting agent number k is send to the environment (coefficient o_d equal to the m_d^k is attributed to this demand).

This configuration of strategies presumes an intervention of system's environment in the liquidation of an agent. The environment calculates the sum of coefficients (for every agent separately) attributed to demands and liquidates all agents which have the maximum sum of coefficients and this sum is larger than constant OU. Periodically, after one constant time period Δt, the calculated sums of coefficients are set to 0. Constant coefficient OU is introduced in order to get tolerance for behavior that is evaluated as "bad" in a short time, or is evaluated as "bad" by a small amount of agents.

4. SIMULATIONS

In order to formulate conclusions about effectiveness of presented security mechanisms there are done simulations. This article presents results of simulations that differ in the way of running from presented among others in [3,4].

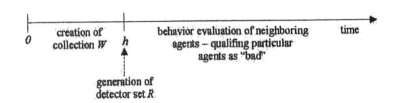

Fig. 1. Stages of division profile functioning

4.1 Synchronously acting agents

Simulations presented in [3,4] assume that agents, existing in the system, act synchronously. There is assumed that in every constant time period Δt every agent executes his full life cycle. It means that in one time period the number of activated agents is equals to the number of agents existing in simulated system. Agents are activated in a constant order that is fixed in the moment of creation of multi–agent system.

Mentioned simulation model with synchronously acting agents enable us to make some conclusions about sense of being engaged in the behavioral security systems. Results of these simulations encourage us to continue work on behavior based detection mechanisms. Next step in our research are simulations (with implemented presented security mechanisms) which model real–world multi–agent systems to a larger degree.

4.2 Asynchronously acting agents

Environment with entities, who act asynchronously is more similar to a real–world multi–agent system. All agents' actions are asynchronized — each agent randomly chooses time of his next activation and length of his next full life cycle. There is assumed that these random values are chosen from a range 1 constant time period Δt to 10 constant time periods Δt. It means that in each constant time period Δt number of activated agents can be different (whereas in 10 constant time periods Δt each agent is activated at least once). In this article presented are results of simulations with asynchronously acting agents.

Asynchronization of agents' acting forced some modification in behavior estimation of neighboring agents and configuration of strategies of agent's division profile. If an agent a tries to undertake any action, the system's environment asks neighboring agents for their "opinion" about him. Each agent in the environment modifies his division state m_d and sends back to the environment the coefficient m_d^a. The environment sums gained coefficients. If final sum is larger than constant OU, agent a is liquidated. In order to make possible behavioral analysis for all agent in the environment (according to stages of division

profile functioning presented onto Fig. 1), agents act synchronously till they are able to generate set of detectors R (only first 18 constant time periods Δt). Afterwards all agents' actions are asynchronized.

5. BEHAVIORAL ANALYSIS EXPERIMENTS

In the computer system there are some operation which must be executed in couples, for example: open and close a file, connection request and disconnection request. There are a lot of attack techniques which consist in doing only one from a couples (or trios...) of obligatory operations (for example so–called SYN flood attack [9]). There is simulated a system with two types of agents:

- type g=0 agents — good agents which perform some operations in couples (e.g. open, open, open, close, open, close, close, close);
- type g=1 agents — bad agents (intruders) which perform only one from a couples of some operations (e.g. open, open, open, open, open, open, open, open).

In all simulations there is no possibility of distinguishing the type of an agent on the basis of the agent's structure. So the only possibility is to observe the agent's behavior and process the observed actions (actions–objects) to distinguish whether the agent is good or bad.

5.1 Unsecured multi–agent system

First a case was simulated which in only good agents exist in the multi–agent system — initially there were 80 type g=0 agents. Next a case was simulated which in good and bad agents exist in the system — initially there were 64 type g=0 agents and 16 type g=1 agents. In both cases simulations is run to 3000 constant time periods Δt and there were 10 simulations performed. Diagram onto Fig. 2 shows the average numbers of agents in separate time periods.

If there are not any type g=1 agents in the simulated system, all type g=0 agents can exist without any disturbance. The existence of intruders in the system causes problems with executing tasks by good agents which die after some time periods. Bad agents still remain in the system that

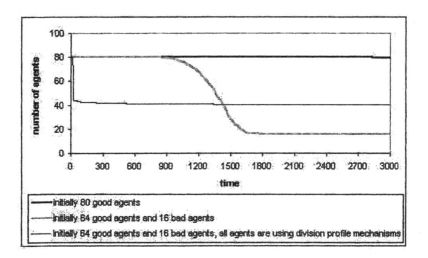

Fig. 2. Number of agents in separate time periods, agent using or not using division profile mechanisms

is blocked by those bad agents. Comparison of presented simulations with simulations described in [3,4] let us to make a conclusion that changing way of agents' acting from synchronous to asynchronous do not influence on final number of agents remained in the unsecured multi–agent system and phenomenon that happen in simulations are identical. More deductions about phenomenon in executed simulation are presented in [3].

5.2 Secured multi–agent system

A case was simulated which in good and bad agents exist in the system — initially there were 64 type g=0 and 16 type g=1 agents. All agents in the system were equipped with the division profile security mechanisms. Simulations is run to 3000 constant time periods Δt and there were 10 simulations performed. Diagram onto Fig. 2 shows the average numbers of agents in separate time periods.

In the environment there are stored last 18 actions undertaken by every agent. During first 18 periods of time all agents act synchronous. After 18 actions have been undertaken by every agent, detectors of length $l = 5$ are constructed and all agents' actions became asynchronous. The environment of multi–agent system asks neighboring agents for their "opinion" about some agents, for example agent a. Each agent in the environment calculates his division state m_d and sends back to the environment the coefficient m_d^a. The environment sums these coefficients and eliminates the agent when final sum of coefficients is larger than constant OU. The constant OU is set up to 480.

After series of tests it turned out that there were eliminated intruders as well as good agents. All agents, who chose moment of their activation at the beginning of simulated system, were eliminated. The highest coefficient sent by bad agent

to the environment causes elimination of good entities. The final number of agents decreased to the level, when the sum of coefficients could not be above the constant OU. So liquidation of agent only on the base of the sum of coefficients is not sufficient. In presented in [3,4] simulations, every one agent choose only "the worst" agent (or agents) which they want to eliminate and sends information about this agent (or agents) to the environment. That mechanisms enable to make proportional behavior evaluation — sending information to the environment an agent compares coefficients in his division state.

Strategy of agent's division profile was reconfigured and additional requirements of liquidation of agents were introduced. An agent in a case of receiving a request (from the environment) of evaluation of an agent a sends:

- the coefficient m_d^a,
- the information about goal of his division profile: true if his goal is to eliminate agent a; false in other cases.

The environment sums these coefficients and eliminates the agent a when final sum of coefficients is larger than constant OU and more than 50 per cent of agents have a goal to eliminate agent a. Diagram onto Fig. 3 shows the average numbers of agents (with modified strategy of division profile) in separate time periods.

After detectors had been constructed in 18 constant time period Δt, process of distinguishing "bad" agents, according to the division profile mechanisms, was started. Accurately analysis of obtained results indicates that undesirable entities were deleting successively from 19 constant time period Δt to 28 constant time period Δt. Delay of deleting of some bad agents is caused by the asynchronous model of agents' functioning (in 10 constant time periods Δt each agent is activated

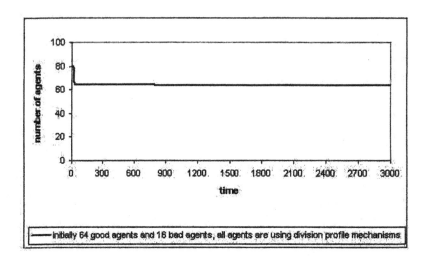

Fig. 3. Number of agents in separate time periods, agents using division profile mechanisms, system with proportional behavior evaluation

at least once, but in 1 constant time periods Δt an agent does not have to be activated). This results are very similar to results of the simulations with synchronous model of agents' functioning presented in [3,4].

6. CONCLUSION AND FURTHER RESEARCH

This paper presents several mechanisms, derived from some ethically–social mechanisms, for solving security problem in the multi–agent system. All these security mechanisms were called a division profile. The mechanisms have been implemented and tested in the multi–agent system with agents acting asynchronously. The results obtained in these simulations confirm earlier research's conclusion about the effectiveness of presented solution.

Presented mechanisms can remove a lot of disadvantages of present centralized security systems which base on looking for some fragments of code (signatures) rather than on activity observation. However multitude of operation each agent have to undertake effect a huge slowness of the entire multi–agent system, due to frequent actualization of division state.

Further research in behavior based detection of unfavorable activities in multi–agent systems should include methods (such as periodical and/or partly division state actualization) for reducing number of operation undertaken by each agents. It will be also useful to set precisely the value of constant OU. This could eliminate the amount of required comparisons of detectors by each one agent with actions undertaken by agents functioning in multi–agent system.

REFERENCES

[1] K. Cetnarowicz, E. Nawarecki and M. Żabińska, M–agent Architecture and its Application to the Agent Oriented Technology, in: *Proceedings of the DAIMAS'97* (St. Petersburg, 1997).

[2] K. Cetnarowicz, G. Rojek, J. Werszowiec-Plazowski and M. Suwara, Utilization of Ethical and Social Mechanisms in Maintenance of Computer Resources' Security, in: *Proc. of the Agent Day 2002* (Belfort, 2002).

[3] K. Cetnarowicz and G. Rojek, Unfavourable Beahviour Detection with the Immunological Approach, in: *Proceedings of the XXVth International Autumn Colloquium ASIS 2003* (MARQ, Ostrava, 2003) 41–46.

[4] K. Cetnarowicz and G. Rojek, Behavior Based Detection of Unfavorable Resources, in: *Lecture Notes in Computer Science, Proceedings of Computational Science - ICCS 2004: 4th International Conference* (Springer-Verlag, Heidelberg, 2004) 607–614.

[5] S. Forrest, A. S. Perelson, L. Allen and R. Cherukuri, Self-nonself Discrimination in a Computer, in: *Proc. of the 1994 IEEE Symposium on Research in Security and Privacy* (IEEE Computer Society Press, Los Alamitos, 1994) 202–212.

[6] S. A. Hofmeyr and S. Forrest, Architecture for an Artificial Immune System, *Evolutionary Computation*, vol. 7, no. 1 (2002) 45–68.

[7] M. Ossowska, *Normy moralne* (Wydawnictwo Naukowe PWN, Warszawa, 2000).

[8] F. Ricken, *Etyka Ogólna* (Wydawnictwo AN-TYK — Marek Derewiecki, Kety, 2001).

[9] E. Schetina, K. Green and J. Carlson, *Bezpieczeństwo w sieci* (Wydawnictwo HELION, Gliwice, 2002).

ELSEVIER

IFAC

PUBLICATIONS
www.elsevier.com/locate/ifac

PROACTIVE/REACTIVE APPROACH FOR TRANSIENT FUNCTIONING MODE OF TIME CRITICAL SYSTEMS

Soizick Calvez[*], Pascal Aygalinc[*], Patrice Bonhomme[**]

(*) LISTIC, ESIA, Université de Savoie, BP 806. F 74016 Annecy Cedex
E-mail: Soizick.Calvez@univ-savoie.fr, Pascal.Aygalinc@univ-savoie.fr

(**) LI, University of Tours, 64, Avenue Jean Portalis, F 37200 Tours
E-mail: Patrice.Bonhomme@univ-tours.fr

Abstract: Time-critical systems are characterized by operation times included between a minimum and a maximum value. A transgression of these specifications causes a deficient quality of service. As the policy consisting in using the earliest functioning mode leads generally to a potential constraint violation, it is necessary to control the starting of these systems. In this paper, a proactive/reactive control of the starting period is proposed to obtain a desired periodic functioning mode and to increase the robustness of this transient phase. P-time Petri Net is used as modeling tool to generate temporal and logical constraints to be considered. To update the tasks scheduling, flexibility on time and on the task order is obtained by permitting negative instants in the firing instant approach. The illustration of the method is made on a graphite ceramic cell.

Keywords: Time Petri-Net, Robustness, Proactive/Reactive Control, Time Critical Systems.

1. INTRODUCTION

The distinctive feature of time critical systems is that their operation times are represented by time intervals. These time intervals may correspond to either a temporal tolerance to complete a task in regard of a typical value or an uncertainty on its effective duration due to the used resource (hardware or human one). They can also express the drift of the process parameters because of the deterioration of the production equipment. So, time is essential in these systems because an acceptable behavior depends not only on the logical correctness of the sequence of the tasks, but also on the time of their completion.

Such systems can be found in computer systems as supervision applications requiring consistency of data, in manufacturing systems of the chemical industry where the chemical reagents are efficient during a time interval or in food industry where handling and delivery of the products are subjected to requirements of freshness.

As these systems may enter a forbidden state if a result or an event occurs too early or too late, this article focuses on the robust control problem of these systems, and more precisely on their starting phase in a disturbed environment. The disturbances considered here generate an uncertainty on the effective task durations. To update the planned scheduling when disturbances occur, and consequently to improve the robustness, we suggest using a proactive/reactive control.

The proposed approach necessitates three steps:

- The first one consists of modeling the process flow and the needed resources. It is the purpose of section two. Thanks to their graphical nature, their ability to model parallel and distributed processes and their firm mathematical foundation, Petri nets (PN on short) (David and Alla, 1994; Diaz, 2001) are used as modeling tool. PN have been extended and modified in several ways in order to match requirements of specific application areas. As the operation times are specified here by time intervals, the chosen PN are p-time Petri nets (p-time PN) (Khansa *et al.*, 1996; Calvez *et al.*,1997). Their firing rules impose a firing compulsion whatever the activities considered are. This modeling tool provides the set of temporal and logical constraints that the transient scheduling and the repetitive one must ensure.

- The second step, described in section three, aims at computing a static reference schedule. When the system is described by a p-time PN, the scheduling problem consists in setting the firing instant of each transition of the model in order to obtain a desired specification as a definite cycle time, makespan, ... A firing instant approach (Boucheneb and Berthelot, 1993; Bonhomme *et al.*, 2001) can be used if the admissible firing sequences have been previously determined by the enumeration phase. So, in the proposed approach, the logical functioning of the model is considered to build the temporal constraints. Via linear programming techniques, these constraints are used to evaluate the performances of the modeled system as well as to determine the exact firing order and instant of each transition for a chosen resource allocation.

- The third step consists in actualizing the static scheduling computed in the second step, to take into account the task duration variations. These disturbances can be viewed as advances or delays on the planned firing instants of the p-time PN model transitions. The firing instants must be actualized to inhibit as quickly as possible the effect of disturbances. In the fourth section, we propose to take advantage of partial order to enlarge performances and to increase the control robustness. The benefits of such a method are practically of great interest since determining a new resource allocation can be of high computational cost.

During the run time of a system using a repetitive mode, some different phases appear: from the free state of the system (no work in process and free equipments), there is a transient phase to start up the system leading to a periodic functioning mode (and conversely from the periodic functioning mode to unload the system). The principle presented here to calculate a transient schedule can be applied to other transient phases linked to the diminishing or to the increasing of the available resources due to preventive maintenance while the system is operating.

An illustration is made on a graphite ceramic cell in the fifth section. Some conclusions are given in the last one.

2. MODELING STEP

2.1 p-Time Petri Net

The formal definition of a p-time Petri Net (Khansa *et al.*, 1996) is given by a pair $< N ; I >$ where:

N is a marked Petri Net (Diaz, 2001) $\qquad$ (1)

$I: \boldsymbol{P} \to (\boldsymbol{Q}^+ \cup 0) \rtimes (\boldsymbol{Q}^+ \cup +\infty)$

$p_i \to I_i = [a_i, b_i]$ with $0 \le a_i \le b_i$
where $\boldsymbol{P}$ is the set of places of net N, $\boldsymbol{Q}^+$ the set of positive rational numbers.

I_i defines the static interval of the operation duration of a token in a place p_i. A token in place p_i will be considered in the enabledness of the output transitions of this place if it has stayed for a_i time units at least and b_i at the most. Consequently, the token must leave p_i, at the latest, when its operation duration becomes b_i. After this duration b_i, the token will be "dead" and will no longer be considered in the enabledness of the transitions. A dead token is not removed from the place; this token state indicates that a temporal violation has occurred.

The particularity of this model requires analysis techniques, allowing taking account efficiently of the various functionalities associated with the modeled system, as well as its temporal features. It leads ineluctably to the need for having formal methods ensuring the system control. Indeed, the policy consisting in firing a transition as soon as it becomes enabled is not always feasible and usually leads to a potential constraint violation.

All the constraints of the associated scheduling problem can be extracted from the model. This is the purpose of the following section.

3. STATIC SCHEDULING

3.1 Firing instant approach

Enumerative analysis (or exhaustive simulation) aims at exhibiting all the admissible functioning of the modeled system. The selection of one of them is not evoked here, only the constraints extraction is developed. This one is based on the evaluation of the firing conditions for the first, the second,..., and q^{th} firing instant (Bonhomme *et al.*, 2001).

Definition 3.1: a fired transition denoted by t_j will be associated with the jth firing instant (i.e. the firing sequence considered is $t_1 t_2 t_3 ... t_q$). A variable x_i represents the elapsed time between the $(i - 1)^{th}$ and the ith firing instant.

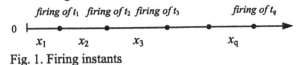

Fig. 1. Firing instants

For instance (see figure 1), $(x_2 + x_3)$ is the time elapsed between the first firing instant and the third one.

In a p-time PN, the sojourn time (*i.e. the amount of time that a token has been waiting in a place*) is counted up as soon as the token has been dropped in the place as seen previously. Thus, quantitative (*i.e. performance*) considerations take precedence over qualitative (*i.e. logical*) ones, in opposition to Merlin's time PN model (Merlin and Faber, 1976). To compute the firing instants, this approach requires that a token is identified by three parameters: the place that contains it, the information of its creation instant and the information of its consumption one. Function *TOK* is defined with this purpose. If the weight of the p-time PN arcs is element of {0, 1}, then:

$$TOK: N \times N^* \to \wp(P),$$ (2)

$TOK(j, n) = \{p \in P \mid p$ contains a token created at the jth firing instant and consumed at the nth one$\}$.

with: $\wp(P)$ the set of subsets of P.

When a place contains several tokens, they are differentiated by the values j and n associated with them. So, it is possible to impose any token management, but in the sequel a FIFO mode will be considered. Moreover, the determination of these sets depends on the firing sequence considered.

Using these sets, the minimal and maximal effective sojourn times of each token in its place are evaluated by:

$$Dsmin(j, n) = \begin{cases} \max(a_i), i \mid p_i \in TOK(j,n) \\ \text{else } 0 \text{ if } TOK(j,n) = \varnothing \end{cases},$$ (3)

$$Dsmax(j, n) = \begin{cases} \min(b_i), i \mid p_i \in TOK(j,n) \\ \text{else } +\infty \text{ if } TOK(j,n) = \varnothing \end{cases}$$

3.1.1 Firing space at the qth firing instant:

To simplify the firing condition for the first, the second, ..., and q^{th} firing instant, the definition of the following coefficients is required (Bonhomme *et al.*, 2001):

$$c_{uq} = \begin{cases} Dsmin(u,q) \text{ if } u \in SEN(q) \\ \text{else } 0 \end{cases}$$ (4)

$$d_{jk} = \begin{cases} Dsmax(j,k) \text{ if } TOK(j,k) \neq \varnothing \\ \text{else } +\infty \end{cases}$$

where $SEN(q) = \{u \mid TOK(u, q) \subset ({}^\circ t_q)\}$ represents the creation instants set of tokens consumed by the qth firing instant.

Theorem 3.1(Bonhomme *et al.*, 2001): A sequence of transitions $\sigma = t_1, t_2, ..., t_q$ may be fired respectively

at firing instants 1, 2,..., q if and only if there exist $x_1 \geq 0, x_2 \geq 0, ..., x_q \geq 0$ such that:

$$S_\sigma(q) \begin{cases} \underset{k=1,n}{c_{0k}} \leq x_1 \leq \underset{k=1,n}{d_{0k}} \\ \underset{k=2,n}{\max} (c_{0k}, c_{1k}+x_1) \leq x_1+x_2 \leq \underset{k=2,n}{\min} (d_{0k}, d_{1k}+x_1) \\ \dots \\ \underset{\substack{j=0,q-1 \\ k=q,n}}{\max} (c_{jk}+\sum_{s=0}^{j} x_s) \leq \sum_{s=0}^{q} x_s \leq \underset{\substack{j=0,q-1 \\ k=q,n}}{\min} (d_{jk}+\sum_{s=0}^{j} x_s) \end{cases}$$ (5)

Each left inequality member can be interpreted as the availability of the tokens taking part in the firing of the transition considered, and the right one can be viewed as the "no token(s) death" constraint.

Definition 3.2: The firing space at the qth firing instant denoted by $FSP(q)$ is the set of non negative vectors $(x_1, ..., x_q)$ such that the first, the second, ... and the qth firing conditions are satisfied.

3.1.2 Conditions on the state of the initial marking

The evaluation of the firing conditions for the first, the second and the qth firing instant was made on the basis of an initial marking where each token duration is included in the static interval of the place containing it (see the definition of *Dsmin* and *Dsmax* (3)). Another assumption corresponds to the hypothesis usually made for the timed models where the initial marking is available for the firing of the initially enabled transitions.

This consideration leads to systems $S_\sigma(q)$ where quantities c_{jk} and d_{jk} satisfying $j = 0, \forall k \in [1, q]$ verify $c_{jk} = 0$ and $d_{jk} = +\infty$, these systems will be referred to as relaxed systems as opposed to non-relaxed ones. For relaxed systems the question of the steady state reachability starting from the initial state (the marking and its temporal state) does not arise (initial tokens are considered as virtual work in process (Aygalinc *et al.*, 2003)).

Whatever the assumption made on the initial marking state, for a given firing sequence on the basis of the system $S_\sigma(q)$, the firing instant of each transition can be easily determined and the firing sequence performance evaluation can be achieved with no difficulty.

As this work concerns the transient functioning mode, performance evaluation and static reference schedule are formulated only for non-relaxed systems.

3.2 Periodic Reference Schedule and Performance Evaluation

Theorem 3.2 (Laftit, 1991): The behavior of the periodic mode is fully determined by:

$\forall k \geq 1$, $s_i(k) = s_i(k-1) + \pi$, where $s_i(k)$ is the kth (6) firing date of the transition t_i and π the functioning period.

The benefit of this optimal functioning mode (Calvez *et al.*, 1997) is that it suffices to compute the firing instants of the transitions on one cycle and the functioning period to build a repetitive static schedule.

Let us consider a firing sequence $\sigma = (\sigma_t \, \sigma_r)$ where σ_t represents the transient functioning leading to a particular repetitive one (represented by σ_r) and such that the last transition of σ_t is the same as the last one of the repetitive sequence σ_r. Let us denote by $|\sigma_t|$ (resp. $|\sigma_r|$) the length of σ_t (resp σ_r). The lower (resp. upper) bound denoted by $\mu_{\sigma_r}^{min}$ (resp. $\mu_{\sigma_r}^{max}$) of the cycle time of σ_r can be computed with the linear programs stated as follows:

$$\mu_{\sigma_r}^{min} = min(\pi) \; / \; \mu_{\sigma_r}^{max} = max(\pi), \qquad (7)$$

$$\text{subject to} \begin{cases} \pi = \sum_{i=|\sigma_t|+1}^{|\sigma_t|+|\sigma_r|} x_i, \\ \text{the set of constraints } S_\sigma(|\sigma_t|+|\sigma_r|). \end{cases}$$

And the transient mode duration μ_{σ_t} is given by:

$$\mu_{\sigma_t} = \sum_{i=1}^{|\sigma_t|} x_i \qquad (8)$$

A reference periodic schedule of period π_{obj}, with $\pi_{obj} \in [\mu_{\sigma_r}^{min}, \mu_{\sigma_r}^{max}]$ and the associated transient schedule σ_t are obtained via the resolution of the following linear system if there exist $x_1 \geq 0$, $x_2 \geq 0$, ..., $x_i \geq 0$ ..., such that:

$$\sum_{i=|\sigma_t|+1}^{|\sigma_t|+|\sigma_r|} x_i = \pi_{obj}, \text{ subject to the set of constraints} \qquad (9)$$

$$S_\sigma(|\sigma_t|+|\sigma_r|)$$

4. ROBUST TRANSIENT CONTROL

The previous firing instant approach requires a strictly ordered sequence. Indeed, this order is necessary to build the sets *TOK* and the associated firing inequalities system $S_\sigma(|\sigma_t|+|\sigma_r|)$. Consequently, a solution of $S_\sigma(|\sigma_t|+|\sigma_r|)$ is a set of non negative instants and performance evaluation is restricted to the sequence considered. The sequence used to establish the associated firing inequalities system $S_\sigma(|\sigma_t|+|\sigma_r|)$ is called in the sequel the reference sequence.

In this section, an extension of the previous approach is proposed to evaluate other possible functioning modes by means of introducing partial order on the reference sequence. The partial order is obtained by permitting negative firing instants. Indeed, as variable x_i represents the elapsed time between the $(i-1)^{th}$ and the ith firing instant, if negative values are permitted then this order may be changed.

4.1 Static Reference Schedule

Whatever the order generated from the reference sequence $\sigma = (\sigma_t \, \sigma_r)$, new constraints must be added to the set of constraints $S_\sigma(|\sigma_t|+|\sigma_r|)$ to preserve the following properties:
  i) all of the firing absolute dates must be non negatives.
  ii) the semantic of the expression of the cycle time $\pi = \sum_{i=|\sigma_t|+1}^{|\sigma_t|+|\sigma_r|} x_i$ for the repetitive sequence σ_r. As the firing instants are expressed as relative quantities, and as a cycle time is a positive value, this amounts imposing that the last fired transition must be the last transition of σ_r, and consequently of σ_t.
  iii) maintaining the same firing occurrences of each transition in the repetitive reference sequence σ_r, because the generated schedule must be a repetitive one. In others words, the firing instants of the repetitive reference sequence σ_r are relevant if they are greater than the transient mode duration.

So, the quantities x_i with $x_i \in Q$ must verify:

$$\begin{cases} \forall j \in \{1,...,|\sigma_t|+|\sigma_r|\}, \sum_{i=1}^{j} x_i \geq 0 & (10) \\ \forall j \in \{1,...,|\sigma_t|-1\}, \sum_{i=1}^{j} x_i \leq \sum_{i=1}^{|\sigma_t|} x_i \\ \forall j \in \{|\sigma_t|+1,...,|\sigma_t|+|\sigma_r|-1\}, \sum_{i=1}^{j} x_i \leq \sum_{i=1}^{|\sigma_t|+|\sigma_r|} x_i \\ \sum_{i=1}^{|\sigma_t|} x_i \leq \sum_{i=1}^{j} x_i, \forall j \in \{|\sigma_t|+1,...,|\sigma_t|+|\sigma_r|\} \end{cases}$$

4.2 Dynamic Transient Control

Let us consider that a disturbance occurs on the ith firing instant x_i in the transient schedule (the ith fired transition in σ_t). So, the disturbed firing instant will correspond to x_i^{new} with $x_i^{new} = x_i + \Delta x_i$ ($\Delta x_i < 0$ representing an advance and $\Delta x_i > 0$ a delay, on the previously scheduled instant x_i).

Deciding whether a disturbance (a delay or an advance on a previously planned transition firing instant) is admissible or not, will be tantamount to determining the system ability to tolerate the disturbance effect with regard to the preceding chosen firing instants. Indeed, the future is dependent on the past events. As in (Aygalinc *et al.*, 2003), the transitions considered for the sequence order revision are only the ones following the disturbed transition. So, additional constraints may be added to obtain that the actualized firing instants of the transitions are relevant if they are greater than the disturbed firing instant of t_i. The sequence $\sigma = (\sigma_i \, \sigma_r)$ may be reordered except the last transition of σ_r and of σ_i.

5. ILLUSTRATIVE EXAMPLE

The line described here (see Figure 2) is a workshop manufacturing graphite ceramic electrodes. It is composed of six mixing tanks, two conveyors, two conditioners and one press. The raw materials (primarily coke and binder pitch) are first mixed together at an elevated temperature in order to obtain an homogeneous mass. The obtained paste is then conditioned (the paste is refreshed to bring down its temperature). Finally, the paste is extruded into the shape and size of end-products. A time interval is associated to each of these operations. This interval represents an uncertainty on the duration (for example, the duration of the conditioning depends on the temperature of the ambient air, of the paste and of the conditioner). This interval may also correspond to the duration in which the good properties of the pitch are preserved (transfer, extrusion...).

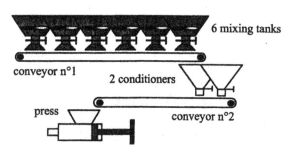

Fig.2: The graphite ceramic cell considered

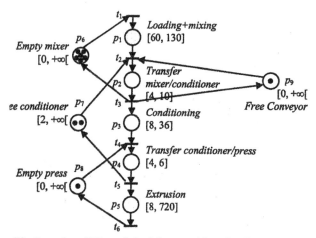

Fig.3: p-time PN model of the considered cell

The modeling is made in two steps (figure 3). First, the precedence constraints provided by the process flow (linear part of the graph) are modeled, considering that in a p-Time PN, the places are

representing the tasks and the transitions are associated with beginning and ending events. Then, the model is completed with the resources needed for the execution of the tasks. For instance, the synchronization structure associated with t_2 describes a synchronization in a constrained time: it implies that the conveyor n°1 (place p_9) and one of the two conditioners (place p_7) must be available in a time compatible with the completion of the heating and mixing operation (place p_1).

The initial marking corresponds to the inactive workshop, where all resources are available. The marking of place p_6 (6) represents the number of mixers, the one of place p_7 (2) is the number of conditioners and the one of p_8 (1) corresponds to the single press. Notice that only the conditioners need a set-up time before re-using.

Using the logical behavior of this model, the reference firing sequence $\sigma = (\sigma_i \, \sigma_r)$ is the following one: $\sigma_i = (t_1 \, t_1 \, t_1 \, t_1 \, t_1 \, t_1 \, t_2 \, t_3 \, t_1)$ and $\sigma_r = (t_2 \, t_3 \, t_4 \, t_5 \, t_6 \, t_1)$, and the associated firing instant vector is: $(x_1 \, x_2 \, x_3 \, x_4 \, x_5 \, x_6 \, x_7 \, x_8 \, x_9 \, x_{10} \, x_{11} \, x_{12} \, x_{13} \, x_{14} \, x_{15})$.

The system $S_\sigma(|\sigma_i| + |\sigma_r|)$ completed by the added constraints allowing the negative instants is:

$$
\begin{cases}
0 \le x_1 \le +\infty \\
0 \le x_1 + x_2 \le +\infty \\
0 \le x_1 + x_2 + x_3 \le +\infty \\
0 \le x_1 + x_2 + x_3 + x_4 \le +\infty \\
0 \le x_1 + x_2 + x_3 + x_4 + x_5 \le +\infty \\
0 \le x_1 + x_2 + x_3 + x_4 + x_5 + x_6 \le +\infty \\
60 \le x_2 + x_3 + x_4 + x_5 + x_6 + x_7 \le 130 \\
4 \le x_8 \le 10 \\
0 \le x_9 \le +\infty \\
4 \le x_{11} \le 10 \\
4 \le x_{13} \le 6 \\
8 \le x_{14} \le 720 \\
0 \le x_{15} + x_{10} + x_{11} + x_{12} \le +\infty \\
0 \le x_{12} + x_{13} + x_{14} + x_{15} \le +\infty \\
60 \le 5 \cdot (x_{10} + x_{11} + x_{12} + x_{13} + x_{14} + x_{15}) + x_{10} \le 130 \\
2 \le x_{14} + x_{15} + x_{10} \le +\infty \\
8 \le x_{12} + x_{13} + x_{14} + x_{15} + x_{10} + x_{11} + x_{12} \le 36 \\
\pi = \sum_{i=10}^{15} x_i \\
60 \le x_3 + x_4 + x_5 + x_6 + x_7 + x_8 + x_9 + x_{10} \le 130 \\
60 \le x_4 + x_5 + x_6 + x_7 + x_8 + x_9 + \pi + x_{10} \le 130 \\
60 \le x_5 + x_6 + x_7 + x_8 + x_9 + 2.\pi + x_{10} \le 130 \\
60 \le x_6 + x_7 + x_8 + x_9 + 3.\pi + x_{10} \le 130 \\
60 \le x_7 + x_8 + x_9 + 4.\pi + x_{10} \le 130 \\
8 \le x_9 + x_{10} + x_{11} + x_{12} \le 36
\end{cases}
\tag{11}
$$

$$
\bigcup
\begin{cases}
\sum_{i=1}^{j} x_i \ge 0, \, \forall \, j \in \{1,...,15\} \\
\sum_{i=1}^{j} x_i \le \sum_{i=1}^{9} x_i, \, \forall \, j \in \{1,...,8\} \\
\sum_{i=1}^{j} x_i \le \sum_{i=1}^{15} x_i, \, \forall \, j \in \{10,...,14\} \\
\sum_{i=1}^{9} x_i \le \sum_{i=1}^{j} x_i, \, \forall \, j \in \{10,...,15\}
\end{cases}
$$

So, the linear programs to evaluate performance are:

$\mu_{\sigma_r}^{min} = \min(\pi)$, ($\mu_{\sigma_r}^{max} = \max(\pi)$), with $\pi = \sum_{i=10}^{15} x_i$,

subject to the constraints (11).

Performance evaluation gives $[\mu_{\sigma_r}^{min}, \mu_{\sigma_r}^{max}] = [12, 26]$ and the associated selected orders are the following ones:

for $\mu_{\sigma_r}^{min}$: $\sigma_f = (t_1\ t_1\ t_1\ t_1\ t_1\ t_1\ t_2\ t_3\ t_1)$ $\sigma_r = (t_2\ t_4\ t_3\ t_5\ t_6\ t_1)$

for $\mu_{\sigma_r}^{max}$: $\sigma_f = (t_1\ t_1\ t_1\ t_1\ t_2\ t_3\ t_1\ t_1\ t_1)$ $\sigma_r = (t_2\ t_3\ t_4\ t_5\ t_6\ t_1)$

These results illustrate the benefits of the proposed approach. Permutations in the reference sequence established on the associated underlying un-timed net are observable in the transient mode as well as in the repetitive one. The schedule corresponding to the performance $\mu_{\sigma_r}^{max}$ is given on the Gantt Diagram (see figure 4).

6. CONCLUSION

In this paper, a proactive/reactive approach to obtain a robust control of time critical systems during their starting phase was proposed. The disturbances considered here generate an uncertainty on the effective task durations. The presented work uses p-time Petri Net as modeling tool and extends firing instant approach by permitting negative instants. So, from a sequencing based on the logical constraints, a system of inequalities is built. This system is completed by additional constraints that allow possible re-ordering of firing instants. Via linear programming, performance evaluation and static and dynamic schedule in transient and repetitive mode are evaluated. This approach is also applicable when a transient schedule becomes necessary due to the diminishing or to the increasing of the available resources.

REFERENCES

Aygalinc, P, S. Calvez, P. Bonhomme (2003) Using Firing Instants Approach and Partial Order to Control Time Critical, Proceedings of the 9th IEEE ETFA03, Vol. 1, Lisbonne, Portugal, 2003, pp. 82-89

Bonhomme, P, P. Aygalinc, S. Calvez (2001) Control and Performance Evaluation Using an Enumerative Approach, *Proceedings of the 5th World Multi-Conference on Systemics, Cybernetics and Informatic, SCI'2001*, Orlando, USA, July 22-25, 2001

Boucheneb, H, G. Berthelot (2001) Towards a Simplified Building of Time Petri Nets Reachability Graph, *Proceedings of the PNPM'93*, September. 1993, pp. 46-55

Calvez, S, P. Aygalinc., W. Khansa (1997) W.:P-Time Petri Nets For Manufacturing Systems with Staying Time Constraints, *Proceedings of the* CIS 97, Belfort (France), May 20-22 1997, pp. 495-500

David, R, H. Alla (1994) Petri Nets for Modeling of Dynamic Systems - A survey, *Automatica, 30(2)*, 1994, pp. 175-202

Diaz, M (2001) *Les Réseaux de Petri*, Hermès (2001)

Khansa, W, J.P. Denat, S. Collart (1996) P-Time Petri Nets for Manufacturing Systems", *Proceedings of the Wodes'96*. Edinburgh (UK), August 19-21 1996, pp 94-102

Laftit, S (1991) *Graphes d'Evénements Déterministes et Stochastiques : Application aux Systèmes de Production*, Thèse de Docteur en Mathématiques, Université Paris- Dauphine.

Merlin, P, D. Faber (1976) "Recoverability of communication protocols - implications of a theoretical study", *IEEE Trans. Communications*, COM-24 (9), September 1976, pp. 381-404

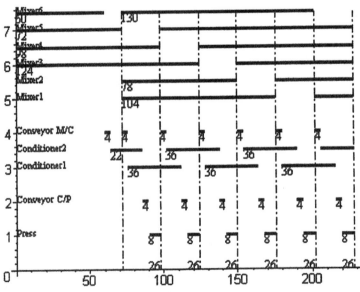

Fig.4: schedule corresponding to the performance $\mu_{\sigma_r}^{max} = 26$

www.elsevier.com/locate/ifac

A PROJECT MANAGEMENT METHOD AND TOOL: BUSINESS COMMUNICATION ENGINEERING

Kenneth Brown
Jean-Claude Tarondeau

Laboratoire Ermite, Université Louis Pasteur, Strasbourg, France
Université de Paris X-Nanterre et ESSEC Business School, France
B. P. 105
95021 Cergy-Pontoise Cédex France
tarondeau@essec.fr

Abstract: Business Communication Engineering (BCE) is a knowledge oriented method of mastering communication among members of project teams in the area of new product design projects. It has been developped to better managed complex projects caracterized by high levels on uncertainty on resource deployment and, at the same time, strong commitment to organizational objectives. It has been mainly tested in R&D environnements. In this paper, we present the BCE method followed by a demonstration toolkit.

Keywords: Project management, Planning, Communication Control Applications, Decentralized control, Concurrent Engineering

The BCE method is the result of several years of research and testing[1]. It has been first developed in a research project on simultaneous engineering called SICPARI. Further refinements and a tool prototype were added in Brown's doctoral dissertation. It has been experienced in german industrial firms and has resulted into management tools called Communigram developed within the Primavera, SAP and Microsoft environments.

This presentation goals is to present the theoritical and practical basements of the BCE method and to demonstrate the use of Communigram in the management of a R&D project.

1. A NEW PROJECT MANAGEMENT METHOD

Business Communication Engineering (BCE) is actually quite simple in its principles. The principles of BCE can best be described by the "Four Ws":

- Who has to communicate with Who?

- When?

- and about What?

[1] Brown, K., H. Schmied and J.-C. Tarondeau (2002). Success factors in R&D : A meta-analysis of the empirical literature and derived implications for Design Management, Design Management Journal, **2**, pp. 72-87

BCE is primarily about getting the right people to talk together (Who with Who?). However, it is difficult to determine Who should be communicating with Who (and When) before we have established What results we are expecting. For this reason, we will begin our description of BCE by addressing the "What?" question.

1.1 What has to be achieved?

Without doubt, the most important step in organizing a project or process is getting its objectives straight. How could a team possibly be effective, let alone efficient, before goals are appropriately set and agreed upon? This point is often dismissed as "obvious", yet alas often ignored in practice. In BCE, we carry this principle to the next level by breaking down an overall objective into a series of sub-objectives which can be treated individually—and keeping these objectives in sight even at the task level. The result is that we begin the conception of our projects by defining every single task in terms of the results it is to produce rather than by the work to be done.

A knowledge orientation.

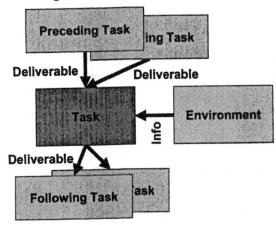

Figure 1: Definition of a Task: A task produces an outcome, a deliverable, and has inputs from other tasks and the environment.

Indeed, the aim of every R&D project is primarily to produce new knowledge. This statement is surprising at first sight since most design projects seem to be aimed at producing some piece of work, such as a set of drawings or even a prototype. Whatever the hardware may look like, it will only reach its stated objectives if the underlying knowledge necessary to produce it was elaborated in the course of the project. For instance, if a prototype of a new product falls into the hands of a competitor, chances are that its designers will not be able to redesign the same product with similar performance. "Reverse Engineering" is far less effective than most people think, and in fact only makes sense in the context of a

longer-term strategy of learning and improving. Simple copying seldom leads to satisfactory results, since the lack of background knowledge will make it impossible for the imitator to improve the initial design to reach economies of scale or to prepare for the next production generation. It the deeper understanding of *why* a certain material, shape or value was chosen which determines the true value created in a R&D project. This understanding, this *knowledge*, is difficult to obtain, because difficult to articulate and transfer. For this very reason, knowledge (be it artistic, technical or other, such as market intelligence or process excellence) is difficult to match and can therefore be of considerable importance in gaining a competitive edge over competitors.

Knowledge comes in many forms. It may be contained in a report, a letter or an e-mail, a telephone conversation, computer software, or a drawing be it technical or not. Even a machined piece of plastic is nothing else than knowledge transformed into hardware. The reason why a R&D project benefits so strongly from result-orientation is because knowledge is created through a chain of creative combinations of intermediate results, made possible through the competence (i.e. the profound background knowledge in a certain discipline) of the people involved in the elaboration of the individual deliverables. Project desgns therefore reach a much higher quality of planning if their activities are defined in terms of the knowledge to be created rather than by the work to be done. In other words, organizing the project around knowledge rather than operations is a way of reducing uncertainty in Project Management because you are using the right language from the outset.

Experience in applying this approach has shown that it offers very clear guidance for the planning of projects. This not only improves the quality of the project plan, but also the execution of the work. Implementing this approach implies that not only the project objectives must be formulated as "knowledge objectives", but that *all activities* to attain these objectives must also be designed, planned and managed in terms of the knowledge they are expected to produce or acquire.

The Deliverable Breakdown Structure. Experience has shown that it is rather difficult to impose the definition of tasks based exclusively on the knowledge creation method. Project staff has a longstanding habit of expressing tasks in terms of work to be done, including the use of abbreviations, code words and so on. They love developing their own language and in this context short task titles correspond much better to their way of working than questions. That is why Business Communication Engineering foresees both task definitions side by side: the definition of the deliverable in terms of knowledge production, often expressed as a question, plus the classic task definition by task title. The corresponding diagram, the Communigram, facilitates this work in a simple way by foreseeing a column reserved for the description of the activities' result.

		Objective (Deliverable)	Task Description
1		"Go / NoGo" Decision	Preliminary phase
1.1		"Buy or Make" Decision	State-of-the-Art
	A1000	Detailed overview of currently available technologies	Literature analysis
	A1010	Is an appropriate product available?	Contact suppliers
	A1020	Are there currently technologies available that solve the problem?	Contact knowledge suppliers
	A1030	Are there other products or technologies which question the viability o	Technology Watch review
	A1040	Continue with project?	Milestone "State-of-the-Art"
1.2		Specifications for system	Design Specification
1.2.3		Specifications for component 1	Component 1
	A1050	Which designs are currently available in the company?	Review current designs
1.2.3.2		Design specifications for "Plug"	Plug
	A1060	Mechanical specifications for development	Mechanical design
	A1070	Electronic specifications for development	Electronic curcuit and components
	A1080	Exact specifications of the interface configuration	Interface definition
	...		

Figure 2: The tool for BCE contains specific columns to explicitly define both the DBS and the WBS

Once it is understood that projects are carried out to create knowledge, the next obvious question is: "What kind of knowledge?".

In the context of a R&D project, this question would probably be answered like this: "Well, the aim is of course the production of a new product set of specifications." But why not include economic aspects? Using these questions, you could ask for knowledge which could save your organization a lot of money and trigger first market contacts:

- Is there anybody who wants the product once it is finished?

- Is there anybody prepared to pay for it and how much?

- How many paying customers could there be?

Include these aspects into the project objectives and you have finally found the way to introduce market and economic orientations in your new product designs. Not that researchers should answer these questions themselves. But the project leader needs to make sure that answers to these questions become available. To do so, he (or she) needs to organize his (or her) collaboration with people proficient in marketing, production or cost control so that they can provide the answers.

Besides the direct effect that asking these tough questions finally gets Design, Production, R&D and Marketing to communicate, it also helps project leaders (and project members) develop a sense of entrepreneurship, making them even more useful for the organization.

All of these concerns can be brought together in an arborescence of objectives which begins from the overall objective of the project. In contrast to the usual Work Breakdown Structure (WBS), our breakdown of the overall project objective into concrete deliverables is called the "Deliverable Breakdown Structure" (DBS).

1.2. Who needs to communicate with Who?

When figuring out how to set up a project or process in which men and women work together, exchange ideas, look for solutions, warn each other when things go wrong and pass on results as they come available, it would seem only logical to put these same people into the center of all preoccupations. Strangely enough, in most project planning techniques, the people are hidden in ominous tables and databases, and they are called "resources". In Communication Engineering, people are put back where they belong: right in the middle of the project plan.

One of the first questions every project leader will ask himself is: "Who is going to do the work?" Consequently, arduous negotiations concerning project staffing begin instantly, even when just a faint possibility of a new project shows at the horizon. Once the battles between competing projects for the most competent people are more or less over, the project manager will set up his project plan and the precious experts he had fought so dearly for disappear in the resource list of a computer program. Communication Engineering proposes a completely different way of going about planning, and illustrates this new way of doing things in the so-called

"Communigram". The Communigram is a two-dimensional matrix in which the rows are the activities of the project, and the columns are the people doing the work. Between the tasks and the columns, a series of very simple symbols show Who should be communicating with Who, and about What.

The Deliverable Breakdown Structure (DBS) is reproduced on the left hand side of the Communigram. Each row is either an individual task, or a headline for a group of tasks that we shall call a work section. By convention, the order of the rows is roughly chronological, i.e. the deliverables on the top are elaborated before those on the bottom. The left hand side of the Communigram is rather traditional inasmuch as a Gantt chart would display a similar table there: the Work Breakdown Structure (WBS). The main difference is that the DBS is oriented around knowledge production rather than work.

System Functions. Similar to the DBS, there is a whole philosophy behind naming the columns in the Communigram. You could of course simply give each person you think should be on your project a column and start defining Who does What. However, before thinking about individuals, it makes sense to first determine which *competencies* are required to elaborate each deliverable. Such competencies can be artistic, technical, such as electronics, biology, information technology, to name but a few, or non-technical such as marketing, accounting, public relations, or quality control. In order to prevent misunderstandings and be independent from various industrial structures and naming conventions, we collectively call the competencies needed in a project "System Functions".

System Functions are generic descriptors for what type of competence, scientific background, experience, or other is required to elaborate a certain deliverable. They allow us to set up a project without immediately worrying about whether particular individuals are available. Thinking in System Functions terms also allows us to describe what type of person we need in a certain task without worrying right away in which department we can find this competence, or whether a person with such competency profile is available at all within the organization.

The reason why we choose the term "System" Function is because we see a project as a system in its own right which is not necessarily a subsystem of a department, or even of the organization. It is all a question of how we define the system boundaries. It is therefore very well possible that a part of the competencies are outside our own organization, such as in a joint project with a partner, a supplier, or even a client.

Taking into account the different types of objectives which should be taken into account when planning a project, it is obvious that *at least* the following five basic System Functions need to be taken into consideration:

- Top Management
- Design
- R&D
- Production
- Marketing

Some projects may not make extensive use of some of these basic System Functions. However, one of the aims of Business Communication Engineering is to provide universal tools that make life easier by helping you avoid forgetting important things. The idea here is that once all the important System Functions are included in your Communigram, it will look awfully awkward if one of these columns remains inactive for a longer period within the project. This would mean that this System Function is not communicating with the others. Either there is a good reason for this, or there is a mistake in your planning.

Resources. Once the necessary System Functions are named and the project gets "kicked off", the System Functions are of course replaced by individuals. In project management, this is called "staffing". This means that through BCE, each person in the company gets their own column in the project plan. In terms of motivation and transparency, this is a very important point.

Having dealt with "Who are the people we need on the project team?", the next step consists in figuring out "Who is collaborating and communicating with Who, and on which occasion?" By now, the Communigram shows the project team as columns and the deliverables they are supposed to produce as rows. All we need to do now is determine the exact persons who work on each task and simply draw a dot at each intersection between a deliverable and an involved person as shown in figure 4. If there is more than one person involved in elaborating a deliverable, this obviously means that they work out the deliverable together by communicating and cooperating. To indicate this intra-task communication and cooperation, we connect the dots with a horizontal bar.

Although it is certainly nice to see people working together to create results, and indeed the results are usually all the better the more different competencies come together, we should not forget that this can also lead to certain difficulties. These inevitably occur when things go wrong, such as when the delivery of results is running late, or when the result of a task is not satisfactory. Mutual finger-pointing between the task participants and endless discussions why the results could not be delivered on time are the usual result. This kind of conflict can often be avoided by appointing a single person responsible for the task and its deliverable.

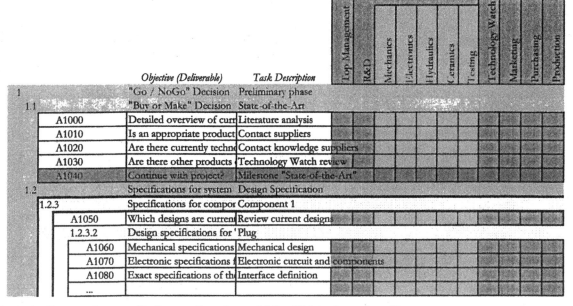

Figure 3: Communigram with System Functions which will later be replaced by persons

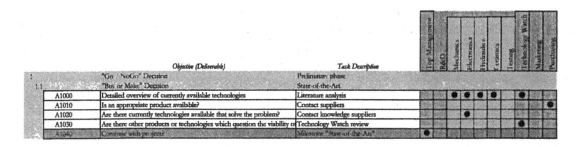

Figure 4: The Participants of each task are designated by a dot

The idea of designating responsibilities is as such not new at all. But how come so many projects go astray simply because responsibilities had not been defined clearly enough? The reason for this phenomenon is that while the concept as such is clear, the methods we dispose of to put it into practice are too complicated or too time-consuming to be implemented effectively. Examples for tools to define responsibilities include:

- o Responsibility matrices

- o Interface matrices

- o Flowcharts,

in which the boxes representing the activities are located in a horizontal or vertical column assigned to

the person responsible or with an additional matrix next to it containing resource columns

- o Cost-Time-Resource sheets,

which define each activity, the responsibilities as well as dependencies with other activities

.

The Communigram integrates all the information contained in these sheets into the planning data by a simple mouse click. In the Communigram, the deliverable responsible is designated simply by a thicker dot as shown in figure 5. Through a simple mouse click, the information is not only confined in the database but also visible in the most intuitively comprehensible diagram that can be used during planning work.

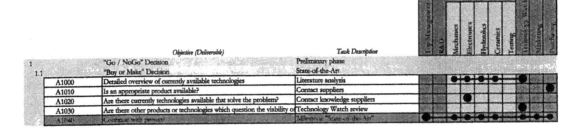

	Objective (Deliverable)	Task Description
1	"Go / NoGo" Decision	Preliminary phase
1.1	"Buy or Make" Decision	State-of-the-Art
A1000	Detailed overview of currently available technologies	Literature analysis
A1010	Is an appropriate product available?	Contact suppliers
A1020	Are there currently technologies available that solve the problem?	Contact knowledge suppliers
A1030	Are there other products or technologies which question the viability of	Technology Watch review
A1040	Conclude with project	Synthesis on "State of the Art"

Figure 5: Each task has a single Responsible

The responsible of a deliverable is in charge of:

- *ensuring* that the knowledge (deliverable) is produced as defined and on time,

- *transmitting* the deliverable to all those who need the knowledge to be able to start their work, and

- *exercising* bilateral control by regularly setting a green, yellow or red status indicator concerning the anticipated timely delivery of the result, as will be explained further on.

In the project plan, the activities are treated as "black boxes". We are concerned only with the inputs and outputs (deliverables) of each task, not in how the people elaborate the deliverable. The Communigram defines a "task team" for each particular task, simply by the dots which appear in each row. It also defines a responsible as a *primus inter pares* to make sure that the results are elaborated in time. As long as the team does what is in its power to produce the deliverable, nobody should interfere. Not the project manager, and certainly not senior management. This is very motivating for the people working in the task, and also an extremely effective way of training future managers.

1.3. When does the Deliverable need to be communicated?

Now that we have defined *Who* works together *with Who* to elaborate certain deliverables *(What)*, the question which remains is *When* to do *What*. Apart from priority considerations, the *When* in a project is primarily defined through the dependencies between tasks.

The Time Axis. The vertical axis of the Communigram is—roughly—its time axis. Tasks which happen first are by convention listed first in the DBS. But of course, since a project will have many tasks which are carried out in parallel, the order in the DBS cannot always be representative for the execution order. Nonetheless, information and knowledge flows more or less from top to bottom in the Communigram, much like a waterfall. "When" deliverables actually need to be created and delivered is primarily determined by their logical order, i.e. the dependencies between deliverables.

Task Dependencies. We call a task "dependent" from another when a certain task needs the deliverable produced by another task before the task team can start working on it. In the Communigram, each time a task needs the deliverable from another task as an input, this is drawn as an arrow. By convention, this arrow is drawn between the responsible of the preceding task to the responsible of the following task.

The same kind of sequencing can of course also be done using the traditional project management methods Gantt and CPM/PERT. The arrows in activity-on-node network diagrams are in fact entirely compatible with the arrows in the Communigram. The same is true for the arrows in the Gantt chart, only that the time bars in this diagram also help understand how long the entire process is going to take. Gantt chart, CPM/PERT and Communigram are thus entirely compatible. They are also entirely complementary. While the Gantt chart shows when certain activities take place in time, CPM/PERT and flow charts help display the logic between tasks. However, neither are much good at explaining to people what their role in a project or process is or who they should be communicating with. The Communigram in fact defines a third, entirely new view which explicitly shows the people and the communication links between them, combined with the logic of the project. To a large extent, Communigram even uses the same data as the other diagrams, and can be set up to draw on the same database. Like looking through a new pair of glasses, the Communigram allows us to focus on the communication and cooperation between people, shows how information flows within the project or process, and illustrates how knowledge and other results are produced step-by-step. Figure 6 shows how the Communigram presents the information flow of a project much better than any other graphical representation.

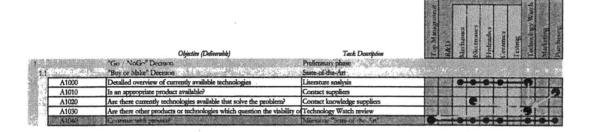

Figure 6: Communigram Showing Task Participants, Responsibles, and the Transmission of Deliverables

The convention made in the Communigram that transmission of deliverables takes place from responsible to responsible reflects a systematic means of handling information flow within the project. It ensures that communication is clearly planned in a formal way and not left to random processes. This does of course not mean that people are not allowed to talk to each other informally. The Communigram addresses only the problem of modeling the minimal flow of information needed in the project.

2. BCE: A "DECENTRALIZED[2] LIVING PLANNING" METHOD

Before even going into the merits of orienting projects around knowledge production and communication, it is important to understand that

BCE provides a means of getting a decent project plan together and, even more importantly, keeping it "alive" despite the inevitable changes which become necessary.

---

[2] By centralized planning we understand a planning procedure in which few people plan the work many others have to execute. This type of planning is quite effective in large construction projects which know have low levels of uncertainty, although it is actually not the size of these projects which distinguishes them from the non-repetitive and knowledge elaboration projects we treat here. Ever since Taylor, large projects have been broken down into smaller pieces. For instance, building a new airplane is of course much easier if the wings, the fuselage and the engines are developed by more or less independent groups of specialists rather than putting everybody into an enormous project team. These teams, which may actually be quite small, will plan their work individually. This is not what we mean by decentralized planning. Here, few people still plan for the others. Decentralized planning is when as many project members as possible participate also in the planning of the project.

2.1. A reluctance to planning

Besides this general attitude against planning, we have the problem that even if we can manage to set up a detailed initial project plan, it is usually not updated regularly, if at all. The trouble is that a project by definition changes course constantly. As soon as the project starts, the beautiful chart is instantly outdated as unforeseen events occur and additional tasks need to be woven in.

The only way we see to get out of this situation is to do what we called "decentralized planning" to change the project planning data at its source by getting the people working in the project to enter the data themselves. Unfortunately, the available project management tools based on Gantt and PERT/CPM come from a time in which centralized project planning was the rule. If we want the team members of projects to participate in keeping the project plan "alive"—we call this "Living Planning®"—then there are a few very demanding rules to respect. Decentralized planning is not easy to implement. Most researchers, engineers and technicians consider planning administrative work which is outside their interests and duties. They have hence developed a number of arguments explaining that new product design cannot be planned anyway and that it is much more important to do the work rather than producing useless paperwork. What is more, design staff also deeply mistrusts controlling mechanisms which may tie down their flexibility in reaching objectives. For this reason, they will fight hard not to give away too much information about where they are going and where they currently stand. After all, if you say what you are doing, people may start asking unpleasant questions.

The obvious reaction to technical staff's reluctance to providing such insight is of course to use coercive measures. Threatening highly competent project staff to participate in project planning may however have disastrous effects. The only way to get project staff to participate in planning activities is therefore through motivation.

The first step in persuading project members to participate in project planning consists in proposing methods which correspond better to their way of thinking and working. If you start planning a design project by making Gantt charts, you have already got it all wrong from the perspective of your creative staff. The Gantt chart forces them to think about task durations and deadlines while they are still trying to figure out if they can treat the problem at all.

Here, the Communigram is of invaluable help because it complies with the way creative people intuitively plan a project. When you ask a designer to make a project proposal, he first tries to understand what you want from him. Then he thinks how he can break down the big problem into smaller ones, which can be handled easier such as to get an overview over the complexity and to calm his fears. At this stage he will think about the knowledge he must elaborate or the information he must collect in order to find a solution to the problem.

The Communigram is therefore a tool which comes much more natural to the people working in R&D projects. Not only does it show which knowledge is to be produced at each stage of the project, it also exhibits how bits of knowledge are combined as they flow from task to task.

2.2. Bilateral Control With Status Indicators

Communigram helps monitor exchange of information between members of the project team. Each deliverable responsible must indicate the status of the operation under his control using traffic-ligth colors.

Status indicators in the traffic-light colors green, amber and red are quite common in project management. They are habitually used to indicate when something in a project is going wrong, such as non-respect of deadlines or budget overruns. In modern project management software packages, these indicators are automatically set through calculations comparing the current project data with the project plan. This is however not the type of status indicators we need for Bilateral Control

What we use are very simple indicators which are not at all calculated. They are simply set by hand by the responsibles of each task—the large dots in the Communigram. The very fact that they are so simple makes them very powerful in application. Using these hand-set indicators, each responsible of a task predicts—once a week—whether the deliverable resulting from this task will be delivered in time. If so, then a green indicator is assigned. If however the task is experiencing difficulties which endanger the timely delivery of the result, the task responsible must set an amber indicator, or even a red indicator if the task is experiencing difficulties which endanger the entire project.

2.3. Planning the Unforeseeable

A further obstacle to project planning in R&D projects is that by nature, such projects have a lot of inherent uncertainty to cope with.

Planning however does not mean that everything must happen as planned. This only makes it all the more important to plan. Yet for these project plans to remain useful, they need to be updated regularly. Everybody in a project will instinctively change course when events occur which question the validity of their current direction. The only problem is that these changes are hardly ever put on record.

The only feasible answer to this difficulty is keeping the project data up to date through "Living Planning®". In this case, the project plan will reflect the current state of knowledge concerning the project. In order to make Living Planning® operational, everybody working in the project must introduce the inevitable and frequent changes of plan into a central planning database of the project. It should be obvious that this can only be obtained if the software used has a user interface which is extremely simple to use and if the system is available at any given time for data entry. Any even minor difficulty to update the data will be a most welcome excuse for a designer not to participate in planning next time.

2.3. A Reliable and Simple Software

Extremely reliable and extremely simple software is an indispensable hygienic motivation factor to get project members to adhere to project planning practices. Hygienic means in this context that, in itself, it will not motivate people to do the planning. But if these conditions are not fulfilled you may as well forget getting your people to participate in project planning. The planning tools must be "easier to handle than paper and pencil".

3. BUSINESS COMMUNICATION ENGINEERING TOOLKIT DEMONSTRATION

Communigram is now in use in several industrial settings. It has been developped within several environments like Primavera and SAP.

To present more concretely these powerfull method and tool of project planning, a R&D project planned with Communigram will be simulated.

Those willing to get the demonstration package should mail their request to tarondeau@essec.fr.

REFERENCES

Allen, T. (1977). Managing the Flow of Information, Cambridge, MA: M.I.T. Press

Baker, B., Murphy, D. and Fisher, D.(1988). Factors Affecting Project Success, in: Cleland, D. & King, W. (eds.): Project Management Handbook, 2nd edition, New York, NY: Van Nostrand Reinhold

Brown, K., Schmied, H. et Tarondeau J. C.(2002). Success factors in R&D : A meta-analysis of the empirical literature and derived implications for Design Management, Design Management Journal, Juillet

Brown, K. (2002). La Gestion de projets de R&D : Une Méthodologie orientée information pour améliorer la rentabilité de la R&D. Doctoral Thesis, Université Paris X, Nanterre

Cooper, R. et Kleinschmidt, E. (1996). Winning Businesses in Product Development: The Critical Success Factors, Research Technology Management, 39, 4

Cooper, R. et Kleinschmidt, E. (1990). New Product Success Factors: A comparison of 'kills' versus successes and failures, R&D Management, 20, 1

Couillard, J. et Navarre, C. (1993). Quels sont les facteurs de succès des projets ?, Gestion 2000, 2

Gerhardt, A., et Schmied, H. (1997). Externes Simultanes Engineering – Der neue Dialog zwischen Kunden und Lieferanten. Springer, 1997

Gerstenfeld, A. (1976). A Study of Successful Projects, Unsuccessful Projects, and Projects in Process in West Germany, IEEE Transactions on Engineering Management, 23, 3, Août

Haase, E. et Leroux, C. (1993). Simultaneous Engineering – Executive Summary. Literature study, IAR Institute for Automation and Robotics. Université Louis Pasteur, Strasbourg

Katz, R. and Allen, T. (1985). Project Performance and the Locus of Influence in the R&D Matrix, Academy of Management Journal, 28, 1

Maidique, M. et Zirger, B. (1984). A Study of Success and Failure in Product Innovation: The Case of the U.S. Electronics Industry, IEEE Transactions on Engineering Management, 31, Novembre

Myers, S. et Marquis (1969). Successful Industrial Innovations: A Study of Social Factors Underlying Innovation in Selected Firms, Report for the National Science Foundation, National Science Foundation, 69-17, Washington, D.C

Pinto, J. et Mantel, S. (1990). The Causes of Project Failure, IEEE Transactions on Engineering Management, 37, 4

Schmied, H. (1995). R&D Management in Europe, Productivity, Performance, International Co-operation. Gabler

Tarondeau, J.-C. (1994). Recherche et développement, Paris: Vuibert

ELSEVIER
IFAC
PUBLICATIONS
www.elsevier.com/locate/ifac

CONCURRENT DESIGN OF PRODUCT FAMILY: FROM THE FUNCTIONAL MODEL TO THE MATERIAL MODEL

Blaise MTOPI FOTSO*, **Maryvonne DULMET, Eric BONJOUR**

*Laboratoire d'Automatique de Besançon (LAB, UMR CNRS 6596). ENSMM /
UFC, 24 Rue Alain Savary 25000 Besançon, France, phone : +33(0)
381 40 28 00, fax : +33(0) 381 40 28 09
Email: bmtopi@ens2m.fr / mdulmet@ens2m.fr / ebonjour@ens2m.fr*
*** Department of Mechanical and Production Engineering, Fotso Victor Institute
of Technology of Bandjoun. Dschang University, P.O. Box: 134 Bandjoun,
Cameroon.*

Abstract: The worry to control the design processes of new products has been growing
these recent years. The process of product design is often considered as the transformation
of numerous and various kind of data of nature and permit to clarify a certain number of
principle designs. The crucial problem is to specify the articulation between the functional
domain and the organic domain. Therefore, a functional survey of the product by a
progressive sophistication is necessary. This paper proposes the use of a hierarchical
approach. We describe an innovative design process model for matrix transformation
between constant generic constituent, variant generic constituent, partial generic constituent
and functions.

Keywords: Concurrent engineering, design, life cycle, products, modeling, models, assembly

1. INTRODUCTION

During the last fifteen, some essential
characteristics of the market have changed. In order
to remain competitive, companies must meet
consumer's needs and expectations at best.
Companies must offer product diversity and low
costs, high quality and environmentally friendly
product to the market. In addition to cost and quality
criteria, a tendency to decrease the life cycle of
product and increase product diversity is observed. In
response to these market changes, enterprises have
developed, among other solutions, Flexible
Production Systems. The main intent is to obtain
production systems easily adaptable to a mix of
products and to change in production demand.

Our research is concentrated on the design of
Product Families and it is founded on previously
developed works (Bourjault, 1984; Henrioud, and
Bourjault, 1991; Djemel, et al., 1992). One of the
strategies to face these questions is referred to
Flexible Assembly System (FAS). A FAS is a
automated system, constituted of robots, feeders,
transfers and control facilities, among other
resources, where a set distinct product (a family of
product) is assembled. Thus, a set of different

assembly processes, sharing the same production
resources, is carried out in FAS.

Traditionally, the design of FAS is developed
sequentially and it involves the following phases:
specification of functional requirements from market
analysis, product design, assembly planning and
production system design. This approach for design
development takes into accounts, mainly, the causal
relation function → product → process → system.
The effects of product design decisions on system
complexity on product design imposed by the
characteristics of the production processes and
system, although implicit, are not systematically
considered. This means that important decisions,
representing a great portion of the goal incurred
costs, may be taken without evaluating all their
effects on the production system.

Growing commercial competition as well as
evolution in management, engineering and computer
domains have motivated the development of a new
design approach called Concurrent Engineering (also
named Simultaneous or Integrated Engineering)
(Eynard, 1999; Stadzisz and Henrioud, 1999).
Concurrent Engineering (CE) is an organization and
methodological approach for the integration of the
different activities concerning the design of products

and their production systems. This integration permits to overcome existing barriers between the various services in the enterprise, and achieve a more effective synergy in the design process. The main goals of CE are to reduce time – to – market, reduce design development cost and improve designs quality. Research on CE focuses on three main, non independent aspects: organizational, technical and methodological issues. Organizational issues concern changes in the structure of the services, improvement of communication between services and horizontal integration (Hadj-Hanou, 2002). Technical issues involve technical data management, communication, software integration, standards for exchange of data, and project management support. Methodological issues concentrate on the support of the design development (e.g., decision support, constraint checking, design evaluation, design rules). This research addresses methodological aspects of models in the design of products family. It focuses on the integration of the model from the assembly point of view. Section 2 presents the problematic of the modelling of products by particular points on the concept and functional analysis of the product family. Section 3 in last, propose different nomenclature support of product family; an accent is put on the typologies of generic constituent as well as the geometric link.

2. PROBLEMATIC OF THE PRODUCT MODELING

The identity of a product is determined by the set of his features that is to say, by the definition of his functionalities, of his behaviour, of his physicochemical constitution, of hi structure and morphology, among others. The knowledge of a product supposes the knowledge of the determining part of these data. The registration, the treatment and the transmission of this knowledge necessarily require a means of representation that named (Stadzisz, 1997) the model of the product. When one speaks of industrial product, it is important to observe that every product (sense unit of a physical object) is, in fact, a instance of a same product abstracts basis: the one describes by the model of the product. The industrial product concept suppose the interchangeability of the products thus or, in other words, that the entities and the process of production are as the exits (the instance of the product) are sufficiently similar to offer undifferentiated manner the same services (functions) to the consumers.

The model of a product describes the knowledge therefore on an industrial product. It can happen, however, that a product is described by more of a model according to the properties or characteristic that one wishes to put in evidence. Thus, the knowledge on a product could be described by a functional model, a behavioural model, a structural

model and / or geometric model. Each of its models captures a part of the knowledge on the product.

2.1 The Concept of Product Family

In the context of the product multi-product assembly, we use the term type of product and "family of product" to designated the products susceptible of production. These terms, take here a particular significance given by the definition follow.

A type of product designate all instances (physical product) interchangeable generated named the model product. Thus, two products are said of a same type if the have been created according to a same model. A family of product is a whole $F = \{P_1, \dots P_{np}\}$ of np $\geq$ 2 types of composed products possessing a function equivalence (at least partial) and differentiated by their shape, their structure, their materials, their properties and / or the means of production of their components, and satisfactory the following conditions:

Condition of similarity: $\forall P_i \in F, \exists P_j$ as P_i and P_j present some similarities concerning the process of assembly of their components.

Condition of production volume: The individual volume of production estimated for the different types of product $P_i \in F$ as well as the orders in the time don't justify the development of specific assembly system to every type of product P_i.

Condition of life cycle: The life cycle of every type of product $P_i \in F$ on the market doesn't justify the development of a specific assembly system to each.

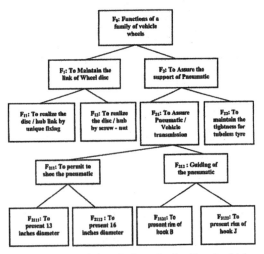

Fig.1. Functional description of a family of vehicle wheels

The functional model of a family describes the organization of the functions of each one of its products. A product function (also called functional requirement) is the product aptitude for fulfilling an user's desired task or for satisfying an user's expectation. We refer to this kind of function as *main functions* (or service function) to distinguish from internal functions of the products. Internal functions and specify the intent of components and subassemblies in the product. In the other word, internal functions represent the decomposition of the main functions into sub-functions. Thus, products in a family have a same base set of main functions with eventual, optional main functions which are specific to some product. Subsequent functional decomposition may produce differentiable or similar sub-functions for products in the family. The functional model is a hierarchical description of the function decomposition. At the highest level there are the main functions that define which are the product intents, and at the other levels there are the sub-functions which establish how main functions will be achieved by components and subassemblies (Pomian, et al. 1997).

Associated with functions, these can be functional constraints. These constraints established the boundary on acceptable solutions and describe dependencies between functions and / or sub-functions. In the multi-product context, constraints also specify diversity of the functions.

The figure 1 is a synthetic illustration of the functional description of a family of vehicle wheels. This family should offer the choice of rim diameter 13 and 16 inches, and of B type and J. This different level that are bound by a relation belonging between the functions.

Relation belonging between functions: Either E_F, the set of the service functions and technique contained in the functional model of a type of product given. The binary relation $R_f(F_i, F_j)$ definite in $E_F X E_F$ whose elements verify the predicate below definite P_f, is called relation belonging between the function of a product.

P_f: The F_i function belong in the F_j function, and is called one sub function of F_j therefore, so F_i is attached hierarchically over to the F_j representing a component of F_j thus. The relation belonging is transitive since $\forall F_i, F_j and F_k \in E_F$ as

$R_f(F_i, F_j)$ and $R_f(F_j, F_k)$ one necessarily has $R_f(F_i, F_k)$. This relation belonging allows us to introduce the abstract functions and the elementary functions. A function any F_i is qualified of elementary functions if another F_j function doesn't exist as $R_f(F_j, F_i)$. In the example of the figure 1, we have five abstract functions (F_1, F_2, F_{21}, F_{211} and F_{212}) and seven elementary functions (F_{11}, F_{12}, F_{22}, F_{2111}, F_{2112}, F_{2121} and F_{2122}). We can now establish a typology of the functions in the setting of the families of products, what permits to distinguish three types of functions: the constant functions, the variant functions and the partial functions [9]. As applying has the example of vehicle wheels, we get the configuration below:

Constant functions: the F_{11} functions, F_{12} and the F_{22} apply indifferently to the four types of product of the family.

Variant Functions: the F_1 functions, F_{21} F_{211}, F_{212}, apply to the four types of products but their parameter (type of assembly, diameter, and type of hook) change. Partial functions: the F_{2111} function only applies to the diameter of rim 13 inches. The F_{2112} function only applies to the diameter of rim 16 inches. The F_{2121} function only applies to the rim of hook B. The F_{2122} function only applies to the rim of hook J. The F_{11} function only applies to the wheel disc non removable. The F_{21} function only applies to the disc wheel disc removable.

3. MATERIAL ANALYSIS OF A FAMILY OF PRODUCT

All family of product contains a set of products possessing some similarities between them. These similarities express themselves as well to the functional level (as us presented it to the previous paragraph) that to the level of their material constitution (components and common or similar links). In this section, we are interested in the generic nomenclature concept and the establishment of a typology of the component and link between components for the family to which the belong. This typology, as we will see it in the continuation is indispensable to the realization of a model of representation of the families of products.

3.1. Nomenclature of a family of product

A composed product is a set of united components following a certain process of assembly. Independently of the other of assembly of the component, a composed product can be considered like a functional arrangement of components and sets of components. (Djemel, 1994) qualified the description of such an arrangement of *nomenclature of product*.

(Stadzisz and Henrioud, 1996) clarified well the definitions of nomenclature (functional) of product, of functional subassembly, of functional equivalence relation, of generic constituent of product family and generic nomenclature. The elements of the generic nomenclature are presented according to an increasing hierarchy. The root of this hierarchy corresponds to the generic constituent representing the whole products family whiles the knots and the leaves correspond to generic constituent representing subassembly. In order to illustrate these concepts, let's consider our family of vehicle wheels whose generic functional model is the one of the figure 1. Four types of products exist in the family of which the graph of links and the nomenclatures are represented to the figure 2.

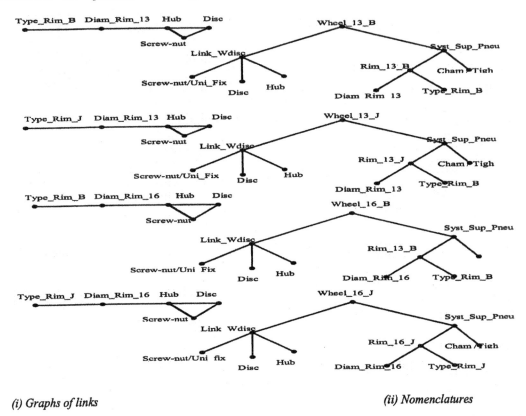

(i) Graphs of links   *(ii) Nomenclatures*

Figure2. Description individualized of four types of vehicle wheels

The analysis of the functional relations of equivalence of the element of the four types of wheels duct to the constitution of thirteen classes of equivalence materialized by the following matrix relation (where S is a diagonal matrix):

$$\begin{bmatrix} F_0 \\ F_2 \\ F_{22} \\ . \\ . \\ F_{12} \end{bmatrix} = \begin{bmatrix} S_1 & & & & \\ & S_2 & & & \\ & & S_3 & & \\ & & & \ddots & \\ & & & & S_{13} \end{bmatrix} \begin{bmatrix} C_1 \\ C_2 \\ C_3 \\ . \\ . \\ C_{13} \end{bmatrix} \quad (1)$$

For example S_1 = {Wheel_13_B, Wheel_13_J, Wheel_16_B, Wheel_16_J} and Cl_1 {S_1} achieve the F_0 function: function of a wheel family. The F_i functions are those presented to the figure 1.

In order to structure better and those for a better exploitation of the matrix, we identified one generic constituent by functional equivalence class. We baptize WHEEL for Cl_1, SYS_SUP for Cl_2, CHAM_TIGH for Cl_3, RIM for Cl_4, DIAM for Cl_5, DIAM_13 for Cl_6, DIAM_16 for Cl_7, HOOK for Cl_8, HOOK_B for Cl_9, HOOK_J for Cl_{10}, LINK_WDISC for Cl_{11}, UNI_FIX for Cl_{12} and SCREW_NUT for Cl_{13}.

Still in the worries to have an advanced sophistication degree, we establish a relation of belonging between the constituent whose hierarchy is presented in equation (1).
The exam of this relation gives:

{LINK_WDISC, SYS_SUP} ∈ WHEEL;
{SCREW_NUT, UNI_FIX} ∈ LINK_WDISC;

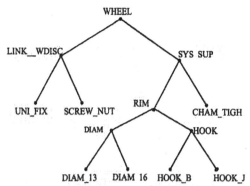

{RIM, CHAM_TIGH} ∈ SYS_SUP;
{DIAM, HOOK} ∈ RIM;
{DIAM_13, DIAM_16} ∈ DIAM;
{HOOK_B, HOOK_J} ∈ HOOK

$$\begin{bmatrix} C \\ G \\ C \end{bmatrix} = \begin{bmatrix} SCREW \\ NUT \\ & & UNI \\ & & FIX \\ & & & CHAM \\ & & & TIGH \end{bmatrix} \begin{bmatrix} F_{12} \\ F_{11} \\ F_{22} \end{bmatrix} \quad (2)$$

$$\begin{bmatrix} V \\ G \\ C \end{bmatrix} = \begin{bmatrix} WHEEL \\ & LINK \\ & WDISC \\ & & SUP \\ & & PNEU \\ & & & RIM \\ & & & & DIAM \\ & & & & & HOOK \end{bmatrix} \begin{bmatrix} F_0 \\ F_1 \\ F_2 \\ F_{21} \\ F_{211} \\ F_{212} \end{bmatrix} \quad (3)$$

$$\begin{bmatrix} P \\ G \\ C \end{bmatrix} = \begin{bmatrix} DIAM \\ 13 \\ & DIAM \\ & 16 \\ & & HOOK \\ & & B \\ & & & HOOK \\ & & & J \\ & & & & SCREW \\ & & & & NUT \\ & & & & & UNI \\ & & & & & FIX \end{bmatrix} \begin{bmatrix} F_{2111} \\ F_{2112} \\ F_{212} \\ F_{2122} \\ F_{21} \\ F_{11} \end{bmatrix} \quad (4)$$

Figure3. Generic nomenclature of family wheel

3.2. Typology of the generic constituent – Typology of the geometric links

The concepts introduced in the previous paragraphs allow us to establish a typology of the generic constituent of products family. Three types of generic constituent are identified: constant, variable and partial.

Constant Generic Constituent: Besides the fact to be constituent where all functions that he achieves are constant, the hypothesis that all differentiation material in one family of product is justified amply in this functional model.

Variant Generic Constituent: In this case of face, at least one of the functions that he achieves are variable, the other capable to be that of the constant functions. This type of constituent is present in all of product of the family.

Partial Generic Constituent: Here, at least one of the function that he achieve are partial, the other capable be indifferently constant or variable. They are not present in all types of products of the family because the partial functions that they achieve only apply to a part of the types of products.

The nomenclature presented to the figure 3 famous the previous definitions well. The matrix transformations of equations (2), (3) and (4) present the different typologies.

The links geometric between the components are not always identical for the set of the types of product. The components capable to vary from a type of product to the other, some links can be also different according to the type product. We introduce the generic link concept below and we present a typology.

Elementary generic link:
Either
- $GC_F = \{GC_1, ..., GC_n\}$ the set of the generic constituent in the generic nomenclature of the family of F products.

- $EGC_F \subset GC_F$ the set of the elementary generic constituent of the nomenclature of F

- $AGC_F \subset GC_F$ as

$EGC_F \cup AGC_F = GC_F$ the set of the abstract generic constituent in the nomenclature of F.

An elementary generic link, noted EGL (GC_p, GC_q), is a fictional link between two generic constituent GC_p and $GC_q \in EGC_F$, definite by the whole non empty set $\{l_1, ..., l_k\}$ of the generic links between the components of GC_p and those of GC_q on the whole of type of product of the family F. The generic constituent belonging in GCE_F represents classes of correspondence of components only having one element, every elementary generic link (EGL) is defined by a geometric link.

Abstract generic link:
An abstract generic link, noted AGL (GC_p, GC_q) is a fictional link between two generic constituent GC_p and GC_q belonging in the set GC_F of the generic constituent of the nomenclature of products family of F, as (Jagou, 1993):

$$GC_P \in GCN_F, \exists ELG_F(C_r, C_s) \quad (5)$$

as

$$C_r \in \bigcup_{i=1}^{ngc}$$

$$\lfloor GC_i \, as \, GC_i \in EGC_F \, and \, R_{ac}\left(GC_i, GC_q\right)\rfloor \quad (6)$$

$$if \, GC_q \in EGC_F \, then \, C_s \in GC_q \quad (7)$$

$$if \, no \, C_s \in \bigcup_{i=1}^{ngc}$$

$$\lfloor GC_i \, as \, GC_i \in EGC_F \, and \, R_{ac}\left(GC_i, GC_q\right)\rfloor \quad (8)$$

Where, ngc is the number of generic constituent in GC_F. $AGL_F(GC_p, GC_q)$ is defined therefore by the set of links elementary EGL_F (C_r, C_s) between GC_p and GC_q.

Equation (5) precise on the one hand, that one of the generic constituent concerned by link is necessarily a elementary generic constituent in generic nomenclature of F; on the other hand that at least exist an generic elementary link between the elements of the two generic constituent concerned by the link. The binary relation noted $R_{al}(GL_t, GL_u)$, definite in $GL_F \times GL_F$, whose elements verify the PR_{al} predicated, as presented to the paragraph 2.2, is called relation of adherence of the generic links.

4. CONCLUSION

The objective of this paper has been to specify the articulation between the functional domain ad the organic domain. Since product and variability imposed assembly process differentiation and the need for flexibility in the assembly system, we proposed different models. In particular: the matrix relation between functions and classes of equivalence, the generic nomenclature, matrix of transformation between the constant generic constituent and the function.

This work is complementary to a set of others. In particular, in Mtopi et al. (2004a), we have established an interrelationship between the abstract generic links and the elementary generic links. On the other hand, in (Mtopi et al. 2004b), we went in depth in the design process while putting in inscription the notions of web grammars for a good understanding of the combinative explosion phenomenon.

REFERENCES

Bourjault, A. (1984). Contribution à une Approche Méthologique de l'Assemblage Automatisé : Elaboration Automatique des Séquences Opératoires. *Thèse d'Etat de l'Université de Franche – Comté*. pp.200, France.

Djemel, N. (1994). Contribution à la conception des systèmes flexibles dans le cas multi – produits. *Thèse de l'Université de Franche – Comté* pp. 200, France.

Djemel, N., P. Lutz and A. Bourjault (1992). Flexible Assembly System Design: Modeling of a Product Family with a Web Grammar. *Proceedings of 10th IPSE / IFAC International Conference on CAD/CAM Robotic and Factories of the future*. Canada, pp. 63 – 68.

Eynard, B. (1999). Modélisation du produit et des activités de conception : Contribution à la conduite et à la traçabilité du processus d'ingénierie. *Thèse de l'Université de Bordeaux 1, pp.205*. France.

Hadj-Hamou, K. (2002). Contribution à la conception de produits à forte diversité et leur chaîne logistique : une approche par contrainte. *Thèse de l'Institut National Polytechnique de Toulouse, pp.182*, France.

Henrioud, J.M. and A. Bourjault (1991). LEGA: a *Computer–Aided Mechanical Assembly Planning, Kluwer Academic Publisher*, **Chapter 8**, Norwell, USA.

Jagou, P. (1993). Concurrent Engineering, *editions Hermes, Paris*, France.

Mtopi, B., M, Dulmet and E. Bonjour (2004a). Different models to support concurrent design of product family. *Proceedings of 5th International Conference on Integrated Design and Manufacturing in Mechanical Engineering (IDMME), Apris 5th – 7th, Bath*, UK.

Mtopi, F. B., M Dulmet and E. Bonjour (2004b). Product family in concurrent design. Preprints of the 10th IFAC/IFORS/IMACS/IFIP Symposium on Large Scale Systems. **Vol. 1**, pp. 118-123, July, Osaka, Japan.

Pomian, J.L., T. Pradere and I. Giallard (1997). Ingénierie & Ergonomie. *Cepadès – Edition* pp. 259, France.

Stadzsiz, P.C. (1997). Contribution à une méthodologie de conception Intégrée des Familles des produits pour l'assemblage. *Thèse de l'Université de Franche – Comté*. pp .204. France.

Stadzsisz, P.C., J.M. Henrioud and A. Bourjault (1995). Concurrent Development of Product Families and assembly Systems. *Proceedings of the 1995 IEEE International Symposium on Assembly and Task Planning ISATP* pp. 327 – 332. Pittsburg. USA.

Vargas, C. (1995). Modélisation du processus de conception en ingénierie des systèmes mécaniques. Application à la conception d'une culasse automobile. *Thèse de l'Ecole Normale Supérieure de Cachan*, France.

ELSEVIER
IFAC
PUBLICATIONS
www.elsevier.com/locate/ifac

USING INTERPRETABLE NEURO FUZZY MODELS FOR COST ESTIMATION IN HIGHLY DYNAMIC INDUSTRIES

M. Camargo, B. Rabenasolo, J-M. Castelain, A-M Jolly-Desodt

Laboratoire Génie et Matériaux Textiles (GEMTEX, EA–2461)
École Nationale Supérieure des Arts et Industries Textiles
9 rue de l'Ermitage, France, BP30329, 59056 Roubaix Cedex01, France
Tel : + 33 (0) 3 20 25 64 91
Fax : + 33 (0) 3 20 27 25 97
e-mail : mauricio.camargo@ensait.fr

Abstract: early economic evaluation of the design alternatives for new products is one of the key concept in order to optimize the conception activity. This helps to attain strategic goal such as design-to-cost or target costing. For this objective, various cost estimation methods and models have been widely applied in several industrial fields (aeronautics, automotive, software industries, etc.) This paper studies the implications of using various modelling techniques to develop Cost Estimation Relationships (CER). The application of fuzzy based modelling techniques is carried out in order to integrate the product features . It is then compared to other techniques, namely regression-based and neural networks models. Their respective modelling capabilities are commented through a case study taken from the textile printing manufacturers.

Keywords: Cost estimation, Hybrid Neuro-Fuzzy Model, Neural Networks

1. INTROUDCTION

The textile industry has suffered from structural changes in terms of production facilities re-localisation, more demands of small lot size, a higher pressure to reduce time to market. At the same time, there is also an increasing complexity resulting from the new filed of application of new fibres and materials. As a result, the manufacturers and designers are forced to improve the product cost estimation in order to optimise their investment. Rapid and accurate cost estimation becomes a complex but strategic task. For this reason, our goal is to develop a cost estimation tool to be used in the current textile environment:

- a high degree of product diversity and frequency of change
- a variable lot size with a tendency to decrease further
- a great pressure for a quick response of the supply chain
- a decreasing profit margin and product price.

Using the parametric cost estimation method, the authors propose a cost modelling approach which is based on traditional and advanced modelling techniques, that has been applied in other industries but new to the textile industry at the design stage. The proposed approach allows to improve the interactions between designers and manufacturers by providing a better (mutual) understanding of the two worlds . The final objective is to increase the percentage of successful products, and to reduce development lead time.

Because of the cost of increased product complexity and shorter lead times, a cost model may also help textile and garment producers to provide a rapid and accurate quotation price to the client, so that they do not endanger their profitability.

As it has been mentioned in (Rosel, et al, 2003) the designers often find that they are confident about the performance of some design alternatives and uncertain about others. Currently, the designers spend an enormous amount of time to review new products, as design alternatives may differ substantially in uncertainty, potential impact, and cost. At this early stage, decisions about materials, dimensions, surface, touch of textile and the production methods have to be made. Since all factors are inter-dependent to each other, lowering the cost of one aspect may increase the cost of another.

For example, the selection of a less expensive material may require extra operation steps in order to achieve a target colour or quality level. Several of these factors are known but certainly not all. Very often, from the designers point of view, many parameters such as colour intensity or design complexity are not included in their set of constraints since they do not have any possibility to estimate the impact of these decision on the final cost. A way to solve this problem is to use cost information from the past products. These data allow us to develop fast and accurate estimation tools especially built for the design stage, where technical descriptions of a product and the possible production plans are not known or uncompletely decided.

In previous papers the authors have studied this problem in terms of domains of potential application of cost estimation in the textile industry and proposed the use of Cost Estimation Relationships (CER) or cost functions obtained from linear and non-linear least square regression (Camargo et al, 2003; Camargo et al, 2004). The results were satisfactory in term of model precision. However, the model selection may not be based only on precision criteria. As studied some authors such as (MacDonell et al, 1997; Mair et al, 2000), there are model performance criteria other than accuracy, which include the model capability to integrate qualitative variables, or the reproduction of the results to new cases, especially when there are scarce number of historical cases. In (MacDonell, et al, 1997) the model choice is related to several criteria: the ability of the modelling technique to determine its own structure, the model robustness, easy output interpretation, the model capability to suit small data sets, the adjustability to model new data, the visibility of the reasoning process and finally the capability to include known facts or to add expert knowledge. In (Farineau, et al, 2002), after having emphasised the characteristics of the different quality criteria, the authors proposed practical selection methods for selecting the adequate cost estimation model. All these methods intend to measure the general quality of CERs accounting of both technical and statistical points of view, where the technical coherence measures and qualifies the matching between the cost model generated automatically and the *a priori* knowledge of the technical expert.

This paper deals with the use of interpretable fuzzy cost estimation model for the design data in the textile industry. The main goal is to take advantage of the available information during the processes of the product development, through the implementation of new software tools such as Product Data Management (PDM). This information will improve the integration of the product design features (quantitative and qualitative) and the final cost, to enable the development of a quick costing tool. The main difficulty is the integration of economical issues before the product process plan and the bill of materials have been completely defined.

2. FUZZY LOGIC AND COST MODELLLING

Fuzzy logic is a technique that models vagueness, ambiguity and imprecision, and enables complex systems with many parameters to be effectively modelled using principally expert knowledge but can also take advantage of 'learning' from input output data pairs to improve the approximation (Zadeh, 1972). Previous research has shown that against the other modelling techniques, the hybrid neuro-fuzzy models provide the combination of strengths of both fuzzy and neural networks models by avoiding their weaknesses (Gray, et al, 1997). Thus the main feature is an adaptive system able at the same time to extract rules and knowledge from historical data bases and express it in a comprehensive way through linguistic rules, to be easily understood and applied by product designers. This approach allows a more efficient treatment of the inputs, to reduce the number of rules needed and to obtain a more clear and interpretable output response surface. In addition to accuracy, these properties are the most relevant for our specific model.

The central part of our work is to apply fuzzy rule induction methods to improve the interpretation of cost models. In order to simplify the set of rules resulting in the last model, the clustering techniques seems to be effective, specially for data bases with high number of variables or scarce cases.

2.1 Hybrid Neuro Fuzzy Cost Estimation Model

We use the well known model named Adaptative Network Based Fuzzy Interference System — ANFIS (Jang, 1993). The learning process builds a model connecting input variables to one output variable, using a set of rules. The process of fuzzification builds a certain number of fuzzy sets represented by membership functions for each variable. The inference systems uses these fuzzy sets and the rules to build an output value which will be translated into real number ("defuzzified").

The ANFIS use a hybrid-learning algorithm in order to identify the fuzzy sets by using the following steps.

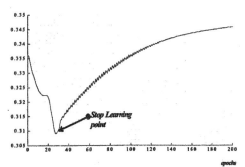

Fig 1. Early stopping training method.

- *Fuzzyfication and rule identification*: we use the sub-clustering method as mentioned in (Chiu, 1994), in order to identify the inputs of the fuzzy model. It allows the building of input-output functions, under the form of IF-THEN sentences.

- *Training*: this step optimises the parameters in the ANFIS model which has been generated by the previous step. In order to avoid the overfitting problem, the early stopping technique was used, by minimising the estimation on the validation dataset while training the model on another data set, as shows the figure. 1. The estimation score of this model can be found in the next recapitulative tables.

2.2 Simplified Hybrid Neuro Fuzzy Cost Estimation Model

In order to simplify the set of rules resulting from the ANFIS model, the clustering techniques seems to be effective, especially for largest data bases or scarce cases. Many authors have studied the problem of the model simplification and the equilibrium between an accurate and an interpretable model (Guillaume *et al*, 2001; Casillas *et al*, 2003). According to (Guillaume *et al*, 2001), there are three main features which make the rules set interpretable:

- The partitions must be legible to the experts; the membership functions must be linguistic terms, related to the system in order to make the rules comparable.
- The number of rules must be minimal. This rule simplification will result in less accurate models, but improves their robustness and their generalisation capability.
- The rules must be defined only by the more influential variables.

The authors simplify the previous ANFIS model, by applying a method which is composed by two main steps: the first simplification concerns the reduction of the number of membership functions (MF) represented by μ_l. This step uses the aggregation of several MFs into a representative MF through a distance notion $\Delta_k(i,j)$ between two MFs μ_l and μ_j. The pair (i,j) is chosen by minimizing the impact on the estimationaccuracy

when using the remaining MFs set as a new model. Many MF distance or similarity measures are possible for this aggregation step. In this paper, we use the distance defined as:

$$\Delta_k(i,j) = \int_{-\infty}^{+\infty} |\mu_i(x) - \mu_j(x)| \cdot dx \qquad (1)$$

The second step looks for the simplification of the rule set in the ANFIS model. The basis here is the search of a rule set satisfying a performance criterion which has been previously defined. The root means squared error (RMSE) is employed in our case, and the simplification algorithm is iteratively run until we find the most accurate and interpretable rule set.

3. APPLICATION EXAMPLE IN THE TEXTILE INDUSTRY

The design processes in textile printing starts from the concept design to the product process plan definition. This dynamic process begins with a set of ideas proposed by the designer that are classified by a complex evaluation system before the product arrives to the market as well explained in (Moxey, et al, 2000). The first designer inspiration sources are for example, the intent to create new forms or the use of elements coming from the social or natural environment. At this stage, the designer decisions depend mainly on the following factors:

Aesthetical factors: colour, texture, brightness, touch, and pattern
Functional factors: isolation, chemical resistance, heat transfer and dissipation etc.
Commercial factors: delay, quality and price.

The classical idea is that the designers make choices based only on aesthetical parameters. But in practice the product definition process takes into account the economical and functional constraints. For a designer the paradigm of aesthetic conventions that determine creative solutions within printed fabric design are constrained by technological and market factors. The creative design must be new related to the previous collections and must be validated by the members of the system.

For the specific case of the textile printing industry, we have carried a research study about the process of selecting design. The results show it is highly speculative and the financial investment is extremely high in comparison to the potential product commercial success. Unfortunately, all the esthetical paradigms and customer requirements are apparent only when the project is at the product freeze point, somewhere between three and four months after the project started. In contrast, the typical sales period may be only of two-four weeks. Thus, it is very important to have reliable economical evaluation of the product changes during the phase where the product is being defined. For manufacturers, this kind of highly dynamic environment requires an immediate response to the customer, as well as an optimisation of the capital investment.

As a matter of fact, the product total cost depends on several components related to the direct and indirect cost. The time and resources spent in the product development process, the technological capabilities (production infrastructure, human knowledge and skills) and of course the product features. The combination of those factors results in a specific product cost.

A theoretical analysis of the actual cost components and the main design decisions were conducted. By using principal component analysis technique, we found that most of the decisions taken in the design stage are related to four main variables. the number of product original designs with a focus on a specific theme (x_1), the base cloth colour options (x_2), the number of design colour options (x_3), the expected amount of production (x_4). As for the experts in printing textile industry the product complexity was related to number of design colours.

4. EXPERIMENTAL RESULTS

We collected first hand production data from the case company. The data concerning the indirect cost have been confirmed by (Moxey, et al, 2000) for the textile printing industry. The database includes 70 cases. The inputs of the CER are the four above-mentioned design parameters. The output of the model is the product total cost which is composed of derect ans indirect costs. The direct cost includes raw material an dproduction cost; indirect cost includes development and promotional cost.

In order to avoid bias in the learning process, we have randomly partitioned the data : 75% of them are allocated to the training set used for the optimisation of the model parameters; and 25% of the data are in the validation set.

Two criteria are used for the evaluation of the quality of the models. The first one is the well-known Root Mean Square Error (RMSE):

$$RMSE = \sqrt{\frac{1}{(n-p)}\sum_{i=1}^{n}e_i^2} \qquad (2)$$

where

e_i is the estimation error for the product i,
n the number of observations in the data set,
p the number of parameters in the model.
($n-p$) represents the degree of freedom of the model. The second criterion is the Normalized Sum of Residues (SNER) :

$$SNER = \frac{\sum_{i=1}^{n}|e_i|}{\sum_{i=1}^{n}c_i} \qquad (3)$$

where c_i is actual cost for the product i.

The SNER which can be read as a percentage error, has some advantages over the usual mean relative absolute errors $MRAE = \sum_{i=1}^{n}|e_i|/c_i$. While this latter gives indication on the mean precision in percentage, the SNER does not treat errors on big values of cost the same way as errors on small values. What is emphasized here is the percentage of the *global* volume of errors on the *global* volume of costs (the total budget).

The table 1 shows the impact of the simplification step. The structures of the models (ANFIS and simplified ANFIS) are given in the columns 3 and 4: '#MF' gives the number of membership functions for each input, and '#Rules' gives the number of rules in the fuzzy inference systems . The column 'Training' (resp. 'Validation') shows performance values measured on the training dataset (resp. validation data set).

Table 1. Hybrid Neuro-Fuzzy Cost Estimation Model Simplification results

Cost Model	Inputs	#MF	#Rules	Training		Validation	
				RMSE	SNER	RMSE	SNER
ANFIS	1	11	11	2.058	7.17%	2.912	13.74%
	2	11					
	3	11					
	4	11					
SANFIS	1	3	7	3.257	8.83%	4.488	16.43%
	2	4					
	3	3					
	4	4					

The results show that the accuracy of the simplified model (SANFIS) remains acceptable when compared with the automatically generated ANFIS model. In this case, the number of MFs is reduced from 11 to 3 or 4 for each input variables as a result of the aggregation algorithm.

As an illustration, the figure 2 shows the impact of the first step in the simplification algorithm for the input variable 'design complexity (number of colours)'. Here the simplified membership functions are more easily interpretable as it allows to assign more readable linguistic markers ('easy', 'mean' and 'complex' fuzzy sets instead of eleven fuzzy sets).

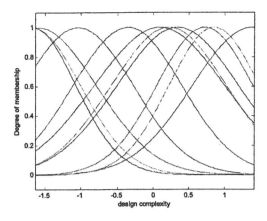

Fig 2 (a) automatically generated ANFIS model

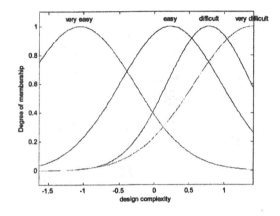

(b) simplified ANFIS cost model
Fig 2 Comparison of the membership functions for the input variable "number of colours".

The final step consists in rule base simplification. The simplification process also improves the rule base interpretation (7 instead of 11 inference rules). The reduced rule base makes possible to build adaptable systems able to: predict the outcome of a decision-making process and provide an understandable explanation of a possible reasoning.

Between global variable removal and selection done at rule level, from (Guillaume, et al, 2004) we introduce an intermediate selection level. It allows for evaluating the influence of a given variable within a context defined by the other inputs, and represented by a group of rules. The implementation requires the definition of a rule distance function.

The simplification stage tolerates some loss of accuracy while being guided by new indices complementary to the usual numerical performance index. Figure 3 shows the estimated cost versus the actual cost, the estimation errors versus the actual cost, and the distribution (histogram) of these errors.

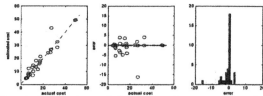

fig. 3 Evaluation of the estimation error on the validation dataset

In addition to these, we also developed a tool which allows us to compare simultaneously a set of various CERs by using Linear Regression (LR), Non-Linear Regression (NLR) and Neural Networks. Table 2. shows the performance and robustness.

These results put in evidence that in terms of model precision the Non-linear (NLR) and the Neural Networks (NN) are the most accurate models, because of their capability to modelling non linearity and interactions in the data. Against that, the SANFIS model allows us to improve the interpretability and knowledge recover from the cost database.

Table 2. Comparison of the different cost models

Model	Training dataset			Validation datase		
	SNER$_{100}$	SNER$_{90}$	SNER$_{75}$	SNER$_{100}$	SNER$_{90}$	SNI
Linear	34.43%	24.51%	16.74%	30.04%	20.81%	11.!
NL	3.81%	2.82%	2.02%	3.26%	2.48%	1.5
NN	2.81%	2.04%	1.38%	3.25%	2.46%	1.4
ANFIS	7.17%	3.80%	2.13%	13.74%	8.90%	5.2
SANFIS	8.83%	3.97%	1.42%	16.43%	10.63%	6.7

This first test allow us to infer that the NN and NLR models will be more desirable for situations with relatively well know cost drivers and stable product features, and when the accuracy is the most important expected result. And the SANFIS models are more convenient when the system to model is highly variable, and we need to extract knowledge by using linguistic interpretation of the simplified rule set. This fact is very important to enhance the use of cost as a design variable in industrial activities where very often the cost drivers are aesthetical features with soft attributes (Leneau, et al, 2003).

5. CONCLUSION AND PERSPECTIVES

The proposed model allows to improve the product development process by taking advantage of the product and process information available in the design and process software tools. The most important fact for highly dynamic environments as textile industry is the availability of multi model tool reach a trade-off between accuracy, interpretability and time pressure in a highly dynamic business environment

More exactly, the methodology developed in this paper can improve the textile product development process by:

- A better comprehension of the system to be modeled by extracting knowledge from the simplified hybrid neuro-fuzzy interpretation.
- The possibility of integration of aesthetical features as cost drivers that could act as a resource for designers.
- The improvement of the designer visibility, in order to have a better perception of the economical implication of his decisions taking into account the functional, aesthetical and structural parameters.
- A better communication between the designer (Stylist or Engineer) with the production process and the supply chain partners.
- Product development and validation time reduction in order to make better and faster customer response.

These results obtained by using this kind of specific cost estimation relationship (CER) have laid milestone to develop better models. However, it is important to recognize that some limitations still exist because of the structural elements used and their application field. Besides, As mentioned in (Leneau, et al, 2003), the way as the interactions between soft and hard cost drivers of the product should be studied.

REFERENCES

Camargo M, Rabenasolo B., Castelain J-M., Jolly-Desodt A-M (2004). Parametric cost estimation in the textile supply chain: a comparison of modelling techniques. *Thirteenth International Working Seminar on Production Economics*. IGLS'2004- Austria. February 2004.

Camargo M, Rabenasolo B., Castelain J-M., Jolly-Desodt A-M (2003). Application of the Parametric Cost Estimation in the Textile Supply Chain. *Journal of Textile and Apparel, Technology and Management*, vol. 3, Issue 1, Summer 2003.

Casillas J, Cordón O., Herrera F., Magdalena L. (2003). Accuracy Improvements to Find the Balance Interpretability-Accuracy in Linguistic Fuzzy Modeling: An Overview, *Studies in Fuzziness and Soft Computing*, Springer, Heidelberg, Germany, pp. 3-2

Chiu S. Fuzzy Model Identification Based on Cluster Estimation (1994). *Journal of Intelligent & Fuzzy Systems*, vol 2, No 3.

Farineau T., B. Rabenasolo, J-M Castelain (2002) Choice of cost-estimation functions based on statistical quality criteria and technical coherence, *The International Journal of Advanced Manufacturing Technology*, no 19, pp 544-550.

Gray, A and MacDonell S. (1997) A comparison of Techniques for Developing Predictive Models of Software Metrics. *Information and Software Technology*, vol 39, pp 425-437

Guillaume S. (2001) Designing Fuzzy interference systems from data: an interpretability-oriented review *IEEE Transactions on Fuzzy Systems*, vol 9, no 3, pp 426-443.

Jang J.-S. R. (1993) ANFIS: Adaptive-Network-based Fuzzy Inference Systems,' *IEEE Trans. on Systems, Man, and Cybernetics*, vol 23, pp. 665-685.

Lenau T, Boelskifte P. (2003) Soft and hard product attributes in design. *Working paper for the Nord code seminar "Semantic & aesthetic functions in design"*. Helsinki October 2-3.

McDonell S., Gray A. (1997) Applications on Fuzzy Logic to Software Metric Models for Development Effort Estimation. *Annual Meeting of the North American Fuzzy Information Processing Society- NAFIPS*. Syracuse NY, USA, IEE pp 394-399.

Mair, C., Kadoda G, Lefley M, Phalp P., Schofield C., Shepperd M., and Webster S. (2000) An Investigation of Machine Learning Based Prediction Systems. *Journal of Systems and Software*, vol 53, pp23-29.

Moxey and J Studd, R. (2000) Investigation Creativity in the Development of Fashion Textiles. *Journal of Textile Institute*, vol 91 part 2.

Roser C., Kazmer D. and Rinderle J. (2003) An Economic Design Change. *Journal of Mechanical Design*, vol 125. p 233..

Zadeh L.A. (1972) A fuzzy-set theoretic interpretation of linguistic hedges. *Journal of Cybernetics*, vol 2, no 2, pp 4-34.

PUBLICATIONS
www.elsevier.com/locate/ifac

A MAX PLUS ALGEBRA APPROACH FOR MODELLING AND CONTROL OF LOTS DELIVERY: APPLICATION TO A SUPPLY CHAIN CASE STUDY

I. Elmahi, O. Grunder, A. Elmoudni

Laboratoire Systèmes et Transport (SeT), UTBM, 90010 BELFORT Cedex, France
Phone (+33) 3.84.58.33.47 Fax :(+33) 3.84.58.33.42
Corresponding author: ilham.elmahi@utbm.fr

Abstract: Timed event graphs in lots are introduced as a new tool for modelling and control of the lots delivery problem in freight transportation systems. The model is developed by enhancing timed event graphs with transitions in lots that have particular dynamic. We develop a linear mathematical model under constraints in the (max,+) algebra. This leads us to develop a just-in-time control for the lots delivery using the residuation theory. The proposed approach is illustrated with a supply chain subject to a non constant logistic demand. The modelling and the simulation-based optimization approach allow evaluating the just-in-time control and the model performances in term of respecting the time criterion while minimizing the supply chain costs.

Keywords: Max-Plus Algebra, Timed Event Graph, Modelling, Lots Delivery, Just-in-Time Control, Supply Chain, Genetic Algorithm.

1. INTRODUCTION

Supply Chain (*SC*) networks are formed out of complex interconnections amongst various manufacturing companies and service providers such as raw material vendors, warehouses, transporters and customers. Modelling and control of such a complicated system is crucial for improving performances and for comparing competing supply chains.

In the recent supply chain (SC) literature, different works are proposed. On one hand, some models based on Petri nets (PN) [17] are proposed to study, the materials accumulation within a framework of a high-speed production [8], the inventory management [2] or to deal with groups (batches) of goods with a given size [4]. Logistic systems have been also modelled using coloured Petri nets [16]. Although different models are developed, few works deals with the supply chain control. On the other hand, a broad category of works, mainly based on operation research, has targeted to techniques to deal with batch transport of goods. Thus, [11] studied the economic lot delivery scheduling problem (*ELDSP*), which deals with deterministic and continuous-time production. [12] developed a policy with equal and unequal sized shipments from a vendor to a buyer. [13] introduced a dynamic programming model to select the load patterns for trucks to move components from a supplier to an assembly company. However, constant demand rate, equally spaced deliveries and constant delivery quantity were restrictive assumptions of these analyses.

For the SC management, it is imperative to master two essential processes. The first one is the production scheduling of goods. The second one is the control of the transport of these goods to their destination. However, very few works deal with a SC model combined to a powerful tool of control. The problem studied in this paper is to propose a suitable model equipped with a sound tool of analysis which is easily exploitable to develop the SC control (just-in-time control for example).

In the aim to achieve this objective, the modelling using the Timed Event Graph (TEG) (subclass of PN) seems to be an interesting solution. First, because a logistic system, and particularly a SC system, can be seen as a discrete event systems (DES). Second, because the TEGs have the interesting property to be associated to a linear dioïd algebra called (max,+) algebra. Indeed, the linearity of the "max" facilitates the resolution of the complex models of the DES ([6], [1]). However, unlike public transport systems, where the (max,+) algebra is initially used ([3], [14] and [15]), in logistic systems, a means of transport (truck) can wait in a production centre (or warehouse) until that lots (batches) of goods with given sizes can be collected. Then, as they are defined and used with the (max,+) algebra, the classical TEGs do not allow a correct modelling of the freight transport with different lots of materials. The information about the capacities using of the means of transport can not be precise. Therefore, to overcome inconvenience we enrich the formalism of the classical TEG in order to improve the quality of the TEG representation for the lots transport.

Section 2 presents the studied *SC* problem. Section 3 recalls *TEG* and some dioid theory results, and introduces the proposed model. The Just-in-Time (JiT) control for the lots delivery is developed in section 4. The application is illustrated in section 5 and completed with an optimization approach. Finally, conclusions are presented in section 6.

2. STUDIED SYSTEM

We define a supply link (*SL*) as a structure connecting one producer to one consumer by the way of a Transport System (*TS*) that transports materials from the producer to the consumer. We consider a supply link (fig.1) that has a *supplier* M_1, a *customer* M_2 and a vehicle *vh* which has a limited capacity and is initially in M_1.

The scenario of this *SL* is as follows: M_2 requires to M_1 a set of products according to a vector of due dates. Each product is needed at one due date. The demand rate may be constant or not constant. Then, while taking into account the vehicle capacity, the supplier has to schedule one or several shipments in the aim to deliver the due products, at last, at their due dates.

Transport system

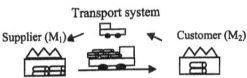

Supplier (M₁)   Customer (M₂)

Figure 1: Transport system of a supply link

When it is not possible to meet all the due dates, the deliveries can be made earlier to deliver the products before their due dates. Thus, advanced deliveries are allowed whereas delays are strictly prohibited. This is the JiT criterion to be fulfilled along this work.

The set of requested products can be divided into several lots that are not inevitably equal. One product is loaded at time and each lot corresponds to one shipment. Thus, we have to compute the optimal dates for the transport start. The following notation is used to formulate the problem:

- n = number of due goods required by M_2, $n \in IN$.

- z = due dates such as $z = z(k)_{k=1,\ldots,n}$.

- c = loading capacity of *vh*, $c \geq 2$.

- t_i = travelling time from M_1 to M_2, $t_i \in IR^+$.

- t_r = travelling time from M_2 to M_1, $t_r \in IR^+$.

- $T = t_i + t_r$, time of a shipment (M_1 to $M_2 + M_2$ to M_1) .

- C_S = holding cost at the supplier per unit of time.

- C_C = holding cost at the customer per unit of time.

- C_T = transport cost per delivery.

- $Seq = \{l_1,\ldots,l_i,\ldots,l_\beta\}$ = sequence of lots to be shipped by *vh* from M_1 to M_2. Thus, β is the number of lots (shipments).

- $\{s_1,\ldots,s_i,\ldots,s_\beta\}$ = sizes of these lots such as: $\sum_{i=1}^{\beta} s$

the sum of all the lots sizes must be equal to the demand size.

- l_i = lot with s_i products $\{p_{i,1},\ldots,p_{i,j},\ldots,p_{i,s_i}\}$ transported in the ith shipment. Thus, $\forall 1 \leq i \leq \beta$, $\forall 1 \leq j \leq s_i$, $p_{i,j}$ indicates the rank of the jth product in the ith lot.

3. MODELING

Before developing the proposed model we give a brief overview of the dioid theory and the *TEG*.

3.1. (Max,+) algebra and *TEG* overview

In the (max,+) algebra, the timed behaviour of a net can be represented by dater functions which provide the occurrence time of all the possible events in the system. More exhaustive details are given in [6], [5] and [1].

3.1.1. Dioid Theory

Definition 1 A dioid D is a set endowed with two operations, sum "$\oplus$" and product "$\otimes$". "max" and "+" are these two binary operations, respectively. The sum is associative, commutative, idempotent ($\forall a \in D$: $a \oplus a = a$) and admits a neutral element denoted ε . The product is associative, distributes over the sum and admits a neutral element denoted e . The element ε is absorbing for the product. ∎

Definition 2 A dioid D is complete if it is closed for infinite sums and if the products distributes over infinite sums $\forall c \in D, A \subseteq D : c \otimes (\bigoplus_{x \in A} x) = \bigoplus_{x \in A} (c \otimes x)$. ∎

Theorem 1 Over a complete dioid, the implicit equation $x = a \otimes x \oplus b$ admits $x = a^* \otimes b$ as least solution, where $a^* = \oplus_{i \in IN} a^i$ (Kleene star operator) with $a^0 = e$. ∎

Our complete dioid is $\overline{ZI}_{max} = ZI \cup \{-\infty\} \cup \{+\infty\}$. Then, $\forall a, b \in \overline{ZI}_{max}$, "$a \wedge b$" is the lower bound of a and b , whereas "a / b", respectively, "$b \backslash a$" which, means the right, respectively, the left division of a by b , is equivalent to the usual subtraction of a and b .

3.1.2. Residuation Theory

The residuation theory provides optimal solutions to inequalities such as $f(x) \leq b$, where f is an order preserving mapping defined over ordered sets.

Definition 3 Let $f : E \to F$ an isotone mapping, where $(E, \leq)$ and $(F, \leq)$ are ordered sets. Mapping f is said residuated if for all $y \in F$, the least upper bound of subset $\{x \in E / f(x) \leq y\}$ exists and lies in this subset. ∎

Theorem 2 Let $f : E \to F$ where E and F are complete dioids of which bottom elements are denoted ε_E and ε_F . Then, f is residuated iff $f(\varepsilon_E) = \varepsilon_F$ and $\forall A \subseteq E$, $f(\oplus_{x \in A} x) = \oplus_{x \in A} f(x)$. ∎

Corollary 1 Mapping $x \to a \otimes x$ and $x \to x \otimes a$ defined over a complete dioid are both residuated. Their residuals are usually denoted respectively $x \to a \backslash x$ and $x \to x / a$ in (max,+) literature. ∎

3.1.3. Timed Event Graphs (*TEG*)

Timed Event Graphs *TEG* are a particular class of *PN* in which there are a single transition upstream and a single transition downstream in every place [7]. With each transition X_i of a *TEG*, we associate a dater x_i which is a non decreasing mapping from ZI into $\overline{ZI}_{max}$, where $x_i(k) = t$ means that the firing numbered k of X_i occurs at time t. In [9], a first *TEG* model for a supply link has been proposed (fig.2). The arcs weights are all equal to one and one token is supposed representing different lot sizes of products.

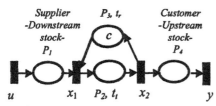

Figure 2: 1st *TEG* for the transport system in a supply link

However, to represent the lots movement of tokens, attributing weights (≥ 1) to the arcs seems to be the solution. Unfortunately, this solution has a significant disadvantage which is complicating the recurrence dater equations between input and output daters. Indeed, we will have dates of fractions of transition firings (events), whereas, we are essentially interested to the dates of the beginning and the end of an entire event. Consequently, it will be difficult to compute a JiT control. To overcome this inconvenience, we enrich the *TEG* to model and control the transport of different lots of tokens (products). This *TEG* model is endowed with a new transition in order to represent the transport of different lot sizes of products with a graphical and analytical interface.

3.2. Timed event graph in lots

We present a complete description of the proposed approach and the formalism of the particular transition that makes the particularity of the proposed model.

3.2.1. Definitions and formalisms

Definition 4

A *Transition in Lots* (*TL*) is a specific transition that allows the passage of lots of tokens by the means of fixed number of close firings during a time interval. After these firings, this *TL* is inhibited during a time interval.
Dynamic description
The *TL* is characterized with two different dynamics that occur in two distinct time intervals:
1. High level: When a transition in lots is validated (tokens exist in upstream places), this transition can be fired several times, almost, at the same time. The number of these firings corresponds to the desired lot size of products. These firings do not occur exactly at the same time but a small amount of time separates them. In practice, this small amount of time represent the handling time of a product in the vehicle. We call this event a *gust* of firings of the transition in lots.

With regard to the *classical* dater $x(k)$ of x, this high level dynamic occurs during a time interval that corresponds to tokens belonging to the same lot. This dynamic can be mathematically formulated as follows:

$$\forall k, \quad x(k) = dt \otimes x(k-1) \qquad (1)$$

dt represents the small amount of time that separates two successive firings in the same gust (same lot of tokens).
2. Low level: After the last firing of a gust, the *TL* can not be fired any more until, a time, at least, equal to T, can be elapsed. This corresponds, in practice, to the delivery travelling time that takes each shipment.
With regards to the *classical* dater $x(k)$ of x, this low level dynamic occurs during a time interval that corresponds to the inhibition period of the transition in lots. This dynamic is formulated as follows:

$$\forall k \quad x(k) \geq T \otimes x(k-1) \qquad (2)$$

In two dynamics, we assimilate the birth, respectively, the dead date of a gust of firing to its first, respectively, last firing date. We represent a transition in lots with double bars as follows:

Figure 3: Transition in lots

Upstream structural conditions A transition in lots can have several input places coming from input transitions. These places can represent for example arrivals of products and/or arrival of loading/transport orders. A transition in lots has only one upstream place coming from an internal transition. The initial marking of this place represents the available capacity of the transporter.

Downstream structural conditions A transition in lots can have one or several downstream places of which the marking represents the information about the transported quantity and the consumed transporter capacity.

Definition 5 *(Timed Event Graph in Lots (TEGL))*

A timed event graph in lots is a *TEG*, which contains at least one transition in lots. We define a *TEGL* by a 5-tuple (P, T_o, T_l, M, C):
- P: finite set of places.
- M: initial marking of places.
- T_o: finite set of ordinary discrete transitions.
- T_l: finite set of transitions in lots.
- C: finite set of arcs connecting places and transitions: $C \subseteq (P \times T_o \cup T_l) \cup (T_o \cup T_l \times P)$.

3.2.2. Supply link model

To illustrate our modelling approach, we consider the supply link (fig.4) with a vehicle for which, the capacity of loading is c. It has to drive the due goods from the supplier to the customer according to the same scenario explained in section 2. After their processing time in the supplier, the goods arrive at the stock (P_1) and wait to be driven to the stock P_4 before being consumed by the

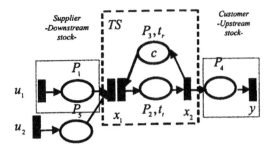

Figure 4: *TEGL* for the transport system of a supply link

customer. The transition x_1, which is a transition in lots, allows loading lots of goods into the vehicle. The firing dates of x_1, respectively, x_2 indicate the transport start, respectively, end dates. The firing dates of u_1 represent the dates when the goods have to arrive at P_1. The firing dates of u_2 provides the order dates of firings of x_1. Each gust of firings of x_1 corresponds to one lot of tokens. The marking of P_2 shows the number of full positions in the vehicle. It represents the number of loaded products in the vehicle. The marking of P_3 shows the number free positions of the vehicle when a lot size is less than c. Finally, the firing dates of y indicate the arrival of the due goods at the customer.

Even if (1) and (2) with the dater $x(k)$ provide two constraint to characterize the transition in lots dynamic, it does not provide a precise idea about which lot of tokens is in-process. For this reason, we enhance the information in the dater $x(k)$ to take into account this addition of notion of lots in the dater system representation.

Let x a transition in lots, $x(k)$ is its the dater, l_{i-1} and l_i are two lots of products to be loaded and transported successively from the supplier to its customer. As defined in section 2, $p_{i,j}$ is the rank of the jth transported product in the ith lot. Therefore, we note $x(p_{i,j})$ the dater that provides the firing date of transition x corresponding to the jth product to be transported in the lot l_i.

This dater expression allows us to give an appropriate representation for the firings dates of x while taking into account the information about the number of a lot and the rank of a product within this lot. Subsequently, the dynamic of the transition in lots x related to two lots l_{i-1} and l_i can be formulated as follows:

$$\forall 1 \leq i \leq \beta, \forall 1 < j \leq s_i.\ x(p_{i,j}) = dt \otimes x(p_{i,j-1}) \tag{3}$$

$$\forall 2 \leq i \leq \beta, \quad x(p_{i,1}) \geq T \otimes x(p_{i-1,s_{i-1}}) \tag{4}$$

The constraint (3) concerns l_i and allows loading two successive products of this lot. Thus, products that compose the same in-progress lot are loaded one after the other while dt time units separate their loading.
The daters in (4) correspond to two products in successive lots l_{i-1} and l_i. Thus, T time units separate the transport of the last token in l_{i-1} and the first token in l_i.

Hence, $x(p_{i-1,s_{i-1}})$ is the dead date of the gust of firings corresponding to l_{i-1}, whereas, $x(p_{i,1})$ is the birth date of the gust of firings corresponding to l_i.

3.2.3. Dater analysis

The system has to evolve under state constraints ((3) and (4)). Then, the state equations can be written as follows:

$$S_1 \begin{cases} x_1(k) = u_1(k) \oplus u_2(k) \oplus t_r \otimes x_2(k-c) & (5) \\ x_2(k) = t_t \otimes x_1(k) & (6) \\ y(k) = x_2(k) \\ subject\ to \\ \forall i, \forall 1 < j \leq s_i\ \ x_1(p_{i,j}) = dt \otimes x_1(p_{i,j-1}) & (7) \\ \forall 2 \leq i \leq \beta \quad x_1(p_{i,1}) \geq T \otimes x_1(p_{i-1,s_{i-1}}) & (8) \end{cases}$$

Let $u(k) = [u_1(k)\ \ u_2(k)]^T$ the input transitions vector. For each lot l_i, we consider $u(p_{i,j})$, the vector of input daters (arrival of product and transport order), corresponding to the product represented by $p_{i,j}$. Thus $u(p_{i,j})$ contains $u_1(p_{i,j})$ and $u_2(p_{i,j})$.

Proposition 1

The system S_1, is equivalent to the following one:

$$S_2 \begin{cases} x_1(k) = u_1(k) \oplus u_2(k) \oplus t_r \otimes x_2(k-c) \\ x_2(k) = t_t \otimes x_1(k) \\ y(k) = x_2(k) \\ s.t \\ h = 1,2, \forall i, \forall 1 < j \leq s_i\ \ u_h(p_{i,j}) = dt \otimes u_h(p_{i,j-1}) & (9) \\ h = 1,2, \forall 2 \leq i \leq \beta\ \ u_h(p_{i,1}) \geq T \otimes u_h(p_{i-1,s_{i-1}}) & (10) \end{cases}$$

∎

Proof : Let Σ_1 the set of solutions of S_1 and Σ_2 the set of solutions of S_2.

1. $\forall u \in \Sigma_2$, using (5) and (6), we have:

$$x_1(p_{i,j}) = dt \otimes u_1(p_{i,j-1}) \oplus dt \otimes u_2(p_{i,j-1}) \oplus t_r \otimes t_t \otimes x_1(p_{i,j-c})$$
$$= dt \otimes (u_1(p_{i,j-1}) \oplus u_2(p_{i,j-1})) \oplus t_r \otimes t_t \otimes x_1(p_{i,j-c})$$

And (5) gives:

$$x_1(p_{i,j-1}) = u_1(p_{i,j-1}) \oplus u_2(p_{i,j-1}) \oplus t_r \otimes t_t \otimes x_1(p_{i,j-c-1}).$$

Since we have $\forall i : 1 \leq j \leq s_i$, and $\forall i : s_i \leq c$ (a lot size cannot exceed the vehicle capacity), we obtain:

$\forall i : x_1(p_{i,j-c}) = x_1(p_{i,j-c-1}) = -\infty$, Then:

$\forall i, \forall 1 \leq j \leq s_i : x_1(p_{i,j}) = dt \otimes (u_1(p_{i,j-1}) \oplus u_2(p_{i,j-1}))$

Using (5), we can write :

$x_1(p_{i,1}) = u_1(p_{i,1}) \oplus u_2(p_{i,1}) \oplus t_r \otimes t_t \otimes x_1(p_{i,1-c})$ and

$x_1(p_{i-1,s_{i-1}}) = u_1(p_{i-1,s_{i-1}}) \oplus u_2(p_{i-1,s_{i-1}})) \oplus t_r \otimes t_t \otimes x_1(p_{i-1,s_{i-1}-c})$

Where $x_1(p_{i,1-c}) = x_1(p_{i-1,s_{i-1}-c}) = -\infty$, which allows us, using (10), to write:

$x_1(p_{i,1}) = u_1(p_{i,1}) \oplus u_2(p_{i,1}) \geq T \otimes (u_1(p_{i-1,s_{i-1}}) \oplus u_2(p_{i-1,s_{i-1}}))$
$= x_1(p_{i-1,s_{i-1}})$

2. Let u a solution of S_1, which means that $u \in \Sigma_1$.

In (7), the expressions (5) are substituted for $x_1(p_{i,j})$ and $x_1(p_{i,j-1})$, then:

$$u_1(p_{i,j}) \oplus u_2(p_{i,j}) \oplus t_r \otimes t_t \otimes x_1(p_{i,j-c}) = dt_2 \otimes (u_1(p_{i,j-1}) \oplus u_2(p_{i,j-1}) \oplus t_r \otimes t_t \otimes x_1(p_{i,j-c-1}))$$

For the same raisons as above, we have: $\forall i : x_1(p_{i,j-c}) = x_1(p_{i,j-c-1}) = -\infty$.

Then, $\forall j, \forall i$ $u_1(p_{i,j}) \oplus u_2(p_{i,j}) = dt \otimes (u_1(p_{i,j-1}) \oplus u_2(p_{i,j-1}))$

Since we look for an optimal synchronization between two actions related to $u_1(p_{i,j})$ (the jth product in the ith lot is ready to be loaded) and $u_2(p_{i,j})$ (start loading this product), we obtain: $u_1(p_{i,j}) = u_2(p_{i,j})$, and particularly: $u_1(p_{i,j}) = dt \otimes u_1(p_{i,j-1})$ and $u_2(p_{i,j}) = dt \otimes u_2(p_{i,j-1})$.

The same reasoning can be developed for the second constraint (10). ∎

Let $x(k) = [x_1(k)\ \ x_2(k)]^T$ the state vector of the *TEGL* model (fig.4), we obtain:

$$S_3 \begin{cases} x(k) = A_0 \otimes x(k) \oplus A_c \otimes x(k-c) \oplus B_0 \otimes u(k) \\ y(k) = C \otimes x(k) \\ s.t \\ \forall i, \forall 1 < j \le s_i \ \ u(p_{i,j}) = dt \otimes u(p_{i,j-1}) \\ \forall 2 \le i \le \beta \ \ \ u(p_{i,1}) \ge T \otimes u(p_{i-1,s_{i-1}}) \end{cases}$$

Where: $A_0 = \begin{bmatrix} \varepsilon & \varepsilon \\ t_t & \varepsilon \end{bmatrix}, A_c = \begin{bmatrix} \varepsilon & t_r \\ \varepsilon & \varepsilon \end{bmatrix}, B_0 = \begin{bmatrix} e & e \\ \varepsilon & \varepsilon \end{bmatrix}$ $C = [\varepsilon\ \ e]$

$T = \begin{bmatrix} T & \varepsilon \\ \varepsilon & T \end{bmatrix}$ and $dt = \begin{bmatrix} dt & \varepsilon \\ \varepsilon & dt \end{bmatrix}$.

Using the properties of the *Kleene star*, we obtain: $A = A_0^* \otimes A_c$ and $B = A_0^* \otimes B_0$. Then S_3 becomes:

$$S_4 \begin{cases} x(k) = A \otimes x(k-c) \oplus B \otimes u(k) \\ y(k) = C \otimes x(k) \\ s.t \\ \forall i, \forall 1 < j \le s_i \ \ u(p_{i,j}) = dt \otimes u(p_{i,j-1}) \\ \forall 2 \le i \le \beta \ \ \ u(p_{i,1}) \ge T \otimes u(p_{i-1,s_{i-1}}) \end{cases}$$

And we obtain a linear representation in (max,+) algebra under equality and inequality constraints as follows:

$$S_5 \begin{cases} \tilde{x}(k) = \tilde{A} \otimes \tilde{x}(k-1) \oplus \tilde{B} \otimes \tilde{u}(k) \\ \tilde{y}(k) = \tilde{C} \otimes \tilde{x}(k) \\ s.t \\ \forall i, \forall 1 < j \le s_i \ \ \ u(p_{i,j}) = dt \otimes u(p_{i,j-1}) \\ \forall 2 \le i \le \beta \ \ \ \ \ u(p_{i,1}) \ge T \otimes u(p_{i-1,s_{i-1}}) \end{cases}$$

Where, $\tilde{x}(k) = [x(k)\ \ x(k-1)\ \ ...\ \ x(k-c+1)]^T$ is the new state vector, $\tilde{u}(k) = [u(k)\ \ \varepsilon\ \ ...\ \ \varepsilon]^T$ is the new inputs vector and $\tilde{y}(k) = y(k)$ is the new output vector. The new system characteristic matrices are:

$$\tilde{A} = \begin{bmatrix} \varepsilon & \varepsilon & ... & \varepsilon & A \\ E & \varepsilon & ... & ... & \varepsilon \\ \varepsilon & E & \varepsilon & ... & \varepsilon \\ \vdots & \ddots & \ddots & \ddots & \vdots \\ \varepsilon & ... & \varepsilon & E & \varepsilon \end{bmatrix}, \tilde{B} = \begin{bmatrix} B \\ \varepsilon \\ \varepsilon \\ \vdots \\ \varepsilon \end{bmatrix} \text{ and } \tilde{C} = [C\ \ \varepsilon\ \ ...\ \ \varepsilon]$$

where: $E = \begin{pmatrix} e & \varepsilon \\ \varepsilon & e \end{pmatrix}$ and $\varepsilon = \begin{pmatrix} \varepsilon & \varepsilon \\ \varepsilon & \varepsilon \end{pmatrix}$.

4. CONTROL

When the vector of due dates z, at which we have to obtain output, is given, and we are asked to provide the latest model input dates that would meet this objective, we speak about the "inverse" problem [1]. Due to the problem inversion, recurrent equations in the event domain for daters proceed backwards in event numbering and 'algebra' $(\Lambda, \backslash)$ can be substituted for $(\oplus, \otimes)$ in these backward equations system ([1]). Moreover, as mappings of S_5 are invertible (corollary 1), the residuation provides the optimal solution $\bar{u}$ for the backward equations. But, under constraints (9) and (10), the purpose will be to find the optimal solution u^* of the whole constrained problem:

$$S_6 \begin{cases} \xi(k) = \tilde{A} \backslash \xi(k+1) \Lambda \tilde{C} \backslash z(k) \\ \tilde{\bar{u}} = \tilde{B} \backslash \xi(k) \\ s.t \\ \forall i, \forall 1 < j \le s_i \ \ \ u(p_{i,j}) = dt \otimes u(p_{i,j-1}) \\ \forall 2 \le i \le \beta \ \ \ \ \ u(p_{i,1}) \ge T \otimes u(p_{i-1,s_{i-1}}) \end{cases}$$

Remark *(Initial conditions)*

As the due dates vector is limited, we have to determine the JiT control only until the nth product, such as $\forall k > n : z(k) = \xi(k) = +\infty$. This means that beyond n events, we have no information about the desired output.

As soon as a lot is entirely loaded, the departure order is assimilated to the dead date of the corresponding gust of firings. This process is repeated β times.

The following proposition provides the JiT control that allows these lots transport. It provides the optimal input dater vector u^* that includes the JiT control (u_1^*) of the products arrivals at the supplier downstream stock, and the JiT control (u_2^*) of the transport start of the same products. In the same way as $x(p_{i,j})$, the daters $u^*(p_{i,j})$, $u(p_{i,j})$ and $z(p_{i,j})$ correspond to the product represented by $p_{i,j}$.

Proposition 2

The JiT control that provides the birth date of the gusts of firings: $\forall 1 \le i \le \beta, \forall h = 1,2$:

$$u_h^*(p_{i,1}) = (\tilde{u}_h(p_{i,1})) \Lambda (\overset{s_i}{\underset{j=1}{\Lambda}} (z(p_{i,j})/(t_t \otimes dt^j)))$$

$$\Lambda (u_h^*(p_{i+1,1})/(dt^{s_i-1} \otimes T))$$

Where:

- $\tilde{u}(p_{i,1})$ is the input model obtained by the backward system (S_b) for the first product in the ith lot.

- $u^*(p_{\beta+1,1}) = +\infty$ since there is no more firing after the death date of last gust. ∎

Proof : Constraint (9) can be rewritten as follows:
$$\forall i, \forall 1 \leq j \leq s_i : u_1(p_{i,j}) = u_1(p_{i,1}) \otimes dt^{j-1}$$

Let Ω the set:
$$u_1(p_{i,j}) \in \overline{ZI}_{\max} / \begin{cases} \tilde{x}(k) = \tilde{A} \otimes \tilde{x}(k-1) \oplus \tilde{B} \otimes \tilde{u}(k) \\ \tilde{y}(k) = \tilde{C} \otimes \tilde{x}(k) \end{cases}$$

Subject to : $\forall i, \forall 1 < j \leq s_i \quad u_1(p_{i,j}) = u_1(p_{i,1}) \otimes dt^{j-1}$

To find solutions of Ω, we are interested to compute the optimal birth dates of the whole gusts $u_1(p_{i,1}), \forall i$.

We have: $\forall i, \forall 1 \leq j \leq s_i \quad u_1(p_{i,1}) = u_1(p_{i,j})/dt^{j-1}$ (11)

Due to the JiT criterion, we must have:
$$\forall i, \forall 1 \leq j \leq s_i \quad u_1(p_{i,j}) \otimes t_t \otimes dt \leq z(p_{i,j}) \quad (12)$$

Consequently, we have: $\forall i \quad u_1(p_{i,1}) \leq \bigwedge_{j=1}^{s_i} (z(p_{i,j})/(t_t \otimes dt^j))$

Then, the greatest solutions of Ω corresponding to the birth dates of the gusts are the least upper bound of two sub-solutions given by the backward system and (9):
$$\forall i \quad u_1^{*-}(p_{i,1}) = \bigwedge_{j=1}^{s_i} (z(p_{i,j})/(t_t \otimes dt^j)) \wedge \tilde{u}_1(p_{i,1})$$

Let Ω' the set : $\begin{cases} u_1^{*-}(p_{i,j}) \in \overline{ZI}_{\max} / u_1^{*-}(p_{i,j}) \in \Omega \\ s.t \\ \forall 2 \leq i \leq \beta, \ u_1(p_{i,1}) \geq u_1(p_{i-1,s_{i-1}}) \otimes T \end{cases}$

Then, the greatest solution $u_1^*(p_{i,1})$ of Ω' (birth date of the gusts of firing) is the least upper bound of two sub-solutions: $u_1^{*-}(p_{i,1})$ of Ω, and the solution of (10). We have then: $\forall 2 \leq i \leq \beta, u_1(p_{i-1,s_{i-1}}) = u_1(p_{i-1,1}) \otimes dt^{s_{i-1}-1}$.

And using (10), we obtain:

$\forall 2 \leq i \leq \beta, u_1(p_{i,1}) \geq u_1(p_{i-1,1}) \otimes dt^{s_{i-1}-1} \otimes T$

And $\forall 2 \leq i \leq \beta : u_1(p_{i-1,1}) \leq u_1(p_{i,1})/(dt^{s_{i-1}-1} \otimes T)$.

Therefore, the greatest solution of Ω is:

$u_1^*(p_{i-1,1}) = u_1^{*-}(p_{i-1,1}) \wedge ((u_1(p_{i,1})/(dt^{s_{i-1}-1} \otimes T))$

$= \bigwedge_{j=1}^{s_i} (z(p_{i,j})/(t_t \otimes dt^j)) \wedge \tilde{u}_1(p_{i,1}) \wedge (u_1(p_{i,1})/(dt^{s_{i-1}-1} \otimes T))$

In the same way, we proof that:
$u_2^*(p_{i-1,1}) = u_2^{*-}(p_{i-1,1}) \wedge (u_2(p_{i,1})/(dt^{s_{i-1}-1} \otimes T))$ ∎

Corollary 2

The JiT control that gives the firing dates corresponding to the products inside all the lots is:
$$\forall 1 \leq i \leq \beta, \forall 1 \leq j \leq s_i, \forall h = 1, 2 \quad u_h^*(p_{i,j}) = u_h^*(p_{i,1}) \otimes dt^{j-1} \quad ∎$$

In a first study, we apply the proposed control based TEGL approach to a supply link with one vehicle. In order to have an optimal sequence of batches of products to make deliveries from the supplier to its customer, we develop a genetic algorithm (section 5). The evaluation of a sequence of batches is based on a cost function. This latter is closely related to the advanced time of the deliveries, the holding cost at the supplier and the holding cost at the customer.

5. APPLICATION

We concentrate on the modelling of the lots transport problem under the (max,+) algebra. Let a supply link modelled using a TEGL (fig.4), where $c = 10$, $t_t = t_r = 5\,(t.u)$ and $dt = 0.2\,(t.u)$. We recall that the primarily criterion to be respected is temporal. Indeed, it is quite allowed to make deliveries before their due dates whereas, any shortage throughout the supply chain is not accepted.

The vector of due dates is $z = [120\ 122\ 129\ 130\ 139\ 147\ 150\ 171\ 175\ 188\ 190\ 195\ 200\ 210\ 211\ 212\ 217\ 220\ 222\ 250\ 320\ 322\ 329\ 330\ 339\ 347\ 350\ 371\ 375\ 388]$ (n=30). We assume, without less of generality, that z is a non decreasing mapping from IN (set of integers) to IR (set of reals).

While using Prop.2 to compute the JiT control for the supply link for a given sequence of batches, we run our GA in order to test several sequences of lots that will constitute possible solutions. The GA evaluation function is based on the proposed JiT control. In other words, it is based on the computed input dates of the TEGL model.

Therefore, the developed GA deals with a population of individuals. Each individual is one potential solution. A solution is a possible sequence of lots to ship from the supplier to the customer. We encode each solution (fig.5) as a chromosome of n dimension.

4	2	3	1	0	0	0	0	..	0

Figure 5: Solution encoding example for 10 due products.

The largest number of shipments is equal to n ($s_i = 1, \forall i$). When the number of shipments performed by the vehicle is less than n, we can distinguish two different parts within the individual: a *useful part* (genes with non null value) and an *ineffective part* (genes with null value). The solutions are improved iteratively through genetic operations, crossover and mutation, evaluation and selection. An initial population of solutions is randomly generated. After, selection of the individuals using the roulette wheel principle, we use a 2-point crossover operator to divide each individual (solution) into 3 segments. We exchange some genes of two parent chromosomes inside the useful part. Sometimes, after the crossover operation, we may achieve two types of infeasible solutions: the sum of genes value is higher than n or the sum of genes value is lower than n. A repair process is used to correct the infeasible individuals. For the first case, it consists to add, according to the vehicle capacity, the missing amount of products to the individual in lack. In the second case, we cut off the overflowed quantity. The repair starts from the left hand side of individual in the aim to keep genes with null values within the ineffective part. The mutation operator chooses randomly two genes inside the useful part of a same solution and exchanges their positions.

The evaluation of each individual is realized by the simulation of the *TEGL* model using the following fitness function which minimizes the Total Cost (TC) of the supply link:

$$Min\ TC = \sum_{k=1}^{n}[(A_k \times C_C) + (W_k \times C_S)] + (\beta \times C_T)$$

with $A_k = y_r(k) - z(k)\ \forall k = 1,..,n$, where $y_r(k)$ is the real arrival date of the kth due product. Then, A_k is the advanced delivery time of the same product. The amount of time W_k is the waiting time of the kth due product at the supplier. In other words, the difference time between the date of its production end and the date of its transport start.

The algorithm is terminated if a given number of iteration is reached or no improvement is observed for a given number of iterations. The reader can refer to our paper [10] for more algorithm details.

To test the performance of our GA, we compare its computational time (table 1) to an enumeration procedure that identifies the optimal solution. The enumeration procedure was used because there are no other algorithm for solving the problem when dealing with a fitness function which contains the holding cost at the supplier and the holding cost at the customer in addition to the transport cost.

Number of due products	Method	
	GA	ENUM.
10	8s 632ms	16ms
20	8s 625ms	1s 46ms
30	11s 254ms	11h 41mn
40	11s 125ms	> 2 days
100	11s 406ms	--
200	13s 984 ms	--

Table 1: Average run time for enumeration methods vs. EA.

It is interesting to note the difference in the running time of the GA as opposed to the enumeration procedure. As the table shows, as the problems become larger, the GA quickly becomes much faster than the enumeration procedure. For example, for an average *n=30* problem, the EA took 11s254ms to find a solution whereas the enumeration procedure took 11h41mn.

To verify that the GA performs well in term of achieving an optimal, or near optimal, solution, several experiments with various *GA* parameters (population size, number of generations and crossover probability) was established. We considered also several values of n, t_i, t_r, dt and c. During each experiment (table 2), the *GA* is simulated 100 times in the aim to compare its performance and robustness. We limited these experiences to $n \le 40$ because beyond this problem size, it is impossible to check the optimality using the enumeration procedure (too long).

Test number	1	2	3	4	5	6
n	20	20	20	30	30	40
Optimal fitness Value	498	72	0	10 78	21	3
Percentage to achieve optimum (%)	95	82	98	89	91	98

Table 2: Genetic algorithm performance

The *GA* provides the optimal solution in more than 82% of cases. Indeed, in test number 3 and 6, the *GA* achieves the optimal solution in 98% of simulations. A *B & B* procedure, that takes much time, has been used to check the optimality for small problem sizes.

A comparison between the TEGL simulations results of the supply link in case of the optimal sequence (of lots) and another sequence is presented hereafter (figure 5). The optimal sequence for this problem is $Seq = \{l_1 = 10, l_2 = 10, l_3 = 5, l_4 = 5\}$ with an advanced time equal to 638 units of time. The second sequence is:

$Seq = \{l_1 = 8, l_2 = 6, l_3 = 2, l_4 = 5, l_5 = 5, l_6 = 4\}$ with an advanced time equal to 710 units of time.

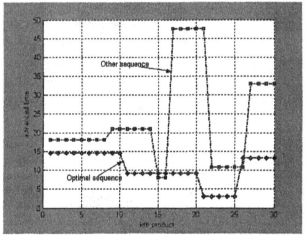

Figure 5: Comparison of average advanced times of deliveries.

We notice that some products arrive at their due date whereas the others arrive before their due dates. Therefore, we ensure that the JiT criterion is widely respected when using the (max,+) algebra control.

6. CONCLUSION

A (max,+) algebra approach that allows modelling and control of the transport of lots (of goods) in a supply link is proposed. The modelling approach is developed based on timed event graphs (*TEGs*) and residuation theory. This new model is developed by enhancing the *TEG* with a new transition that we call transition in lots.

This approach is illustrated with a supply link. We developed a dater model which is a linear state equations system under state constraints, and we computed the JiT control for the lots delivery problem when sequences of lots are given. The application also illustrates that our model using simulation based-performance evaluation and optimization approach can solve supply chain issues such as evaluation and optimization of delays of deliveries. Moreover, the proposed approach can be applied to larger *SC* that has more than two production stages. Further work includes more complex transport systems with fleets of several vehicles and other costs.

REFERENCES

[1] F. Bacceli, G. Cohen, G.J. Olsder, J. P Quadrat. "Synchronization and linearity, an algebra for discrete event system". Wiley, New York, 1992.

[2] F. Balduzzi, Has. Giua and G. Menga. "First-order hybrid Petri net: In model for optimisation and control" IEEE trans. On Rob. And Aut. 16 (4):382-399, 2000.

[3] J. G. Braker. "Max-algebra and analysis of the time-table dependant transportation networks" ECC, July, France, 1991.

[4] A. Chen, L. Amodeo, F. Fallen. "Modelling and performance evaluation of supply chain with batch deterministic and stochastic Petri net," MOSIM, April 25-27, Troyes France, 2001.

[5] G. Cohen, P. Moller, J.P. Quadrat, M. Viot. "Algebraic tool for the performance evaluation of discrete event systems". IEEE, n° 1, pp. 39-58, 1989.

[6] R. Cuninghame-Green. Minimax Algebra. Number 166. Springer Verlag, Berlin, 1979.

[7] R. David, H. Alla. "Du Grafcet au réseaux de Petri". Hermes, Paris, 1997.

[8] I. Demongodin. "Generalized batches Petri nets: Hybrid model for high speed systems with variable delays". DEDS: Theory and Applications, Vol. 11, no. 1, pp. 137-162, 2001.

[9] I. Elmahi, O. Grunder, A. Elmoudni, "A Max plus algebra approach for modeling and control a supply chain". IEEE CCA. June, Istanbul, Turkey, 2003.

[10] I. Elmahi, O. Grunder, A. Elmoudni, "A genetic algorithm approach for the batches delivery optimization". IEEE ICNSC. March, Taipei, Taiwan, 2004.

[11] J. Hahm, C. A. Yano, "The economics lot delivery scheduling problem: The single item case". Journal of production Economics, n° 28, pp. 235-252, 1992.

[12] M. A. Hoque, Goyal, "An optimal policy for a single-vendor single-buyer integrated production-inventory system with capacity constraint of the transport equipment". JPE.65, pp. 305-315, 2000.

[13] J.B. Kim, K.H. Kim, "Determining load pattern for the delivery of assembly components under JIT systems". JPE, 77, 25-38, 2002.

[14] G. J. Olsder, Subiono. M. Mc Gettric, "On large scale max-plus algebra model in railway systems". 26ième école de printemps d'informatique. Algèbre Max-Plus et application en informatique. Mai. Ile de Noirmoutier, Vendée, pp. 177-192, France, 1998

[15] B. D. Schutter, R. D. Vries and B. D. Moor. "On max-algebraic models for transportation networks", WODES. Tech. rep., SISTA, Liège, Belgium, 1998.

[16] Van der Aalst, W. M. P., "Timed coloured Petri nets and their application to logistics". Ph-D thesis, Technical University of Einddhoven, 1992.

[17] N. Viswanadaham, N. R. S Srinivasa Raghavan. "Performance analysis and design of supply chain: a Petri net approach," Newspaper of the Operational Research Society, 51 (10): 1158-1169, 2000.

Proposal for The Optimization of Feed Forward Structure Neural Networks Using Genetic Algorithms

Joldeş Remus "1 Decembrie 1918" University of Alba Iulia, Romania, rjoldes@uab.ro

Corina Rotar "1 Decembrie 1918" University of Alba Iulia, Romania, crotar@uab.ro

Abstract: Artificial neural networks (RNA) are an interesting approach in solving some difficult problems like pattern recognition, system simulation etc. The artificial neural networks have the specific feature of "storing" the knowledge in the synaptic weights of the processing elements (artificial neurons). There are a great number of RNA types and algorithms allowing the design of neural networks and the computing of weight values. In this paper we present several results related to optimization of feed-forward neural networks structure by using genetic algorithms. Such a network must satisfy some requirements: it must learn the input data, it must generalize and it must have the minimum size allowed to accomplish the first two tasks.

Keywords: feed/forward neural networks, layers, backpropagation, genetic algorithms.

1. INTRODUCTION

Artificial neural networks (RNAs) are an interesting approach in solving some difficult problems like pattern recognition, system simulation, process forecast etc. The artificial neural networks have the specific feature of "storing" the knowledge in the synaptic weights of the processing elements (artificial neurons). There are a great number of RNA types and algorithms allowing the design of neural networks and the computing of weight values.

In this paper we present several results related to optimization of feed-forward neural networks structure by using genetic algorithms. Such a network must satisfy some requirements: it must learn the input data, it must generalize and it must have the minimum size allowed to accomplish the first two tasks. The processing element of this type of network is shown in figure 1.

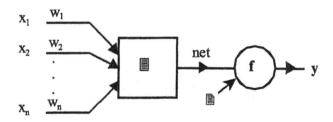

Fig. 1. The model of artificial neuron used

In this figure x_1, x_2,... x_n are neuron's inputs, w_1, w_2, ...w_n are interconnection weights, 2 is neuron's threshold, f() is activation function and y is neuron's output. We shall denote $\mathbf{x}=[x_1, x_2, ..., x_n]^T$ input vector, $\mathbf{w}=[w_1, w_2, ..., w_n]^T$ synaptic connections vector, 2 thresholds vector,

$$net = \sum_i w_i x_i = \mathbf{w}^T \mathbf{x} \qquad (1)$$

The output of the neuron may be written:

$$y = f(net-2) = f(\mathbf{w}^T \mathbf{x}-2) \qquad (2)$$

In practical applications, the neural networks are organized in several layers as shown in figure 2.

2. DESIGN APPROACHES FOR FEED-FORWARD NEURAL NETWORKS

Implementing a RNA application implies three steps [3]:
- Choice of network model;
- Correct dimensioning of the network;
- The training of the network using existing data (synaptic weights synthesis).

The present paper is dealing with feed-forward neural network so we concentrate on steps 2 and 3.

Network dimension must satisfy at least two criteria:

- The network must be able to learn the input data ;
- The network must be able to generalize for similar input data that were not in training set.

The accomplishment degree of these requirements depends on the network complexity and the training data set and training mode. Figure 3 shows the dependence of network performance in function of the complexity and figure 4 shows the same dependence of the training mode.

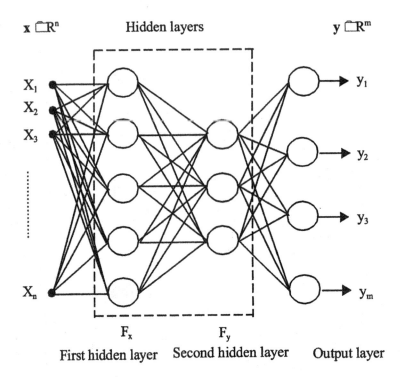

Fig. 2. Multilayer feed-forward neural network.

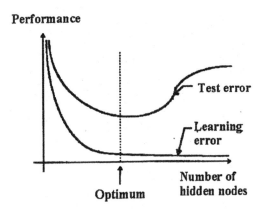

Fig. 3. Performance's dependence of network complexity. Source: [3], p. 74.

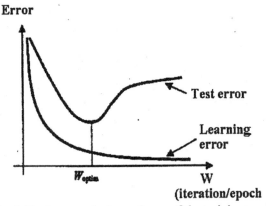

Fig. 4. Performance's dependence of the training mode. Source: [6], p. 40.

One can notice that the two requests are contradictory and that the establishment of the right dimension is a complex matter. We generally wish to diminish the complexity of the model, which leads to a better generalization, to an increased training speed and to lower implementation cost.

The designing of a feed-forward structure that would lead to the minimizing of the generalization error, of the learning time and of the network dimension implies the establishment of the layer number, neuron number in each layer and interconnections between neurons. At the time being, there are no formal methods for optimal choice of the neural network's dimensions.

The choice of the number of layers is made knowing that a two layer network (one hidden layer) is able to approximate most of the non linear functions demanded by practice and that a three layer network (two hidden layers) can approximate any non linear function. Therefore it would result that a three layer network would be sufficient for any problem. In reality the use of a large number of hidden layers can be useful if the number of neurons on each layer is too big in the three layer approach.

Concerning the **dimension of each neuron layer** the situation is as follows:

- Input and output layers are imposed by the problem to be solved;

- The dimension of hidden layers is essential for the efficiency of the network and there is a multitude of dimensioning methods (table 1).

Table 1. Source: [3] p. 86.

Hidden layer dimensioning methods		Type of method
Empirical methods		Direct
Methods based on statistic criteria		Indirect
Ontogenic methods	Constructive	Direct
	Destructive	Direct
	Mixed	Direct
	Based on Genetic Algorithms	Direct

Most methods use a constructive approach (one starts with a small number of neurons and increases it for better performances) or a destructive approach (one starts with a large number of neurons and drops the neurons with low activity).

As a general conclusion, there are a multitude of methods for feed-forward neural networks designing, each fulfilling in a certain degree the optimization requests upper presented.

3. THE PROPOSED APPROACH

As we have previously seen, optimal designing of feed forward neural networks is a complex problem and three criterions must be satisfied:

- The network must have the capacity of learning
- The network must have the capacity of generalization
- The network must have the minimum number of neurons.
Next we shall present one method of optimizing the network's structure, by the use of Genetic Algorithms (GA). Genetic algorithms proved their efficiency in solving optimization problems and, moreover, many evolutionary techniques have been developed for determining multiple optima of a specific function.

According to the evolutionary metaphor, a genetic algorithm starts with a population (collection) of individuals, which evolves toward optimum solutions through the genetic operators (*selection, crossover, mutation*), inspired by biological processes [2].

Each element of the population is called chromosome and codifies a point from the search space. The search is guided by a fitness function meant to evaluate the quality of each individual. The efficiency of a genetic algorithm is connected to the ability of defining a "good" fitness function. For example, in real function optimization problems, the fitness function could be the function to be optimized. The standard genetic algorithm is illustrated in the figure 2.

Multicriteria optimization

The multicriteria optimization (multi-objective optimization or vector optimization) may be stated as follows: the vector:

$$\overline{x^*} = \begin{pmatrix} x_1^* \\ x_2^* \\ \vdots \\ x_n^* \end{pmatrix},$$

that satisfies the constraints:

$$g_i\left(\overline{x}\right) \geq 0 , i = 1,2,\ldots,m \qquad (3)$$

$$h_i\left(\overline{x}\right) = 0 , i = 1,2,\ldots,p \qquad (4)$$

and optimizes the vector function:

$$\overline{f}\left(\overline{x}\right) = \begin{pmatrix} f_1\left(\overline{x}\right) \\ f_2\left(\overline{x}\right) \\ \vdots \\ f_k\left(\overline{x}\right) \end{pmatrix} \qquad (5)$$

must be determined, where $\overline{x}$ represents the vector of decision variables.

In other words we want to find from the set F of the values that satisfy (1) and (2), the particular values $x_1^*, x_2^*, \ldots, x_n^*$ that produce optimal values for the objective function. In this situation the desired solution would be $\overline{x^*}$, but there are few situations in which all the $f_i\left(\overline{x}\right)$ have minimum (or maximum) in F in a common point $\overline{x^*}$. It is necessary to state what "optimal solution" is.

Pareto optimum

The most popular optimal approach was introduced by Vilfredo Pareto at the end of XIX century: One vector $x^* \in F$ of decision variables is *Pareto optimal* if there isn't another vector $x \in F$ with properties:

$$f_i\left(x\right) \leq f_i\left(x^*\right), \text{ for every } i = 1,2,\ldots,k, \text{ and}$$

$$\exists j = 1,2,\ldots,k, \text{ so that } f_j\left(x\right) < f_j\left(x^*\right).$$

This optimality criterion will supply a set of solutions denoted *Pareto optimal set*. The vectors x^* corresponding to the solutions included in Pareto set will be denoted non-dominants. The area from F formed by the non-dominant solutions makes *Pareto front*.

Evolutionary approach

For solving the above problem we used two evolutionary approaches: one approach that doesn't use the concept of Pareto dominance when evaluating candidate solutions (non Pareto method) - weights method - and one Pareto approach recently developed inspired by endocrine system.

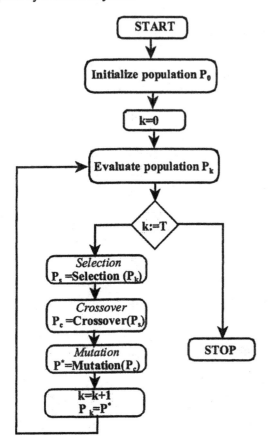

Fig. 2. Standard Genetic Algorithm

Non-Pareto method (Weights method)

The technique of combining all the objective functions in one function is denoted the *function aggregation method* and the most popular such method is the *weights method*.

In this method we add to every criterion f_i a positive sub unity value w_i denoted weight. The multicriteria problem becomes a unicriteria optimization problem

If we want to find the minimum, the problem can be stated as follows:

$$\min \sum_{i=1}^{k} w_i f_i\left(\overline{x}\right) \ (1), \ 0 \le w_i \le 1, \qquad (6)$$

$$\text{usually} \sum_{i=1}^{k} w_i = 1 \qquad (7)$$

One drawback of this method is the determination of weights if one don' know many things about the problem to be solved.

Pareto method

We have applied a recently developed method inspired by the natural endocrine system ([8], [9], [10]). The main characteristics of this method are:

1. One maintains two populations: one active population of individuals (hormones) H, and one passive population of non-dominant solutions T. The members of passive T population act like an elite collection having the function of guiding the active population toward Pareto front and keeping them well distributed in search space.

2. The passive T population doesn't suffer any modifications at individuals' level through variation operators like recombining or mutation. In the end the T will contain a previously established number of non-dominating vectors supplying a good approximation of the *Pareto front*.

3. At each T generation the members of the H_t active population are divided in *st* classes, where *st* represents the number of tropes from the current T population. Each hormone class is supervised by a correspondent trope. The point is that each h hormone from H_t is controlled by the nearest a_i trope from A_t.

4. Two individuals' evaluation functions are defined. The value of the first one for an individual represents the number of individuals of the current population that are dominated by it. The value of the second function for an individual represents the agglomerate degree from the class corresponding to that particular individual.

5. The recombining selection takes into consideration the values of the first function as well as the values of the second. The first parent is selected through competition taking into consideration the values of the second performance function. The second parent is selected proportional from the first parent's class taking into consideration the second performance function.

4. EXPERIMENTAL RESULTS

We have applied the optimization method presented for a feed-forward neural network synthesized to approximate the function illustrated in figure 5. The training and testing data were different; even more the data used for testing were outside training set.

The chromosomes will codify the network dimension (number of layers, number of neurons on layer) and the fitness function will integrate the training error (after a number of 100 epochs), the testing error (as a measure of the generalization

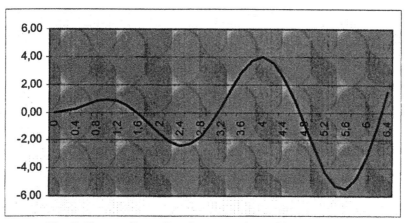

Fig. 5. The non linear function approximated by the network

capacity) and the entire number of neurons of the network. To simplify we considered only two or three layers networks. The maximum number of neurons on each layer is 10.There is one output neuron. The results obtained are shown in tables 2 and 3.

Table 2.

Weights method	
Population dimension	20
Number of generations	100
Criteria	F1 – training error, F2 – test error, F3 – inverse of neuron number
W1	1
W2	1
W3	0.005
Solutions:	(1,4), (2,8)

The structure obtained by the use of the optimizing algorithm was tested in Matlab and the results are shown in figures 6 and 7.

Table 3.

Technique inspired by endocrine system	
Population dimension	20
Number of generations	100
Criteria	F1 – training error F2 – test error
Solutions:	(2,8)

5. CONCLUSIONS

The optimal dimensioning of a feed-forward neural network is a complex matter and literature presents a multitude of methods but there isn't a rigorous and accurate analytical method.

Our approach uses genetic computing for the establishment of the optimum number of layers and the number of neurons on layer, for a given problem. We used for illustration the approximation of a real function with real argument but the method can be used without restrictions for modeling networks with many inputs and outputs.

We intent to use more complex fitness functions in order to include the training speed.

References:

1. Bäck T., *Evolutionary Algorithms in Theory and Practice*, Oxford University Press, 1996.
2. Coello C. A.. C., *An Updated Survey of Evolutionary Multiobjective Optimization Techniques : State of the Art and Future Trends* , In *1999 Congress on Evolutionary Computation*, Washington, D.C., 1999.
3. Dumitraş Adriana (1997): *Proiectarea reţelelor neuronale artificiale*, Casa editorială Odeon, 1997.
4. Dumitrescu D., Lazzerini B., Jain L.C., Dumitrescu A., *Evolutionary Computation*, CRC Press, Boca Raton London, New York, Washington D.C., 2000.
5. Goldberg D.E., *Genetic Algorithms in Search, Optimization, and Machine Learning*, Addison-Wesley Publishing Company, Inc., 1989.
6. Năstac Dumitru Iulian: *Reţele neuronale artificiale. Procesarea avansată a datelor*, Editura Printech, 2002.
7. Kavzoglu Taskin: *Determining Optimum Structure for Artificial Neural Networks*, Proceedings of the 25-th Annual Technical Conference and Exhibition of Remote sensing Society, Cardiff, UK, pp. 675/682, 8-10 September 1999.

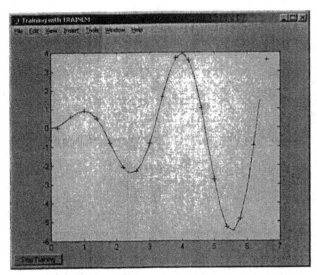

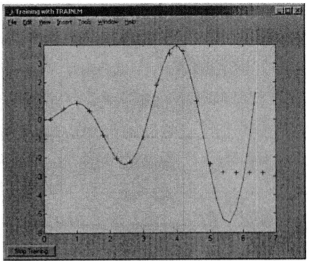

Fig. 6. Output of the 2:8:1 network (+) compared with desired output (line)

Fig. 7. Output of the 1:4:1 network (+) compared with desired output (line)

8. Rotar C., Ileană I., *Models of Population for Multimodal Optimization. A New Evolutionary Approach*, Proc. 8th Int. Conf. on Soft Computing, Mendel2002, Czech Republic, 2002.

9. Rotar Corina, *A new evolutionary algorithm for multicriterial optimization based on endocrine paradigm*, ICTAMI 2003, publicat in Acta Universitatis Apulensis, seria Matematics-Informatics.

10. Rotar Corina, Ileana Ioan, Risteiu Mircea, Joldes Remus, Ceuca Emilian, *An evolutionary approach for optimization based on a new paradigm, Proceedings of the 14th international conference* Process Control 2003*, Slovakia*, 2003, pag. 143.1 – 143.8.

ELSEVIER
IFAC
PUBLICATIONS
www.elsevier.com/locate/ifac

FORMALISATION OF QUANTITATIVE UML MODELS USING CONTINUOUS TIME MARKOV CHAINS

Nawal Addouche, Christian Antoine, Jacky Montmain

URC CEA/EMA-LGI2P Research Center of Ales School of Mines
Parc scientifique - Georges Besse, 30035 Nîmes, Cedex 01 France
nawal.addouche@ema, christian.antoine@ema.fr, jacky.montmain@ema.fr

Abstract: In this paper, a quantitative modelling with UML is presented and a translation from concurrent extended UML statecharts into continuous time Markov chains (CTMCs) is proposed. These are widely used in the context of performance and reliability evaluation of various systems. Our aim is to carry out a probabilistic model checking of properties related to dependability of probabilistic systems. A semantics based on stochastic clocks is used to easily translate from UML statecharts augmented with stochastic time to CTMCs. Temporal probabilistic properties related to system dependability are specified with continuous stochastic logic (CSL).

Keywords: Probabilistic systems, UML, CTMCs, Dependability properties, Probabilistic model checking.

1. INTRODUCTION

Recently, the analysis of functional system requirements in combination with quantitative aspects of system behaviour have come into focus. Several approaches have already been explored to introduce a quantitative information in the dynamic UML models. A stochastic extension of UML statechart diagrams is proposed in (Gnesi *et al.*, 2000). It is based on a set of stochastic clocks which can be used as guards for transitions. The clock value is given by a random variable with specified distribution function. A probabilistic extension of UML statecharts is presented in (Jansen *et al.*, 2002). The probabilistic UML statecharts describe probabilistic choice and nondeterministic system behaviour. Their formal semantics is given in terms of Markov decision process as defined in (Kwiatkowska, 2003). To evaluate system performance, other approaches are also proposed. Dynamic UML models are formalised with stochastic Petri nets in (King and Pooley, 1999; Merseguer and Campos, 2002) or with stochastic process algebra in (Pooley, 1999; Canevet *et al.*, 2002). In order to take into account either real-time constraints and probabilistic behaviour (undesired signals, lost signals, etc.), a profile called DAMRTS (Dependability Analysis Models for Real-Time Systems) is defined by (Addouche *et al.*, 2004). A translation from UML statecharts used in this profile to probabilistic timed automata is also proposed.

In this paper, another approach intended to quantitative dependability analysis is presented. Adding to evaluate system performance using the several existing tools, the formalisation of combined collaboration statechart diagrams using continuous time Markov chains contains tree fault information. Introduction of this type of information allows to verify formally properties related to system dependability. In our works, we focus on UML statechart diagrams, which allow to describe dynamic aspects of system behaviour. In section 2, semantics of extended UML statecharts using stochastic clocks is presented. Dependability information excerpted from faults trees analysis (failures with their causes) are included in these statecharts. A set of extended UML statecharts describing the behaviour of an assembly chain is presented in section 3. In this example, the activities duration are considered as distributed exponentially. That make possible to translate UML models into continuous time Markov chains as given in section 4. To verify formally temporal probabilistic properties, we finally propose to use the probabilistic model checker Prism. In section 5, the formal model which represents the chain example is given with CTMCs. Dependability properties are specified with CSL and some results are presented before the conclusion.

2. EXTENDED UML STATECHARTS

In UML, each class of the class diagram has an optional statechart which describes the behaviour of its instances (the objects). This statechart receives events from other ones and reacts to them. The reactions can include the sending of new events to other objects and the execution of internal methods on the object. Communications between system components are generally modelled as events. In proposed UML statecharts, exchanged signals, orders and random events (e.g. undesired and lost signals) are represented as events. Syntax of UML statecharts defined in the standard UML of the (OMG, 2003) is extended with integrating rates on transitions. A dependability information related to faults trees analysis are also introduced as defined in (Addouche et al., 2004). Semantics of extended statecharts is presented in section 2.2.

2.1 Syntax

This section describes informal interpretation of extended UML statecharts. The graphical representation is based on a set of nodes and a set of edges. An edge is presented by the following syntax:

Edge: = Event [Guard] / Action.
Event: = Event name.
Guard: = Boolean Expression.
Action: = Operation name[Rate].

Event represents either received signals or orders or random events. Guard is a boolean expression that represents AND-composition and OR-composition states of different objects. The compositions can be a particular qualitative information which represent the causes of undesirable events. In this case, when guard is TRUE, a mode of failure appears on the system. This type of information is available in dependability analysis based on faults trees. Action expresses operation execution or sending messages to other objects. The considered time on duration is stochastic and exponentially distributed.

2.2 Semantics

The semantic model used in our extended UML statecharts is inspired by stochastic automata defined in (D'Argenio et al., 1999). These are a variant of timed automata proposed by (Alur and Dill, 1994). Stochastic automata are defined with a set of random clocks and a clock setting function that determines for each state, called location, with clocks are set to which value. The transitions of stochastic automata, called edges, are labelled by an action and a finite set of clocks. A stochastic automaton can perform a transition from location s to location s' labelled by action a and clock set C by performing action a as soon as all the clocks in the set C have expired. A global time is assumed and all clocks are decreased at the same speed. Immediately after the transition takes place all clocks associated to s' by the clock setting function are randomly set according to their probability distributions.

Stochastic automata has been chosen by (D'Argenio et al., 1999) to express semantics of stochastic process algebra and by (Jansen et al., 2003) to extend UML statecharts with randomly varying duration associated to arbitrary probability distribution (e.g. exponential, uniform, etc.). In our case, the random variables are exponentially distributed. Translation from proposed UML statecharts to continuous time

Markov chains requires this choice. UML statechart diagrams are a variant of classical statecharts of (Harel, 1987). Semantics of UML statecharts is defined in (Latella et al., 1999). The associated extension is defined by adding to each action name, a distribution function that determines the stochastic timing of the transition. A single extended UML statechart consists of :

- A finite set of nodes with a tree structure, described by the function $children_i$: $Nodes_i \rightarrow \mathbf{P}(Nodes_i)$
- A finite set of events,
- A set of guard expressions. Guard expressions are boolean combinations,
- A set of actions. A set of clocks is defined on actions. We consider that actions are executed in their correspondent states,
- A E-edge, an extended edge with event e, guard g, action a and set of clocks is denoted as an arrow $\xrightarrow{e[g]/a,C}$. C is as set of stochastic clocks assigned to action a.

As in the case of stochastic automata, when entering a state, the clocks listed in the state are set to values which are determined by random variables associated to the clocks. We also assume a global time and all clocks decreased at the same time.

3. EXAMPLE : ASSEMBLY CHAIN OF MICRO-MOTORS

This example presents an automated chain for assembly of electrical micro-motors. It is excerpted from a European project named PABADIS (Plant Automation Based on DIstributed Systems) and presented in (Lüder et al., 2004). This one deals with flexible and reconfigurable system designed for production of different types of micro-motors. Fig 1 presents the controlled system. Micro-motors consist to stators and rotors. The firsts are transported to assembly robots, on pallets via a conveyor system and seconds are available into stocks near each robot. A set of pallets containing stators moves along the conveyor belt. These are detected by pallet sensors PSi at different levels of the conveyor system.

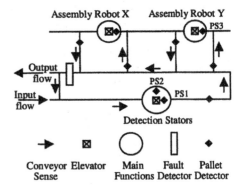

Fig. 1. Assembly chain

When assembly of micro-motors is completed, the pallets then move into a fault detection station where a camera detects the possible assembly faults. Set of PLC (Programmable Logic Controller) and PC composes the control system. Let us consider the mode of failure "*Assembly fault*". Among the causes of this mode of failure, there are undesired signals sent by the sensors, undesired orders sent by the controller and material failures of elevator.

To represent dynamic aspects of the system, extended UML statecharts and collaboration diagrams are used. Combination of these two diagrams allows to represent all system interactions. Indeed, the collaboration model describes external interactions between objects whereas UML statecharts diagrams represent how an instance of a class reacts to an event occurrence.

3.1. Collaboration diagram

Interactions between objects of classes are presented in fig 2. Exchanged messages describe signals sent from *Sensors* (S.PPi for *sensor* i) and *Fault detector* objects (S.Fault) to *controller*. They also represent orders sent from *Controller* to *Robot* and *Elevator* objects. Our example presents a distributed system such that several PLC interact to control the system functioning. To simplify, one *controller* object is presented in the collaboration diagram.

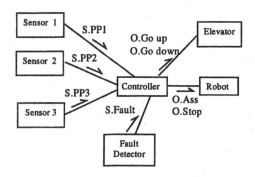

Fig. 2. Collaboration diagram

3.2 Statechart diagrams

Robot and *Controller* objects behaviour are respectively modelled as given in fig 3 and 4. In *Robot* statechart, orders are modelled as events. In the example: "O.Ass[PS1.Ds OR PS2.Ds]/ Assembly()[0,03] ", guard expresses that the edge is enabled if one of sensors PS2 or PS3 is in degraded state *Ds*. The rate of executing assembly tasks represents constant of exponential distribution function associated to the clock C=EXP (0,03). The guard "S.Active AND E.Active" represents the condition to leave the state *Emergency stop*: sensors and elevator must be in the state *Active* of their respective UML statecharts. Among the malfunctions of controller, sending of undesired

orders or lost orders are modelled in *Controller* statechart as a fault which arrives randomly.

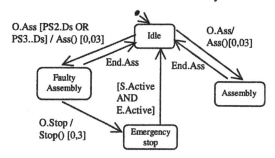

Fig. 3. Robot statechart

In the example "UO/ Send.O ()[0.001]", an undesired order (UO) can arrives randomly. After which an order with a rate of 0.001 is sent to the corresponding component of controlled system. The received sensor signals are presented as events and the sending of orders as actions. The edge "SPP1/ O.Go up()", expresses that when PS1 detects a pallet, the order *go up* is sent from controller to elevator.

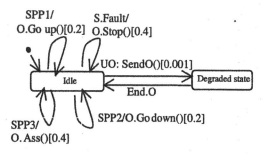

Fig. 4. Controller statechart

4. TRANSLATION OF UML MODELS WITH CTMCs

System behaviour is composed by a set of scenario. A scenario is composed by signal, order and executing tasks. It is presented with a set of combined collaboration and extended statecharts diagrams. Formalisation of this behavioural UML models is given as follow:

- Each scenario is modelled by a set of combined collaboration extended statecharts diagrams. Number of combined diagrams varies according scenario,
- At any time the combined diagram must have each of its objects in exactly one state. These combinations of active states are called "marking" and represent in a continuous time Markov chains the reachable states,
- It is assumed that the rates associated to sent messages are exponentially distributed.

Let us consider the following scenario : when a pallet arrives in the detection station and sensor PS2 send an undesired signal to controller. This make erroneous, position stator data sent to the controller.

Then a fault assembly will appear and will be detected in fault detection station.

Translation of behavioural UML models to its correspondent continuous time Markov chain is explained in fig 5 and 6 such that first one presents a scenario of assembly chain example described above with behavioural UML models and second one shows a possible translation to CTMCs.

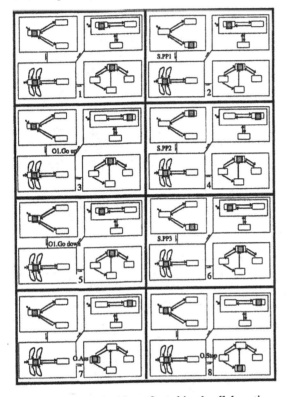

Fig. 5. Direct marking of combined collaboration statechart diagrams

4.1 Application

The fig 5 describes a faulty assembly due to sending of undesired signal by sensor 2. This scenario is described by eight combined diagrams. In each one from up to down and from right to left are respectively presented, statecharts of *sensor i, elevator i, controller* and *robot* objects. To simplify fig 5, *Default detector* object is deliberately omitted, one *sensor* statecharts and one elevator statecharts are shown, index (i) used on signals and orders gives which sensor or elevator is modelled in combined diagrams. Transition from a marking to another results of exchanging signals or orders between objects. Active states of each marking are grey tented in the figure.

The scenario of faulty assembly begins in marking 1 and takes end in marking 8. From marking 1 to marking 5 tasks of stators detection station (go up, go down) are executed with undesired signal sent from sensor 2 to controller. Malfunctioning of sensors 2 leads erroneous detection of stator position

on pallet. Consequently, robot do not put rotors in good pallet compartments. From marking 5 to marking 8, the faulty assembly and the emergency stop are modelled.

If we assume an initial marking, such as all objects are in *Idle* state and *elevator* in *Down* state, it is possible to derive all markings by following the exchanging messages between objects. The CTMC reachable states are thus formed. Fig 6 shows the continuous time Markov chain derived from model of fig 5. The numbers on reachable states are related to numbers on markings of fig 5.

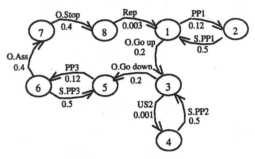

Fig. 6. Continuous time Markov chain

4.2 Principle of translation

A continuous time Markov chain is composed by set of states and set of edges subject to an exponentially distributed random delay. Each state of CTMC represents a combination of active states of extended UML statecharts. A CTMC edge is fired when an action is executed by an extended UML statechart. Rates associated to actions are directly translated as rate of firing edges. In fig 6, the rates are related either to orders sent by controller (O.Stop, O.Ass, etc.) or to signals sent by sensors (SPP1, SPP2 and SPP3). They are also related to stochastic events such as random passage of pallets (PP1 and PP3), operator intervention or repairing system (Rep). Guards are used to model AND-composition and OR-composition of degraded or normal states. Abnormal or normal functioning scenarios are obtained according to whether these guards are true or false. So, Guards are implicitly expressed in the CTMC.

4.3 Comments

The global system behaviour is obtained from the set of continuous time Markov chains related to different scenario. Combination of these models gives global continuous time Makov chains. This method represents an intuitive proposition for formalisation. It is not applied in this paper because it is not necessary to construct it while Prism tool is used for probabilistic model checking. This one is based on a modular modelling of the system. It does not require a global CTMC but rather a set of CTMCs, where each represents one "module" of the system, as given in section 5.1.

5. PROBABILISTIC MODEL CHECKING

The model checking is an algorithmic method designed to check if a system satisfies the specifications. A model checker is a software tool which introduces a system model (e.g. automata) and specifications (e.g. logic formula), as well as the return *yes* or *no* depending on whether the system satisfy or not the specifications. In a probabilistic model checking, the return *yes* or *no* are replaced by probabilities evaluations. The probabilities are introduced in two cases:

- The first case is related to probabilistic systems, i.e. which include probabilistic information such as the mean time between failures of a material device or in transmission protocols, the probability of message transmission,
- The second case concerns non-probabilistic systems whose complexity makes exhaustive verification practically impossible. So, a probabilistic description is used for unavailable information or for the information whose the too complex representation would be not exploitable.

To verify dependability properties, the model checker Prism is adopted (Kwiatkowska *et al.*, 2002). This tool is designed for analysis of probabilistic models and supports various models such as Markov decision processes, discrete time Markov chains and continuous time Markov chains. Prism is a tool developed at the University of Birmingham which supports the model checking described before. The tool takes as input a description of a system written in Prism language. It constructs the model from this description and computes the set of reachable states. It accepts specification in either the logic PCTL or CSL (Kwiatkowska, 2003) depending on the model type. It then determines which states of the system satisfy each specification.

5.1 Model analysis

In order to model check a system with Prism tool, it must be specified in the Prism language, based on the Reactive Modules formalism of (Alur and Henzinger, 1996). This formal model is designed for concurrent systems and represents synchronous and asynchronous components in a uniform framework that supports compositional and hierarchical design and verification.

The fundamental components of Prism language are *modules* and *variables*. A system is composed of a number of modules which can interact with each other. A module contains a number of *local variables*. The values of these variables at any given time constitute state of the module. Global state of the system is determined by local states of all modules.

The translation from UML statechart diagrams to reactive modules is given as follow :

- The *modules* are defined for each UML object,
- The states of extended UML statecharts are modelled by one or several local variables (integer or boolean),
- The signals, orders and random events are given by one local boolean variable that determine their presence,
- The rates associated to actions (as given in our extended statecharts) are modelled by a set of constants proposed in reactive modules to assign stochastic information to the transition,
- The guards of extended UML statecharts are expressed with constraints. These are predicates over the local variables of other modules and are proposed in Prism language in order to condition the transition firing,
- The actions of UML models are also defined as actions in reactive modules.

The behaviour model of assembly chain is proposed such that each presented object in collaboration diagram of section 3.1 is taken into account. Part of the model is presented in fig. 7.

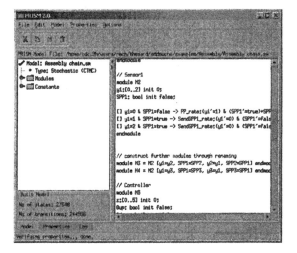

Fig. 7. Prism interface: Editing a model

5.2. Properties in Prism specification language

Some probabilistic properties related to our example are presented. Their informal specifications is given as follow:

Property 1: "In the long run, the probability the robot carries out a faulty assembly is less than 1%".

Property 2: "In initial state, the probability that the robot remains in emergency stop until the elevator and the sensors are reactivated is at least 0,95".

Property 3: "Elevator remains in down position less than k units of time until the sensor1 detects a pallet presence with a probability $\geq p$ ".

These dependability requirements are formally specified with the temporal logic CSL. The following are their Prism specification language.

Property 1: S<0.1 [(r=2)]

Property 2: "init" ⇒ P>0.95[(r=3) U (e=0) & (y1=0)
      & (y2=0) & (y3=0)]

Property 3: P=? [(e=0)U$^{\geq k}$ SPP1= true]

5.3. Experimental results

The results of our experiments are shown in fig. 8. Dependability properties 1 and 2 are verified (true). The verification of property 3 is presented with a curve. The CSL requirement are evaluated for increasing time points k and the boundary probabilities p at which the requirement turns from being true to being false are calculated. We plotted a graph generated by Prism where a pair (t, p) above a plot the requirement is FALSE, while for pairs below it is TRUE.

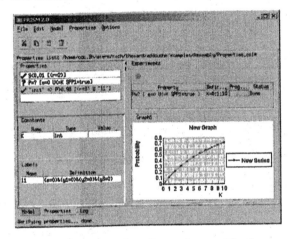

Fig. 8. Prism interface: Properties specification

6. CONCLUSION

The paper presents UML models to analyse the system dependability. The aim is to verify formally dependability properties of probabilistic systems. Extended UML statecharts are proposed with semantics related to stochastic automata. Duration of activities are expressed with stochastic clocks. The time is exponentially distributed, that make possible translation to continuous time Markov chains. As example, an assembly chain is described with extended statecharts. A translation method from combined collaboration statechart diagrams to CTMCs is proposed. Using the Prism tool, some dependability properties related to the example are specified with the temporal logic CSL and verified by probabilistic model checking.

REFERENCES

Addouche, N, C. Antoine and J. Montmain. (2004). UML Models for Dependability Analysis of Real-Time System. In: *Proc of SMC'04*, The Hague, The Netherlands.

Alur, R and D. Dill. (1994). A theory of timed automata. *Theoretical Computer Science*, **126(2)**, 183-235.

Alur, R and T. Henzinger. (1996). Reactive Modules. In: *Proc of LICS'96*, pp. 207-218, IEEE Computer Society Press, New Jersey.

Canevet, C, S. Gilmore, J. Hillston and P. Stevens. (2002). Performance modelling with UML and stochastic process algebra. In: *Proc of UKPEW'02*, pp. 16. The Univ of Glascow, UK.

D'Argenio, P.R, J-P. Katoen and E. Brinksma. (1999). Specification and Analysis of Soft Real-Time Systems: Quantity and Quality. In: *Proc. of RTSS'99*, pp.104-114. IEEE Society Press, Phoenix, Arizona, USA.

Gnesi, S, D.Latella and M. Massink. (2000). A Stochastic Extension of a Behavioural Subset of UML Statechart Diagrams. In: *Proc of HASE'00*, Albuquerque, New Mexico.

Harel, D. (1987). Statcharts: A visual formalism for complex systems. *Science of Computer Programming*. Elsevier, **8(3)**, 231-274.

Jansen, D.N, H. Hermanns and J-P Kaoten. (2002). A Probabilistic Extension of UML Statecharts: Specification and Verification. In: *Proc of FTRTFT'02*, pp. 355-374. Oldenburg, Germany.

Jansen, D.N, H. Hermanns and J-P Kaoten (2003). QoS-oriented Extension of UML Statecharts. In: *UML 2003*, (Perdita Stevens et al.), pp. 76-91., LNCS, 2863, San Fransisco, USA.

King, P and R. Pooley. (1999). Using UML to derive stochastic Petri nets models. In : *Proc of UKPEW'99*, pp. 45-56, The Univ of Bristol.

Kwiatkowska, M, G. Norman and D. Parker. (2002), "Prism: Probabilistic Model Checker", In: *Proc of TOOLS'02*, (T. Field et al), pp. 200-204, London, UK.

Kwiatkowska, M. (2003). Model Checking for Probability and Time : From Theory to Practice. In: *Proc of LICS'03*, pp. 351-360, IEEE Computer Society Press, Ottawa, Canada.

Latella, D, I. Majzik, and M. Massink. (1999), Towards a formal operational semantics of UML statechart diagrams. In: *Proc of FMOODS'99*, pp. 331-347, Florence, Italy.

Lüder, A, J. Peschke, T. Sauter, S. Deter and D, Diep. (2004), Distributed intelligence for plant automation on multi-agent systems: the PABADIS approach, *Production Planning and Control*, **15(2)**, pp. 201-212.

Merseguer, J and J. Campos. (2002). A Compositional Semantics for UML State Machines Aimed at Performance Evaluation. In: *Proc of WODES'02*, pp. 295-302, IEEE Computer Society Press, Zaragoza, Spain.

OMG (2003) Unified Modeling Language Specification. v.1.5, *OMG Document Formal / 03-03-01.*

Pooley, R. (1999). Using UML to Derive Stochastic Process Algebra Models. In: *Proc of UKPEW'99*, (J.T. Bradley and N.J. Davies), pp. 23-33. The Univ of Bristol.

PROJECTED ORIENTATION MAPS BY SPACE REDUCTION FOR INTELLIGENT MONITORING OF INDUSTRIAL PROCESSES

Justo Matheus, António Dourado, Jorge Henriques
matheusj@dei.uc.pt, dourado@dei.uc.pt, jh@dei.uc.pt

Center for Informatics and Systems
University of Coimbra Portugal
Departamento de Engenharia Informática
Pólo II Pinhal de Marrocos
3040-290 Coimbra Portugal

Abstract.. The simplified visualization of operational data in industrial plants might guarantee to the process operators take quick decisions for convenient supervision and monitoring of complex industrial systems. A space reduction technique is implemented to visualize and identify different work zones in the production of multiple grade products. The analysis of a historical database helps to create a bi-dimensional (2-D) representation of the operational regions of a Hydro Treating Process which works in 7 different modes. The space reduction technique keeps the relevant and informative geometric characteristics of the original space and builds a 2-D map, called Projected Operational Map (POM), where the operational regions of the process under a specific feed rate or load can be identified. In the POM can be detected some points where the system works away of its traditional work zone, indicating a probable abnormal operation of the process. A strategy to monitor dynamic systems based on POM is proposed in order to increase the productivity, quality and safety of the industrial plant.

Keywords: supervision and monitoring; space reduction; data mining; synthetic visualization; refinery

1. Introduction

Modern Industrial plants must produce a wide range of products to serve other plants or the consumer market. For these reasons the processes are submitted to inevitable changes in the feed rate or load, these changes leading to modifications of the operational points of the production systems. Even though the developments in the field of multivariable control and real time optimization, the transition between grades of operation is still performed by human operators, who use their knowledge and expertise about the process to find the new operating points.

Jaeckle and MacGregor (1988) developed an approach using multidimensional statistics to analyze historical operational data, for the case of a high-pressure tubular reactor producing low density polyethylene, with data generated from computer simulations. Chen and Wang proposed the use of Principal Component Analysis (PCA) as a reduction technique to analyze data collected from a refinery Fluid Catalytic Cracking process. Nomikos and MacGregor (1994) show that the PCA just performs a good bi-dimensional representation

when the 85% of the accumulative variance is concentrated in the first two principal components. Mao and Jain (1995) showed how the PCA projection overlaps classes than are linearly separable by other projection methods; according to Borg and Groennen (1997), this basically occurs because the criteria to perform the reduction of the space intends to keep the maximum amount of variance from the original data in the reduced space, without any attention to the geometric structure of the data. The term geometric structure means here the distance among the points or in general the intrinsic characteristics of the original space that will be kept in the reduced one. The concept of geometric structure will be used here in the sense that distance relationships are important (Michael Tipping (1996)).

In this work the original multidimensional space is reduced to a bi-dimensional one where a POM is represented. The Euclidean distance among the points in the original space is preserved in the 2-D space. A data base of 186 cases and 7 operational modes collected from a refinery hydro treating process is used to create the POM. The interpretation of the POM can be used to detect some operational points where the process works away of its proper zone, points that represent potential abnormal operation of the process.

The remaining part of this paper is organized as follows: Section 2 describes the process and the data characteristics. In Section 3 is presented the concept of space reduction for process monitoring. Section 4 presents the Projected Operational Map and discusses the results obtained. Section 5 concludes with the final remarks.

2. Process Description

Hydro-Treating (HDT) is a generic name for a general refinery process for treating fuels (Kerosene, Diesel, Gasoline) or lubricant base stocks at high temperatures in the presence of hydrogen and a catalyst. Our case study is a HDT process located in a lubricant plant; this process is used to eliminate the sulphur, to improve the colour and odour and to increase the resistance for oxidation for the lubricating base stocks. The relevant sections of the process are shown in Fig. 1.

The lubricant feedstock is reacted with hydrogen in the presence of a catalyst at a very high temperature ($\approx 400°$ C) and high pressure (≈ 3000 psig). Under these severe conditions, the impurities such as sulphur, nitrogen, or oxygen are removed. The HDT process produces very high quality ("semi-synthetic") lubricant bases. Due to the multiplicity of specifications for consumer market, the process has to change continuously of operation mode according to the characteristics of the lubricant base to be processed.

The data collected from the refinery historical database is composed by 189 cases, and summarizes the production of 7 different feed stocks (Table 1). The process is characterized by 8 variables (listed in Table 2) and will be reduced to a 2-D representation. The selection of these variables was made by process engineers using their knowledge and expertise about the plant.

3. Data Visualization for Process Monitoring and the POM

Process plant operators must monitor a great amount of data collected from a multitude of sensors every few seconds, minutes, or hours. This data sometimes overloads the interpretation capabilities of the operators, who can miss the beginning of an abnormal situation, leading probably to a process fault. Statistics show that about 70% of the industrial accidents are caused by human errors (Venkatasubramanian V,

Rengaswamy R(2003)). Under this scenario, the processing of operational historical databases using intelligent techniques is a decisive task to improve the productivity, quality, safety and reduce environment accidents in the industrial plants.

The idea behind the data visualization algorithms is to generate a lower dimensional data, preserving the main characteristics of the original higher dimensional data space. This can be researched by finding a mechanisms or function which preserves the geometric structure of the original space in the reduced space. This concept is illustrated by Fig. 2.

Points bellowing to the original space of characteristics $X \in \Re^n$ are transformed into a reduced space $Y \in \Re^q$ · $q \ll n$, using the transformation $F(X)$. The reduced representation is found when the dissimilarity matrix of the original space (D) is approximately equal to the dissimilarity matrix of the reduced one (L). This formulation guarantees that any particular characteristic of the original space is kept in the reduced one. Appendix A presents the algorithm used to perform the reduction of the dimensionality for the data.

4. Projected Operational Map (POM)

The information about the structure, clusters, tendency and shape of the data is very relevant for process monitoring; however this information is not directly accessible for high dimensional data. In the present case study the Hydro- treating plant has to process 7 different feed stocks, with different operational conditions in the process variables; so the process has 7 modes of operation that should be separable in the original space. This separation should be visible in a good bi-dimensional projection.

The projected operational map is the bi-dimensional projection of the 8 process variables, using the Euclidean distance as the property to be preserved from the original space in the reduced one. A mathematical formulation can be seen as (1).

$$X \in R^8 \xrightarrow{\ F(X)\ } Y \in R^2$$
$$\text{Original} \qquad 2\text{-D Operational Map} \tag{1}$$
$$\text{Space}$$

Figure 3 shows the projection of the operational database, each operation mode is represented by a different icon. Each of the points in the 2-D plot has a correspondence with a point in the original eight dimensional space, under the following objective:

Points very close (far) in the original space should be very close (far) in the 2-D plot.

4.1 Analysis of the Projected Operational Map (POM)

In figure 3 can be observed how the data is grouped; forming clusters for each of the operation modes. The projected operational map shows that the modes LVI-40 and NB0-170 are the most distant. The points bounded by a rectangle show a point of the LVI-40 feed stock processed in the region of the NBO-170, and a point of the NBO-170 processed in the region of the LVI-40.These points could indicate a bad or abnormal operation of the process in the treatment of these feeds. The points bounded by a triangle are far from their traditional working zones and could indicate some problem during the operation of the process.

By omitting the points bounded by the rectangle and triangle, a frontier can be drawn to separate the operation of the HDT process when it is processing the feed LVI and NBO. The frontier line is shown in figure 3 (hand drawn). The use of pattern classification algorithms or support vector machines to find the best separation function is an open research topic being worked out. Above the frontier are the LVI cases and below are the NBO.

The working zone can be delimited using geometric forms according to the metric used to define the cluster. Figure 4 shows a definition of the operation zones using ellipses (hand drawn in the case). The use of pattern classification algorithms to define the best contour is a present research topic.

A strategy to monitor industrial systems based on the construction of projected operational maps (POM) is proposed in figure 5. It takes an operational database and performs a non-linear transformation of the original space into a reduced space. The POM identifies regions of operation indicating the performance or behavior of the process in a certain time. The relative position in the cluster may give information about relative quality.

5. Conclusion

This paper briefly proposed a strategy to create Projected Operational Map (POM) using space reduction techniques to visualize work conditions in a database of 189 cases collected from a Hydro- Treating process. The projected

data forms clusters that could represent the operational modes of the process. From the interpretation of the POM two possible abnormal situations might have been discovered when two feeds were processed with a wrong specification in their operational conditions. A new process monitoring scheme was proposed using the Projected Operational Map as a base to find a simplified representation of real time process data, allowing the quick identification of operational regions.

The development of such advanced monitoring system requires further research of pattern recognition techniques for on-line clustering as well as knowledge extraction from the cluster forms and volumes. The main goal is to obtain a decision support system for engineers to improve quality, productivity and safety by early detection of abnormal operation conditions leading probably to a bad operating trajectory and eventually to faults.

References

Borg,I.; P.Groenen (1997), Modern Multidimensional Scaling. Theory and Applications. Springer-Verlag.

Chen F.Z.; X.Z. Wang (2000), Discovery of Operational Spaces from Process Data for Production of Multigrades of Products. Ind Eng Chem Res., 39.

Eckart, C; Young,G (1936) Approximation of one matrix by another of lower rank, Psychometrika,1,211-218

Jaeckle, C.M.; J.F.MacGregor (1988), Product design through multivariate statistical analysis of process data. AICHE J.,44.

Mao, J.; Jain, K. Anil (1995), Artificial Neural Networks for Feature Extraction and Multivariate Data Projection IEEE Trans On Neural Networks, 6, 2.

Nomikos,P.; J.F MacGregor (1994), Monitoring Batch Process Using Multiway Principal Component Analysis, AICHE J.,40.

Salager, J. L. (1986), Refinación Petróleo II, Universidad de los Andes.

Tipping M. (1996), Topographic Mapping, PhD Dissertation. Univ. of Aston in Birmingham.

Torgerson,W.S (1958) Multidimensional Scaling I. Theory and Methods. New York Wiley.

Venkatasubramanian,V.,R. Rengaswamy(2003), A review of Fault Detection and Diagnosis. Part I: Process. History based methods. Computers and Chemical Enginnering, 27.

Young, G; A.S Householder(1938), Discussion of a set of points in terms of their mutual distances, Psychometrika, 1938,3,19-22

Appendix A

Classical Scaling is also known under the names of Togerson Scaling (Torgerson(1958)) and Togerson-Gower Scaling. It is based on the pioneer theorems by Eckart and Young (1936) and by Young and Householder(1938). The basic idea of classical scaling is to assume that the dissimilarities are distances and then find coordinates that explain them.

The classical scaling algorithm is as follows:

1. From the dissimilarity matrix Δ, generate the double-centered inner product matrix B_Δ

$$B_\Delta = \frac{-1}{2} J\Delta^2 J \qquad (A.1)$$

With Δ^2 the matrix whose elements are the squares of those of Δ. J. The matrix J is the *centering* matrix, given by $I - (M-1)11'$, where $11'$ is the matrix whose elements are all 1 and M is the number of rows or columns of the matrix Δ.

2. Compute the eigendecomposition of B_Δ by (A.2)

$$B_\Delta = Q\Phi Q' \qquad (A.2)$$

The matrix Φ is the diagonal matrix of eigenvalues of B_Δ. Q is the corresponding matrix of eigenvectors.

3. Let m be the dimensionality of the solution. Denote the matrix of the first m eigenvalues greater than zero by Φ_+ and Q_+ the first m

columns of Q. Then the coordinate's matrix of classical scaling is given by (A.3)

$$Y = Q_+ \Phi_+^{1/2} \qquad (A.3)$$

If Δ happens to be a Euclidean distance matrix, then B_Δ is a positive semi-definite matrix and so negative eigenvalues can not occur.

Mapping New Data Points.

The classical scaling does not allow the reduction of new data points. Given this difficulty the following idea is proposed to obtain a way to perform the reduction of new data points:

- Apply the classical scaling algorithm and obtain the value Y.

- Let the following parametric model be (A.4)

$$Y = X * C \qquad (A.4)$$

  with X the original data to calculate the dissimilarity matrix Δ.

- Compute the matrix C using the following expression (A.5).

$$C = (X'X)^{-1} X'Y \qquad (A.5)$$

The reconstruction error can be obtained using the following equation:

$$e = Y - \bar{Y} \qquad (A.6)$$

$$\bar{Y} = X * C$$

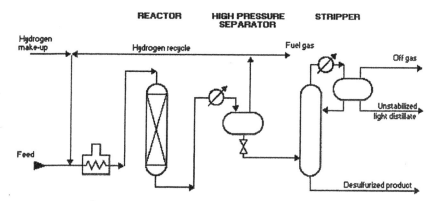

Figure1. The Hydro-Treating Process

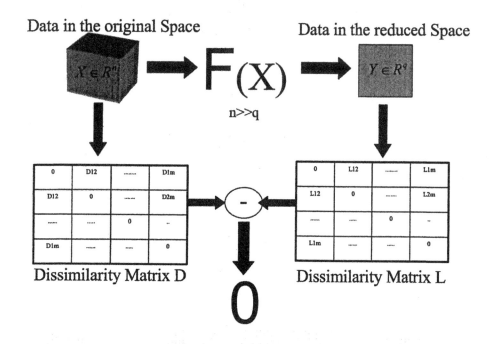

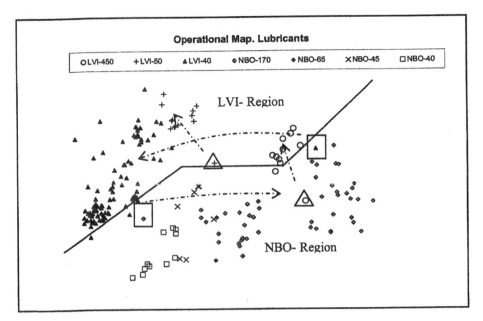

Figure 3. Projected Operational Map (POM), for the Hydro Treating.

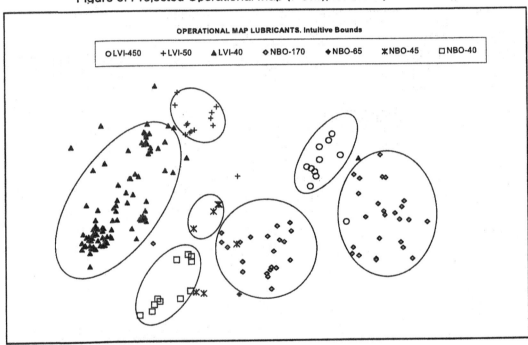

Figure 4. Projected Operational Map with operational bounds.

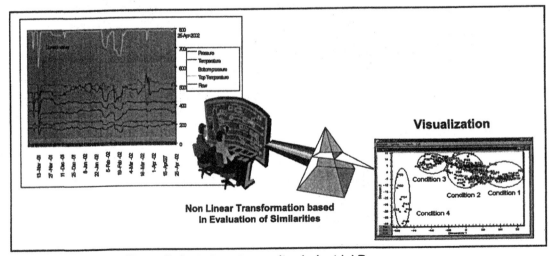

Figure 5. A strategy to monitor Industrial Processes.

ELSEVIER
IFAC
PUBLICATIONS
www.elsevier.com/locate/ifac

IDENTIFICATION OF CHUA'S CHAOTIC CIRCUIT BASED ON AN ARTIFICIAL NEURAL NETWORK AND LEVENBERG-MARQUARDT OPTIMIZATION METHOD

Leandro dos Santos Coelho [1]   and   **Fábio A. Guerra** [2]

[1] *Pontifícia Universidade Católica do Paraná, Grupo Produtrônica*
Programa de Pós-Graduação em Engenharia de Produção e Sistemas
Rua Imaculada Conceição, 1155, CEP 80215-901, Curitiba, PR – Brazil
E-mail: leandro.coelho@pucpr.br

[2] *Artificial Intelligence Division*
TECPAR –Paraná Institute of Technology
St. Algacyr Munhoz Mader, 3775, 81350-010 - Curitiba, PR - Brazil
E-mail: guerra@tecpar.br

Abstract: In recent years, the interest in the development of identification strategies of nonlinear systems with chaotic behaviour reappeared, many of these using neural networks. This paper presents the implementation and study of a multilayer perceptron neural network for identification of a dynamic system, this with nonlinear and chaotic behaviour, called Chua's circuit. The construction of Chua's circuit is composed for a network of linear passive elements connected to a nonlinear active component (Chua's diode). The Chua's circuit consists of an electronic chaotic oscillator that has been used as platform of tests in many areas related to the study of the chaos, telecommunications, cryptography and physics.

Keywords: chaotic circuits, neural networks, Chua's circuit, systems identification, soft computing, chaotic systems, nonlinear systems.

1. INTRODUCTION

One of the relevant areas of science is the forecast ("Gave the past, as I can foresee the future"). The time series analysis is a relevant area in different diverse fields of the knowledge, mainly in the areas of engineering and economy. The main motivation for research about time series is to provide a forecast when the mathematical model of a phenomenon is complex, incomplete and unknown. A time series consists of measures or observation of previously obtained of a phenomenon that is realized sequentially under a time interval. If the consecutive observations are dependents one of the others, then is possible to obtain the system identification.

The conception of mathematical models for the representation of complex systems of time series is an excellent procedure and with practical applications. However, in general, the construction of adjusted mathematical models for the engineering intentions is not a simple task. In the last decades, diverse conceptions of algorithms for modelling and

identification of complex dynamics systems have been proposed in literature, such as: frequency methods, techniques based on estimates of Wiener models, Hammerstein, bilinear and Volterra, nonlinear regression, wavelets and recursive identification (Ljung 1987; Haber and Unbehauen 1990). An excellent boarding, between much others, for mathematical representation of dynamics systems with complex or chaotic behaviour it is of the neural networks.

The mathematical models based in some boarding of neural networks have received attention recently of scientific community and of the professionals that work in industry, for being about project tool that offer to promising solutions to identification problems of complex dynamic system (McLoone *et al.*, 1998; Hovakimyam *et al.*, 2000; Pan *el al.*, 2001).
This paper presents the foundations, analysis and project of a multilayer perceptron neural network applied to a identification of the nonlinear dynamic behaviour of a Chua's circuit. The boarded stages, in

this paper, for the processes identification through of neural networks include the determination of: (i) a data set of inputs and outputs of the process; (ii) a classroom of models candidates (structure of the neural model); (iii) a criterion for verification of the approach between the real data and the mathematical model obtained by the neural networks; and (iv) of the routines of validation of the resultant mathematical models.

The paper is organized as follows. The description Chaos' theory and Chua's circuit is presented in section 2. An introduction about system identification, multilayer perceptron neural network and the identification procedure using neural network is boarded in section 3. The simulations and the analysis of the obtained results of neural network application are presented and argued in section 4. The conclusion is presented in section 5.

2. CHAOS' THEORY AND CHUA'S CIRCUIT

Chaos' theory studies pertinent phenomenon to the nonlinear dynamic systems (any process that evolve with the time) and that present complex behaviour to be treat mathematically. The theory of the chaos studies the unexpected phenomenon apparently, in the search of hidden standards and simple laws that conduct the complex behaviours. However, this study if became effectively reasonable from the decade of 1960, when the computers had started to possess reasonable graphical capacity and of processing, giving to the physicists and mathematicians the power to discover answers for basic questions of the science in general way, that before were obscure.

The nonlinear systems, had appeared from the chaos' theory that supplies to an explanation, many times adequate, (through formulas and equations), to many behaviours current in the nature, such as: natural phenomenon (populations, turbulence, fluid movement, and cloud formation), complex behaviours in electric circuits (converting buck, oscillators) (Iu and Tse, 2001; Lian and Liu, 2000), behaviour of stock exchange and economy, (Aguirre and Aguirre, 1997), nonlinear systems and variant in the time (Ljung et al., 1987), telecommunications (Schweizer and Schimming, 2001), control of processes (Lim and Mareels, 2000), dynamic behaviour of the cardiac beatings (Pei et al., 1999), among others.

The behaviour of the chaotic systems can present the great sensitivity in relation to the initial conditions that are applied. Although, to be a difficult task to describe the behaviours of a chaotic system, in probabilistic terms this situation can be treated and some paradigms have been presented in literature for this purpose.

The Chua's circuit (Aguirre et al., 1997; Tôrres and Aguirre, 1999, 2000; Thamilmaran et al., 2000; Itoh

et al., 2001) is one of the systems used for the study of nonlinear and chaos dynamics. The Chua's circuit is illustrated summarized in Fig. 1 and its characteristics of current-voltage are illustrated in Fig. 2, presented to follow. The construction of this circuit can be realized by the composition of a network of linear passive elements connected to a nonlinear active component called Diode of Chua, as Fig. 3 illustrates. This conception of electronic chaotic oscillator can be used as platform of tests in diverse areas related to the study of the chaos.

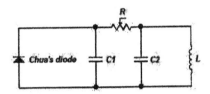

Fig. 1. Chua's circuit.

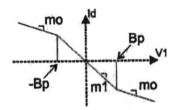

Fig. 2. Characteristics of current-voltage.

The equations that describe the circuit are:

$$C_1 \frac{dv_{C_1}}{dt} = \left[\frac{v_{C_2} - v_{C_1}}{R} \right] - i_d(v_1) \quad (1)$$

$$C_2 \frac{dv_{C_2}}{dt} = \left[\frac{v_{C_1} - v_{C_2}}{R} \right] + i_L \quad (2)$$

$$L \frac{di_L}{dt} = -v_{C_2} \quad (3)$$

where v_{ci} is the voltage on capacitor C_i, i_L is the current through the inductor. The equations that describe the current of Chua's Diode are:

$$i_d(v_1) = \begin{cases} m_o * v_1 + B_p(m_0 - m_1), & v_1 < -B_p \\ m_1 * v_1, & |v_1| \le B_p \\ m_o * v_1 + B_p(m_1 - m_0), & v_1 > -B_p \end{cases} \quad (4)$$

The idea of this oscillator is not only to measure values of voltage on the capacitor, but, yes, to observe the oscillation chaotic, as illustrated in Fig. 4. In Fig. 4 is presented a simulation realized in the software Electronics Workbench (version 5.0c). The oscillation chaotic of experimental result is illustrated in Fig. 5. An important detail is that different values for the components of the electric circuit result in attractors with different geometry. In this application, resulted in a double-scroll attractor, as illustrated in Fig. 6, where vc1 is the voltage in capacitor 1.

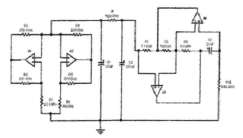

Fig. 3. Chua's circuit (electronic chaotic oscillator).

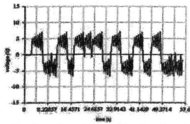

Fig. 4. Oscillation chaotic graphic simulated in EWB.

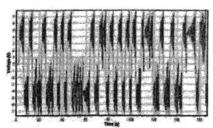

Fig. 5. Oscillation of experimental chaotic graphic.

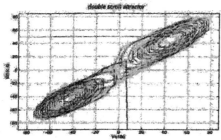

Fig. 6. Double-scroll attractor resulted of Chua's circuit.

3. SYSTEM IDENTIFICATION AND NEURAL NETWORKS

The processes identification is a procedure to identify a model of an unknown process, for intentions of forecast and/or understanding of the dynamic behaviour of the process. A model describes reality in some way, and system identification is the theory of how mathematical models for dynamical systems are constructed form observed data. Typically, a parameterized set of models, a model structure, is hypothesized and data is used to find the best model within this set according to some criterion. The choice of model structure is guided by prior knowledge or assumptions about the system that generated the data. When little prior knowledge is available it is common to use a black-box model. A black-box model is a standard flexible structure

and it can be used to approximate a large variety of different systems (Sjöberg, 1995). Neural network models have proven to be successful non-linear black-box model structures in many applications.

The neural networks consist of elements of processing highly interconnected called neurons. Each neuron has inputs and one output. The output of each neuron is determined as a nonlinear function of weighed sum of the inputs, however mathematical operations more complex could be included. The neurons if interconnect through weights, which are adjusted during the period of training. Between the excellent characteristics of the neural networks had: parallel processing, learning, associative and distributed memory. These characteristics are inspired in the biological neural networks, exactly that rudimentarily.

The neural networks provide, usually, not-parametric quantitative knowledge and are adjusted for systems identification, learning and adaptation of complex processes. The neural network to be boarded in this project is the multilayer perceptron neural network (MLP-NN). The MLP-NN is applied to decide the most diverse and complex problems, had its capacities of "universal approximations" of a function with one given precision (Cybenko, 1989), providing a promising tool to the identification of nonlinear systems and boarding of intelligent, predict and adaptive control.

The MLP-NN is a generalization of perceptron of a layer. The MLP-NN contains three types of layers: the input layer, hidden layer(s) and the output layer. Any neuron of a layer can be established connection with another neuron of the following layer. The neurons of the input layer receive the signals from the external world and they transmit them for the neurons of the next layer. While this, the neurons of the output layer send the information of the hidden layer(s) neurons for the external world, as presented in Fig. 7.

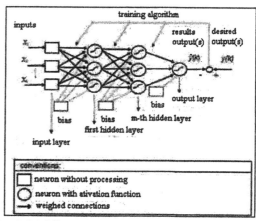

Fig. 7. Multilayer perceptron neural network.

The training of the MLP-NN is realized of way supervised and usually realized with the back-propagation algorithm (BP). BP algorithm is based on the learning rule of the error correction that can be seen as a generalization of the adaptive filtering algorithm or a same special case of the squared minimums algorithm. The BP algorithm is an iterative method of projected gradient to minimize the addition of the quadratic error, between the current output and the desired output. To minimize this function-objective, the BP algorithm uses one technique of search based on gradient, the generalized delta rule. However, the BP presents limitations as low learning speed. The increase of the learning speed, through of the heuristic procedures, for the adaptive modification of the parameters, prescription of a specific selection of the initial weights of the MLP-NN or same for utilization of the second derivative (use of the Hessian matrix) are alternatives viable. The performance of the BP is sensible to the configuration of the learning coefficient and only uses the first derived (the bending/curvature) from the error surface. With the utilization of the second derivative (the change tax of the bending), the training time can be reduced of significant form.

The used algorithm in this paper is the algorithm of Levenberg-Marquardt (Levenberg, 1944; Marquardt, 1963) (LM) that it consists of an approach of the Newton method and frequently consists of an efficient alternative to the algorithms of type BP. The LM considers the function bending to establish the size of step to the long one of its bending. The strategy consists of the function expansion $f(x,w)=t$ in a series of Taylor. The intention is to minimize the function cost, $J(w)$, with relation to weight w. The LM presents high computational cost and the required high operation number. Suppose that have a function $V(x)$ which want to minimize with respect to the parameter vector x, then Newton's method would be:

$$\Delta \underline{x} = -\left[\nabla^2 V(\underline{x})\right]^{-1} \nabla V(\underline{x}) \quad (5)$$

where $\nabla^2 V(\underline{x})$ is the Hessian matrix and $\nabla V(\underline{x})$ is the gradient. If $V(\underline{x})$ is a sum of squares function:

$$V(\underline{x}) = \sum_{i=1}^{N} e_i^2(\underline{x}) \quad (6)$$

then it can be shown that:

$$\nabla V(\underline{x}) = J^T(\underline{x}) e(\underline{x}) \quad (7)$$

$$\nabla^2 V(\underline{x}) = J^T(\underline{x}) J(\underline{x}) + S(\underline{x}) \quad (8)$$

where $J(x)$ is the Jacobian matrix:

$$J(\underline{x}) = \begin{bmatrix} \dfrac{\partial e_1(\underline{x})}{\partial x_1} & \dfrac{\partial e_1(\underline{x})}{\partial x_2} & \cdots & \dfrac{\partial e_1(\underline{x})}{\partial x_n} \\ \dfrac{\partial e_2(\underline{x})}{\partial x_1} & \dfrac{\partial e_2(\underline{x})}{\partial x_2} & \cdots & \dfrac{\partial e_2(\underline{x})}{\partial x_n} \\ \vdots & \vdots & \ddots & \vdots \\ \dfrac{\partial e_N(\underline{x})}{\partial x_1} & \dfrac{\partial e_N(\underline{x})}{\partial x_2} & \cdots & \dfrac{\partial e_N(\underline{x})}{\partial x_n} \end{bmatrix} \quad (9)$$

and

$$S(\underline{x}) = \sum_{i=1}^{N} e_i(\underline{x}) \nabla^2 e_i(\underline{x}) \quad (10)$$

For the Gauss-Newton method it is assumed that $S(x) \approx 0$ and the update becomes:

$$\Delta \underline{x} = \left[J^T(\underline{x}) J(\underline{x})\right]^{-1} J^T(\underline{x}) e(\underline{x}) \quad (11)$$

The Levenberg-Marquardt modification to the Gauss-Newton method is:

$$\Delta \underline{x} = \left[J^T(\underline{x}) J(\underline{x}) + \mu I\right]^{-1} J^T(\underline{x}) e(\underline{x}) \quad (12)$$

The parameter μ is multiplied by some factor (β) whenever a step would result in an increased $V(x)$, μ is divided by β (this can start with $\mu=0.01$ and $\beta=10$). Notice that when μ is large the algorithm becomes steepest descent (with step $1/\mu$), while for small μ the algorithm becomes Gauss-Newton. The Levenberg-Marquardt algorithm can be considered a trust-region modification to Gauss-Newton.

The key step in this algorithm is the computation of the Jacobian matrix. For the neural networks mapping problem the terms in Jacobian matrix can be computed by a simple modification to the BP algorithm. It easy to see that this is equivalent in form to the sum of squares function ($V(\underline{x})$), where $\underline{x} = [w^1(1,1) \ w^1(1,2) \ ... \ w^1(S_1,R) \ b^1(1) \ ... \ b^1(S_1) \ w^2(1,1) \ ... \ b^M(S_M)^T]$, and $N=Q \times S_M$. Standard BP calculates terms like:

$$\frac{\partial \hat{V}}{\partial w^k(i,j)} = \frac{\partial \sum_{m=1}^{S_M} e_q^2(m)}{\partial w^k(i,j)} \quad (13)$$

For the elements of the Jacobian matrix that are needed for the Marquardt algorithm is necessary to calculate terms like:

$$\frac{\partial e_q(m)}{\partial w^k(i,j)} \quad (14)$$

These terms can be calculated using the standard BP algorithm with one modification at output layer:

$$\Delta^M = -F^M\left(\underline{n}^M\right) \quad (15)$$

Note that each column of the matrix in this equation is a sensitivity vector that must be back propagated through the network to produce one row of the Jacobian.

The training of the MLP-NN through the LM method can be described for the following procedure: (i) to initiate the MLP-NN and to present all inputs and outputs; (ii) to calculate the reply of the MLP-NN; (iii) to calculate the addition of the quadratic errors; (iv) calculation of the Jacobian matrix; (v) to resolve the equation of Gauss-Newton to update of the weights of the MLP-NN; (vi) to recalculate the addition of the squares of the errors being used update of the weights, w, through $w + \Delta w$. S, is the addition of squares is minor that the previous (step iii), to reduce the factor μ for β, to take $w = w + \Delta w$, and to return to the step (i). In contrary case, to increase the factor μ for β, and to return for the step (v); and (vii) to stop when a number of epochs are realized or when the addition quadratic error found is below of a priori definitive value for the designer.

4. RESULTS

The Table 1 presents a summary of the experiments realized for identification of the Chua's circuit using MLP-NN. In this in case were evaluated the quadratic average error to the validation and estimation phases for diverse configurations of MLP-NN, that is, different number of neurons in the hidden layer of the MLP-NN.

Table 1 Results for a number of neurons in the hidden layer of the MLP-NN (full connected) to identification of Chua's circuit.

number of neurons	estimation phase (learning phase) quadratic mean error	validation phase (test phase) quadratic mean error
1	16.4900	7.6316
2	16.3878	7.5316
3	**3.0563**	**3.3218**
4	3.1417	3.3551
5	3.2210	3.3493

Observing the Table 1, is noticed that the results of the identification of the dynamic behavior of the Chua's circuit are promising using a MLP-NN, mainly with more than 3 neurons in the hidden layer. It must be emphasized that the LM algorithm of the MLP-NN presents advantages of being a fast algorithm, however, it presents limitations due requires much memory and sensible being to it the initial conditions of the optimization procedure, or either, to the random procedure of generation of the initial weights of the neural network. This last factor the generation of the initial weights of the network was excellent in the attainment of the boarded results in Table 1 and must be analyzed in future works of form to get one better commitment between the neurons of the hidden layer and its optimization.

The best one application resulted of the dynamic behavior identification procedure of the Chua's circuit using a MLP-NN with 3 neurons in the hidden layer and with 100 epochs of training is presented in Fig. 8. The obtained maximum error for this configuration was –6.4166, i. e., a maximum error of 8% (see Fig. 9).

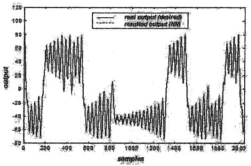

Fig. 8. Identification results of Chua's circuit using MLP-NN.

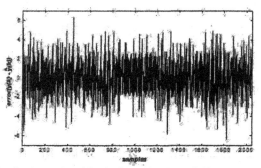

Fig. 9. Error signal in the identification of Chua's circuit using MLP-NN.

5. CONCLUSION

In the last years of interest in the development of strategies of nonlinear identification of chaotic systems reappeared. This interest is motivated by diverse factors, such as: (i) advances of the nonlinear systems theory, causing applicable methodologies of project to an extension of control nonlinear problems; (ii) development of efficient identification methods for the treatment of empirical nonlinear models; (iii) continued development of the capacities of software and the hardware, becoming possible the incorporation of complex nonlinear models in control systems design.

The behaviour of the chaotic systems can present the great sensitivity in relation to the initial conditions that are applied. Although, to be a difficult task to describe the behaviour of a chaotic system, in probabilistic terms this situation can be treated and some paradigms have been presented in literature for this purpose. The behaviour of a chaotic system can be evaluated through the configuration of nonlinear identification methodologies. The methodologies that can be used include the fuzzy systems, neural networks, genetic algorithms, among others approaches.

The case study of this paper it aimed the nonlinear identification and validation of the Chua's circuit dynamic behaviour using a neural network. The artificial neural networks are motivated by biological neural systems, with intention to simulate the form as the brain learns, remembers and processes information. The preliminary results presented in this paper had shown that the MLP-NN is a powerful tool for forecast of time series and the studies of systems with complex or chaotic behaviour.

A negative aspect of the MLP-NN is that they can suffer of the curse of dimensionality, therefore when the dimension of the input vector is increased the dimensionally of the activation functions of the hidden layer of the MLP-NN has, generally, exponential increase to represent the process with a permissible precision. In this context it is desired to use procedures of optimization with multiple objectives for attainment of one better commitment between interpolation, generalization and learning of the MLP-NN aiming at itself identification applications.

REFERENCES

Aguirre, L.A., Rodrigues, G.G. and Mendes, E.M.A.M. (1997), "Nonlinear identification and cluster analysis of chaotic attractors from a real implementation of Chua's circuit," *International Journal of Bifurcation and Chaos*, vol. 7, no. 6, pp. 1411-1423.

Cybenko, G. (1989), "Approximation by superpositions of a sigmoidal function," *Mathematics of Control Signals and Systems*, vol. 2, pp. 303-314.

Haber, R. and Unbehauen, H. (1990), "Structure identification of nonlinear dynamic systems — a survey on input/output approaches," *Automatica*, vol. 26, no. 4, pp. 651-677.

Hovakimyan, N.; Lee; H. and Calise, A. (2000), "On approximate NN realization of an unknown dynamic system from its input-output history," *Proceedings of the American Control Conference*, Philadelphia, Pennsylvania, vol. 2, pp. 919-923.

Itoh, M.; Yang, T. and Chua, L. O. (2001), "Conditions for impulsive synchronization of chaotic and hyperchaotic systems," *International Journal of Bifurcation and Chaos*, vol. 11, no. 2, pp. 551-560.

Kollias, S. and Anastassiou, D. (1989), "An adaptive least squares algorithm for the efficient training of artificial neural networks", *IEEE Transactions on Circuits and Systems*, vol. 36, no. 8, pp. 1092-1101.

Levenberg, K. (1944), "A method for the solution of certain non-linear problems in least squares," *Quart. Applied Mathematics*, vol. 2, no. 2, pp. 164-168.

Lian, K.-Y. and Liu, P. (2000), "Synchronization with message embedded for generalized Lorenz chaotic circuits and its error analysis," *IEEE Transactions on Circuits and Systems I: Fundamental Theory and Applications*, vol. 47, no. 9, pp. 1418-1430.

Ljung, L. (1987), *System identification: theory for the user*, Prentice-Hall: New York.

Ljung, L. (2001), "Black-box models from input-output measurements," *Proceedings of IEEE Instrumentation and Measurement Technology Congress*, Budapest, pp. 138-146.

Marquardt, D. W. (1963), "An algorithm for least-squares estimation of nonlinear parameters," *SIAM Journal of Applied Mathematics*, vol. 11, no. 2, pp. 431-441.

McLoone, S., Brown, M.D., Irwin, G. and Lightbody, A. (1998), "A hybrid linear/nonlinear training algorithm for feedforward neural networks," *IEEE Transactions on Neural Networks*, vol. 9, no. 4, pp. 669-684.

Pan, Y.; Sung, S.W. and Lee, J.H. (2001), "Data-based construction of feedback-corrected nonlinear prediction model using feedback neural networks," *Control Engineering Practice*, vol. 9, pp. 859-867.

Pei; W.; Yang, L. and He, Z. (1999), "Identification of dynamical noise levels in chaotic systems and application to cardiac dynamics analysis," *Int. Joint Conference on Neural Networks*, vol. 5, Washington, DC, pp. 3680-3684.

Petrick M.H. and Wigdorowitz, B. (1997), "A priori nonlinear model structure selection for system identification," *Control Engineering Practice*, vol. 5, no. 8, pp. 1053-1062.

Schweizer, J. and Schimming, T. (2001), "Symbolic dynamics for processing chaotic signal. ii. communication and coding," *IEEE Transactions on Circuits and Systems I: Fundamental Theory and Applications*, vol. 48, no. 11, pp. 1283-1295.

Sjöberg, J. (1995), *Non-linear system identification with neural network*, Dept. of Electrical Engineering, Linköping University, Sweden.

Thamilmaran, K.; Lakshmanan, M. and Murali, K. (2000), "Rich variety of bifurcations and chaos in a variant of Murali-Lakshmanan Chua circuit," *Int. Journal of Bifurcation and Chaos*, vol. 10, no. 7, pp. 1781-1785.

Tôrres, L.A.B. and Aguirre, L.A. (1999), "Extended chaos control method applied to Chua's circuit," *Electronics Letters*, vol. 35, no. 10, pp. 768-770.

Tôrres, L.A.B. and Aguirre, L.A. (2000), "Inductorless Chua's circuit," *Electronics Letters*, vol. 36, no. 23, pp. 1915-1916.

Wilamowski, B.M.; Chen, Y. and Malinowski, A. (1999), "Efficient algorithm for training neural networks with one hidden layer neural networks," *International Joint Conference on Neural Networks*, vol. 3, Washington, DC, pp. 1725 –1728.

PUBLICATIONS
www.elsevier.com/locate/ifac

Performance comparison for the methods of 'in stock' components optimal replacement

Eduardo de Oliveira Pacheco

Pontifical Catholic University of Parana (PUCPR)

Rua Imaculada Conceição, 1155 Prado Velho

Curitiba, PR, Brazil

Tel.: +55412711333

Fax: +55412711345

edupache@terra.com.br

Gustavo Henrique da Costa Oliveira

Pontifical Catholic University of Parana (PUCPR)

Rua Imaculada Conceição, 1155 Prado Velho

Curitiba, PR, Brazil

Tel.: +55412711333

Fax: +55412711345

gustavoc@rla01.pucpr.br

Abstract: This article approaches the problem of optimal replacement of components in stock, using the dynamic systems analysis in closed-loop, more specifically the control theory. The referred productive environment is mono-product and has only one level. It can operate in several known strategies (ATO, MTO, MTS, among others). The level is composed by the manufacturing and the assembly stages with an intermediate stock, which should be minimized by the components' replacement. The system considers capacity constraint in the manufacturing, lead-time in the stages and uncertainties in the forecasted future demand. A performance comparative analysis is realized among three methods for the optimal replacement of components in stock. In the first algorithm, one deals with the replacement problem, using the linear programming technique. In the second, the replacement algorithm uses the optimal control technique with the minimization of one step. In the third, the replacement algorithm uses the optimal control technique, based on the minimization of the anticipation horizon.
Key Words: Dynamic System; Stock Management; Optimal Control; Demand Uncertainties.

1. Introduction

The control systems have been explored since the fifties due to their relevance in various applications (Aström & WIttenmark, 1995). Productive environments and manufacturing systems, which involve stock management are also another area for the application of the control theory (Vassian, 1954; Towill et al., 1997). Control systems with feedback are of vital importance, since they allow the compensation of unpredictable factors as systems errors, uncertainties in lead times for delivery/manufacturing/assembly or for the demand of products in stock. Within this area, the control systems are related to the methodology for the components' replacement, in a given moment in time, based on the information from the productive environment. The modeling of a productive environment using the dynamic systems concepts in closed-loop for a mono-product/mono-level productive environment is approached in Towill (1982) under the jargon of replacement system in stock based upon the production order, or

IOBPCS (Inventory and Order Based Production Control System). Within this modeling, the objective was to verify the quantity of components in stock, whenever the system was under the influence of the demand flotation. This work is extended in Towill & Del Vecchio (1994) for a supply chain composed by three productive levels (three-echelon). Throughout this proposal, the demand increase among productive levels is discussed. Other works that deal with dynamic systems, optimal stock replacement and closed-loop systems can be designed. For instance, the works of Vassian (1954), Pacheco et al. (2002), Oliveira & Pacheco (2003), Pacheco & Oliveira (2003 a, b).

In Vassian (1954), the analysis and the design of an algorithm for the stock control, that is, the components' replacement in a warehouse, it is made using a control project for servomechanisms viewpoint. The dynamic systems theory is, then, used due to the variables 'essentially discrete' characteristic, involved in the analyzed productive environment, i.e., production orders, components position, demand, etc. The productive and stock environment is modeled through the difference equations and based on the hypothesis that the system alternates stages in set-time pauses (data collection and periodical components' replacement), a stock control strategy is approached. In a more recent work, Pacheco et al. (2002) deals with the stocks optimal replacement problems inside an one level productive environment, mono-product, that attends the demand management *ATO*. This productive environment is composed by two stages. The first represents the components' manufacturing and the second the components' assembly, generating the final product. There is an intermediate stock between stages. One proposes a management solution for the determination of the production orders to minimize the variance in the stock position, regarding a reference value. One analyzes the seasonal demand case, the uncertainties' consideration in the demand forecast and the non-existence of capacity constraints. In Pacheco & Oliveira (2003 a, b), one approaches the problem of demand peaks not forecasted with exactness and the manufacturing capacity constraints. Within this situation, the manufacturing capacity may not be able to supply the demand, considering high-demand peaks, and there is the need for a strategic stock arrangement. One proposes, then, a division in the management system in two stages, the control stage and the supervision stage, which create a strategic stock, related to the future forecast of the demand peaks. Both stages, acting together, avoid the lack of components' delivery for the supply chain's next level. Oliveira & Pacheco (2003) present an analogous proposal to these works. Within their work, the components' in stock optimal replacement uses the control technique, based on the receding horizon approach. This strategy is characterized by the determination of the production order and consequent stocks' replacement through the determination of a forecast horizon, regarding the system's future behavior. This approach considers production capacity limits and it is capable of dealing, automatically, with the future demand peaks without the need for additional stages in the control algorithm. Throughout the current article, three methods for the components' in stock optimal replacement are presented. The first is based on the classic modeling and uses the methodology of linear programming in order to replace the components. The other two are proposed in Oliveira & Pacheco (2003) and Pacheco & Oliveira (2003b). The methods are implemented and the results are submitted to a performance analysis, measured through the total cost of production and the components' in stock average replacement.

The current article is divided in 5 (five) sections. In section 2, the main characteristics of the stocks'

management model are described. Section 3, contains the stock optimal replacement methods. Section 4 presents the minuscule analysis of the replacement algorithms performance. Finally, in section 5, the article is concluded.

2. Description of the productive environment

One assumes a mono-level and a mono-productive productive system, composed by the manufacturing and the assembly stages with an intermediate stock between both. One considers the capacity constraint and the delay (*lead-time*) in the manufacturing stage. In the assembly stage, one assumes the lead-time is negligible and presents assembly infinite capacity. The modeling of the stock dynamic behavior within this productive level is made using the following differences equation:

$$y(k+1) = y(k) + p(k) - d(k) \qquad (1)$$

Where k is a discrete variable, which represented the time k (k is integer), $y(k)$ is the stock position in the beginning of the time interval k; $p(k)$ is the components' in stock replacement in the time interval k; $d(k)$ are the components removed from the stock to the assembly in the same time interval, it is the market demand. This structure is illustrated by Figure 1. In the Figure, $u(k)$ is the order for the components' manufacturing in the beginning of the time interval k.

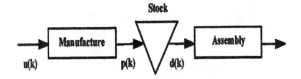

Figure 1: The productive environment

The manufacturing stage can be modeled with a pure delay d, the equation (1) can be re-written as:

$$y(k+1) = y(k) + p(k) - \hat{d}(k) + \zeta(k+1) \qquad (2)$$
$$p(k) = u(k-d)$$

Within the equation, therefore, d is the delay between the production order $u(k)$ and the replacement in the stock $p(k)$; $\hat{d}(k)$ is the market demand forecasted in the time interval k, $\zeta(k+1)$ is a random variable related to the error anticipation in k.

3. Methods for the products in stock optimal replacement

Figure 1 contains a productive environment, which represents a sub-problem of a supply chain composed by several productive levels (for instance, manufacturing, distribution and retail market). The control management aims to calculate the optimal replacement orders in the time k, that is, to calculate $u(k)$, which minimizes a cost criterion, considering the productive environment constraint and the uncertainties of the forecasted future demand. Within this section, three methods for the optimal in stock replacement are reviewed: the method based upon the linear programming and the methods proposed in Oliveira & Pacheco (2003) and Pacheco & Oliveira (2003b).

3.1 The replacement algorithm based upon linear programming

The linear programming (LP) has being applied in order to solve various types of problems that involve manufacturing (Moreira, 1993). One can name the work of Barchet et al., 1993, which discusses the production aggregated planning problem. Within this work, the goal is to minimize the total production cost in order to attend the forecasted demand. The unknown variables (stock position, productive capacity, quantities produced in regular regime, in extra hours and in sub-contracts, etc) are included in the cost function, representing the solution for the LP problem. Within this section, this concept is used in the problem of optimal 'in stock' replacement and it is related to the cost function, which congregates the storage costs and the 'in stock' replacement. The constraints regarding the operational limits of the productive environment are included to compose the LP problem. The methodology is presented below:

Be $J(k)$ a cost function related to the stock system in the time moment k. The optimization problem is:

$$\min_{\mathbf{u}(k)} \quad \sum_{k=1}^{N} \hat{y}(k) + \sum_{k=1}^{N} \mathbf{u}(k) \tag{3}$$

$s.to$

$$\hat{y}(k+1+i) = \hat{y}(k+i) + u(k-d+i) - \hat{d}(k+d+i)$$

$$y_{\min} \leq \hat{y}(k+i) \leq y_{\max}$$

$$u_{\min} \leq u(k+i) \leq u_{\max}, \quad i = 0, \cdots, N-1$$

Where:

$$\hat{y}(k) = \left[\hat{y}(k+1|k) \quad \hat{y}(k+2|k) \quad \cdots \quad \hat{y}(k+N|k) \right] \tag{4}$$

$$\mathbf{u}(k) = \left[u(k|k) \quad u(k+1|k) \quad \cdots \quad u(k+N-1|k) \right]$$

$y_{\min}$, $y_{\max}$, $u_{\min}$ and $u_{\max}$ are the operational constraints within the productive environment. In the problem, N is the end of the productive time, $\mathbf{u}(k)$ and $\hat{y}(k)$ contain, respectively, the optimal values for the production future orders and the forecasted values for the components' in stock future position, calculated in the time moment k. This algorithm is applied in a time-period N times k, i.e., the N replacement optimal values $u(k)$ are calculated in a sole time and, later on, used in the system until the end of the productive period. The solution for the problem is found in Bazaraa (1979).

3.3 The replacement algorithm based on the optimal control with forecast of one step ahead

Within this section, one deals with the problem of components' in stock optimal replacement proposed in Pacheco & Oliveira (2003b), using the control technique based on the forecast of one step ahead. The control rule is defined from the minimization, related to the variable $u(k)$ in the time instant k of the following cost function:

$$J(k) = (\hat{y}(k+j|k) - w(k+j))^2 \tag{5}$$

Where $w(k+j)$ and $\hat{y}(k+j|k)$ are, respectively, the wanted value for the stock in the instant $k+j$ and the stock position forecast in the instant $k+j$ realized in k. This function minimization related to $u(k)$ in the time k assures the orders optimal replacement. In order to solve the problem, it is necessary to foresee the stock position j steps ahead, considering the information until the time k. The forecast equation is presented in Pacheco & Oliveira (2003b) and it is given by:

$$\hat{y}(k+j|k) = u(k) + \hat{y}_l(k+j|k) \tag{6}$$

Where $\hat{y}_l(k+j|k)$ it is:

$$\hat{y}_l(k+j|k) = (1 + q^{-1} + \ldots + q^{-d-1})u(k-1) - \tag{7}$$

$$(q^d + q^{d-1} + \ldots + 1)\hat{d}(k|k) + y(k)$$

This equation $\hat{y}_l(k+j|k)$ represents all the terms of the stock position forecast, which are known in the instant j time k. Replacing this equation in the cost criterion and minimizing, regarding the components replacement, one defines a quadratic optimization problem of a variable given by:

$$\min_{u(k)} \quad u^2(k) + 2u(k)(\hat{y}_l(k+j|k) - w(k+j)) \tag{8}$$

$$s.a. \qquad u_{\min} \leq u(k) \leq u_{\max}$$

Where $u_{\min}$ and $u_{\max}$ are the operational constraints of the productive environment. This algorithm implementation is made according to: to each time moment k the algorithm is calculated and generates an optimal $u(k)$ and in the following instant is repeated until the end of the productive period.

3.4 The replacement algorithm based on the optimal control with forecast j steps ahead

Within this section, one treats the components in stock optimal replacement proposed in Oliveira & Pacheco (2003), using the control technique known as receding horizon. The generated control signs are based on the system forecasted information j steps ahead. The procedure for the stocks' management is based on the representation of the stock system performance. Through a function cost and in this function minimization regarding the future production orders of components, that is, the vector $u(k)$. Being the cost' function given by the equation

(3) and redefined in the following manner:

$$J(k) = \left\| \hat{y}(k) \right\| + \left\| u(k) \right\| \qquad (9)$$

In order to solve the problem, it is necessary to foresee the stock's future position j steps ahead. The prevision equation is presented in Oliveira & Pacheco (2003) and it is given by:

$$\hat{y}(k+j \mid k) = H_j(q^{-j})u(k+j-1) + \hat{y}_l(k+j) \qquad (10)$$

Re-writing the prevision equation (10) of a vectorial manner, one has:

$$\hat{\mathbf{y}}(k) = \mathbf{H}\mathbf{u}(k) + \hat{\mathbf{y}}_1(k) \qquad (11)$$

Within this equation $\hat{\mathbf{y}}(k)$ and $\hat{\mathbf{y}}_1(k)$ are vectors, which contain the elements $y(k+j \mid k)$ and $\hat{y}_l(k+j \mid k)$, respectively.

The criterion minimization of the cost equation (9) regarding the future orders of components' production is similar to the equation (3) and it is given by:

$$\min_{\mathbf{u}(k)} \quad J(k) \qquad (12)$$
$$s.to \quad \mathbf{y}_{min} \le \hat{\mathbf{y}}(k) \le \mathbf{y}_{max}$$
$$\mathbf{u}_{min} \le \mathbf{u}(k) \le \mathbf{u}_{max}$$

Where y_{min}, y_{max}, u_{min} and u_{max} are vectors, whose components are $y_{min}, y_{max}, u_{min}$ and u_{max}.

Replacing the equation (11) in the cost function (9) and (12), and minimizing regarding the components replacement, one reaches the classical problem of the quadratic programming, whose solution is largely studied in the literature (Bazaraa, 1979). This optimization problem is equivalent to:

$$\min_{\mathbf{u}(k)} \quad \mathbf{u}^T(k)\left(\mathbf{H}^T\mathbf{H} + \lambda\mathbf{I}\right)\mathbf{u}(k) + \left(2\mathbf{H}^T + \mathbf{y}_1(k)\right)\mathbf{u}(k) \qquad (13)$$
$$s.to \quad \mathbf{y}_{min} - \mathbf{y}_1(k) \le \mathbf{H}\mathbf{u}(k) \le \mathbf{y}_{max} - \mathbf{y}_1(k)$$
$$\mathbf{u}_{min} \le \mathbf{u}(k) \le \mathbf{u}_{max}$$

Where $\mathbf{I}$ is the identity matrix. This algorithm implementation is made according to: to each time instant k the algorithm is calculated and generates an optimal $u(k)$, based on the system future information and in the subsequent instant is repeated until the end of the productive period.

4. Comparative study among the replacement algorithms

In this section, one makes a comparative study among the designed three optimal replacement algorithms. The result is measured through the total production cost (*Ct*), composed by the costs, the lack of stock (*Cfe*) and the manufacturing (*Cp*). An average replacement value (*Rme*) identifies if the system performance reaches, approximately, the average value of the forecasted demand throughout the productive period. The productive system

illustrated by Figure 1 is subject to the uncertain demand and the capacity constraint.

The actual demand describes the behavior of the consumer-market and it is presented by Figure 2. One observes through the figure that the demand peaks are seasonal and represent periods of high consumption. These periods occur, often, when there are commemorative dates. For instance, Mother's Day, Fathers' Day, Valentine's Day and Christmas Holidays. The forecast error represented by $\zeta(k+1)$ demonstrates the uncertainty in the forecasted demand.

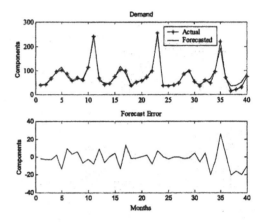

Figure 2: Actual and forecasted demand and forecast error

4.1 Comparative study without constraint in the productive system

Within this section, the objective is to make a performance's minuscule analysis of the replacement algorithms presented in section 3, when there isn't operational constraint in the productive environment. The stock position is determined through the average value of the forecasted demand, that is, within the 75 components, the initial stock position is of 60 components and the productive lead-time assumes integer values between 0 and 2. The algorithms are implemented in the productive environment exactly as presented, i.e., one implements in the system the N (productive period) stock replacement optimal values, i.e., only $u(k|k)$ is implemented in the system and, in the subsequent time instant $k+1$, the procedure is repeated. The nomenclature used in the replacement algorithms is the following: *LP* (*Linear Programming*), *MV* (*Minimum Variance* or replacement based on the forecast of one step ahead) and *GPC* (*Predictive Control* or replacement based on the forecast j steps ahead). Table 1 presents the system's performance.

In Table 1, one observes that in all the algorithms, independently of lead-time, the average stock replacement index (*Rme*) comes near to the designed value for the stock initial position. The *LP* algorithm presents a good performance for environments without lead-times in the

Algorithms	Lead-time	Cp	Cfe	Ct	Rme
LP	0	3029	0	3029	75
LP	1	3047	1540	4587	76
LP	2	3055	3698	6753	76
MV	0	3043	0	3043	76
MV	1	3080	0	3080	77
MV	2	3093	0	3093	77
GPC	0	3043	0	3043	76
GPC	1	3080	0	3080	77
GPC	2	3116	0	3116	77

Table 1: Performance of the optimal replacement algorithms without operational constraint

Algorithms	Lead-time	Cp	Cfe	Ct	Rme
LP	0	3029	0	3029	75
LP	1	3047	1425	4472	76
LP	2	3055	3525	6580	76
MV	0	3043	0	3043	76
MV	1	3080	0	3080	77
MV	2	3097	0	3097	77
GPC	0	3043	0	3043	76
GPC	1	3080	0	3080	77
GPC	2	3064	13	3077	77

Table 2: Performance of the optimal replacement algorithms with operational constraint

manufacturing, i.e., the market demand is supplied. However, when there is a lead-time increase, the *Cfe* tends to increase also, making the stock rupture inevitable (*stock-out*). Incorporating to the productive environment penalties related to lower productivity and leading to the clients' non-supply, this is illustrated by the lead-times 1 and 2. The *MV* algorithm presents an excellent performance for all lead-times, managing to assure the manufacturing with the lowest possible operational value and minimizing the stock position. The *GPC* presents a similar performance to the *MV* algorithm in the observed results with little variations in the lead-time 2, that is, due to the adjustments produced in the forecast and the control horizons.

4.2 Comparative study with constraint in the productive system

Within this subsection, the goal is to make a performance minuscule analysis for the replacement algorithms presented in section 3, when there is operational constraint in the productive environment and under the same initial conditions defined in subsection 4.1. The operational constraints are rigid if compared to the demand peaks and have strong influence in the productive environment performance. The capacity constraint in the manufacturing is defined in 220 components and the stock position in 150 components. In the replacement algorithm *MV*, one utilizes the criterion of prevention in the lack of components in stock presented in Pacheco & Oliveira (2003b), with a 4 (four) months forecast for the market peak reach, i.e., one anticipates the manufacturing start. In the *GPC* algorithm, one changes the forecast and the control horizon in order to improve the system's performance. In Table 2, the system's performance with constraints in the productive environment is presented.

In Table 2, one observes the replacement average index (*Rme*) presents similarity to the comparative analysis made in subsection 4.1. The *LP* replacement algorithm presents a similar performance when there isn't operational constraint. In the productive environment without manufacturing lead-time, the market demand is supplied and the manufacturing operates with a

minimum cost. However, when there is a lead-time increase in the manufacturing, the algorithm cannot supply the market demand, and in consequence the stock rupture is inevitable (*stock-out*). Incorporating to the productive environment penalties related to lower productivity and leading to the clients' non-supply, that is illustrated for the lead-times 1 and 2. The *MV* algorithm presents excellent performance, even when there is a lead-time increase for the manufacturing. In all cases, the strategy of the lack of components can supply the demand peak reach and the manufacturing operational capacity is not surpassed, thus the stock rupture risk is removed. The *GPC* algorithm presents similar performance to the presented through *MV* algorithm, when the lead-time is 0 and 1. In the lead-time 2, the manufacturing cannot supply the demand, consequently a negligible cost related to the lack of components is observed, even when adjustments have been made in the forecast and the control horizon.

5. Conclusions

This article discussed the performance of three algorithms for the 'in stock' components optimal replacement presented in the works of Oliveira & Pacheco (2003) and Pacheco & Oliveira (2003b). Two analysis were accomplished and the performance was measured through the total cost of production and the 'in stock' average replacement, in the first analysis the productive environment doesn't present operational constraints and in the second analysis rigid operational constraints are imposed. The *MV* algorithm and the *GPC* presented excellent performance in both analyses even when there is a lead-time increase. The *LP* algorithm presented excellent performance, when there isn't lead-time in the manufacturing, however, to the lead-time increase the manufacturing cannot supply the market demand, in consequence costs and penalties are applied, demonstrating its weak performance. Thus, one concludes the algorithms proposed in the designed works, based upon the forecast of one step and *j* steps ahead, obtain superior performance related to the linear programming, within this comparison.

References

ASTROM, K., J. & WITTENMARK, B. (1995). Adaptive control. Addison Wesley Publishing Company, Inc. 3° Edição.

BARCHET, V., M., F., LARA, C., S. & SAIBT, E. (1993) – Métodos para Planejamento Agregado. Anais do XIII Encontro Nacional de Engenharia de Produção. Vol. 2, pp. 951-956.

BAZARAA, M., S. & SHETTY, C., M. (1979) - Nonlinear Programming. John Wiley & Sons, Inc.

MOREIRA, D., A. (1993) – Administração da Produção e Operações. Ed. Pioneira.

OLIVEIRA, G., H., C. & PACHECO, E., O. (2003). Estratégia Baseada em Horizonte Rolante para Gestão de Estoques. XXIII - Encontro Nacional de Engenharia de Produção (ENEGEP).

PACHECO, E., O., OLIVEIRA, G., H., C. & PACHECO, R., F. (2002) - Utilização de Conceitos de Variância Mínima na Gestão de Estoques. XXII - Encontro Nacional de Engenharia de Produção (ENEGEP).

PACHECO, E., O. & OLIVEIRA, G., H., C. (2003a) – Using MV control concepts in inventory management subject to uncertain demand. Annual meeting of the Production and Operations Management Society (POMS'03)

PACHECO, E., O. & OLIVEIRA, G., H., C. (2003b). Controle Ótimo de Estoques Sujeitos a Picos de Demanda do Mercado. XXIII - Encontro Nacional de Engenharia de Produção (ENEGEP).

TOWILL, D., R. (1982) Dynamic analysis of an inventory and order based production control system. International Journal of Production Research; 20; pp 671-678.

TOWILL, D., R. & DELVECCHIO, A., L. (1994) - The application of filter theory to the study of supply chain dynamics. Production Planning and Control, pp 82-96; 5.

TOWILL, D., R., EVANS, G., N. & CHEEMA, P. (1997). Analysis and Design of an Adaptive Minimum Reasonable Inventory Control System. Production Planning and Control. Vol. 8, n. 6, pp. 545-557.

VASSIAN, H., J. 1954. Application of Discrete Variable Servo Theory To Inventory Control. Operations Research Society of America, p. 272-282. Chicago, Illinois.

HYDROCYCLON SYSTEM FAILURE PREDICTION USING ARTIFICIAL NEURAL NETWORK

Pedro Palominos[*]
Claudio Carrasco
Ruben Galarce

Industrial Engineering Department
University of Santiago of Chile
3769 Ecuador Ave., Santiago, Chile
* ppalomin@usach.cl

ABSTRACT

This research presents a predictive failure model based on Artificial Neural Networks (ANN) developed for a mineral classification process. The classification process uses hydrociclons and it is performed at the semi-autonomous grinder plant of Codelco's Division El Teniente. The developed model is based on a Perceptron Multilayer and the nnarx function from Matlab6p5, which gives a correlation factor of 99.11 % with he real data series, and a MAPE of 0.0031%.

Keywords: Artificial Neural Networks Artificiales, Failure Prediction, Hydrocyclones

1.- INTRODUCCIÓN

Codelco's Division El Teniente, the Chilean largest copper producer, has three main production processes: the underground mine, in which mineral is extracted by the block sinking method; the concentrator, which operates by copper sulfurs flotation; and the foundry which produces copper of 99.98% purity by traditional concentrates fusion. This copper is then sent to international markets.

Raw material from the underground mine is first processed at the primary grinding stage and then sent by conveyor to the stock pile. The stock pile has a maximum storage capacity of 60,000 tons, 80% of which is less than 200,000 microns in size.

Raw material is taken from the stock pile by four feeders with a linear arrangement and variable speed conveyors. Feeders take the raw material to conveyor 461 (60 inches wide and 120 meters long from pulley to head) which takes the material to the Semi Autonomous Grinder (SAG).

The grinding plant processes the raw material that comes from the mine and which has been through several comminution stages. The key aspect of this grinding stage is the delivery of ground mineral ready to be beneficiated in the subsequent flotation stage. The classification system in the SAG plant consists of filters for the mineral particles, allowing only the desired caliber to continue to the next process stage. Due to this classification process, it is very important to maintain the classification systems stabile and minimize the affecting disturbances upon them.

Natural disturbances in the feeding of mineral to this stage generate instabilities of the circulating load at the comminuting circuit. The instabilities generate an overload in the pulp flow in the internal pumping circuits, which results in thrust duct obstruction by exceeding the ducts capacity (saturation) and load spills on the bay floor. Mineral saturation at the hydrociclons normally takes a considerably large amount of non-programmed maintenance hours, forcing the complete stop of the grinding process of the corresponding line. Annual production losses due to this reason adds to US$280.000, which makes attractive to minimize the failure events per year.

One of the techniques commonly used for predictive purposes is Artificial Neural Networks (ANN), for example in finance [1], weather forecast [2], air quality [3], or market forecast [4] among others. Even more, the comparison study of ANN to statistical methods by Hill et al. [5], concludes that ANN applications result in better performance. In predictive maintenance, particularly in mining industry, few research can be found, and only Flament et al. [6] propose control strategies based on ANN for a grinding plant, but not applied to predictive maintenance.

2.- DEVELOPMENT

In this section, the key variables affecting the obstruction of the thrust ducts of the mineral classification process are identified. Additionally, several ANN architectures are evaluated by simulation for identifying a model capable of predicting failures (obstructions) in a timely and reliability fashion.

2.1.- Key Process Variables

Identification of the key variables involved in a failure event was obtained with assistance of metallurgical and grinding process company experts, considering measuring instruments reliability.

Variable selection considers the fact that mineral saturation in a battery of hydrociclons occurs mainly because of either water excess in the mineral download receiving box or mineral overfeeding to the SAG. Both of these situations generate from an uncontrolled increment of fine mineral in the fresh feed to the system. These operational conditions affect in a significant manner the mineral pulp feeding to the classification system. With this reasoning in mind, the selected process variables are:

a) Weight of fresh mineral feeding to the grinder (tons per hour) measured with a four stations weight sensor system.

b) Amount of water added to the mineral download of the SAG (cubic meters per hour), measured with a magnetical flow meter.

c) Amount of mineral pulp feed to the hydrociclons battery (cubic meters per hour), measured with a nuclear densimeter and a flow meter.

Output variable (Control) is the hydrociclons battery feeding pressure (psi), measured with a pressure transmissor. This measurement indicates the occurrence of obstructions.

2.2.- Data Collection

Operational data was obtained for normal and abnormal plant's operation condition. Figure 1 presents a random selection period of 5000 minutes from the historical data series.

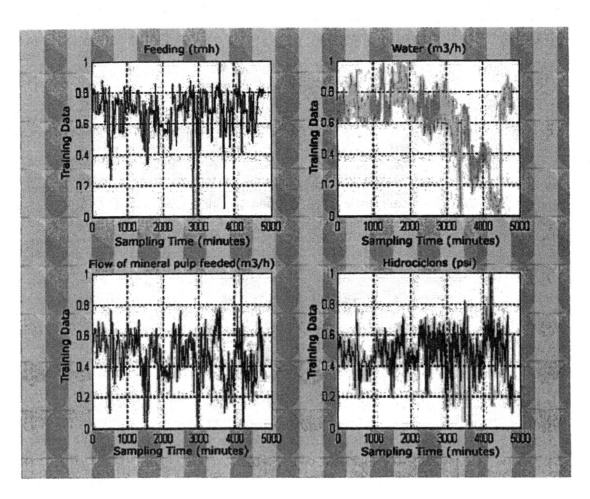

Figure 1. Training Data

316

For training of the ANN, an additional set of data is required. This data set does not take part of the ANN training process and is used to validate the control output generated as result of the ANN learning process. The verification data set also includes normal and abnormal plant operation situations and it includes the plant operation data from October to December 2003. A last set of data is used for ANN generalization and it includes operational data for the period from December 2003 to January 2004.

2.3.- ANN Architecture Design

The ANN architecture design is based on the Multilayer Perceptron Model (MLP). This particular model was chosen because its capacity to model a wide variety of functional relationships, as shown in a number of applications [7]. The MLP ANN class considered here has one hidden layer and only hyperbolic tangents and linear activation functions. The synaptic weights are the adjustable ANN parameters and its values are set through examples during the ANN training process. The training process involves a set of input data u(t), which relate to the process output data y(t) for a time period t (Eq. 2)

$$Z^N = \{ [\, u(t),\, y(t)\,] \,/\, t = 1,\ldots,N\} \quad (2)$$

The objective for the training stage is to determine a mapping of the training data set to the set of weightings, in a way that prediction ŷ(t) generated with the model will be the closest to the real output y(t). Backpropagation algorithm, using the descending gradient is used for the training of the ANN. This process is accomplished with the software product *Neural Network Based System Identification Toolbox*, version 2, developed by *Magnus Norgaard* [8] using *MatLab6p5*.

In order to obtain the failure forecasting model, non-linear identification systems from *MatLab6p5* were used. This system has the Multilayer Perceptron structure and some variations from it such as:

➢ nnarx function, which takes the input signals and predicts the output signal.

Regression Array:

$$\varphi(t) = [y(t-1)..y(t-n_a)u(t-n_k)..u(t-n_b-n_k+1)]^T \quad (3)$$

Predictor:

$$\hat{y}(t/\theta) = \hat{y}(t/t-1,\theta) = g(\varphi(t),\theta)$$

$$(4)$$

➢ nnoe function, which takes the predicted value for the input signal and the input signals and predicts the output signals.

Regression Array:

$$\varphi(t) = [\hat{y}(t-1/\theta)..\hat{y}t-n_a/\theta u(t-n_k)..u(t-n_b-n_k+1)]^T$$

$$(5)$$

Predictor:

$$\hat{y}(t/\theta) = g(\varphi(t),\theta)$$

$$(6)$$

➢ nnarmax2 function, which takes the input signals and the generated error and predicts the output signal.

Regression Array:

$$\varphi(t) = [y(t-1)..y(t-n_a)u(t-n_k)..u(t-n_b-n_k+1)\varepsilon(t-1)..\varepsilon(t-n_c)]$$

$$(7)$$

Predictor:

$$\hat{y}(t/\theta) = g(\varphi(t),\theta)$$

$$(8)$$

$\varphi(t)$: Regression array.

θ: Synaptic weight array.

g: ANN function.

n_a, n_b, n_c, n_k: Input characterization array.

$\varepsilon(t)$: Forecast error, defined as $\varepsilon(t) = y(t) - \hat{y}(t/\theta)$

$y(t)$: Desired output.

$\hat{y}(t)$: Forecasted output.

$u(t)$: Input arrays.

2.4.- ANN simulation and error measurement

To determine the number of neurons per layer, the ANN is simulated with two hyperbolic tangential neurons and one linear neuron at the output, 1,200 iterations. At each iteration, the number of internal neurons is increased by two until it reaches 30, keeping all other parameters fixed. Performance of the ANN is evaluated at each iteration, This process will locate the design for which the training, cross validation, and generalization error is minimum. This design process and evaluation was performed for the nnaex and nnarmax2 functions, while the errors measurements considered were: Correlation Factor (R), network efficiency (PI), and Mean Absolute Percentage error (MAPE).

3.- RESULTS

No significant difference was observed in the correlation factor or efficiency index for the three ANN architectures simulated. However, the nnarx function with architecture 2-2-1,200 (2 delays in input data, 2 neurons per delay, and 1,200 iterations) is the best and its results are presented in Figure 2 and the respective error measures in Table 2.

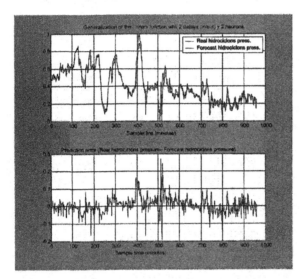

Figure 2. Adjustment and Forecast Error for Architecture 2-2-1,200

2 – 2 – 1,200	Training	Cross Validation	Generalization
R (%)	91,31	96,49	99,11
PI (%)	0,083	0,079	0,033
MAPE (%)	0,0010	0,00084709	0,0031

Table 2. nnarx Function Architecture with 2 Neurons

Forecast quality of the ANN was tested with a new set of data for a period of 10 days of continuous operation (March 1 to 10, 2004). The results are shown in Figure 3 with errors R(%)= 99.71 ; PI(%) = 0.045 y MAPE (%) = 0.00032652.

Regarding its operation, the model is a reliable instrument to predict failures, triggering corrective actions when an alarm signal indicating a negative pressure gradient between times (t-1) and (t) advises. Under that emergency situation, the operator has one minute to take the necessary action in the automatic control system, activating the closing and opening valves for the operating and backup hydrociclons, allowing the pressure to build up until reaching the normal operating level. The system will allow the steady state operation, minimizing the characteristic disturbances in continuous processes, increasing considerably the SAG plant availability.

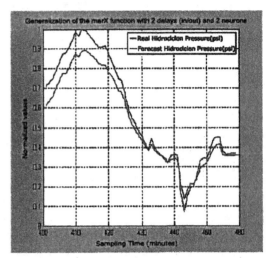

Figure 3. Hydrociclon Real and Forecast Pressure for Architecture 2-2-1,200 for 460 data Points.

4.- CONCLUSIONS

From the obtained results, it can be concluded that ANN based models present fair performance to predict the saturation condition in the SAG grinding process at Codelco's Division El Teniente, with a response time of one minute, considered enough for the operator to take corrective action.

Specifically, a 2-2-1,200 ANN architecture was selected because of its performance (highest correlation factor, 99,11% and the lowest MAPE, 0,0031%). This architecture implements two neurons, two regression elements in the input signals, 1,200 iterations, and the nnarx function.

Finally, implementation of the ANN in the production process, according to preliminary estimation, would cost US$ 56,366 for instruments while the saturation cost per year adds to US$ 279,065. These figures evidence the economic feasibility of the model implementation, because the payback period is 2.4 months with the savings obtained.

5.- REFERENCES

[1] Aiken Milam, "Using a neural network to forecast inflation", Industrial management & Data Systems, Vol 99, N° 7, pp. 296-301, 1999.

[2] Farshad Fred, at all., "Predicting temperature profiles in producing oil wells using artificial neural networks", Vol.17, N°6, pp. 735-754, 2000.

[3] Wang Wenjian, et all., " Three improved neural network model for air quality

forecasting", Enginnering Computations, Vol.20, N° 2, pp. 192- 210, 2003.

[4] Van Wezel and Baets Walter, "Predicting market responses with a neural network: the case of last moving consumer goods", Marketing Intelligence & Planning, Vol. 13, N° 13, pp. 23-30, 1995.

[5] Hill Tim, O'connors Marcus and Remus William, "Neural network Models for time series forecast", Management Science, Vol. 42, N°7, 1996.

[6] Flament Frederick y Thibault James,"Neural Network Based Control Of Mineral Grinding Plants", Mineral Engineering, Vol .6, N°3, pp. 235 - 249, 1993.

[7] Demunth, Hilton y Michael Beale, "Neural Network Toolbox", The Mathworks Inc., USA, 2001.

[8] Norgaard Magnus, "Neural Network Based System Identification TOOLBOX", Department Automation, Technical University of Denmark, January 23, 2000.

www.elsevier.com/locate/ifac

THE CHALLENGE OF DESIGNING NERVOUS AND ENDOCRINE SYSTEMS IN ROBOTS

Felisa M. Córdova and Lucio R. Cañete

*Laboratorio de Concepción e Innovación de Productos, LACIP.
Departamento de Ingeniería Industrial, Universidad de Santiago de Chile.
Santiago de Chile, fcordova@usach.cl*

Abstract: We discuss in conceptual terms the feasibility of designing a Nervous System and an Endocrine System in a robot and to reflect upon the bionic issues associated with such highly complex automatons. The emulation of biological phenomenon in artificial systems, both nervous and endocrine imitation in a mechatronic automaton is an attractive proposition as a mechanism of organic integration. The ability of living organisms to maintain their internal organization in spite of external changes, encourages attempts to imitate such achievements in devices. The complexity of the network sensors-integrators-effectors and the proportion of internal fluids must be increased when seeking the homeostasis of the robot from mechatronic design to bionic design.

Key words: bionics, integration, robots.

1. INTRODUCTION

Biology has developed in a very accelerated way over the last three decades with the 21st Century, widely predicted to be the "Biology Age" (Ovchinnikov, 1987). In this context, a certain interest in emulating biological characteristics in appliances can be seen, for the purpose of recreating the successful behavior of living organisms in artificial systems.

Sensors such as hair and skin, effectors such as feet and fins and integrators such as artificial brains are some examples of bionic projects (Cañete, 2002). However, in order to come close to achieving the behavior exhibited by animals as inspiratory beings, then by necessity (Ofek, 2001), it must be recognized the pertinence of Nervous System and Endocrine System.

Of course, when the Nervous System and the Endocrine System work in unison in animals, they could generate synergic effects in the behavior of these alive beings. The use and integration of both systems in a human being can be seen for example when a person is attacked by a swarm of wasps. The electromagnetic receptors (the eyes), acoustics (the hearing) and those related to pain (the skin) receive information which is interpreted as threatening, so the association neurons and effectors not only order the locomotors effectors (the leg's skeletal muscles) to take flight but also order the glands to release adrenaline and other hormones that designate resources to the task of escaping.

No doubt, animal characteristics like these are interesting to emulate in automatons (Cañete and Córdova, 2003). So, the objects of this work are to conceptually design a Nervous System and Endocrine System in robots as a biological emulation, and to reflect upon the bionic design of those systems.

2. WHY A NERVOUS SYSTEM AND A ENDOCRINE SYSTEM ?

On examining any animal, one observes that the digestive, circulatory, respiratory and excretory systems have specific structures and functions, but none of them can carry out their functions independently of the others; rather they work together in harmony to meet the metabolic needs of the body. Any living being, not only an animal, is an organized entity that behaves as a unit and this unity is the result of the participation of a larger coordinated and integrated system. As an automaton, a robot must also possess such a system, and select animal inhalation as a perfect example of how it must fit together with its environments.

Vertebrates and other multicellular beings whose structures are the result of the evolution (Paccault, 1991), the source of the emulation, posses two coordination systems of varying form and complexity that are closely related: Nervous and Endocrine. Both serve to make the body a unique active entity, individually regulating the functions of it parts and enabling it to adjust to changes.

In biology texts, robots designers are accustomed to seeing both systems compared with each other with the emphasis on the considerable difference that exist between them: in the Nervous System the message is an electrochemical disturbance (a nerve impulse) that travels through a nerve fiber, whereas in the Endocrine System the message is a chemical substance (a hormone) that travels in fluid, like blood.

If a robot had a sort of these system, it behavior in aggressive environments would be better. So, as it shown in Figure 1, the immediate task is to emulate the Nervous System and the Endocrine System from natural stage to artificial stage (animal to robot).

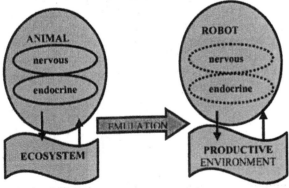

Fig 1: From a scene of origin towards another one of destiny. Considering the contribution of the Nervous System and Endocrine System to the successful performance of the animal facing its ecosystem, such systems conceptually may be copied in a robot to face its productive environment.

3. THE NERVOUS SYSTEM

3.1 The necessity of sensors

It is possible to face the design of basic a Nervous System that allows to make the sensomotor coordination to guide a Load-Haul-Dump (LHD) vehicle inside a street within a tunnel in an underground mine (Córdova, 1996). The vehicle loads mineral from a pit and it dump it in a ore-pass. Then, a set of sensors could be in charge to acquire the relevant data from the tunnel. Among them, is possible to use some of the following ones:
- Distance sensors, that can be an adjustment of ultrasonic sensors, or infrared sensors.
- Laser sensors to measure depth.
- Tact and of proximity sensors.
- Vision sensors, that uses TV cameras and processors of stereoscopic images, that can detect a painted line in the ceiling of the gallery or the profile of the tunnel.
- Angle sensors.
- Torque and acceleration sensors.
- Sensors of virtual surroundings, that can detect them holes of ditches and resentments.
- Sensors of the operational conditions of the vehicle, among them sensorial of humidity, temperature, pressure.
- Sensors of environmental conditions of the tunnel, such as fire detectors.
- Tags, that allows to detect figures of bar codes and it the location of the vehicle inside the gallery.

3.2 A sensomotor coordination design

In this particular design, an array of six located ultrasonic sensors in the flanks of the vehicle is used. The navigation criteria is of homeostatic type, that means that the vehicle maintains a equidistant distance to the walls of the tunnel. In this design, a neuronal network of sensomotor coordination can be used that coordinates the sensors and the actuators. The strategy of sensomotor coordination makes use of a propioceptive network presented in Figure 2.

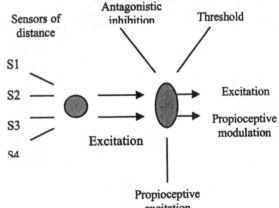

Fig 2: Propioceptive network

In the reflective network of senso-motor coordination of Figure 3, from the general and previous design of a fuzzy neuron, defined the characteristic of design for each neuron in individual, obtaining itself eight types of neurons. In the case of the sensorial neuron a not-linear excitation (sigmoidal) and with different weight for each sensor from the respective sub-network is required (the Agonist-Antagonist sub-networks are symmetrical in their construction).

Modulating Neuron **m** dynamically varies the characteristic of the Excitatory Neuron **e1** and the Threshold Neuron. This activity of modulation is function of the speed of the vehicle and the ratio of turn **Rg**. The rest of the neurons has excitatory and/or inhibitory activities according to its connectivity in the network.

The rank of activity of all the neurons varies between 0 and 1, minimum activity and maximun activity for all the variables of the neurons. The signals of activity are Excitation, Inhibition, Modulation and a Threshold of adjustment for each neuron.

- Excitatory Neuron e1: this neuron receives the activity of the distance sensors whose weight decreases and this is modulated as well by the activity of the Modulating Neuron depending on inverse on the turning ratio and the speed of vehicle v at low speeds and v2 at high speeds (over v2 march).
- Excitatory Neuron e2: this neuron actives the turn actuator, receiving a modulated activity of the previous activity of its sensors, inhibitions of the antagonistic network that are against to him and of its own actuator that restrains it and stabilizes as the networks arrive at an activity balance.
- Inhibitory Neuron i1: this neuron allows the incorporation of the activity from one sub-network to the other, and its action is strong in high ranks of internal activity to allow to a fast compensation forehead to strong stimulus of the network that originates it when the networks are very far from the balance.
- Inhibitory Neuron i2: this neuron restrains the activity of the turn actuator when it is in conditions near the balance, or allows a fast turn (depending on the Threshold Neuron) when a strong sensorial activity of distance exists and the vehicle goes fast.
- Modulating Neuron m: this neuron modulates the activity of sensors of distance in agreement with the speed and the turning ratio of the vehicle, in addition stimulates to the neuron threshold when its own activity is high along with the activity of the distance sensors.
- Threshold Neuron: When this neuron is stimulated by the Modulating Neuron, it inhibits to inhibitor of the actuator to correct in fast form the direction of the vehicle. This is

required in the case that the vehicle has a small turning radius or when the speed is high.
- Sensor Neuron: The activity received by each sensor of the sub-network is heavy in decreasing form while more moved away it is of the advance direction, in addition the sensation one to the activity is not-linear for each sensor.

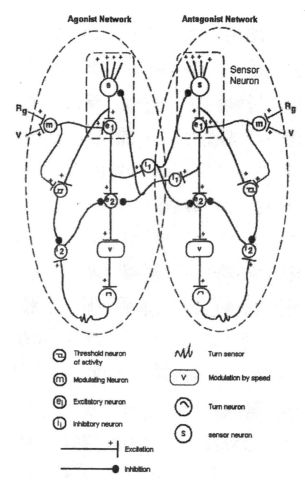

Fig. 3: Sensomotor network

Structurally speaking, nerves can be emulated and in fact they have been, by way of electrical conductors specifically in mechatronic automaton like LHD.

4. THE ENDOCRINE SYSTEM

4.1 The necessity of an internal fluid

In the case of blood, it has not been emulated in robots and the absence of internal fluid in an automaton was one of the reasons why work was halted on the emulation of other animal features dependent on internal chemical transportation.

Of course, the existence of fluids in the body of the robot would not only facilitate the emulation of the Endocrine Systems' own hormones but also the emulation of antibodies for immunity, of bradicins (when animal tissue is broken, enzymes are released that convert certain plasmatic proteins into a substance called bradicin), for nociception (perception of damage) and of solvents for quimosensitivity (to stimulate the sense of taste, the chemical substances must be dissolved).

Furthermore, a fluid not only facilitates the movement of intra-corporal messengers but could also act as a lubricant, combustible and a thermo-regulator.

A classic mechatronic robot contains in percentage terms of volume approximately one sixth part of fluids which are principally lubricants, and substances for hydraulic devices and combustion. An arthropod on the other hand, contains at least three quarter parts of fluid (Cañete and Córdova, 2003).

Therefore, the first challenge in this respect consists of the proportion of internal fluids in the robot. Such an increase would lead to an increase in homeostasis which would bring with it an increase in the design complexity as shown in Figure 4.

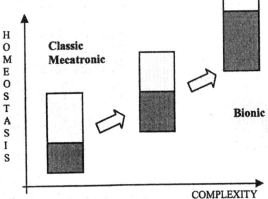

Fig. 4: Complexity versus homeostasis.
An increase in complexity for Endocrine System, comes with increasing the proportion of internal fluids when seeking the homeostasis of the robot from mechatronic design to bionic design.

4.2 A simple design

When a robot works in changing environments, some parts of the surroundings without its control would become dangerous. In such situations, the robot will have to react in urgent form to fit its internal structures and to assure its viability. One of such urgencies can be the repair or reinforcing a part of the body that is being exposed to an external requesting. In order to face such situation it is possible to be resorted to a endocrine emulation as it is explained next and presented in Figure 5.

The internal part of the sensible zone of the body exposed to harmful emboss and ruptures are designed rough and it is covered with little smooth laminas.

The last ones are conductive elements with bad plastic behavior; of such form get fractures and can release them when happens a damage of the outside.

The internal part in addition is bathed by a fluid and slightly energized with electrical current.

An internal sensor to the circuit of the fluid exists somewhere in addition that is able to capture fragments of little laminas conductors by magnetic action. Whenever it captures some piece of laminas, releases a plastic and rough substance in form of grume (small semisolid mass) and it orders to a pump to increase the speed of the fluid circulation. The rough grumes slide by the smooth surface and they are crowded in the rough surface that is being dangerous. Finally the grumes reinforce the exposed zone.

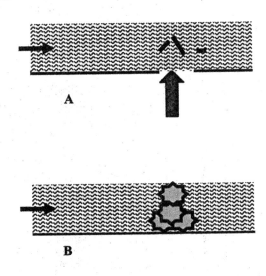

Fig 5: Cross section of the bodywork with phenomenon of endocrine emulation.
A: An external agent breaks or damages the interior where a fluid circulates, conductive fragments are freed in the open leaving one rough surface. The fragments when traveling by the fluid are captured by a magnetic receiver that releases rough grumes.
B: The rough grumes travel by the current and they lodge in the naked zone, cushioning the external action. The effect on the speed flow caused by the grumes is insignificant due to the extension planar of the conduit and to the precise location of the grumes.

Why not to design a hard bodywork and thus to avoid a complex Endocrine System ? For the same reason for which the nature through million years of evolution has not done it (David and Samadi, 2000). Comparatively (Massé and Thibault, 2000), to maintain a hard bodywork is more expensive than to maintain a light bodywork with repair capacity.

5. ORGANIC INTEGRATION IN THE ROBOT

Once the Endocrine and Nervous Systems have been conceived, both systems may be connected to each other and at the same time joined to the functions which they control. It is then that an organization is configured, which for the purpose of the robot being proposed in the current research, would have a internal communication structure as shown in Figure 6.

In animal kingdom, the levels are recursive and thus subordinates: that is to say the Supreme Level contains and governs the Homeostasis Level (Nervous and Endocrine) which in turn does the same with the Sensomotor Level.

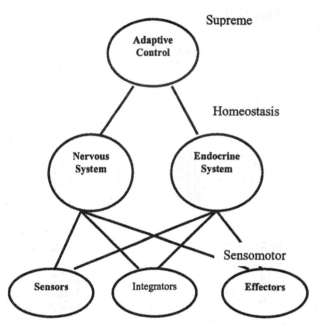

Fig. 6: Three levels in organic integration of bionic robot.The nervous and endocrine reaction corresponds at a Homeostatic Level. Such reactions are subordinated to Supreme Level, but they govern at Sensomotor Level.

Adaptive Control, as the name implies, is the higher level of coordination and functional integration in a robot. Its structure can be a managed by a system that knows the mission the robot and links both information of the inner state of the automata and his surroundings. A proposed future challenge in the present work, is the conceptual design of this Adaptive Control.

6. CONCLUSION

Although the robot has neither nerves nor blood where it may send messages in charge of functionally integrating and coordinating all its parts, they can be emulated.

There is evidence that the emulation with certain restrictions of both Nervous and Endocrine System is feasible. However such an emulation would require the convergence of two sciences: Biology and Robotics.

The convergence of Biology with its different branches (Ecology, Genetics, Evolution among others) and Robotics is no accidental; rather it constitutes an essential issue in as far as abstract terms are concerned, that is: there is little to distinguish automatons and living beings. In reality both beings show behavior that is autopoietic (self produced), telenomic (self initiating) y homeostatic (self regulating with feedback function). Put more precisely: many of the features of a living being are desirable in an automaton.

Aspiring to have such features comes from the observations that man has carried out in the biological world of he himself is a part. The ability of living organisms to maintain their internal organization in spite of external changes, encourages attempts to imitate such achievements in devices, whose internal compositions are quite modest when compared with those of plants and animals.

This motivation, which has further increased on the verification of real successes achieved by man more than three decades ago (in sonar, prosthesis and artificial neuronal networks among others) provides a particular incentive to continue with the current research.

REFERENCES

Cañete, L. (2002). *Ecología cognitiva en robots terrenos para el desierto de Atacama*. Tesis Doctoral, Facultad de Ingeniería de la Universidad de Santiago de Chile.

Córdova, F. (1996). *Guiado autónomo de equipos cargadores frontales LHD en una mina subterránea*. Informe Proyecto FONDEF. CONICYT. Chile.

Cañete, L. and Córdova, F. (2003). *Ecología cognitiva en robots terrenos*. Proceedings of the First IEEE Latin American Conference on Robotics and Automation LCRA, Santiago de Chile.

Cañete, L. and Córdova, F. (2003). *A robot with immunity and perception of damage*. Proceedings of the First IEEE Latin American Conference on Robotics and Automation LCRA, Santiago de Chile.

David, P. and Samadi, S. (2000). *La théorie de l'évolution*. Paris: Champus Université Flammarion.

Glávic, N. and Ferrada, C. (1985). *Biología*. Tercer año de Educación Media. Santiago: Ministerio de Educación..

Ofek, H. (2001). *Second Nature: economics origins of human evolution*. Cambridge: University Press.

Ovchinnikov, Y. (1987). *Basic Tendencies in Physico-Chemical Biology*. Moscow: MIR Publisher.

Massé, G., and Thibault, F. (2001). *Intelligence économique*. Bruxelles: De Boeck Université.

Paccault, I. (1991). *La terre et la vie*. Paris: Larousse.

ELSEVIER
IFAC
PUBLICATIONS
www.elsevier.com/locate/ifac

Branch and Bound Algorithm
for a Transfer Lines Balancing Problem

A. Dolgui * and I. Ihnatsenka *,**

* Ecole des Mines de Saint-Etienne, 158 cours Fauriel Cedex 2, 42023, France

** Grodno State University, 22 Ozshechko St., 230023, Belarus
e-mail: dolgui@emse.fr, ivan@emse.fr, ignatenko@grsu.by

Abstract

In the paper a transfer line balancing problem is considered. Its main difference from the classical simple assembly line balancing problem is in the fact that the operations are grouped into blocks. All the operations of the same block are carried out simultaneously by one piece of equipment (spindle head) and all the blocks at the same workstation are executed sequentially. The set of all available blocks is given beforehand. The constraints on compatibility of operations and blocks at the same workstations are taking into account. The problem is to find the best assignment of blocks to workstations that provides a minimal line cost (estimated by sum of blocks and workstations cost) satisfying the precedence, cycle time and compatibility constraints. A new lower bound based on solving a special set partitioning problem is proposed and a branch-and-bound algorithm is developed. A good quality of the lower bound allows to obtain very promising results.

Key words: Line balancing, optimization, lower bound, branch-and-bound, set partitioning.

1. Introduction

Flow lines are frequently used in manufacturing systems. They allow to increase the production rate and minimize the cost of produced items. The classical line balancing (LB) problem consists of finding an assignment of a given set of operations to workstations taking into account precedence relations between operations. The maximal available work time per workstation is limited by a given *line cycle time*. The objective is to minimize the total idle time for the given line cycle time. Such a problem is a so-called simple assembly LB problem (SALBP) [9].

The SALBP has been intensively studied and a great number of approaches like dynamical programming, branch and bound algorithms, heuristics and metaheuristics are proposed to solve it. Comprehensive surveys on SALBP are given in [9, 8, 10, 14].

A lot of papers are devoted to time oriented assembly line balancing problem and only in a few papers is considered the so-called cost-oriented SALB [15]. For cost oriented SALBP the objective is to minimize the cost per produced unit. In this paper, a cost oriented line balancing problem is considered. The optimal solution may not be time balanced but it must give with a minimal cost.

Most of the works on line balancing are dedicated to manual assembly environment. In this paper, a line balancing problem in machining/process environment for transfer lines is considered. This problem is called transfer line balancing problem (TLBP). The transfer lines are designed for mass production and have high investment cost. However, their high productivity allows to decrease their exploitation cost [6]. In transfer lines, the operations are grouped into blocks. The operations of the same block are executed simultaneously by one spindle head.

At one hand, the high investment cost of transfer line requires an exact solution. At other hand, searching an optimal solution for such a problem is a very complex combinatorial problem.

In literature, there are only a few papers which deal with TLBP [2, 3, 4, 5]. In these papers, it is supposed that the blocks are executed sequentially and the set of all available blocks is not known beforehand. The goal is to minimize a weighted sum of number of workstations and number of blocks (spindle heads). Three approaches for TLBP were proposed: a constrained shortest path, mixed integer programming (MIP) and heuristics.

This paper deals with a TLBP for the case when all the blocks assigned to the same workstation are executed sequentially and the set of all available blocks is fixed. Last assumption signifies that the modular transfer lines composed of "standard" blocks are considered. The objective function is the investment cost for the line design.

The set of all available blocks can be obtained by experience or using an additional study. Therefore, the problem studied in this paper consists of choosing a subset from the set of available blocks which minimise the line cost. The obtained subset must include all operations and respect all the constraints.

There is no buffer permitted in between the workstations and movement of items is synchronized. The workstation time is the sum of block times for the blocks assigned to the workstation. The line cycle time is equal to the maximum of the workstation times.

2. Problem Statement

The line balancing problem, which is considered in this paper, is to assign blocks from a given set to workstations in such a way that:

i) each operation must be assigned once (on one workstation only);

ii) each workstation time is not greater than the given line cycle time;

iii) the precedence constraints which are defined on the set of all operations are not violated. In spite of the fact that the precedence constraints are defined in the same way as for the SALB their usage for the TLBP has some distinctions (due to the fact that the operations are grouped into blocks). For SALB, if operation i precedes directly operation j, then operation i must be executed before operation j. For the TLBP, in some case, a block can contain both operation i and operation j (i.e. in such a block these operations are executed simultaneously). Let operation i precedes directly operation j. The precedence relation signify that: these operations are in different blocks b' and b'' and block b'' is assigned after block b' (to the same workstation or to two different workstations); operations i and j are in the same block (it means that they are executed simultaneously) and this block is chosen and assigned to a workstation. In other words, for different sets of assigned blocks precedence constraints are used differently.

iv) technological constraints cause that some groups of operations must be performed at the same workstation (i.e. they are in different blocks assigned to the same workstation or in the same assigned block). It is supposed, that these groups of operations are given. This type of constraints is called *inclusion constraints for operations*;

v) similarly, groups of blocks which cannot be carried out at the same workstation are given. This type of constraints is called *exclusion constraints for blocks*.

vi) the cost of the transfer line, which can be evaluated as the sum of assigned blocks cost and workstations cost, is to be minimised.

The following notation is used:

$\mathbf{N} = \{1, 2, \ldots, n\}$ is the given set of operations to be assigned;

T_0 is the given transfer line cycle time;

n_0 is the maximal number of blocks for a workstation;

$\mathbf{B} = \{1, 2, \ldots, q_0\}$ is the given set of available blocks;

$\mathcal{N}(b) \subseteq \mathbf{N}$ is the set of operations which are included in block b;

$t(b)$ is the processing time of block $b \in \mathbf{B}$;

$c(b)$ is the cost of block $b \in \mathbf{B}$;

C_0 is the cost of one workstation;

$Pr(b)$ is the set of operations which directly precede (in the ordinary sense) each operation from $\mathcal{N}(b)$;

m is the number of workstations in the obtained solution;

$\sim$ is the following relation between two operations: $i \sim j$, $i, j \in \mathbf{N}$ iff[1] operation i and operation j either must be assigned to two different blocks of the same workstation or must be included in the same assigned block. Clearly, this relation is transitive;

$\nsim$ is the following relation between two blocks: $b' \nsim b''$, $b', b'' \in \mathbf{B}$ iff block b' and b'' can not be assigned to the same workstation;

$I^O(i)$ is a set of operations such that $i \sim j$, for all $j \in I^O(i)$;

$G^r = (\mathbf{N}, D^r)$ is an acyclic digraph representing the precedence constraints for operations. $(i, j) \in D^r$ iff operation i precedes (in the ordinary sense) operation j;

$G^{BE} = (\mathbf{B}, E^{BE})$ is a graph representing the exclusion constraints for blocks, where $E^{BE} = \{(b', b'') \mid$ iff $b' \nsim b'', b', b'' \in \mathbf{B}\}$;

The TLBP with sequentially activated blocks can be formulated as follows: to find a sequence $S = \{s_{11}, \ldots, s_{1q_1}, \ldots, s_{m1}, \ldots, s_{mq_m}\}$, $s_{ku} \in \mathbf{B}$ such that the following conditions are held:

$$\bigcup_{k=1}^{m} \bigcup_{u=1}^{q_k} \mathcal{N}(s_{ku}) = \mathbf{N}, \tag{1}$$

$$\mathcal{N}(s_{ku}) \bigcap \mathcal{N}(s_{rv}) = \varnothing, \ ku \neq rv, \\ k, r = 1, 2, \ldots, m, \ u = 1, 2, \ldots, q_k, \\ v = 1, 2, \ldots, q_r, \tag{2}$$

for all $s_{ku} \in S$,

$$Pr(s_{ku}) \subseteq \Big(\bigcup_{r=1}^{k-1} \bigcup_{v=1}^{q_r} \mathcal{N}(s_{rv})\Big) \bigcup \Big(\bigcup_{v=1}^{u} \mathcal{N}(s_{kv})\Big), \tag{3}$$

$$\bigcup_{u=1}^{q_k} I^O(s_{ku}) \subseteq \bigcup_{u=1}^{q_k} \mathcal{N}(s_{ku}), \ k = 1, 2, \ldots, m, \tag{4}$$

$$(s_{ku}, s_{kv}) \notin E^{BE}, k = 1, 2, \ldots, m, \\ u, v = 1, 2, \ldots, q_k, \tag{5}$$

$$\sum_{u=1}^{q_k} t(s_{ku}) \leq T_0, k = 1, 2, \ldots, m, \tag{6}$$

$$q_k \leq n_0, k = 1, 2, \ldots, m, \tag{7}$$

and the objective function (transfer line cost)

$$C(S) = C_0 m + \sum_{k=1}^{m} \sum_{u=1}^{q_k} c(s_{ku}), \tag{8}$$

is to be minimized.

Constraint (1) and (2) provide that the execution of all operations from $\mathbf{N}$ and that each operation is executed once. Expression (3) defines the precedence constraints between operations. (4) and (5) are operations inclusion and blocks exclusion constraints respectively.

---

[1] if and only if

Constraint (6) allow to obtain workstation times which do not exceed the given line cycle time. Condition (7) means that each workstation can contain at most n_0 blocks.

In next section, a new lower bound for the problem (1)-(8) is proposed.

3. Lower Bound

Let $S' = \{s'_{11}, \ldots, s'_{1q'_1}, \ldots, s'_{m'1}, \ldots, s'_{m'q'_{m'}}\}$, $s'_{ku} \in \mathbf{B}$ be a sequence which satisfies the constraints (2)-(3), (5)-(7) and the following constraint:

$$\bigcup_{u=1}^{q'_k} I^O(s'_{ku}) \subseteq \bigcup_{u=1}^{q'_k} \mathcal{N}(s'_{ku}), \; k=1,2,\ldots,m'-1, \quad (9)$$

Constraint (1) may not be respected.

Let $\mathcal{N}(S') = \bigcup_{s' \in S'} \mathcal{N}(s')$. The set $\mathcal{N}(S')$ can be considered as the state of item after executing all the blocks $s' \in S'$. Denote $\overline{\mathcal{N}}(S') = \mathbf{N} \backslash \mathcal{N}(S')$.

Suppose, there exists a set $\{b_1, b_2, \ldots, b_l\}$, which satisfies the constraints:

$$b_u \in \mathbf{B}, \; \mathcal{N}(b_u) \subseteq \overline{\mathcal{N}}(S'),$$
$$\mathcal{N}(b_u) \bigcap \mathcal{N}(b_v) = \varnothing, \; u \neq v, u, v = 1, 2, \ldots, l, \quad (10)$$

$$\bigcup_{u=1}^{l} \mathcal{N}(b_u) = \overline{\mathcal{N}}(S'). \quad (11)$$

By definition, it is required $m'-1$ workstations to assign subsequence $\{s'_{11}, \ldots, s'_{1q'_1}, \ldots, s'_{(m'-1)1}, \ldots, s'_{(m'-1)q'_{m'-1}}\}$.

How many workstations it is necessary to assign the set of blocks $B = \{s'_{m'1}, \ldots, s'_{m'q'_{m'}}, b_1, b_2, \ldots, b_l\}$? This number can be estimated (lower bound) by taking into account the constraints (5)-(7). Let $\tilde{m}$ be a lower bound of workstations number.

Firstly, constraint (5) is considered. The set B induces in block exclusion graph G^{BE} a subgraph $G^B = (B, E^B)$, $E^B = \{(u,v) : u, v \in B, (u,v) \in E^{BE}\}$. Let $\overline{G}^B$ be the complement of graph G^B and $\overline{G}^B_k = (V_k, \overline{E}^{V_k})$, $k = 1, 2, \ldots, l_B$ are the components of $\overline{G}^B$. Indeed, the blocks which correspond to nodes of V_k may be assigned to the same workstation but if two nodes from B are in different components, then corresponding blocks cannot be assigned to the same workstation. Therefore, the constraint (5) allows to reformulate the problem of $\tilde{m}$ estimation. For estimate $\tilde{m}$ it is sufficient to estimate number of required workstations m_k for each component $\overline{G}^B_k$.

Now, consider component $\overline{G}^B_k = (V_k, \overline{E}^{V_k})$. The set V_k induces in G^B a subgraph (V_k, E^{V_k}). If a subset of nodes $\tilde{V}_k \subseteq V_k$ of this subgraph belongs in totality to the same clique (by report $\overline{E}^{V_k}$), then the blocks which correspond to the nodes of $\tilde{V}_k$ cannot be assigned to the same workstation together (must be assigned to different stations) and the number of these workstations is equal

to $|\tilde{V}_k|$ ($|A|$ is the cardinality of set A). Moreover, taking into account constraint (5), to assign set of blocks V_k it is required at least $\omega(G^B)$ workstations, where $\omega(G)$ is the clique number of graph G. A lower bound for the clique number (see [1]) for component G^B_k can be computed as follows:

$$\omega(G^B_k) \geq \sum_{v \in V_k} \frac{1}{|V_k| - deg(v)},$$

where $deg(v)$ is the degree of node v in a subgraph G^B_k. Taking into account constraint (5), it is required at least

$$\lceil \omega(G^B_k) \rceil \quad (12)$$

workstations to assign all the blocks from set V_k.

Now, consider constraint (6) and (7). Then,

$$m_k \geq \left\lceil \sum_{v \in V_k} t(v) \Big/ T_0 \right\rceil, \qquad m_k \geq \left\lceil \frac{|V_k|}{n_0} \right\rceil.$$

Using last inequalities and (12), a lower bound of minimal number of required workstations for a subgraph G^B_k can be obtained as follows:

$$\tilde{m}_k = \max \left\{ \lceil \omega(G^B_k) \rceil, \left\lceil \frac{\sum_{v \in V_k} t(v)}{T_0} \right\rceil, \left\lceil \frac{|V_k|}{n_0} \right\rceil \right\}. \quad (13)$$

Finally, a lower bound of minimal number of required workstations for the set B is:

$$\tilde{m} = \sum_{k=1}^{l_B} \tilde{m}_k. \quad (14)$$

Therefore, to obtain a lower bound for the problem (1)-(8), for a given sequence S', it is necessary to find a set $\{b_1, b_2, \ldots, b_l\}$ which satisfies the constraints (10), (11) and minimises the objective function:

$$LW(B) = \sum_{s' \in S'} c(s') + \sum_{u=1}^{l} c(b_u) + (m'-1+\tilde{m})C_0. \quad (15)$$

In other words, to obtain a lower bound, the constraints (1)-(2) are taken into account, the constraints (3)-(4) are ignored and (5)-(7) are use partially only.

The problem (10), (11), (15) is a so-called set partitioning problem. It was studied in [12, ?, 13]. All these papers focused on tree-search algorithms used directly or combined with linear programming. Considering this paper case, some difficulties appear for the model (10), (11), (15) due to the fact that the goal function (15) is nonlinear and does not have an analytical expression (it is defined by an algorithm). But, this function (15) is isotonic (if sets $B' = \{b_1, b_2, \ldots, b_l\}$ and $B'' = \{b_1, b_2, \ldots, b_l, b_{l+1}\}$ satisfy (10), (11), then $LW(B') \leq LW(B'')$). This fact allows to solve the problem (10), (11), (15) by an well-known tree-search algorithm using the two following propositions:

i) Let B' and B'' be two sets such that:

$$\bigcup_{b' \in B'} \mathcal{N}(b') = \bigcup_{b'' \in B''} \mathcal{N}(b''),$$

and constraint (10) is respected.
If $LW(B') \leq LW(B'')$ then the set B'' is unpromising (dominated by B') and it can be eliminated from search.

ii) If B' is promising (not dominated) and there is an operation

$$n' \in \mathbf{N} \backslash \left(\left(\bigcup_{b' \in B'} \mathcal{N}(b') \right) \bigcup \mathcal{N}(S') \right) \quad \text{such that}$$

there exists only one block $b'' \in \mathbf{B}$, $n' \in \mathcal{N}(b'')$ and

$$\mathcal{N}(b'') \bigcap \left(\left(\bigcup_{b' \in B'} \mathcal{N}(b') \right) \bigcup \mathcal{N}(S') \right) = \varnothing,$$

then only the set $B' \bigcup \{b''\}$ is feasible (i.e. it satisfies to (10, 11)). All other sets $B'' \supseteq B'$, $b'' \notin B''$ are infeasible and can be eliminated from the search. If there is no block which satisfies explained above constraints, then the set B' is infeasible.

Proposition i) is a classical dominance rule. Proposition ii) was firstly mentioned in [12] as a pre-processing procedure for the solving of the set partitioning problem. However, because the blocks for transfer line balancing problem (as a rule) are already constructed by respecting some constraints, then the set $\mathbf{B}$ has some "good" properties. Therefore, these rules are useful not only as a pre-processing procedure but also for any node of the tree-search. Computational experiments (see table 1 and 2, column "average time of getting lower bound") are shown that the lower bound calculation is enough fast.

Let $\mathbf{F}$ and $\mathbf{F'}$ be a family of sets satisfied (10). Then, a tree-search algorithm for solving the problem (10), (11), (15) may be described as follows:

Input: a sequence S';
Output: an element $B^* \in \mathbf{F}$ which satisfies (10), (11).
Start: $\mathbf{F} = \{\varnothing\}$, $\mathbf{F'} = \{\varnothing\}$, $u = 1$;
Step 1: for each set $B \in \mathbf{F}$, check if set $B \bigcup \{u\}$ satisfies (10), if true then $\mathbf{F'} = \mathbf{F'} \bigcup (B \bigcup \{u\})$;
Step 2: $\mathbf{F} = \mathbf{F} \bigcup \mathbf{F'}$, $\mathbf{F'} = \{\varnothing\}$;
Step 3: for each pair $B', B'' \in \mathbf{F}$ check dominance rule i) and delete dominated sets from $\mathbf{F}$;
Step 4: for each set $B \in \mathbf{F}$ check proposition ii) and delete infeasible elements from $\mathbf{F}$, $u = u + 1$;
Step 5: Repeat steps 1-4 while $u \leq q_0$;

At the end of algorithm, family $\mathbf{F}$ must contain only one element, this element satisfy constraint (11). If family $\mathbf{F}$ is empty, then sequence S' is infeasible.

In next section, branch-and-bound procedure for solving the problem (1)-(8) optimally is proposed.

4. Branch-and-Bound Algorithm

Obviously, the problem (1)-(8) is NP-hard because the simple line balancing problem is already NP-hard.
Let $S' = \{s'_{11}, \ldots, s'_{1q'_1}, \ldots, s'_{m'1}, \ldots, s'_{m'q'_{m'}}\}$, $s'_{ku} \in \mathbf{B}$ be a sequence which satisfies the conditions of Section 3. It corresponds to a node in the search tree, i.e. the blocks of sequence S' are already assigned to m' workstations, first $m' - 1$ workstations are determined according to S'.

Consider the last wokstation (with index m'). If a block $b \in \mathbf{B}$ satisfies constraints:

$$\mathcal{N}(b) \bigcap \mathcal{N}(S') = \varnothing, \quad Pr(b) \subseteq \mathcal{N}(S') \bigcup \mathcal{N}(b) \quad (16)$$

then there are two possible sequences:

i) $\tilde{S} = \{s'_{11}, \ldots, s'_{1q'_1}, \ldots, s'_{m'1}, \ldots, s'_{m'q'_{m'}+1}\}$,
   $s'_{m'q'_{m'}+1} = b$ which satisfies (6)-(7);

ii) $\hat{S} = \{s'_{11}, \ldots, s'_{1q'_1}, \ldots, s'_{m'1}, \ldots, s'_{m'q'_{m'}}, s'_{(m'+1)1}\}$,
   $s'_{(m'+1)1} = b$ which satisfies (9) for $k = m'$.

In other words, block b can be added to last occupied workstation (according to S') or assigned to next workstation. In the case i) only block exclusion constraints must be checked, in second case only operation inclusion constraints must be checked (due to the fact that the last workstation is already determined).

Let $FB(S')$ be a set of blocks which satisfy both (16) and, at least, one of following constraints: i) or ii).

A multi-choice search tree is built. At each node, set $FB(S')$ is obtained. As stated above, each element from $FB(S')$ can generate at most two new nodes. If for $b \in FB(S')$ only one of conditions either i) or ii) is true, then only one new node can be generated (according to i) or ii)). Otherwise, if both i) and ii) are true and $I^O(b) = \varnothing$ or $I^O(b) \subseteq \mathcal{N}(b)$, then the node of subtree corresponding to sequence $\hat{S}$ can be pruned (due to the Bellman principle of optimality).

Moreover, a large number of nodes can be pruned (deleted) using the following procedure. Due to the fact that the set $\mathcal{N}(S')$ can be considered as a *state* after sequential assignment of elements from S', then during the search each pair $(\mathcal{N}(S'), C(S'))$ can be considered as a candidate to be stored in an appropriate data structure $\mathbf{DS}$ (such as list, array, balanced tree, etc.). There are three kinds of processing for each pair which can be stored: *adding, checking, updating*. Suppose $C^{\mathbf{DS}}_{S'}$ is a value which is corresponded to state $\mathcal{N}(S')$ which is stored in $\mathbf{DS}$. If a sequence S' is considered and $\mathbf{DS}$ contains pair $(\mathcal{N}(S'), C^{\mathbf{DS}}_{S'})$ and in addition $C(S') \geq C^{\mathbf{DS}}_{S'}$, then the node corresponding to S' can be pruned. Pair $(\mathcal{N}(S'), C(S'))$ can be stored in the $\mathbf{DS}$ if there exists a block $b \in \mathbf{FB}$ such that there exists a sequence $\hat{S}$. The same rule must be applied for checking and updating. Namely, if previous conditions are true and $C(S') \geq C^{\mathbf{DS}}_{S'}$, then the node corresponding to $\hat{S}$ can be pruned. Otherwise (i.e. $C(S') < C^{\mathbf{DS}}_{S'}$), $C^{\mathbf{DS}}_{S'}$ can be updated to $C(S')$. Finally, it should be noted that if the branch-and-bound procedure is organized as iterative one, then each node of enumeration tree should use an indicator (i.e. the node is marked by the indicator if and only if new pair has been added or the cost has been updated).

Described above procedure is a simple dominance rule and such approach is widely used for the enumeration tree reducing [11]. This procedure will be called in the rest of paper as a *operations dominance rule checking*.

Similarly, to increase the efficiency of the search procedure, an auxiliary dominance rule can be defined, i.e. *blocks dominance rule checking*. In particular, a set of blocks can be also used as a state for sequence S' (and not only a set of operations). In spite of the fact that this state is a special case of outlined above state, it can be useful. It is possible to define a cost function (in previous case, it was $C(S')$) for checking and updating of dominance property in another (more simple) way:

$$T(S') = T_0(m' - 1) + \sum_{u=1}^{q'_{m'}} t(s'_{m'u}).$$

If a sequence S' satisfies the constraints mentioned in Section 3 and constraint (9) for $k = m'$, then for any sequence S'' corresponding to the constraints mentioned in Section 3 which are obtained by reordering of elements of S', constraint (9) for $k = m'$ is also held. This proposition can be proved by contradiction. It allows carry out checking and updating more frequently.

Let $\hat{DS}$ be an auxiliary data structure. Therefore, if $\hat{DS}$ contains a pair such that their state is the same that the state from sequence S' and in addition $T(S') \geq T_{S'}^{\hat{DS}}$, then the node corresponding to S' can be pruned. For updating, if $T(S') < T_{S'}^{\hat{DS}}$, then $T_{S'}^{\hat{DS}}$ can be changed to $T(S')$. Conditions for adding has no changes (except for the term *state*).

Node having minimal value of lower bound is considered as most promising, it must be branching first.

Let Stack be a LIFO data structure. Notation Stack.Push(S') (S'=Stack.Pop()) denotes that the node corresponding to sequence S' is placed in Stack (extracted from Stack). Proposed branch-and-bound algorithm is the following:

Input: sets N, B, $\mathcal{N}(b)$, $I^O(i)$, graphs G^r, G^{BE}, number $c(b), t(b), C_0, T_0$, $b \in$ B, $i \in$ N;
Output: an optimal sequence S^* and corresponding value of object function C_{S^*};
Step 1: $S^* = \{\varnothing\}$, $C_{S^*} = \infty$, for sequence $\{\varnothing\}$ solve the problem (10), (11), (15) . If optimal solution does not exist then goto Step 16, otherwise, Stack.Push($\{\varnothing\}$);
Step 2: if Stack is non-empty then S'=Stack.Pop() else goto Step 16;
Step 3: if $LW(S') \geq C_{S^*}$ or for given sequence S' solution of the problem (10), (11), (15) does not exist then goto Step 2;
Step 4: if $\mathcal{N}(S') =$ N, (9) for $k = m'$ and $C(S') < C_{S^*}$, then $S^* = S'$, $C_{S^*} = C(S')$ and goto Step 2;

Step 5: compute a set FB; if FB $= \varnothing$ goto Step 2, otherwise select first block from FB;
Step 6: List=$\{\varnothing\}$;
Step 7: if a sequence $\tilde{S}$ exists (according to selected block), and it is not dominated by blocks and $LW(\tilde{S}) < C_{S^*}$, then add the node to List in decreasing order of lower bound;
Step 8: check does a sequence $\hat{S}$ exist and check the condition $LW(\hat{S}) < C_{S^*}$, if false, then goto Step 14;
Step 9: if the node associated with S' has indicator, then goto Step 13;
Step 10: check for S' the dominance rule for operations; if DS does not contain a pair $(\mathcal{N}(S'), C_{S'}^{DS})$, then goto Step 12;
Step 11: if $C(S') < C_{S'}^{DS}$, then update in the pairs $(\mathcal{N}(S'), C_{S'}^{DS})$, and $(S', T_{S'}^{\hat{DS}})$ second component to $C(S')$ and $T(S')$ respectively, mark node associated with S' by indicator and goto Step 13; otherwise goto Step 14;
Step 12: add to DS and $\hat{DS}$ pairs $(\mathcal{N}(S'), C_{S'}^{DS}$ and $(S', T_{S'}^{\hat{DS}})$ respectively; goto Step 13;
Step 13: add to List a node with associated sequence $\hat{S}$ in decreasing order of lower bound;
Step 14: if FB is not exhausted, then select next block from FB and goto Step 7;
Step 15: move all nodes from List to Stack and goto Step 2;
Step 16: The end.

5. Numerical results

Let $p(G)$ be a ration between the number of (arcs) edges in the (digraph) graph G and amount of (arcs) edges in the complete (digraph) graph with the same number of vertices.

Test instances were generated randomly for different values of $|$N$|$, $|$B$|$ (10 instances per value) and for two values of $p(G^r)$. For test instances, the following values of parameters are used: $C_0 = 2000$, $n_0 = 4$, $80 \leq T_0 \leq 100$, $200 \leq c(b) \leq 280$, $10 \leq t(b) \leq 45$, $b \in$ B, $p(G^{BE}) = 0.01$. Tests have been carried out on Compaq nx9010, Intel IV with 2.8Ghz and obtained results are presented in table 1 and 2.

The computational results show that the value of $p(G^r)$, has a great influence on the calculation time. It should be noted that in real transfer line balancing problem $p(G^r) \geq 0.15$ and then the time of solution of the problem (1)-(8) does not exceed three hours (in average) for $|$N$| \leq 80$.

Denote $\tilde{c} = 1 - \hat{C}/C^*$ (relative impovement of object function), where $\hat{C}$ is objective function of first feasible solution by branch-and-bound algorithm and C^* is a best (optimal, for solved instances) objective function. Table 5 and 5 show that average (for all instances) objective function impovement takes 16%. It must be noted that this value depends on a station cost C_0.

Table 1. Numerical results for $p(G^r) = 0.1$

|N|	50	60	70	80
Min |B|	94	114	132	172
Max |B|	99	118	139	178
Average |B|	99.1	115.8	136.4	174.4
Min number of iteration	410	613	1119	1234
Max number of iteration	277	23120	27156	192599
Average numb. of iteration	1332.8	5606.6	10959.4	49509.2
Min running time (sec)	2.5	6.62	52.9	15.9
Max running time (sec)	42.1	317.5	842.5	21757.5
Average running time (sec)	20.3	135	459.2	10175.3
Average time of getting lower bound (millisec)	11.3	15	42	178.2
$\bar{c}$ in percents	13	15	18	19

Table 2. Numerical results for $p(G^r) = 0.05$

|N|	50	60	70	80*
Min |B|	94	116	126	
Max |B|	98	118	138	
Average |B|	96	116.1	134.5	
Min number of iteration	169	1609	18602	
Max number of iteration	17137	71133	3326865	
Average number of iteration	3287.6	12701	477300.2	
Min running time (sec)	0.45	14.2	79.2	
Max running time (sec)	153.2	697.8	40287.3	
Average running time (sec)	31.9	192.6	10402.5	
Average time of getting lower bound (millisec)	11	19.1	35	145.3
$\bar{c}$ in percents	14	14	18	18

* - 4 instances are not solved in 16 hours.

6. Conclusion

In this paper, a transfer line balancing problem with blocks of parallel operations has been investigated. The set of available blocks is given. The blocks of the same workstation are executed sequentially. By report with SALBP, the investigated problem has many additional constraints. For solving the problem exactly, an original lower bound based on set partitioning problem has been proposed and a branch-and-bound algorithm has been implemented.

As the computational results show, the proposed lower bound is enough efficient and the transfer line balancing problem with the number of operations smaller than 80 can be solved in the time of three hours in average. The average improvement of objective function takes 16%. This value justifies a searching for optimal solution althought it is time-consuming.

Acknowledgements

Second author is supported by a NATO grant.

References

[1] Aigner, M. (1995) Turán's Graph Theorem. *Amer. Math. Monthly 102*, pp. 808-816.

[2] Dolgui, A., Guschinsky, N., Levin, G. (1999) On Problem of Optimal Design of Tranfer Lines with Parallel and Sequential Operation. *Proceedings of the 7th IEEE International Conference on Emerging Technologies and Factor Automation (ETFA'99)*, Oct. 18-21, Bracelona, Spain. J.M. Fuertes (Ed.), Vol. 1, pp. 329-334.

[3] Dolgui, A., Guschinsky, N., Levin, G. (2000) *Approaches to balancing of transfer line with block of parallel operation*. Institute of Engineering Cybernetics/ Minsk, Preprint No. 8, 42 pages.

[4] Dolgui, A., Guschinsky, N., Harrath, Y., Levin, G. (2002) Une approache de programmation linéaire pour la conception des lignes de transfert. *European Journal of Automated System (JESA)*, 36 (1), pp. 11-33.

[5] Dolgui, A., Finel, B., Guschinsky, N., Levin, G., and Vernadat, F. (2004) An heuristic approach for transfer lines balancing. *Journal of Intelligent Manufacturing*, 30 pages (accepted, to appear)

[6] Groover, M.P. (1987) *Automation, Production Systems and Computer Integreted Manufacturing*, Prentice Hall, Eaglewood Cliffs, New Jersey.

[7] Rekiek, B., De Lit, P. Delchambre, A. (2000) Designing Mixed-Product Assembly Lines. *IEEE Transactions on Robotic and Automation*, 16(3), pp. 268-280.

[8] Rekiek, B., Dolgui, A., Delchambre, A., Bratcu, A. (2002) State of art of assembly lines design optimisation. *Annual Reviews in Control*, 26(2), pp. 163-174.

[9] Scholl, A. (1999) *Balancing and sequencing of assembly lines*. Heidelberg Physica.

[10] Erel, E., Sarin, S.,C. (1998) A survey of the assembly line balancing procudures. *Production Planning and Control*, 9(5), pp. 414-434.

[11] Christofides, N. (1975) *Graph Theory: An Algorithmic Approach*, Academic Press, New York.

[12] Garfinkel, R.,S., Nemhauser, G.,L. (1969) The Set Partition Problem: Set Covering with Equality Constraints, *Operations Research*, 17, pp. 848-856.

[13] Pierce, J., F., Lasky, J.,S. (1973) Improved combinatorial programming algorithms for a class of all-zero-one integer programming problem, *Management Science*, 19, pp. 528-536.

[14] Ghosh, S., Gagnon, R., J. (1989) A comprehensive literature review and analysis of the design, balancing and scheduling of assembly line systems. *International Journal of Production Research*, 27, pp. 637-670.

[15] Amen, M. (2002) Heuristic methods for cost oriented assembly line balancing: A survey. *International Journal of Production Economics*, 68, pp. 1-14.

NO-WAIT FLOWSHOP PROBLEM WITH TWO MIXED BATCHING MACHINES

A. OULAMARA *

* MACSI Project LORIA - INRIA Lorraine, Ecole des Mines de
Nancy, Parc de Saurupt, 54042 Nancy, France.
Ammar.Oulamara@loria.fr

Abstract: We study a problem of scheduling n tasks in a no-wait flowshop consisting
of two mixed batching machines. Each task has to be processed by both machines.
All tasks visit the machines in the same order. Batching machines can process several
tasks in batch so that all tasks of the same batch start and complete together. The
processing time of a batch on the first batching machine is equal to the maximal
processing time of the tasks in this batch, and on the second batching machine is
equal to the sum of the processing time of tasks in this batch. We assume that the
capacity of any batch on the first machine is bounded, and when a batch is completed
on this machine should immediately transferred to second machine. The aim is to
make batching and sequencing decisions so that the makespan is minimized.

Keywords: No-wait flowshop problem, batch processing machines, complexity.

1. INTRODUCTION

The problem of scheduling a no-wait flowshop
with mixed batching machine can be formuled
as follows. There are n tasks to be batched and
scheduled for processing on two machines. Each
task has to be processed on each of the machines
1 and 2 visiting them in this order. The processing
time of a task j on machines 1 and 2 is at least $p_{1,j}$
and $p_{2,j}$ time units respectively. Each machine is a
batching machine such that it can process several
tasks in batch. All tasks in same batch start
and finish their processing on the same batching
machine at the same times (batch availability).
The first batching machine is called max-batch
machine, it treats several tasks simultaneously in
which the batch processing time on this machine

is determined as the maximum processing time of
tasks in this batch on this machine. The capacities
of the batches on this machine is bounded by
some integer value $b < n$. The second machine is
called sum-batch machine, it treats several tasks
sequentially, in which the batch processing time
on this machine is determined as the sum of
processing time of tasks in this batch on this
machine. A setup time is required on such machine
before each execution of a tasks. We distinguish
two variantes depending on the nature on setup
times. In the anticipatory model, the setup can be
processed before the associated batch is available
on the machine and for the non-anticipatory setup
time model, the setup can only be processed
when all tasks assigned to the corresponding batch
are available on the machine. The batches are

processed in a no-wait fashion, such that the completion time of each batch on machine 1 should coincide with its starting time on machine 2.

A schedule is characterized by partition of tasks into batches on each machine and batch sequences on the machines. Due to the no-wait constraints, batch partition and batch sequences are the same on both machines. Therfore, we shall not distinguish the notion of a schedule and batch sequence. The objective is to find a schedule such that the completion time is minimized. By using the general notation for scheduling problems introduced by Graham et al. (Graham *et al.*, 1979), we write $F2 \mid max - batch(1), sum - batch(2), b, no - wait \mid C_{max}$ to refer to this problem.

Motivation for the formulated problem comes from processing products in iron and steel industry. A metal sheet first passes through a multi-head machine for a hole punching procedure. Depending on the position of a hole on the sheet, batches of holes are punched simultaneously in each working phase. At the second stage, after short preparation period, the metal sheet undergoes a series of sequential bending operations.

Reviews on batch scheduling were provided by Potts and Van Wassenhove (Potts and Wassenhove, 1991), Webster and Baker (Webster and Baker, 1995), Potts and Kovalyov (Potts and Kovalyov, 2000). A flowshop system with two batching machines was studied by Potts, Strusevich and Tautenhahn (Potts *et al.*, 2001). The no-wait constraints were not imposed in their studies. Hall, Laporte, Selvarajah and Sriskandarajah (Hall *et al.*, 2003) studied problems of partitioning a set of identical tasks into batches in the no-wait flowshop with batching machines. Lin and Cheng (Lin and Cheng, 2001) studied the two sum-batch machines in the no-wait flowshop problem, the objective is to minimize. They prove that the minimization of makespan is strongly NP-hard. Oulamara and Finke (Oulamara and Finke, 2001) studied the flowshop problem with two mixed batching machine and without no-wait constraints. They prove that the minimization of makespan criterion is polynomial for the case were the capacity of batch is unbounded on the max-batch machine. For the bounded case, they prove that the makespan minimization is NP-Hard.

The remainder of this article is organized as follows. In section 2 we prove that the makespan minimization is NP-Hard in the strong sense. Section 3 addresses the case in which all tasks have the same processing times on the first machine, we prove that when the capacity of the first machine is equal to two, the makespan minimization reduces to the maximum weight matching problem and when this capacity is greater or equal to three, the makespan minimization is NP-Hard in the strong sense. In the rest of the paper, we consider that the setup time on the second machine is equal to zero. All results that will be given here can be easily generalized for the case with positive setup times.

2. COMPLEXITY RESULT

In this section, we study the problem $F2|max - batch(1), sum - batch(2), b, no - wait, |C_{max}$, abbreviated in the following as problem $P(b)$.

We show that the problem $P(b)$ is NP-hard in the strong sense. We use the reduction of the NUMERICAL MATCHING WITH TARGET SUMS (NMWTS) problem, which is known to be NP-hard in the strong sense, see (Garey and Johnson, 1979).

The NMWTS decision problem is given as follow:

NMWTS. given two disjoint sets A and B of integers where $A = \{a_1, \ldots, a_n\}$ and $B = \{b_1, \ldots, b_n\}$, and a target vector $E = \langle E_1, \ldots, E_n \rangle$ with positive integers entries, where $\sum_{i=1}^n a_i + \sum_{i=1}^n b_i = \sum_{i=1}^n E_i$. Does there exists a partition of $A \cup B$ into n disjoint sets $N_1, \ldots, N_n$, each N_i $(i = 1, \ldots, n)$ containing exactly one element from each A and B, and $a_{N_i} + b_{N_i} = E_i$.

Theorem 1. Problem $P(b)$ is NP-hard in the strong sense.

Proof. First we establish the proof for $b = 2$, and an extension will be given to any value b with $b < n$.

Given an arbitrary instance I' of NMWTS problem, firstly we modify I' as follow:
Let $w = \max\{a'_1, \ldots, a'_n, b'_1, \ldots, b'_n\}$, we construct a new instance I of NMWTS such that, $A = \{a_1, \ldots, a_n\}$, $B = \{b_1, \ldots, b_n\}$ and a target vector $E = \{E_1, \ldots, E_n\}$ where,

$$a_i = a'_i, i = 1, \ldots, n$$
$$b_i = b'_i + 3w, i = 1, \ldots, n$$
$$E_i = E'_i + 3w, i = 1, \ldots, n$$

Clearly, the instance I has a solution, if and only if, the instance I' has a solution.

We consider an arbitrary modified instance I of NMWTS problem (described previously), we construct the following instance of problem $P(b = 2)$ with $2n^2 + 2(n + 1)$ tasks. The tasks are partitioned into three groups:

- A-tasks: denoted $A_{i,j}$, $(i, j = 1, \ldots, n)$,
- B-tasks: denoted $B_{i,j}$, $(i, j = 1, \ldots, n)$,
- D-tasks: denoted $D_{i,j}$, $(i = 1, \ldots, n, j = 1, 2)$.

The processing times of the tasks on both machines are given in the following table:

		$p_{1,j}$	$p_{2,j}$
A-tasks	$A_{i,j}, i, j = 1, \ldots, n$	E_i	a_j
B-tasks	$B_{i,j}, i, j = 1, \ldots, n$	E_i	b_j
D-tasks	$D_{1,1}$	0	0
	$D_{1,2}$	0	E_1
	$D_{i,1}, i = 2, \ldots, n$	E_{i-1}	0
	$D_{i,2}, i = 2, \ldots, n$	E_{i-1}	E_i
	$D_{n+1,1}$	E_n	0
	$D_{n+1,2}$	E_n	0

We show that the NMWTS problem has a solution, if and only if, the problem $P(b = 2)$ has a feasible schedule S, such that the completion time of S is $C_{max}(S) \le (n+1) \sum_{i=1}^{n} E_i$.

First, suppose that NMWTS has a solution. Let $N_1, \ldots, N_n$ be n disjoint sets such that, each set N_i contains exactly one element of A and one element of B and $a_{N_i} + b_{N_i} = E_i$, $i = 1, \ldots, n$. Then there exists a schedule S for the problem $P(b = 2)$ such that the completion time $C_{max}(S)$ is equal to $(n+1) \sum_{i=1}^{n} E_i$. In this schedule, the tasks are grouped into two classes of batches, and each batch contains exactly two tasks. The classes are constructed as follows:

- C_1 : Class containing the batch of the form $\{D_{i,1}; D_{i,2}\}$, $i = 1, \ldots, n$, with the processing time E_{i-1} on the first machine, except for the batch $\{D_{1,1}; D_{1,2}\}$ with zero processing time on the first machine.
- C_2 : Class containing the batch of the form $\{A_{i,j}; B_{i,k}\}$, $i = 1, \ldots, n$, $j \in N_i$ and $k \in N_i$, with the processing time E_i on the first machine.

The first batch $\{D_{1,1}; D_{1,2}\}$ belonging to C_1 is scheduled on the first position in S. Let r_i be the position of the batch $\{D_{i,1}; D_{i,2}\}$ in S, r_i is given by the following formula:

$$\begin{cases} r_1 = 1, \\ r_i = r_{i-1} + 2(n - i + 2), & i = 2, \ldots, n+1. \end{cases}$$

Also, denote $r_{i,l}$ be the position of the batch $\{A_{i,j}; B_{i,k}\}$, $i, l = 1, \ldots, n$ of the class C_2 in S, where $j \in N_i$ and $k \in N_i$. $r_{i,l}$ is given by,

$$r_{i,l} = \begin{cases} r_i + 2(l - i) + 1, & \text{if } i \le l \\ r_l + 2(i - l), & \text{if } i > l \end{cases}$$

where r_j, $j = 1, \ldots, n$ is the position of the jth batch of the class C_1 in S.

To illustare the sequence S, let consider the following example of NMWTS with $n = 4$. We have $A = \{a_1, a_2, a_3, a_4\}$, $B = \{b_1, b_2, b_3, b_4\}$ and $E = \langle E_1, E_2, E_3, E_4 \rangle$.

Assume taht the NMWTS have the following solution, $N_1 = \{a_1, b_2 \ / \ a_1 + b_2 = E_1\}$, $N_2 = \{a_2, b_1 \ / \ a_2 + b_1 = E_2\}$, $N_3 = \{a_3, b_4 \ / \ a_3 + b_4 = E_3\}$, $N_4 = \{a_4, b_3 \ / \ a_4 + b_3 = E_4\}$. We obtain the classes C_1 and C_2 of the batches as:

$C_1:$ $C_2:$

$\begin{pmatrix} 0, & 0 \\ 0, & E_1 \end{pmatrix}^1$ $\begin{pmatrix} E_1, & E_1 \\ a_1, & b_2 \end{pmatrix}^2$ $\begin{pmatrix} E_1, & E_1 \\ a_2, & b_1 \end{pmatrix}^3$ $\begin{pmatrix} E_1, & E_1 \\ a_3, & b_4 \end{pmatrix}^5$ $\begin{pmatrix} E_1, & E_1 \\ a_4, & b_3 \end{pmatrix}^7$

$\begin{pmatrix} E_1, & E_1 \\ 0, & E_2 \end{pmatrix}^9$ $\begin{pmatrix} E_2, & E_2 \\ a_1, & b_2 \end{pmatrix}^4$ $\begin{pmatrix} E_2, & E_2 \\ a_2, & b_1 \end{pmatrix}^{10}$ $\begin{pmatrix} E_2, & E_2 \\ a_3, & b_4 \end{pmatrix}^{11}$ $\begin{pmatrix} E_2, & E_2 \\ a_4, & b_3 \end{pmatrix}^{13}$

$\begin{pmatrix} E_2, & E_2 \\ 0, & E_3 \end{pmatrix}^{15}$ $\begin{pmatrix} E_3, & E_3 \\ a_1, & b_2 \end{pmatrix}^6$ $\begin{pmatrix} E_3, & E_3 \\ a_2, & b_1 \end{pmatrix}^{12}$ $\begin{pmatrix} E_3, & E_3 \\ a_3, & b_4 \end{pmatrix}^{16}$ $\begin{pmatrix} E_3, & E_3 \\ a_4, & b_3 \end{pmatrix}^{17}$

$\begin{pmatrix} E_3, & E_3 \\ 0, & E_4 \end{pmatrix}^{19}$ $\begin{pmatrix} E_4, & E_4 \\ a_1, & b_2 \end{pmatrix}^8$ $\begin{pmatrix} E_4, & E_4 \\ a_2, & b_1 \end{pmatrix}^{14}$ $\begin{pmatrix} E_4, & E_4 \\ a_3, & b_4 \end{pmatrix}^{18}$ $\begin{pmatrix} E_4, & E_4 \\ a_4, & b_3 \end{pmatrix}^{20}$

$\begin{pmatrix} E_4, & E_4 \\ 0, & 0 \end{pmatrix}$

The schedule S is defined by the sequencement order of the batches given by the power number on each batch. The completion time of the schedule S is equal to $C_{max}(S) = 5 \sum_{i=1}^{4} E_i$.

Suppose now there is a schedule S^* such that $C_{max}(S^*) \le (n+1) \sum_{i=1}^{n} E_i$. Let v be the number of batches in S^*, with $v \ge n^2 + n + 1$ (batch size on the first machine is bounded by 2). Denote by $P_1(B_i)$ the processing time of batch B_i, $i = 1, \ldots, v$ on the first machine. For each B_i, define $P'_1(B_i)$ as follows:

$$P'(B_i) = \begin{cases} 0 & \text{if } B_i \text{ contains exactly one task} \\ \min\{p_{1,j} \ / \ j \in B_i\} & \text{otherwise} \end{cases}$$

The batch size on the first machine is bounded by 2, so

$$\sum_{i=1}^{v} P_1(B_i) + \sum_{i=1}^{v} P_1'(B_i) = \sum_{i=1}^{2n^2+2(n+1)} p_{1,j}$$

$$= 2(n+1)\sum_{i=1}^{n} E_i$$

and $C_{cmax}(S^*) \le (n+1)\sum_{i=1}^{n} E_i$, thus,

$$\sum_{i=1}^{v} P_1(B_i) = \sum_{i=1}^{v} P_1'(B_i) = (n+1)\sum_{i=1}^{n} E_i$$

So we establish that,

1. Each batch B_i, $i = 1, \ldots, v$, contains exactly two tasks, i.e. $v = n^2 + n + 1$.
2. The tasks which have the same indices are grouped in the same batch (i.e. a batch containing the task $A_{i,j}$ contains also a task $A_{i,k}$ or $B_{i,l}$ or $D_{i,t}$).
3. No idle time on the first machine.

On the second machine, the sum of processing times of tasks are given by

$$n\sum_{i=1}^{n} a_i + n\sum_{i=1}^{n} b_i + \sum_{i=1}^{n} E_i = (n+1)\sum_{i=1}^{n} E_i$$
$$= C_{max}(S),$$

we note that there is no idle time on the second machine. Thus, we establish that only the batch $\{D_{1,1}; D_{1,2}\}$ can be scheduled in the first position, with zero processing time on the first machine and E_1 processing time on the second.

Also, because there is no idle time on both machines, and all batches will be sequenced in no-wait fashion, then all batches of the sequence S^* are of the form $\{A_{i,j}; B_{i,k}\}$ and $\{D_{i,j'}; D_{i,k'}\}$, because, in opposite case, we create an idle time on the first or on the second machine (that is related to the choice on the instance I of NMWTS problem).

Without loss of generality, we assume that the batch $B = \{A_{i,j}; B_{i,k}\}$ is sequenced at position r and let B' be the batch sequenced at position $r + 1$. The processing time of B' on the first machine is equal to some E_l, where $E_l \in E$. But we have B' is sequenced in no-wait fashion and there is no idle time on both machines, we have $P_2(B) = p_2(A_{i,j}) + p_2(B_{i,k}) = E_l$ for each $E_l \in E$. Thus, the set $N_l = \{a_j, b_k \, / \, a_i + b_j = E_l\}$ contains exactly one element of A and one element of B.

Repeating this last operation for all value of l, $(l = 1, \ldots, n)$ we obtain a solution for the instance I of NMWTS problem.

To complete the proof, the instance of scheduling problem (constructed at the beginning of the proof) can be extended to the problem $P(b)$ with $2 < b < n$. In order to faclitate the reduction of NMWTS problem to $P(b)$ we add to the scheduling instance $(b-2)n^2 + (b-2)(n+1)$ tasks. Their processing times are given by the following table:

		$p_{1,j}$	$p_{2,j}$
U	$U_{i,j,l}$, $i,j = 1, \ldots, n$ $l = 1, \ldots, b-2$	E_i	0
H	$H_{1,l}$, $l = 1, \ldots, b-2$	0	0
	$H_{i,l}$, $i = 2, \ldots, n+1$ $l = 1, \ldots, b-2$	E_{i-1}	0

By using the same principle, we can show that the problem $P(b)$ admits a schedule S^* with makespan $C_{max}(S^*) \le (n+1)\sum_{i=1}^{n} E_i$, if and only if, the NMWTS has a solution. Indeed, the schedule S^* built before remains feasible for $P(b)$. Only the seize of its batches is changing. For each index i, j, $(i, j = 1, \ldots, n)$ all tasks $U_{i,j,l}$, $l = 1, \ldots, b-2$ are sequenced on both machines in the same batch as $A_{i,j}$. Also, for any index i, $(i = 1, \ldots, n)$ the tasks $H_{i,l}$, $(l = 1, \ldots, b-2)$ are assigned to batch containing $D_{i,1}$ on both machines. It follows that the preceding theorem remains valid for the problem $P(b)$ with $2 < b < n$. $\square$

3. CONSTANT PROCESSING TIMES ON THE FIRST MACHINE

We consider here the problem where all tasks have the same processing times on the first machine, $p_{1,j} = p$, $j = 1, \ldots, n$. We distingush here two cases, namely, the case where the capacity of first machine is equal to two (denoted by $P(p, b = 2)$) and the case where the capacity is greater or eqaul to three (denoted by $P(p, b \ge 3)$).

3.1 Capacity equal to two

Lemma 2. There exists an optimal schedule for the problem $P(p, b = 2)$ in which all batches are saturated except the last one if the number of tasks is not a multiple of 2.

Proof. Denote by S an optimal schedule of the problem $P(p, b = 2)$, containing the batches $(B_1, \ldots, B_u)$. Let B_j be the first unsaturated batch of S (B_j contains only one task). Let B_k be the first unsaturated batch of S, sequenced after B_j. We consider a new schedule S' built from S by completing B_j with task of B_k and shift on the right all batches sequenced after B_k. It is easy to show that $C_{max}(S') \leq C_{max}(S)$ (see figure below). Therefore S' is also an optimal schedule.

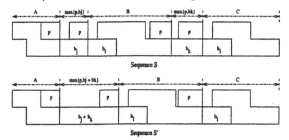

Repeating such modification, we obtain a schedule with saturated batches except for the last one if the number of batches is not a multiple of 2. □

Without loss of generality, we assume here that the number of tasks is a multiple of two, otherwise we can add a dummy task with processing time p on the first machine and zero processing time on the second machine. Thus, for an instance of problem $P(p, b = 2)$ with n tasks the optimal sequence contains $\frac{n}{2}$ batches.

Theorem 3. The problem $P(p, b = 2)$ reduces to the maximum weight matching with cardinality $\frac{n}{2} - 1$

Proof. Consider a complet graph $G = (V, E)$ where the set of vertices $V = \{u_1, \ldots, u_n\}$ corresponds to the tasks and the set $E = \{e_1, \ldots, e_m\}$ of edges corresponds to the possible batching of tasks in the same batch. Let w_j be the weight of edge e_j such that $w_j = \max\{p_{2,u_j} + p_{2,v_j} - p, 0\}$ where $e_j = (u_j, v_j)$, $j = 1, \ldots, m$. Let A be the vertex-edge incidence matrix of the graph $G = (V, E)$ where

$$a_{i,j} = \begin{cases} 1 & \text{if the edge } j \text{ is incident to the vertex } i \\ 0 & \text{otherwise} \end{cases}$$

Let x_j be a decision variable such that

$$x_j = \begin{cases} 1 & \text{if the tasks connected by the edge } j \text{ are} \\ & \text{in the same batch} \\ 0 & \text{otherwise} \end{cases}$$

The linear program corresponding to mimimize the makespan is

$$\min C_{max} = p + \sum_{j=1}^{m} w_j x_j + \sum_{i=1}^{n} \left(1 - \sum_{j=1}^{m} a_{i,j} x_j\right) p_{2,i}$$

$$\text{s.t.} \quad \sum_{j=1}^{m} a_{i,j} x_j \leq 1 \quad i = 1, \ldots, n$$

$$\sum_{j=1}^{m} x_j \leq \frac{n}{2} - 1$$

We have

$$\begin{aligned} C_{max} &= p + \sum_{j=1}^{m} w_j x_j + \sum_{i=1}^{n} \left(1 - \sum_{j=1}^{m} a_{i,j} x_j\right) p_{2,i} \\ &= p + \sum_{j=1}^{m} w_j x_j + \sum_{i=1}^{n} p_{2,i} - \sum_{i=1}^{n} \sum_{j=1}^{m} a_{i,j} p_{2,i} x_j \\ &= p + \sum_{i=1}^{n} p_{2,i} - \sum_{j=1}^{m} \left(\sum_{i=1}^{n} a_{i,j} p_{2,i} - w_j\right) x_j \\ &= p + \sum_{i=1}^{n} p_{2,i} - \sum_{j=1}^{m} (p_{2,u_j} + p_{2,v_j} - w_j) x_j \\ &= p + \sum_{i=1}^{n} p_{2,i} - \sum_{j=1}^{m} (\min\{p, p_{2,u_j} + p_{2,v_j}\}) x_j \end{aligned}$$

$p + \sum_{i=1}^{n} p_{2,i}$ is constant and greater than $\sum_{j=1}^{m} (\min\{p, p_{2,u_j} + p_{2,v_j}\}) x_j$. Then minimizing C_{max} is equivalent to maximizing $\sum_{j=1}^{m} d_j x_j$ where $d_j = \min\{p, p_{2,u_j} + p_{2,v_j}\}$ if the edge e_j is incident to the vertices u_j and v_j. Hence the linear model reduces to the weight matching problem with cardinality $\frac{n}{2} - 1$ in complet graph $G = (V, E)$ with weight d_j on the edge e_j, $j = 1, \ldots, m$. This problem can be solved optimaly with Edmonds algorithm (Edmonds, 1965). □

3.2 *Capacity greater or equal to three*

We consider here the problem where the capacity b of the first machine is greater or equal to three. We proof that minimizing the makespan criterion is strongly NP-Hard by using the reduction of the 3-PARTITION problem which is known to be strongly NP-Hard (Garey and Johnson, 1979).

3-PARTITION. Given $3n$ positive integer $u_1, \ldots, u_{3n}$ where $\sum_{i=1}^{3n} u_i = nW$ for some W and $W/4 < u_i < W/2$ for $i = 1, \ldots, 3n$, does there exists a partition of the index set $N = \{1, \ldots, 3n\}$ into n

disjoint subsets $N_1, \ldots, N_n$ such that $\sum_{i \in N_j} u_i = W$ and $|N_j| = 3$ for $j = 1, \ldots, n$.

Theorem 4. The problem $P(p, b \geq 3)$ is strongly NP-Hard.

Proof. First we consider case where the capacity of first machine is equal to three and we extend the proof for the case where the capacity is greater than three. Given an arbitrary instance of 3-PARTITION, we construct the following instance of problem $P(p, b \geq 3)$ with the set $N = \{1, \ldots, 3n+3\}$ tasks. The processing times of tasks on the first machine are equals to W and on the second machine $p_{2,j} = u_j$ for $j = 1, \ldots, 3n$ and $p_{2,3n+1} = p_{2,3n+2} = p_{2,3n+3} = 0$. The capacity of the first machine is equal to three.

To prove the theorem, we show that for the constructed instance of problem $P(p, b = 3)$ there exists a schedule S such that the makespan $C_{max}(S) \leq (n+1)W$ if and only if 3-PARTITION has a solution.

First, suppose that 3-PARTITION has a solution and $N_1, \ldots, N_n$ are the required subsets of N. The schedule S constructed by following $n+1$ batches, which are defined by $N_1, \ldots, N_n$ and N_{n+1} where N_{n+1} contains the tasks $3n+1$, $3n+2$ and $3n+3$. The tasks of batch N_1 are processed on machine one in the interval $[0, W]$ and on machine two in the interval $[W, 2W]$. In the interval $[jW, (j+1)W]$ for $j = 1, \ldots, n$ the tasks of batch N_{j+1} are processed on machine one and on machine two the tasks of batch N_j are processed. At time $(n+1)W$ machine one has completed all of its processing. Since the tasks of N_{n+1} have zero processing times on machine two, then its processed on machine two in the interval $[(n+1)W, (n+1)W]$. Thus $C_{max}(S) = (n+1)W$.

Suppose now there is a schedule S such that $C_{max}(S) \leq (n+1)W$. Let v be the number of batches in S. Since all tasks have the same processing times W on the machine one it is easy to show that :

1. Each batch B_i, $i = 1, \ldots, v$, contains exactly three tasks, *i.e.* $v = n+1$.
2. No idle time on the first machine.

On the second machine, the sum of processing times are given by $\sum u_i = nW$. Therefore there is no idle time on the machine two, except at the begining witch the idle time W coresponds to the

waiting time of the machine two to process the first batch.

Without loss of generality, we assume that the batch B_l composed by tasks i, j and k is sequenced at position l and let B_{l+1} the batch sequenced at position $l + 1$. The processing time of B_{l+1} on machine one is equal to W. But we know that B_{l+1} is sequenced in no-wait fashion and there is no idle time on both machines before batch B_{l+1}, then $u_i + u_j + u_k = W$. Thus the set $N_l = \{i, j, k\}$ contains exactly three tasks. Repeating this last operation for all value of l ($l = 1, \ldots, n$) we obtain a solution for the instance of 3-PARTITION.

To complete the proof, the instance of scheduling problem can be extended to the problem $P(p, b)$ with $3 < b < n$. In order to facilitate the reduction of 3-PARTITION problem to $P(p, b)$, we by add to the scheduling instance $(b - 3)(n + 1)$ tasks with W processing times on machine one and zero processing times on machine two. By using the same principle, we can show that the problem $P(p, b)$ admits a schedule S with makespan less than or equal to $(n + 1)W$ if and only if 3-PARTITION has a solution. □

4. CONCLUSION

In this paper, we have studied a new extension of flowshop problems on two batch processing machines. We have analyzed the case in which the first machine is a *max-batch machine* and the second is a *sum-batch machine*. We have shown that the problem to minimize the makespan for the bounded batch size of the max-batch machine is NP-Hard in the strong sense. We have studied the particular case in which all tasks have the same processing times on the first machine. It is shown that the makespan minimization for this particular case is polynomial when the capacity b of batch is equal to two and it is NP-Hard in the strong sense when $3 \leq b < n$.

5. REFERENCES

Edmonds, J. (1965). Matching and a polyhedron with 0,1 vertices. *J. Res. N.B.S.B* **69**, 125–130.

Garey, M.R. and D.S. Johnson (1979). *Computers and Intractability: A Guide to the Theory of NP-Completness*. W. H. Freeman.

Graham, R.L., E.L. Lawler, J.K. Lenstra and A.H.G Rinnooy Kan (1979). Optimization and approximation in deterministic machine scheduling: a survey. *Annals of Discrete Mathematics* **5**, 287–326.

Hall, N.G., G. Laporte, E. Selvarajah and C. Sriskandarajah (2003). Scheduling and lot streaming in flowshops with no-wait in process. *Journal of Scheduling* **6**, 339–354.

Lin, B.M.T. and T.C.E. Cheng (2001). batch scheduling in the no-wait two-machine flowshop to minimize the makespan. *Computers & Operations Research* **28**, 613–624.

Oulamara, A. and G. Finke (2001). Flowshop problems with batch processing machines. *International Journal of Mathematical Algorithms* **2**, 269–287.

Potts, C.N. and L.N. Van Wassenhove (1991). Integrating scheduling with batching and lot-sizing: a review of algorithms and complexity. *Journal of the Operational Research Society* **46**, 395–406.

Potts, C.N. and M.Y. Kovalyov (2000). Scheduling with batching: A review. *European Journal of Operational Research* **120**, 228–249.

Potts, C.N., V.A. Strusevich and T. Tautenhahn (2001). Scheduling batches with simultaneous job processing for two-machine shop problems. *Journal of Scheduling* **4**, 25–51.

Webster, S.T. and K.R. Baker (1995). Scheduling groups of jobs on a single machine. *Operations Research* **43**, 692–703.

www.elsevier.com/locate/ifac

FLOWSHOP SCHEDULING PROBLEM WITH BLOCKING AND TRANSPORTATION CONSTRAINTS

Ameur SOUKHAL *

* *Laboratoire d'Informatique, 64 avenue Jean Portalis 37 200, Tours FRANCE. e-mail: {soukhal}@univ-tours.fr*

Abstract: In most manufacturing and distribution systems, semi-finished jobs are transferred from one processing facillity to another and finished jobs are delivered to customers or warehouse by vehicles such as trucks.
We investigate flowshop scheduling problems that explicitly consider constraints on both transportation and buffer capacities. The finished jobs leave the processing facility to be delivered to customer or sent to warehouse by vehicle such as truck. The objective function that is considered is makespan C_{max}. We prove that this problem is strongly $\mathcal{NP}$-hard when the capacity of truck is greater or equal to one within additional constraints, such as blocking. A heuristic with guaranteed performance is proposed and well solvable cases are identified.

Keywords: Scheduling, Blocking flowshop, Transporter, Complexity.

1. INTRODUCTION

A flexible manufacturing system (FMS) is an integrated system composed of automated material handling devices and numerically controlled machines that can process various types of parts (MacCarthy and Liu, 1993). This system has been introduced to give more flexibility by overcoming the traditional hypotheses such as unlimited intermedate storage area. In practice the intermediate stock capacity is limited to k parts. As an other traditional hypotheses the number of available transporters is infinite and the jobs are instantaneously delivered from one machine to another. The resolved problem appears to be far away from the physical workshop. Thus, it is inadequate to make the supervisor apply a sequence produced by scheduling calculation under these hypotheses. So, in most manufacturing distribution systems, finished jobs are transferred from processing facility and delivered to customers or warehouses by vehicles such as trucks. These vehicles have a capacity c. It means that at time t the truck can transport at most c parts (c jobs).

In (Lee and Chen, 2001) the authors investigate machine scheduling models that explicitly consider constraints on both transportation capacity and transportation times. They consider two types of transportation. The first one, the authors consider the intermediate transportation where the semi-finished jobs are transferred from one machine to another. The authors denote this problem "type-1" transportation. The second one denoted "type-2" transportation is the transportation necessary to deliver finished jobs to the customers. So, in (Lee and Chen, 2001) some computational complexity results are given and *"open"* problems are underlined. In their studies, the authors consider an unlimited intermediate buffer area. But in practice, the intermediate buffer areas are often limited.

This paper deals with a blocking two-machine flowshop "type-2" transportation. Here there is no buffer area between the two machines. But at the output of the flowshop there are an unlimited buffer and only one vehicle with a capacity c. This paper does not only aim to identify the complexity status of the blocking two-machine flowshop type-2 transportation scheduling problem as the works presented in (Lee and Chen, 2001) but also to identify well solvable cases.

This problem can be also viewed as a workshop, where jobs can be gathered into batches depending on the capacity of resources: two discrete machines form the processing facility and the truck represents the batch machine with capacity c (Glass *et al.*, 1998) and (Potts *et al.*, 2001). In (Ahmadi *et al.*, 1992) the authors consider a situation in which the manufacturing system is equipped with batch and discrete processors. They analyse the complexity of a class of two-machine batching and scheduling problems.

This paper is organized as follows: In section 2 we describe a problem and commonly used notations. In this section, we present the two-machine flowshop type-2 transportation problem by classifying their computational complexity. In section 3 we study the complexity of the two-machine flowshop problem under transportation and buffer capacities constraints. For a fixed value of c greater or equal to 1, we show that this problem is strongly $\mathcal{NP}$-hard. Section 4 proposes a greedy algorithm to solve the blocking two-machine flowshop type-2 transportation scheduling problem where its guaranteed performance is given. Some special cases that can be solved optimally are presented in the section 5.

2. NOTATIONS AND PROBLEM DESCRIPTION

The blocking flowshop type-2 transportation with limited truck capacity and the intermediate stock capacity equal to zero can be described as follow:

n jobs will be processed on two machines: M_1 and M_2 without preemption in this order. There is no buffer area between the two machines and an unlimited buffer area at the output of the flowshop (i.e. machine $M2$). The absence of an intermediate buffer between the M_1 and M_2 causes the blocking of jobs on M_1 when the machine M_2 is busy. Thus a job cannot leave a machine until the downstream machine becames available. The processing time of a job T_j on machines 1 and 2 is p_{1j} and p_{2j} time units respectively. At a time t one machine can only process one job. After the end of the process on the second machine, jobs must be transported and delivered to the single customer or warehouse using the truck. The transportation time of job T_j is denoted by t_{1j}. The vehicle can be viewed as a batching machine in which all jobs in the same batch (trip) start and finish their transporting at the same time. The batch processing time is determined as the maximum trasporting time of jobs in this batch.

By using the general notation for scheduling problems $\alpha|\beta|\gamma$ and according to the notations of Lee and Chen (Lee and Chen, 2001) and Soukhal et al. (Soukhal *et al.*, 2004) the problem of minimising the makespan for a two-machine flowshop type-2 transportation problem is denoted by $F2 \rightarrow D \mid v, c_i, blocking(1,2) \mid C_{max}$. In this notation, v denotes the number of identical trucks and c_i the capacity of the truck i. One truck ($v = 1$) is considered in this paper and at the time $t = 0$ the truck is located at the output of the flowshop. $blocking(1,2)$ means that there is a blocking constraint btween the M_1 and M_2. The objective is to minimize the makespan C_{max} given by the time to delever the last job plus the necessary time to come back to the output of the processing facility. The necessary time to come back to the output of the processing facility is a constante value t_2.

In (Hall and Sriskandarajah, 1996) the authors investigate $blocking(i, i+1)$ shop scheduling problem. The authors present a survey of complexity results of machine scheduling problems with blocking and no-wait in process. In (Lee and Chen, 2001) the authors study the $F2 \rightarrow D \mid v = 1, c \mid C_{max}$ with unlimited buffers in the system and give some computational complexity results whene $c = 1$ or $c = k$ with any fixed $k \geq 4$, they prove that the problem $F2 \rightarrow D \mid v = 1, c = 1 \mid C_{max}$ is strongly $\mathcal{NP}$-hard. The problem $F2 \rightarrow D \mid v = 1, c \geq 4 \; fixed \; c \mid C_{max}$ is also proved to be strongly $\mathcal{NP}$-hard. The proof is given by reduction from the $\mathcal{NP}$-hard (k-1)-Partition problem (see section 3.3).

In their paper, two *"open"* problems are underlined: $c = 2$ and $c = 3$. Recently, Soukhal et al. (Soukhal *et al.*, 2004) show that the two previous problems with unlimited buffers in the system are $\mathcal{NP}$-hard in the strong sens. The same results are given for blocking two-machine flowshop type-2 transportation scheduling problem in (Soukhal *et al.*, 2004) when the capacity of the truck is equal to two or three.

3. COMPLEXITY RESULTS

In this section and according to the capacity of the transporter c, we analyse the complexity status of the studied problem. Bellow we present complexity results depending on the capacity c of the vehicle.

3.1 *Capacity equal to one*

The $F2 \rightarrow D \mid v = 1, c = 1, blocking(1,2), t_i = p \mid C_{\max}$ abbreviated in the following as problem $P(b)$ is proved to be strongly $\mathcal{NP}$-hard.

Theorem 1. Problem $P(b)$ is $\mathcal{NP}$-hard in the strong sense.

Proof. We use the reduction of 3-Partition problem (Garey and Johnson, 1979) to $P(b)$.

3-PARTITION: Given $3h$ positive numbers $a_1, a_2, \ldots, a_{3h}$ with: $\sum_{i=1}^{3h} a_i = hB$ and $\frac{B}{4} < a_i < \frac{B}{2}$ for $i = 1, \ldots, 3h$

does there exist a partition $I_1, I_2, \ldots, I_h$ of the index set $\{1, 2, \ldots, 3h\}$ such that $\mid I_j \mid = 3$ and $\sum_{i \in I_j} a_i = B$ for $j = 1, \ldots, h$?

We show that a permutation π for the constructed instance with $C_{\max}(\pi) \leq Y$ exists if and only if 3-Partition has a solution.

Given an arbitrary instance of 3-Partition, we construct the following instance of the $P(b)$ with $n = 4h + 1$ jobs. The processing time of the jobs is given by the table 1.

One-way transportation time, $t_{1j} = t_2 = B/2$. So, we ask for a permutation π with: $C_{\max} \leq (4h + 1)B$.

	p_{1j}	p_{2j}
Job T_j $j = 1 \ldots 3h$	0	a_j
Job $T_{3h+1,1}$	0	0
Job $T_{3h+1,2}$	B	3B
Job T_j $j = 3h + 3 \ldots 4h + 1$	4B	3B

Table 1. Jobs processing times

Given a solution to the 3-Partition instance $I_1, I_2, \ldots, I_h$, we can construct a schedule for processing jobs on the two machines where jobs are schedule in the sequence: $(3h + 1), I_1, (3h + 2), I_2, (3h + 3), \ldots, I_h, (4h + 1)$. Based on this schedule, we let the vehicle deliver one job on each delivery trip, and the kth delivery trip departs at time $(k - 1)B$, for $k = 1, \ldots, 4h + 1$; and hence, the vehicle has no idle time after its first depart at time *zero*. Job $3h + j$ $(j = 1, \ldots, h)$, are delivered, respectively, on the 1th and $(4j)^{th}$ trips. It is easy to see that this schedule is feasible and its makespan is $C_{\max} = (4h + 1)B$.

Suppose that there is a schedule for the instance of our problem with a makespan no more than $(4h + 1)B$. We first show that in any feasible schedule for the instance of our problem with makespan no more than $(4h + 1)B$, the following must be true: (i) there is no idle time on M_1 and M_2; (ii) the job $3h + 1$ is proceced first; and (iii) the vehicle starts at time 0 and has no idle time between consecutive deliveries, i.e. the kth delivery trip departs at time $(k - 1)B$, for all $k = 1, \ldots, n$.

The total processing time of jobs on M_1 is $(h - 1)4B + B = Y - 4B$. It takes $3B$ units of time to process the last job on M_2 and $B/2$ units of time to deliver its to the destination and $B/2$ units of time to come back, and hence M_1 should have no idle time. Thus, The total processing time of jobs on M_2 is $4hB = Y - B$. It takes $B/2$ units of time to deliver the last job to the destination and $B/2$ units of time to come back, and hence M_2 should have no idle time. These show (i).

If a job other than $3h + 1$ is processed first, then the completion time of the last job will be greater than $(4h + 1)B$, which implies that the makespan of the problem will be greater than $(4h+1)B$. This shows (ii). Furthermore, since there are $4h + 1$ round trips, the total transportation time is at least $4hB + B$. Hence, the vehicle has to start at time 0 and there is no idle time after it starts. This shows (iii).

Now, we prove that there is a solution to the instance of 3-Partition. Note that we only need to consider permutation schedules because there is no buffer between M_1 and M_2. Given a permutation schedule with a makespan no more than $(4h+1)B$, by (i) and (iii), we can see that the jobs are scheduled in the following sequence: $3h+1, I_1, 3h+2, I_2, \ldots, I_h, 4h+1$ where I_i $(i = 1, \ldots, h)$ is a subset of jobs. Let the number of items in I_1 be q. In the following, we show that $q = 3$, and the total processing time of jobs in I_1 on M_2 is B. By the same argument, it can be shown that each other I_i, for $i = 2, \ldots, h$, contains exactly 3 jobs and $\sum_{j \in I_i} a_j = B$. Hence, the subsets $I_1, I_2, \ldots, I_h$ give a solution to the 3-Partition instance.

Suppose that $q \leq 2$. Then the completion time of job $3h+2$ on M_2 is $3B + \sum_{j \in I_1} p_{2j} > B + qB$. On the other hand, job $3h+2$ is the $(q+2)^{th}$ job delivered and hence it must be completed by time $(q+1)B = B + qB$. This results in a contradiction.

Now, suppose that $q \geq 4$. As the completion time of the last job of I_1 on M_2 is $g = \sum_{j \in I_1} a_j$ then the completion time of job $3h+2$ on $M2$ is $C_{3h+2,2} = g + 3B$. It means that there exists a subset I_k that $|I_k| \leq 2$. Let k be the first subset in the sequence π that $|I_k| \leq 2$. So, the delivered time of the job $3h+k+1$ is equal to $(q+4k-3)B$. But, the completion time of the job $3h+k+1$ on M_2 is equal to $4kB$. So, we have: $(q+4k-3)B - 4kB = (q-3)B > 0$ it means that there is an idle time on machine M_1. This violates the property (i). We conclude that $q = 3$. We now prove that $\sum_{j \in I_1} a_j = B$. Suppose that $\sum_{j \in I_i} a_j < B$ (respectively $\sum_{j \in I_i} a_j > B$) then the completion time of job $3h+2$ on M_1 is equal to B. It means that there is an idle time on M_2 (respectively M_1). This violates the property (i). It must be true that $\sum_{j \in I_1} a_j = B$. $\square$

3.2 Capacity equal to two or three

If the capacity of vehicle is limited to 2 jobs then in (Soukhal *et al.*, 2004), the authors show that the $F2 \rightarrow D \mid v = 1, c = 2, blocking(1,2), t_j = p \mid C_{\max}$ is strongly $\mathcal{NP}$-hard. The proof is given by a reduction from NMTS problem (Garey and Johnson, 1979). In the other hand, if the capacity of vehicle is limited to 3 jobs then in (Soukhal *et al.*, 2004), the authors show that the $F2 \rightarrow D \mid v = 1, c = 3, block(1,2), t_j = p \mid C_{\max}$ is strongly

$\mathcal{NP}$-hard. The proof is given by a reduction from 3-Partition problem.

Although we believe that the problem $F2 \rightarrow D \mid v = 1, c = 2, blocking(1,2), t_i = p \mid C_{\max}$ is also $\mathcal{NP}$-hard, we are not able to prove that now.

3.3 Capacity greater or equal to four

In this section, we are interested to the complexity status of $F2 \rightarrow D \mid v = 1, c \geq 4 \, fixed \, c, block(1,2) \mid C_{\max}$ abbreviated in the following as problem $P(b1)$. In (Lee and Chen, 2001) the authors prove that $F2 \rightarrow D \mid v = 1, c \geq 4 \, fixed \, c, t_j = p \mid C_{\max}$ (abbreviated in the following as problem $P(b2)$) is strongly $\mathcal{NP}$-hard. The proof is given by a reduction from the k-Partition problem:

k-Partition: given kh items, $I = \{1, 2, \ldots, kh\}$, each item $i \in I$ has a positive integer size x_i satisfying $\sum_{i \in I} x_i = hB$, for some integer B, the question asks whether there are h disjoint subsets $I_1, I_2, \ldots, I_h$ of I such that each subset contains exactly k items and its total size is equal to B.

P(b2): we define the instance of P(b2) as follow: Number of jobs, $n = kh + 1$; processing times, $p_{1,j} = 0$, $p_{2,j} = 2(x_j + Q)$, for $j = 1, \ldots, (k-1)h$, where Q is sufficiently large value. And $p_{1,(k-1)h+1} = 0$, $p_{2,(k-1)h+1} = 2$. $p_{1,j} = 2R$ and $p_{2,j} = 2$, for $j = (k-1)h+2, \ldots, kh+1$, where $R = (k-1)Q + (k-3)khB + 1$. One-way transportation time, $t_1 = t_2 = p = R$. Makespan threshold value, $Y = 2hR + R + 2$.

Theorem 2. Problem $P(b1)$ is $\mathcal{NP}$-hard in the strong sense.

Proof. With the same instance of the $P(b2)$ the strongly $\mathcal{NP}$-hard problem k-Partition is reducible to the decision problem of $P(b1)$. It can be proved in the same way as presented in (Lee and Chen, 2001) for the $P(b2)$. Of course, we can verify that the blocking constraints are respected. $\square$

4. HEURISTIC WITH GARANTEED PERFORMANCE

These results mean the absence of polynomial algorithms to solve optimaly these problems. Consequently, heuristic methods must be used. The classical problem $F2|blocking|C_{\max}$ is equivalent to the two-machine flowshop problem that is polynomial (Gilmore and Gomory, 1969). Thus, a greedy algorithm using the Gilmore&Gomory's (G&G) order (Gilmore and Gomory, 1969) is proposed solving the $F2 \rightarrow D \mid v = 1, c, block(1,2) \mid C_{\max}$ problem where its worst-case performance is established.

Let $t = min_j(t_{1j}) + t_2$ be the minimum transportation time of the jobs and the necessary time to come back to the output of the processing facility. Let $P = min(min_j(p_{1,j}), min_j(p_{2,j}))$. Let $C_{(n)}$ be the completion time on M_2 of the last job obtained by applying Gilmore&Gomory's algorithm without taking into account the delivery times. We denote the optimal makespan by $C_{\max}^{opt}$.

Property 1: For the scheduling problem $F2 \rightarrow D \mid v = 1, c, blocking(1,2) \mid C_{\max}$, the makespan given by Gilmore&Gomory's order $C_{\max}^{GG}$, verifies the following formula:

$$\frac{C_{\max}^{GG}}{C_{\max}^{opt}} \le 1 + \frac{(\lceil \frac{n}{c} \rceil - 1)t}{\lceil \frac{n}{c} \rceil t + nP - cP(\lceil \frac{n}{c} \rceil - 1)}$$

Proof. For the proof we should define two lower bounds. Let $LB1$ be the first one. Thus, $LB1 = c_{(n)} + t$. Of course, the blocking two-machine flowshop is equivalent to the no-wait two-machine flowshop. So, the $C_{(n)}$ is the earlier completion time of the last job. And as t is the minimum necessary time to deliver one job then $LB1$ is a lower bound.

We can show that the minimum number of batches (a set of jobs delivered simultaneously within one trip) is equal to $\lceil \frac{n}{c} \rceil$. So, the second lower bound denoted $LB2$ is given by the following formula: $LB2 = t \lceil \frac{n}{c} \rceil + (n - c \lfloor \frac{n}{c} \rfloor)P$. $LB2$ is the necessary time to transport all jobs minus the first idle time. In the ideal situation, there is exactly $\lceil \frac{n}{c} \rceil$ batches where each batch contains exactly c jobs except for the first one contained n mod $c = n - c \lfloor \frac{n}{c} \rfloor$.

Let UB be an upper bound given by $C_{(n)}$ and the necessary time to transport the jobs. Then we have: $UB = C_{(n)} + \lceil \frac{n}{c} \rceil t$.

So we have: $C_{\max}^{GG} \le UB = C_{(n)} + \lceil \frac{n}{c} \rceil t$

$C_{\max}^{GG} \ge C_{(n)} + t$ and $C_{\max}^{GG} \ge (t - Pc) \lfloor \frac{n}{c} \rfloor + Pn + t$

Thus: $C_{\max}^{GG} \le C_{(n)} + \lceil \frac{n}{c} \rceil t \le c_{(n)} + \lceil \frac{n}{c} \rceil t + t - t$

$$\Rightarrow C_{\max}^{GG} \le (C_{(n)} + t) + \lceil \tfrac{n}{c} \rceil t - t$$

$$\Rightarrow C_{\max}^{GG} \le C_{\max}^{opt} + \lceil \tfrac{n}{c} \rceil t - t + (nP - cP(\lceil \tfrac{n}{c} \rceil - 1)) - (nP - cP(\lceil \tfrac{n}{c} \rceil - 1))$$

$$\Rightarrow C_{\max}^{GG} \le C_{\max}^{opt} + C_{\max}^{opt} - t - (nP - cP(\lceil \tfrac{n}{c} \rceil - 1))$$

$$\Rightarrow C_{\max}^{GG} \le 2C_{\max}^{opt} - t - (nP - cP(\lceil \tfrac{n}{c} \rceil - 1))$$

$$\Rightarrow \frac{C_{\max}^{GG}}{C_{\max}^{opt}} \le 2 - \frac{t + nP - cP(\lceil \frac{n}{c} \rceil - 1)}{t \lceil \frac{n}{c} \rceil + nP - cP \lfloor \frac{n}{c} \rfloor}$$

$$\Rightarrow \frac{C_{\max}^{GG}}{C_{\max}^{opt}} \le 2 - \frac{t + nP - cP(\lceil \frac{n}{c} \rceil - 1) + t \lceil \frac{n}{c} \rceil - t \lceil \frac{n}{c} \rceil}{t \lceil \frac{n}{c} \rceil + nP - cP(\lceil \frac{n}{c} \rceil - 1)}$$

Then: $\dfrac{C_{\max}^{GG}}{C_{\max}^{opt}} \le 1 + \dfrac{(\lceil \frac{n}{c} \rceil - 1)t}{\lceil \frac{n}{c} \rceil t + nP - cP(\lceil \frac{n}{c} \rceil - 1)}$ $\square$

If the capacity of the vehicle is equal to n then the property 1 ensures that the optimal schedule can be determined by applying G&G's algorithm. In fact, we have : $\dfrac{(\lceil \frac{n}{c} \rceil - 1)t}{\lceil \frac{n}{c} \rceil t + nP - cP(\lceil \frac{n}{c} \rceil - 1)} = 0$

Then, $\dfrac{C_{\max}^{GG}}{C_{\max}^{opt}} \le 1 + 0 \Rightarrow C_{\max}^{GG} = C_{\max}^{opt}$.

5. WELL SOLVABLE CASES

In the following we assume that the transportation time is a constant $t_{1j} = t_2 = p$. Firstly, we have established the following results:

Property 2: For the scheduling problem $F2 \rightarrow D \mid v = 1, c, blocking(1,2), t_j = p \mid C_{\max}$ there is an optimal solution where each batch contains exactly c jobs except for the first one if the number of jobs n is not a multiple of c.

Proof. If a batch, contains less than c jobs, we can always fill the batch with more jobs from earlier batches without increasing the objective value. $\square$

The property 2 ensures that for a given permutation S the optimal schedule can be determined.

Let's $t = t_{1j} + t_2 = 2p$. So, for a given instance of $F2 \rightarrow D \mid v = 1, c, blocking(1,2), t_j = p \mid C_{\max}$ if one of the two following conditions:

- $t \le k \times min_{j=1}^n (p_{1,j})$
- $t \le k \times min_{j=1}^n (p_{2,j})$

are true then we have the proporty 3:

Property 3: For the scheduling problem $F2 \rightarrow D \mid v = 1, c, blocking(1,2), t_j = p \mid C_{\max}$ there is an optimal solution given by G&G's order where each batch contains exactly c jobs except for the first one if the number of jobs n is not a multiple of c.

Proof. Under the one of these two conditions, it is easy to show that the makespan is equal to $LB1 = c_{(n)} + t$. As G&G's algorithm gives an optimal value of $c_{(n)}$ then $C_{\max}^{GG}$ is optimal. $\square$

An other well solvable cases is identified when: $max_{j=1}^{n}(p_{1,j}) \leq min_{j=1}^{n}(p_{2,j})$. In this case, to solve the cosidered problem we propose the following algorithm:

(1) Determine n solutions $S1, S2, .., Sn$ as follows:
- At the iteration j, the job j is scheduled in the first position of the sequence Sj
- The remaining jobs are scheduled according to SPT rule.

(2) Each batch is composed of c jobs (except for the first one if n is not a multiple of c) and delivered immediately when a batch is available.

(3) The optimal solution is given by the best sequence.

Property 4: If $max_{j=1}^{n}(p_{1,j}) \leq min_{j=1}^{n}(p_{2,j})$ then the previous algorithm gives an optimal solution for the scheduling problem $F2 \rightarrow D \mid v = 1, c, blocking(1,2), t_j = p \mid C_{\max}$.

Proof. The proof is given by using the critical path and pair-wise interchange argument. $\square$

6. CONCLUSIONS

In this paper, we investigate a blocking two-machine flowshop type-2 transportation. In this system there is no intermediate buffer between the two machines but at the output of the processing facility there is an unlimited storage area. New complexity results depending on the capacity c of the vehicle are developed and are summarized in the table 2.

Vehicle capcity c	Complexity	Reference
$c = 1$	SNP	3.1
$c = 2$	SNP	(Soukhal *et al.*, 2004)
$c = 3$	SNP	(Soukhal *et al.*, 2004)
$c \geq 4, \text{ fixed } c$	SNP	3.3

Note: SNP:$\mathcal{NP}$-hard in the strong sense.

Table 2: Summary of complexity status of blocking two-machine flowshop type-2 transportation.

However, in this paper we consider that there is only one transporter in the system. It will be interesting to extend these results to the case $v > 1$. Some works are in progress to extend these results.

7. REFERENCES

Ahmadi, J.H., R.H. Ahmadi, S. Dasu and C.S. Tang (1992). Batching and scheduling jobs on batch and discrete processors. *Op. Res.* **39**(4), 750–763.

Garey, M.R. and D.S. Johnson (1979). Computers and intractability: A guide to the theory of $\mathcal{NP}$-completeness.

Gilmore, P.C. and R.E. Gomory (1969). Sequencing a one-state variable machine: A solvable case of the traveling salesman problem. *Oper. Res* **12**, 655–679.

Glass, C.A., C.N. Potts and V.A. Strusevich (1998). Scheduling batches with sequential job processing for two-machine flow and open shops. *Preprint series No. OR99, University of Southampton, Faculty of Mathematical Studies, England.*

Hall, N.G. and H. Sriskandarajah (1996). A survey of machine scheduling problems with blocking and no-wait in process. *Oper. Res* **44**(3), 510–525.

Lee, C.Y. and Z-L Chen (2001). Machine scheduling with transportation considerations. *Journal of Scheduling* **4**, 3–24.

MacCarthy, B.L. and J. Liu (1993). A new classification sheme for flexible manufacturing systems. *Int. J. Prod. Res* **31**(2), 299–309.

Potts, C.N., V.A. Strusevich and T. Tautenhahn (2001). Scheduling batches with simultaneous job processing for two-machine shop problems. *Journal of Scheduling* **4**, 25–51.

Soukhal, A., A. Oulamara and P. Martineau (2004). Complexity of flowshop scheduling problems with transportation constraints. *Eur. Jou. Oper. Res* **(to appear)**, 1–15.

ELSEVIER
IFAC
PUBLICATIONS
www.elsevier.com/locate/ifac

PROPOSITION OF A DISAGGREGATION METHOD FOR ROBUST TACTICAL PLANNING

V. ORTIZ ARAYA[1*], A. THOMAS[1] and C. HUTT[2]

[1] CRAN –Research Center for Automatic Control of Nancy, CNRS UMR 7039, Henry Poincaré University - BP 239, Faculty of Sciences, 54506 Vandœuvre les Nancy CEDEX – France
Virna.Ortiz@cran.uhp-nancy.fr, Andre.Thomas@cran.uhp-nancy.fr

[2] DynaSys – Society Dynamic Systems, 10, avenue Pierre Mendès France, 67300 Schiltigheim - France
cedric_hutt@dys.com

Abstract

The hierarchical production planning is a structure that provides the possibility to determine the master production scheduling. Different analyses are presented in reviewed literature for applying hierarchical production planning process. Nowadays enterprises look for a method to know when and how much to produce per week, while maintaining coherence between usable stock and forecasted monthly demand.

The purpose of this paper is to show a disaggregation procedure in two levels and so to determine master production scheduling and analyze behaviour of robustness and stability in second level.

Keywords: Hierarchical production planning, Master production scheduling, Linear optimization, Nonlinear optimisation.

1. Introduction

Production can be defined as the process of converting raw materials into finished products. An effective management of the production process should provide the finished products in appropriate quantities, at the desired times, of the required quality, and at a reasonable cost. The production management encompasses a large number of decisions. Such decisions involve complex choices among a large number of alternatives. These choices have to be made by trading off conflicting objectives under the presence of financial, technological, available resources and marketing constraints.

Hierarchical production planning (HPP) systems are developed with the aim of establishing different levels of decision-making and information so that managers are not among irrelevant details of the planning issues. Many great contributions have been made by using operations research to solve the production planning problem, [5] [6] [8]. Many authors take the model proposed by Bitran *et al.* [5] [6] to test the robustness, coherence and feasibility in the disaggregation process (Erschler *et al.* [3], Mercé [7], Özdamar *et al.* [2]). The most important task consists in providing, with HPP, a robust, stable and feasible plan taking into account, capacity allocation and priority management at the level of master production scheduling.

Enterprises look for a method to know when and how much to produce per week, while maintaining coherence between usable stock and forecasted monthly demand. To attain these objectives, APS are useful to calculate production at the master production schedule level. In HPP, once the volume of production quantity is calculated per month for each family, the decision to define the best finished product quantity to produce per week is taken according to the needs of the customers. In this paper, we propose a disaggregation process to obtain an optimal volume to transform a production plan into MPS.

This article is organized as follows: in the first part, an analysis to introduce the chosen methodology is made. Then, we propose two-level disaggregation models. A production plan for product families is transformed into Master Production Scheduling (MPS) for finished products. Finally, the analysis of found results for some months is discussed.

2. Hierarchical Production Planning : context and problems

The Hierarchical Production Planning (HPP) is a concept carried out in many researches. Hax and Meal [14] introduced this concept (HPP) in 1975. Method consists primarily of recognizing the differences between tactical and operational decisions. The tactical decisions are associated with aggregate production planning while the operational decisions are an outcome of the disaggregation

* Corresponding author. Tel.: +33-383684449; fax: +33-383684447.

process. Hax and Meal proposed three levels of disaggregation: items, families and groups. Bitran *et al.* [5] proposed the disaggregation process for an aggregated planning, considering the idea presented in [14]. The model suggested by these authors uses the convex knapsack problems to disaggregate production. The first level consists of allocating production capacity among product types using an aggregate planning model. The planning horizon of this model normally covers a year to take into proper consideration the fluctuation demand requirements for the products. This model is solved with linear programming. In the second level, the disaggregation process takes into account the optimal production found in the first level to issue the production of families. In this stage, the objective is to minimize the total set up costs. Finally, the family production allocation is divided into the items belonging to each family (see figure 1). For the second model, an algorithm is proposed par Bitran and Hax [4]. The analysis of these models in the context of the consistence, coherence and robustness were presented by Erschler *et al.* [3] and Mercé [7].

Özdamar *et al.* [2] propose the heuristic modification to Bitran and Hax [4] algorithm and an alternative filling procedure to improve the infeasibilities in the disaggregation stage for the family production. Giard [1] examines the disaggregation problems in the context of the coherence for cases in which the products are not homogenous, in volume or structure to the interior of the family. Then, the principal problem is to coordinate the different levels to determine an optimal, robustness and instable plan of disaggregation.

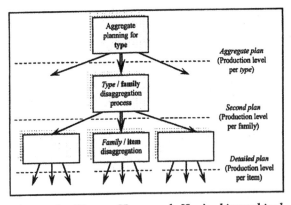

Figure 1. *Bitran, Haas and Hax's hierarchical structure [5]*

The term stability is generally associated, by opposition, with that of nervousness: De Kok and Inderfurth [9] have defined "nervousness" as "a lack of stability in the material requirements planning". Donselaar *et al.* [10] compare the nervousness of the plan generated by MRP with that of their heuristics. The robustness is compared with that of risk and decision-making [11]. The underlying idea of system

robustness is that the measured functions do not diverge significantly from a given value. According to Lee *et al.* [12], robustness is relative to one or several functions results and to their dispersion caused by uncertain parameters and costs.

Taguchi [13] reminds us that minimizing a function by setting control parameters can cause a lack of robustness. He proposes a compromise evaluated by the signal/noise ratio. It can be calculated in different ways depending of the situation: function to minimize, to maximize, and to get to a target. Genin [8] proposed a special SOP (Sales and Operation Planning) procedure to obtain a stable and robust production plan. The figure 2 shows a general idea for this study. The aim of this procedure is to compare different indicators as: service level, signal/noise ratio, stock level, range, standard deviation, to characterize his objective function. He suggested that with some variation in demand plan quantities it is possible to obtain a stable and robust distribution of production planning quantities.

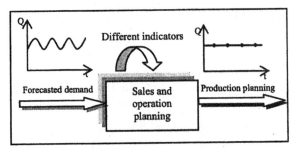

Figure 2. *Concept of procedure for "reference plan" used by [8]*

In this paper, disaggregation process takes place considering a stable production planning. The final objective is to show the effect of the proposed disaggregation procedure on the MPS quantities.

3. Disaggregation models

In this article, the disaggregation procedure proposed is structured only two stages (see scheme 1). First, sales and operations planning is optimized to obtain the production and inventory quantities per family. In the first level, these production quantities are disaggregated into finished product for twelve months. The decision variables of the model are: available inventory, overstock and safety stock. The proposed model is considered as intermediate model. Then, the objective is to apply the disaggregation process to obtain finished product quantities to be analyzed into the second stage.

The second level takes into account finished product quantities found monthly; corresponding safety stock, set up cost, available inventory and overstocks. The MPS quantities are thus obtained at this level.

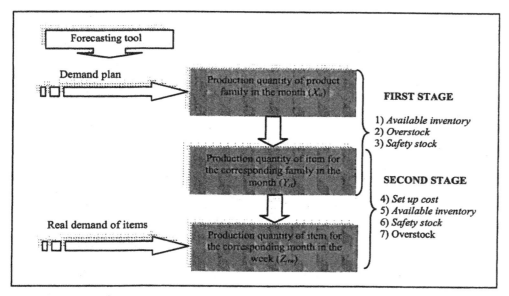

Scheme 1. *Disaggregation stages for the problem*

3.1. Notation used in the model

3.1.1 Parameters

T planning horizon.

d_{it} demand for family i in period t.

c_{it} production cost for unit product of family i in period t (excluding labor).

h_{it} inventory carrying cost for unit product of family i in period t.

r_t, o_t cost per manhour of regular labor and overtime labor in period t.

m_i hours required to produce one unit of family product i.

rm_t, om_t total availability of regular hours and of overtime hours in period t.

$ss_{jt}, as_{jt}, os_{jt}$ safety stock , available inventory and overstock limit for item j in period t.

d_{jt} demand for item j in period t.

lb_{jt}, ub_{jt} lower and upper bounds. Where:

$$lb_{jt} = Max\ (0, \overline{d}_{jt} - ai_{jt} + ss_{jt})\ \text{and}$$

$$ub_{jt} = os_{jt} - ai_{jt}\ .$$

S_{jt} set up cost for item j in moth t.

d_{jtw} demand for item j in moth t and week w.

3.1.2 Variables

X_{it}, I_{it} production and inventory quantities of product family i in period t.

R_t, O_t regular and overtime hours used during period t.

Y_{jt} production quantity of item j in period t.

Z_{jtw} production quantity of item j for moth t and week w.

3.2. Sales and operations planning model

In this model, the objective function (1) consists of minimizing the total cost (production, inventory and hours resources) to seek the optimal production quantity of product families. A complete model can incorporate the hiring and firing costs, backorders, raw materials and others features that can be easy considered.

The equation (3) corresponds to necessary hour constraints for production and equations (4) and (5) consider the limits in utilization of regular and overtime hours.

$$Min \sum_{i=1}^{I} \sum_{t=1}^{T} (c_{it}X_{it} + h_{it}I_{it}) + \sum_{t=1}^{T} (r_t R_t + o_t O_t)$$
(1)

s.t.

$$X_{it} - I_{it} = d_{it} - I_{it-1}\ ,\ t = 1,2,...,T.\ i = 1,2,...,I.\ (2)$$

$$\sum_{i=1}^{I} m_i X_{it} \leq R_t + O_t\ , t = 1,2,...,T.$$
(3)

$$R_t \leq (rm)_t\ , t = 1,2,...,T.$$
(4)

$$O_t \leq (om)_t\ , t = 1,2,...,T.$$
(5)

$$X_{it}, I_{it}, R_t, O_t \geq 0\ ;\ t = 1,2,...,T.\ i = 1,2,...,I.\ (6)$$

3.3. The family disaggregation model in item (intermediate level)

The model presented in this stage corresponds strictly to convex knapsack problem [4]. The central condition to be satisfied at this level is the equality between the sum of finished product and production quantities found in the first model. This equality will assure consistency between the sales and operations planning and the family disaggregation process.

The constraint (9) guarantees to limit finished product quantities to usable volume stock (for upper bound).

$$Min \sum_{j \in J(i)} \left[\frac{X^*_{it} + \sum_{j \in J(i)} \left(ai_{jt} - ss_{jt} \right)}{\sum_{k \in K(j)} d_{jt}} - \frac{Y_j + ai_{jt} - ss_{jt}}{d_{jt}} \right]^2 \tag{7}$$

s.t.

$$\sum_{J \in J(i)} Y_{jt} = X^*_{it} \tag{8}$$

$$lb_{jt} \le Y_{jt} \le ub_{jt} , j \in J(i), t = 1...T \tag{9}$$

3.4. The item disaggregation

The objective function (10) attempts to minimize the set up cost for each item per week. The finished product quantities are proportional to the set up cost and the weekly demand for a given finished product. The presented model is a non-lineal problem. Bitran and Hax [4] proposed an algorithm to solve this problem and Özdamar et al. [2] settled down some modifications for this procedure.

We developed this model with the improved heuristic considered in [2].

$$Min \sum_{w \in W(t)} \frac{\left(s_{jt} * d_{jt} \right)}{Z_{jtw}} \tag{10}$$

s.t.

$$\sum_{w \in W(t)} Z_{jtw} = Y^*_{jt} \tag{11}$$

$$lb_{jtw} \le Z_{jtw} \le ub_{jtw} , w \in W(t),$$
$$j = 1...J; t = 1,2,...,T \tag{12}$$

The calculations lower and upper bounds are made by using the following conditions ((13) and (14)) proposed by Erschler et al. [3].

$$lb_{jtw} = d_{jtw} \text{ and } ub_{jtw} = \sum_{t=1}^{T} d_{jt} , \tag{13) (14}$$

The function (10) is convex and differentiable, so the corresponding Kuhn-Tucker conditions are applied to seek the expression (15). This function (15) allows finding the values of Z_{jtw} that are compared with the lower and upper bounds (constraint (12)). The procedure presented by Özdamar et al. looks for avoiding the possible inconsistencies between the bounds and the values of finished product quantities in the corresponding week.

$$Z_{jtw} = \left(\frac{\sqrt{s_{jt} * d_{jt}}}{\sum_{t \in T(t)} \sqrt{s_{jt} * d_{jt}}} \right) * Y^*_{jt} \tag{15}$$

The application of this procedure is important to test the calculated values and to modify these quantities with the optimal conditions.

4. Results and analysis

The proposed models are tested for three families and fifteen items in a-year planning horizon. In this paper, only the results of the second model for three months are shown. The solutions are founded used CPLEX 9.0 for the first and second model.

According to the Genin's [8] results, the production was leveled at the following quantity: 1200 units for each month of production plan.

Tables (1), (2) and (3) contain the input data for the three corresponding months. The found solutions for the intermediate level model are summarized in the table (4). In this level, it is possible to observe the disaggregation process of three families in fifteen items. For the first month, the production quantity for family 1 is stable (value 1200 units). Nevertheless, once the family is disaggregated, the behavior of finished products (items) changes with respect to demand variation. The final objective is to verify if the quality of the stable plan remains at the end of the disaggregation process.

Table 1. Input data for month 1

Items	Demand			Safety Stock	Available Inventory			Overstock limit		
	Family 1	Family 2	Family 3	Family 1, 2, 3	Family 1	Family 2	Family 3	Family 1	Family 2	Family 3
1	85	100	60	25	100	90	50	5000	7500	8500
2	60	60	80	50	80	40	60	5500	8000	9500
3	80	90	60	40	0	80	40	8000	9500	12000
4	85	65	95	50	50	55	75	7500	9000	12000
5	70	55	120	60	20	45	100	9500	9000	9800
6	80	90	30	45	95	80	20	6500	6500	6500
7	70	70	40	25	60	60	20	7000	7000	4000
8	65	50	60	50	35	40	50	9000	5900	8500
9	80	90	50	40	80	80	40	9000	9500	9500
10	100	30	40	50	10	20	30	10000	5600	6500
11	85	30	60	60	10	20	50	10000	5600	8600
12	90	30	100	45	10	20	60	8500	8500	9800
13	75	150	150	25	50	20	50	8000	10000	5000
14	50	120	85	50	60	20	60	9500	10000	8500
15	60	150	100	40	50	20	40	6500	9500	6500

Table 2. *Input data for month 2 (the overstock limit is the same that for month 1)*

Items	Demand Family 1	Family 2	Family 3	Safety Stock Family 1	Family 2	Family 3	Available Inventory Family 1	Family 2	Family 3
1	125	50	120	25	25	20	50	50	30
2	60	80	80	50	50	10	40	20	20
3	80	60	60	40	55	40	30	100	100
4	100	85	95	50	20	40	20	20	50
5	70	140	120	60	60	30	10	45	120
6	80	90	30	45	45	45	80	30	60
7	70	80	100	30	25	25	40	60	60
8	140	75	60	50	20	60	30	50	50
9	80	100	80	40	20	40	50	120	90
10	30	60	40	50	50	50	60	20	30
11	85	140	50	60	60	60	60	20	60
12	50	80	60	45	20	60	60	30	50
13	75	50	85	30	20	60	30	60	75
14	50	60	85	50	20	60	60	50	60
15	105	50	80	80	30	60	20	60	40

Table 3. *Input data for month 3 (the overstock limit is the same that for month 1)*

Items	Demand Family 1	Family 2	Family 3	Safety Stock Family 1	Family 2	Family 3	Available Inventory Family 1	Family 2	Family 3
1	125	125	55	25	25	20	80	50	30
2	60	80	80	50	50	10	40	20	80
3	65	60	75	40	55	25	60	100	100
4	50	90	85	50	20	40	20	20	50
5	150	70	70	60	60	30	30	45	100
6	80	125	80	45	45	45	50	100	120
7	80	60	90	30	25	25	20	60	60
8	140	90	60	50	20	60	30	50	50
9	90	100	80	40	20	30	50	120	90
10	30	70	40	50	50	50	60	120	30
11	60	60	90	60	50	60	60	20	60
12	50	80	70	45	20	60	50	30	60
13	25	50	150	30	20	60	30	60	75
14	50	60	80	50	20	60	50	50	90
15	50	80	80	80	30	60	60	60	60

Table 4. *Results of disaggregation families in items for three months*

Items	Month 1 Items Family 1	Items Family 2	Items Family 3	Month 2 Items Family 1	Items Family 2	Items Family 3	Month 3 Items Family 1	Items Family 2	Items Family 3
1	130	90	195	120	240	400	150	495	480
2	30	70	70	70	110	70	70	110	10
3	120	50	60	90	15	0	45	15	0
4	85	60	70	130	85	85	80	90	75
5	110	70	80	120	155	30	180	85	0
6	30	55	55	45	105	15	75	70	5
7	35	35	45	60	45	65	90	25	55
8	80	60	60	160	45	70	160	60	70
9	40	50	50	70	0	30	80	0	20
10	140	60	60	20	90	60	20	0	60
11	135	70	70	85	180	50	60	90	90
12	125	55	85	35	70	70	45	70	70
13	50	155	125	75	10	70	25	10	135
14	40	150	75	40	30	85	50	30	50
15	50	170	100	80	20	100	70	50	80
Total	1200	1200	1200	1200	1200	1200	1200	1200	1200

For the third model, the input data are finished product values obtained in the second model (table 4). These quantities are allocated for each week in the corresponding month. The set up cost varies between $20 and $40. In this paper, the analysis is realized on three items (1, 5 and 11) for first family in month 1. The final results are shown in the figure 3. This figure shows the effects on the production plan once the disaggregation procedure is applied obtaining the master production scheduling. In the graph (a), production plan is stable (1200 units for each month) and range for these production quantities is 0.

After carrying out the disaggregation procedure, stability existing in production plan is lost (see graph (b)). For this case, ranges in items 1, 5 and 11 are 20, 30 and 20 respectively. Consequently, this disaggregation procedure does not help to preserve robustness and stability qualities.

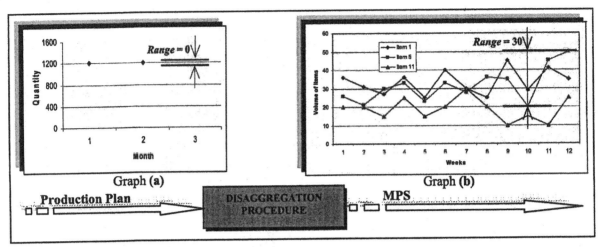

Figure 3. *Comparison between production plan and MPS once applied disaggregation process*

5. Conclusion and future work

We proposed the methodology to disaggregate a production plan into MPS. The initial production planning is established monthly for product families, whereas the resulting MPS is established weekly for finished products. The analysis achieved in this article shows that initial characteristics of production plan (stable and robustness) are not preserved after disaggregation in the MPS.

Future expectations concerning this study are:

- analyze these results using a rolling plan with a longer horizon (50 months, for example),
- apply the methodology reference plan [8] with the purpose to examine behaviour of disaggregation plan,
- implement stochastic optimization for this disaggregation process in order to see the impact that will generate for allocation into MPS.

References

[1] V. Giard, "Gestion de la production et des flux", 3éme éd. *Economica*, 2003.

[2] L. Özdamar, A.Ö. Atli, M.A. Bozyel, "Heuristic family disaggregation techniques for hierarchical production planning systems", *International Journal of Production Research*, 1996, vol. 34, N° 9, 2613-2628

[3] J. Erschler, G. Fontan and C. Mercé, "Consistency of the disaggregation process in hierarchical planning", *Operations Research*, 1986, 34, 464-469.

[4] G.R. Bitran, A.C. Hax, "Disaggregation and resource allocation using convex knapsack problems with bounded variables", *Management Science*, 1981, vol.27, n°4, 431-441.

[5] G.R. Bitran, E.A. Haas, A.C. Hax, "Hierarchical production planning: A single stage system", *Operations Research*, 1981, vol.29, n°4, 717-743.

[6] G.R. Bitran, A.C. Hax, "On design of hierarchical production planning systems", *Decision Science*, 1977, vol.8, 28-55.

[7] C. Mercé, "Cohérence des décisions en planification hiérarchisée". *Thèse de Doctorat d'Etat*, Université Paul Sabatier, Toulouse, 1987.

[8] P. Genin, "Planification tactique robuste avec usage d'un APS", *Thèse de Doctorat de l'Ecole des Mines de Paris*, 2003.

[9] A.G. De Kok, K. Inderfurth, "Nervousness in inventory management: comparison of basic control rules", *European Journal of Operational Research*, 1997, 103, 55-82.

[10] K. Van Donselaar, J. Van Den Nieuwenhof, J. Visschers, "The impact of material coordination concepts on planning stability in supply chains", *International Journal of Production Economics*, 2000, 68, 169-176.

[11] J. P.C. Kleijnen, E. Gaury, "Short-term robustness of production management systems: A case study", *European Journal of Operational Research*, 2003, 148, 452-465.

[12] H.L. Lee, V. Padmanabhan, S. Whang, "Information distortion in a supply chain: the bullwhip effect", *Management Science*, 1997, vol. 43, n°4, 546-558.

[13] Taguchi, "Orthogonal arrays and linear graph", *American Supplier Institute press*, 1987.

[14] A.C. Hax, D. Candea, "Production and inventory management", *Prentice-Hall, Inc.* 1984.

ELSEVIER
IFAC
PUBLICATIONS
www.elsevier.com/locate/ifac

METHODOLOGY TO ANALYZE ORGANIZATIONAL TRAJECTORIES OF SMEs NETWORKS

M. Benali[1], P. Burlat[1]

[1] Ecole des Mines de Saint-Étienne, 158 Cours Fauriel, F 42023 Saint-Étienne, 00 33 4 77 42 66 36, benali@emse.fr, burlat@emse.fr, France

Abstract: This article represents a contribution toward deriving a concept of coordination within networks of firms and their trajectories, with a special focus on SMEs that are virtually linked to achieve goals. First, we develop a methodology for plotting cartographies of the network coordination modes based on the activities and the competencies of each firm. Next, the network evolution is represented by a scenario tree according to the initial position and the possible change of competencies and activities. The methodology presented integrates a global survey of the design of tools for the diagnosis of SMEs network performance.

Keywords: networks, co-ordination, trajectories, transformations, skill, decision making.

1. INTRODUCTION

When an enterprise doesn't possess the necessary resources, skills and competencies to offer competitive solutions, it becomes difficult for it to face up to a more and more volatile competitive market environment and to answer the needs of increasingly demanding customers. To be more competitive, SMEs have responded by initiating extensive industrial co-operations. Thus, new organizational structures are emerging: extended enterprises, virtual enterprises, networks of firms, etc. In this article, we focus our attention on networks of SMEs, i.e., on organizations formed by several independent SMEs virtually linked together to achieve common goals.

Two organizational maps are plotted. First, we map out the actual coordination links within a network, according to the level of exchange and the common goals. A typology of network enterprises is then proposed based on two parameters: complementary activities and competencies similarity. These two parameters are used to identify the preferential

coordination mode between two enterprises within the network (the choice of these two parameters is based on the theoretical works of Richardson (Richardson, 1972). This analysis allows us to draw a map of potential coordination modes, according to the activities and the competencies of each firm. The map is then validated by practical studies and industrial cases (Burlat et. al., 2003).

These studies show that networks live through periods of stability interrupted by a period of transition. The evolution of networks will be thus described as a succession of configurations, linked together by a process of transformation. However, the concepts of configuration (Mintzberg, 1998) and transformation (Pettigrew, 1987) have been created for monolithic firms. Necessary adaptations are proposed to face the case of networks of firms viewed as a whole. In this view, the full process of configuration and transformation is named the 'organizational trajectory' of a network (Burlat, et. al., 2003). The evolution of this trajectory depends on the evolution of the two parameters investigated: complementary activities, similarity of competencies. Therefore, several trajectories can appear, according to the initial position and according to the possible evolution of competencies and activities of each firm inside the network.

In previously work (Peillon, 2001), based on industrial cases, it has been noted that networks are rarely guided by formal and deliberate strategies and that their evolutions and mutations cannot be forecasted with the usual strategic planning tools. In this respect, this work presents a methodology to design the trajectories of networks of enterprises, with the aims of offering a decision-making tool to facilitate the strategic control of the network.

2. CARTOGRAPHY OF ACTUAL COORDINATIONS

In our survey, we treat especially a SME network where the links are often informal and where the parties are somewhat equal, so that relationships are most often non-hierarchical. We distinguish two types of co-operation links between SMEs:

- Non-competing enterprises work together in the same product/market sector, usually at various stages of the value chain, but also within different value chains (they co-operate to add value to the final product or service). This type of coordination forms a *Proactive Network*.

- Potentially competitive companies of different product/market sectors have similar needs and interests (they co-operate to reduce costs by reaching an optimum size). This type of coordination is *Defensive Network* (Peillon, 2001).

Note that if companies compete in the same market, they are coordinated through market forces.

To lay out the map of the actual coordination, it is necessary to identify the coordination modes between each pair of enterprises. To achieve this goal, a questionnaire have been set up, based on the different previous parameters (revolving around the themes: *do the two enterprises belong to the same market sector or not?, are the two enterprises competing or not?*).

Figure 1 illustrates an example of a map of actual coordination modes of a network of ten SMEs working in the mechanical industry: the nature of the link between two companies of the network is identified by using the questionnaire (this example is inspired from industrial cases (Benali, and Burlat, 2003; Benali, and Burlat, 2004)).

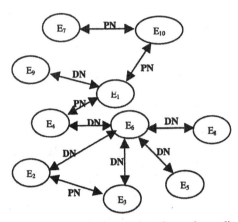

Fig. 1. example of cartography of actual coordination modes within a SMEs network.

PN: Proactive Network (Co-operation to add value)
DN: Defensive Network (Co-operation to reduce costs)

3. CARTOGRAPHY OF POTENTIAL COORDINATION MODES

Inside networks, coordination is not carried out through a hierarchical organization (as in the monolithic firm) or through price mechanisms (as on the market), but trough formal or informal interrelation between independent companies (Richardson, 1972). To identify the type of interrelation between two firms, a networks analysis framework is proposed (Figure 2) (Burlat et. al., 2003). This framework is based on two criteria: complementarity of activities, similarity of competencies. Note that a competency is here an ability to sustain the coordinated deployment of assets in a way that helps a firm achieve its goal. The competencies required to achieve activities are analyzed to define more precisely what should be coordinated by a hierarchical direction inside a firm. Therefore, when activities are complementary and require similar skills, the most efficient coordination mode seems to be the hierarchical direction within a single firm. On the other hand, when activities are complementary and competencies are not similar, the most frequent coordination mode is inter-company cooperation. Two types of inter-companies cooperation are identified:

- When two SMEs having complementary activities require dissimilar skills, the most frequent coordination mode is a proactive network. We note that this type of cooperation often corresponds to an anticipation of the economic environmental constraints.

- When two SMEs have non complementary activities needing similar competencies, they may find it interesting to cooperate in a defensive network. There, we note that such networks are often formed in reaction to environmental changes (for example a purchase networks whose goal is primarily to reduce costs).

Figure 2 depicts a framework to analyze the coordination mode between two companies (Burlat, et. al., 2003):

	Non complementary Activities	Complementary Activities
Non similar competencies	**Market**	**Proactive Network**
Similar competencies	**Defensive Network**	**Firm**

Fig. 2. Network analysis based on activities and competencies.

According to this framework, the map of potential co-operation modes is plotted. To illustrate this methodology, an example is given later. First, it is necessary to clarify the two concepts of competencies and activities and briefly introduce their modeling.

3.1. complementary activities

Activities are called complementary if they correspond to various successive phases of a production process or if they constitute very interconnected steps of an administrative process (for example : a step of the marketing process and R&D mode within an innovation process).

Two activities are complementary if the increase of one of these activities increases (or at least does not decrease) the marginal profitability of the other (Milgrom, 1997). In a network, the complementary activities between companies are modeled by a graph whose vertices represents the enterprises and the arcs (valued between 0 and 1) represent the degree of complementarity. This degree is taken from a questionnaire that permits to identify the relationships between firms via inter-company sales. The methodology to analyze the activities and detecting the sub-groups of enterprises having complementary activities was developed in previous work (Benali, and Burlat, 2004). This methodology is based on three steps: constructing graph of complementary activities, applying a transitive closure to identify the chains of indirect complementarity, applying different algorithms of graph partitioning to identify subsets of complementary firms.

3.2. Similarity of competencies

Competencies are called similar if they correspond to the same profession (mechanical, electricity, stripping, ...). The theory of fuzzy subsets (Zadeh,

1965) has been used for modeling and quantifying the concept of competency (Boucher, and Burlat, 2003).
The proximity between two enterprises in terms of competency is quantified by using the relative generalized Hamming distance (Kaufmann, 1973).
An array of distance is computed and then analyzed, using principal component analysis, to detect sub-groups of firms having a similar competencies (Benali, and Burlat, 2003)

3.3. Example of plotting maps of potential coordination

To continue the previous example, in a network of ten firms E_1, E_2 ..., E_{10}, a graph of complementary activities has been built by consulting each company. After analyzing the graph (transitive closure and partitioning), the following sub-groups of firms having complementary activities have been obtained: $G1 = \{E_5\}$, $G2 = \{E_8\}$, $G3 = \{E_9\}$, $G4 = \{E_1, E_2, E_3, E_4, E_6, E_7, E_{10}\}$
The ten firms have been evaluated on four key competencies of production: machining, cutting, forging and assembly.
The sub-groups having similar competencies are:
$F1 = \{E_5, E_6, E_8\}$, $F2 = \{E_1, E_4, E_9\}$, $F3 = \{E_2, E_3\}$, $F4 = \{E_7, E_{10}\}$
The map of Figure 3 is plotted according to the framework (Figure2) applied to the sub-groups Gi and Fi. We can recommend a preferential coordination mode between two companies inside the same network.

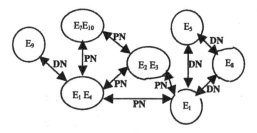

Fig. 3. Example of cartography of potential modes of coordination

4. ANALYSING THE TWO MAPS

In the previous example, the enterprises E_1 and E_4 are currently coordinated by a proactive network (Figure1), but they should find it interesting to merge into one firm (Figure3). However, the theoretical analysis based on activities and competencies shall be refined thanks to parameters of contingency influencing the coordination mode. These parameters moderate the tendencies issued from the framework (Figure2), having either slowing or accelerating effects on the cooperation. Some parameters having significant effects, have been detected in practical studies about SMEs networking: Size of enterprises (Menguzzato, 2003; Mutinelli, and Piscitello, 1998;

Hagedoorn, and Schakenraad, 1994); family culture within SMEs (Menguzzato, 2003; Kets de Vries, 1996; Miles, and Snow, 1994); geographical, technological and organizational proximity (Burlat, and Peillon, 2002); degree of internationalisation (Murray, and Mahon, 1993).

For example, the organizational proximity[1] between E_1 and E_4 has a slowing effect because the organizational structures, the culture and knowledge framework of E_1 and E_4 are different. To work in closer cooperation, E_1 and E_4 should establish common learning processes leading to a deep modification in their decision structure and permitting the development of a common culture and common knowledge framework (Burlat, and Peillon, 2002).

5. ANALYSING THE ORGANISATIONAL TRJECTORY OF NETWORKS

This section presents the preliminary conclusions of our research about SME network evolution. This evolution is represented by a scenario tree. More precisely, the organizational trajectory of the network is considered as a succession of configurations linked together by a process of transformations based on the evolution of activities and competencies of each company. It is possible to influence this trajectory by controlling the attributes making up the two dimensions of the analytic framework of Figure 2: complementary activities and similarity of competencies. For example, if two companies that are coordinated in a defensive network wish to merge into one firm, they will need to reorganize their activities in order to become more complementary.

Activity transformations may either be *internalizing* or *externalizing*. The activity is internalized to become less dependant (and then less complementary) with others: often, enterprises will seek some independence after a time of collaboration in a network where they have acquired sufficient knowledge and know-how to produce in-house. In fact, it has been observed in investigations about networking in the Rhone-Alpes region (France) that knowledge exchanging is one of the main reasons for independent firms to join networks (Hannoun, et. al., 1996). The activity is externalized when it is transformed from internal to external. Then, the firm maybe come more complementary with others (externalization is often necessary when an internal activity needs to be improved or to become competitive).

Competency transformation can be of either nature: develop (acquire, when the competency does not

exist) or abandon competencies in order to focus on key competencies. Its goal is to align them with a partner's competencies.

These transformations operate gradually. After each transformation (internalize/externalize an activity or develop/abandon a competency), a new map of potential coordination modes is plotted.

Evolutions and mutations of SMEs networks are rarely guided by a deliberate collective strategy. It actually appears that evolutions are often the result of a local decision-making process distributed amongst the network.

Often SMEs want to cooperate without merging. The difficult identification of the efficient decisions to stay independent needs to rest on a vision of future mutations according to the different strategic decisions. We suppose here that only one firm proceeds with such transformations.

The scenario tree offers an efficient decision-making aid to the manager. It helps to provide a highly effective structure within which one can lay out options and investigate the possible outcomes of choosing those options, then to form a picture of the risks associated with each possible action and eventually to choose between several courses of strategic decisions .

The initial position in the scenario tree is the map plotted Figure 1, which represents the actual modes of coordination of the network. Before laying out the tree, the effect of the contingency parameters should be reconciled by comparing the two maps. In fact, acting on the contingency parameters can transform a currently coordination mode into a potential one. This step is not easy because it is often quite hard to transform a cultural and social structure in SMEs.

Figure 4 illustrates the organizational trajectory of a network by a scenario tree considering a transformation process (skills and activities) of only one company that looks for best trajectory to attain its objective (Ei Objective). Where:

P_a, P_b: Occurrence probability of the result
A_i: Activity i
C_i: Competency i
p_i: Occurrence probability of the decision line

Each decision arc in the scenario tree (Figure 4) has a cost (financial, strategic or social) and a probability of occurring. In such a stochastic framework, identifying the best decisions to take is often difficult and it is important for the manager to have a vision for the evolution of the company according to the different decisions.

---

[1] The organizational proximity is the result of organized relations between partners. Agent will be close if they accept a common framework to structure their exchanges.

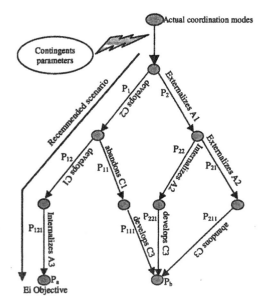

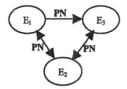

The cartography of potential coordination modes gives:

Fig. 5. Cartography of potential coordination modes

To merge with E_1, E_2 should have the same competencies as E_1. So the decisions to make are: develop C_1; develop C_3; abandon C_2.

Figure 6 illustrates the scenario tree of the decision-making of E_2. Each position represents a map of potential coordination modes after each decision. The order of decisions is important. The recommended trajectory is: develop C1, then abandon C2, then develop C3. Others scenarios have strategic risks: a merger with E_1 may generate irreversible consequences. We note that the probability of merging with E_3 is larger than the probability of merging with E_1. The transformation process of the recommended scenario is longer than the two others scenarios and therefore its cost can be increased. This type of scenario tree can yield additional modeling opportunities.

Fig. 4. Scenario tree.

A strategic and financial auditing of each enterprise in the network to find its objectives and to assess the levels of its future investments and skills transfer will yield a probability for each trajectory and thus for each outcome (the occurrence probability of a given outcome is the sum of the occurrence probability of each trajectory toward this outcome).

We supposed in this section that only one firm proceeds to the activities and competencies transformations. However, in reality, several enterprises within the network proceed to the transformations according to their own objectives and therefore the scenario tree is more complex. An extension of this analysis will be to lay out a second scenario tree allowing a choice of which enterprises proceed to the transformations.

Example

An example of the scenario tree is analyzed to illustrate the methodology. This example is a textbook case of organizational trajectory with tree firms E_1, E_2, E_3. We suppose that only E_2 performs a competency transformation (E_1 and E_3 are static and don't transform) with the aim of merging with E_1. We suppose also that the 3 firms have complementary activities. The competencies C_1, C_2, C_3 are valued at 0 or 1 (0 if the competency does not exist, 1 otherwise):

	C1	C2	C3
E1	1	0	1
E2	0	1	0
E3	0	1	1

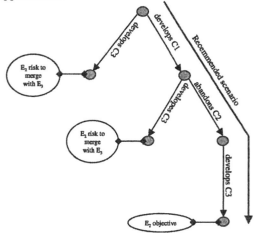

Fig. 6. Example of tree scenario for a network of 3 firms.

6. CONCLUSION AND PERSPECTIVES

Industrial cooperation imposes itself as a necessary resort for firms to be competitive in the modern market. However, one cannot initiate a co-operation without a strategic vision of the evolution of the firm with its partner. Therefore, developing within each firm a strategy for cooperation with other companies, has become a prerequisite to selecting an effective mode of cooperation with some particular partners.

The methodology developed in this paper may help a manager to develop a strategy of inter-organizational cooperation, to evaluate the risks associated with any decision, to assess the pros and cons of an involvement in an enterprise network, and to overcome co-ordination costs.

The industrial validation of this methodology is currently on progress. A questionnaire has been drafted in order to collect necessary information from industrial SME networks. This practical step will allows us to specify the limits of our theoretical procedure. The comparison between the actual coordination modes and the potential coordination modes will permits to identify to what degree the contingency parameters influence SME cooperation.

The detailed steps of the dynamic analysis proposed in Section 5 remain to be investigated. First, how to identify the occurrence probability of the decision lines in the tree scenario. Second, lay out a second tree scenario according to which companies will execute the transformations and interconnect the two trees. Finally, evaluate the risks and costs of each line decision and introduce the duration taken to make each decision.

To become a successful diagnostic tool, the cartography of potential coordination must be connected with elements influencing the performance of the network of enterprises. Different factors have been identified in the literature: planning and control, work structure, organization structure, information technology, risk and reward structure (Cooper, et. al, 1997; Chan, and Qi, 2000). A work is in process to build a typology of I.T. according to the intensity of cooperation and to the integration of the enterprises. Therefore, these interconnections will permit a more complete diagnosis, with the aim of orienting the choice of manufacturing methods within the network according to the potential coordination modes and the evolution of the network.

REFERENCES

Benali, M. and P. Burlat (2003). Une modélisation des relations de coordination dans les réseaux d'entreprises. *Congrès international de génie industriel*, Québec.

Benali, M. and P. Burlat (2004). Une démarche d'analyse de la complémentarité des activités dans un réseau d'entreprises. *5ème Conférence Francophone de Modélisation et SIMulation « Modélisation et simulation pour l'analyse et l'optimisation des systèmes industriels et logistiques » MOSIM'04*, Nantes, France.

Boucher, X. and P. Burlat (2003). Vers l'intégration des compétences dans le système de performances de l'entreprise. *Journal Européen des Systèmes Automatisés (JESA)*, No. 3.

Burlat, P. and S. Peillon (2002). Skills networks and local dynamics. In: *Global Competition and Local Networks* (Rod B. McNaughton and Milford B. Green (Ed)), pp. 133-149. Ashgate Publishing Limited, London.

Burlat, P., B. Besombes and V. Deslandres (July-August 2003). Constructing a typology for networks of firms. *Production Planning and Control (PPC)*, Vol. 14, No. 5, pp. 399409.

Chan, F. and H. J. Qi (2003). Feasibility of performance measurement system for process-based approach and measures. *Integrated Manufacturing System*, Vol. 14, No.3, pp. 179-190

Cooper, M. C., D. Lambert, and J. Pagh (1997). Supply chain management: more than a new name for logistics. *International Journal of Logistics Management*, Vol. 8, No. 1.

Hagedoorn, J. and J. Schakenraad (1994). The effect of strategic technology alliances on company performance. *Strategic Management Journal*, Vol. 15, pp. 291-309.

Hannoun, M. and G. Guerrier (1996). Le partenariat industriel. *Ministère de l'industrie, de la poste et des Télécommunications*, SESSI, Paris.

Kaufmann, A. (1978). Introduction à la théorie des sous-ensembles flous (Eléments théoriques de base), Masson.

Kets De Vries, M. F. R. (1996). Family business: Human dilemmas in the family firm. *International Thomson Business Press*, Boston.

Menguzzato-Boulard, M. (2003). Les accords de coopération : Une stratégie pour toutes les entreprises ?. *XIIème Conférence de l'Association Internationale de Management Stratégique (AIMS)*.

Miles, R. and C. Snow (1994). Fit, Failure, and the hall of fame : how companies succeed or fail. *New York : Free Press*.

Milgrom, P. and J. Roberts (1997). Economie Organisation et Management. *De Boeck Université*.

Mintzberg, H. (1998). Le management. Editions d'Organisations.

Murray, E. A. Jr. and J. F. Mahon (1993), Strategic Alliances: Gateway to the new Europe. *Long Range Planning*, Vol. 26, No. 4, pp. 102-111.

Mutinelli, M. and L. Piscitello (1998), The entry mode choice of MNEs: an evolutionary approach. *Research Policy*, Vol. 27, No. 5, pp. 491-506.

Peillon, S. (2001). Le pilotage des coopérations interentreprises : le cas des groupements de PME. PHD in Economy, University of Jean Monnet, France.

Petitgrew, A. M. (1987). Context and action in transformation of the firm. *Journal Of Management Studies*, Vol. 24, pp. 649-670.

Richardson., G. B. (1972). The Organization of Industry. *Economic Journal*, Vol. 82, No. 327, pp. 883-895.

Zadeh L. A. (1965). Fuzzy sets. *Information and Control*, Vol. 8, pp. 338-353.

ELSEVIER
IFAC
PUBLICATIONS
www.elsevier.com/locate/ifac

Integration of a Management Information System in GRAI decisional modelling

S. Sperandio, F. Pereyrol and J.P. Bourrieres
LAP UMR CNRS 5131
University Bordeaux 1-ENSEIRB
France
name@lap.u-bordeaux1.fr

Abstract – *GRAI Grid and networks are formalisms to model production systems, specifically focusing on the organisation of management. Decision centres are identified and decision-making processes are detailed as well as the information required by decision centres. However, only major information flows are presented, and the process through which the information is provided and adapted to decision centres requirements is not modelled. This paper introduces a management information system (MIS) providing the relevant information to the appropriate decision centre in the enterprise, and defining how these data must be treated through a Data Adjustment Process (DAP) to adapt information to decision requirements. The DAP is made up of several activities: partition, valuation, collection, integration and reduction of data. Modelling concepts are applied to an industrial case.*

Keywords: Management Information System (MIS), production management, decisional modelling

Introduction

Production system engineering and management require to provide decision-makers with a large number of information to let them envisage and select the strategies to be implemented, then to operate the selected strategies. In general terms, a management information system (MIS) can be defined as "a system which provides relevant information to the appropriate persons in the enterprise at the right time. The information provided helps managers to plan their activities in the short and long term, to organize the tasks necessary for the plan and to monitor the execution of the tasks and activities" [1]. Given the mass of information available, the role of information management is moreover "a response to, and a search for new and improved means of controlling the information explosion and the resultant increasing complexity of decision making by improving the flow, the control, the analysis of information for decision makers" [2].

GRAI Grid and networks [3] are formalisms used to model the organisation of systems management through the identification of the decision centres and of the decision-making processes based on the information available. However, only major information types are taken into account, and the upstream process to adapt the information to decision centres requirements is not described: GRAI formalims not include adequate processing for supporting decision making, Consequently, this paper focuses on Enterprise MIS and its integration in GRAI Enterprise Modelling techniques.

In Section 1 the primitives of GRAI modelling (grid and networks) are reminded. In Section 2 the information flows required by GRAI modelling are characterised and

classified. Section 3 is devoted to the MIS, which consists in the development of relevant information her called Data Adjustment Process (DAP), made up of five activities: partition, valuation, collection, integration and reduction. Lastly, the integration of the DAP in the GRAI model is illustrated using an industrial case.

1 GRAI modelling

GRAI grid and networks are focused on decisional aspects of systems management [3].

1.1 GRAI grid

A GRAI grid (see Figure 3) is a functional and temporal decomposition of the management system. The functional decomposition identifies the three basic functions of production management:

- To manage products (MP): product related decisions (purchasing, supplying, inventory control, etc.)
- To manage resources (MR): decisions related to human and technical resources (maintenance, manpower assignment, etc.)
- To plan manufacturing (PL): decisions dealing with the transformation of products by the resources. The main purpose here is to manage the activities by synchronizing MP and MR functions.

The temporal decomposition identifies the decisional levels, characterized each by a Horizon/Period couple, suitable for any type of decision problems.

- The Horizon is the part of the future taken into account by planning, and therefore is relative to the notion of terms (long term, short term, etc.). Here the planning horizons are quantified.

• The Period is the time duration after which a decision must be re-evaluated to take into account the evolution on the system according to the paradigms of system control.

The grid includes moreover two complementary columns gathering information of distinct origins (endogenous vs exogenous). The endogenous information comes from the controlled system itself (production follow-up, resources breakdowns, etc.), whereas the exogenous information comes from the environment (orders, forecasts, etc.).

1.2 Decision centres and GRAI networks

A Decision Centre is in charge of a set of decisions related to one of the functions at one level. GRAI networks detail the decision-making process performed by each Decision Centres identified in the grid. This process may include two types of activities (decisional or executive). An executive activity is a purely procedural and transforms inputs data into outputs data. The duration of such a treatment is generally known. A decisional activity carries out a choice in a set of whole of possible solutions using definite criteria. It is a non procedural activity the duration of which is not systematically known.

2 Information Flows characterisation

Each Decision Centre interacts with other centres by several connections: decision frames and information flows.

2.1 Decision frames

Decision frames are sets of information entities both structural (or qualitative) or operational (or quantitative). A structural decision frame defines the autonomy of a Decision. It is composed of objectives (i.e. expected performances), decision variables (i.e. action means to reach the objectives), constraints (i.e. limitations on possible values of decision variables), performance indicators (i.e. information allowing the comparison of the performances to the objectives). An operational decision frame is the set of quantitative and contextual values of the objectives variables and constraints defined by the structural decision frame.

2.2 Data vs Information

The information flows received by a Decision Centre may be emitted by another centre, by an entity outside the production system or by the physical system. An informational flow is composed of a certain number of data and / or information. In [4] data can be seen as structured and quantitative entities like measures, whereas a number of data parts and their descriptions constitute information [5]. Information can be formal or informal, structured or not structured [6]. Despite the disparity of definitions available in the literature, most of the authors consider data as factual and semantically poor whereas information is semantically more elaborated and eventually subjective. In decision problems, the information provided to the decision makers has to be semantically appropriate, as the result of data processing.

2.3 Decision making data classification

For the purpose of this study, three data families are considered :
• Data relating to resources, products and manner of transforming them: identification of products, know-how of human resources and equipment, etc. These technical data can be seen as parameters, static or slightly evolutionary.
• Interfacing data between the production management system and the external world : orders to suppliers, outsourcing, etc. Such data are dynamic (variable in the time), exact or approximate.
• Interfacing data between the production management system and the physical system : product quantities, data relating to men, machines and tools. Such data are exact or approximate variables.

Table 1 draws up a classification of data handled in production management.

3 Data Adjustment Process (DAP)

Let three decision centres α, β and γ be considered. Centre α receives an informational flow coming from center β and a decision frame coming from γ. The flows resulting from β and γ relate to data classification in Table 1. Nevertheless, all these data are not exploitable as such and require a treatment to be usable by center α. This treatment, here called Data Adjustment Process (DAP), aims to transform data into relevant information allowing decision makers to understand the situation and reason appropriately, in other words to derive knowledge [5] [7] from data. A communication protocol follows this process (DAP). It describes the conventions and rules to be followed for the informational and decisional data exchange between decision-making centres. The "protocol" aspect is not detailed in this paper.

3.1 DAP functions

The functions carried out by a DAP are of two types : *principal* functions and *constraints* functions. The *principal* functions are:
• F_p1: To fall short of relevant information requirement expressed by the decision centre α
• F_p2: To transform data resulting from β into information adapted to α
• F_p3: To transform data resulting from γ into information adapted to α.

decision frames		information flows		
Structural decision frame	Operational decision frame	External data	Internal data	Technical data
Qualitative objectives	Quantitative objectives	<u>Orders</u> Firm orders Estimated orders Urgent orders	<u>Production data</u> Operation carried out Quantity of products Quantity of inventories/ stocks	<u>Products (levels of products)</u> Macro bills of materials Bills of materials Articles Pieces
Qualitative decision variables				
Quantitative structural constraints	Quantitative operational constraints	<u>Partners companies</u> Size Turnover Reputation Financial standing Market share ...	<u>Data relating to machines</u> Production time Pauses Causes of pauses Maintenance Installation	<u>Resources</u> Unit production Cost centre Work station Machines, tools
Qualitative criteria				
Qualitative commands	Quantitative commands			
Qualitative performance indicators	Quantitative performance indicators	<u>Transactions between partners</u> Proposal Agreement Counterproposal Refusal	<u>Data relating to human</u> Assignment at work stations Effective work time Absences (period, duration, causes)	<u>Routings</u> Manufacture Assembly Maintenance
		<u>Material flows</u> Supplies Consignments	<u>Data relating to tools</u> Tools assignment Use time Wear rate Places and modes of storage	(routings levels) Macro routings Activity Operation Phase Gesture

Table 1. Classes of data for decision making environment in production management

The *constraints* functions of the DAP are:

- F_c1: To adapt the DAP to the role of α as identified in the GRAI grid (decisional level, function)

- F_c2: To respect a total time of treatment (adjustment and transfer of information) which must be minimal compared to the decision period ($t_1 - t_0$) of recipient centre α (Figure 1). Moreover, to ensure the continuity of the successive decision-makings, it is necessary to adapt the horizon (H). At t_1, information of this period is removed, and information for the following period is added (Figure 1). Such a representation allows to have always the same vision at the time of the decision-making.

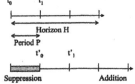

Figure 1: Dynamics of the decision making process

- F_c3: To match the expression of knowledge and know-how of centres ß and γ.

3.2 DAP definition

In [8], a DAP is described as a partially ordered sequence of treatment activities, started by an event. Here, the DAP is defined as a structured organisation, which fulfils principal functions (F_p1, F_p2 and F_p3). This organisation includes a sequence of activities (data treatment for decision-making), and links between these activities. A DAP is moreover characterised (see Figure 2) by :

- Inputs i.e. starting points of the process. The inputs are i) the relevant information required by a Decision Centre, and ii) the data received by the same centre (external data, internal data, technical data, and decision frame) coming from other decision centres in the GRAI grid.

- Outputs i.e. relevant information produced by the DAP.

- Controls necessary to match the constraints functions (F_C1, F_C2, F_C3 and F_C4), i.e the objectives, horizon and period of the decision centre which receives the information.

- Technical or human resources allowing to carry out the activities : experts, know-how, etc.

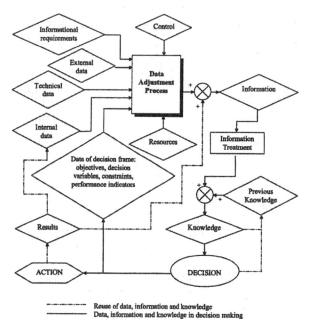

Figure 2. Integration of the DAP in decision-making

3.3 DAP activities

Data processing includes various activities whose classification can be found in [9] as i) classification, ii) rearranging/ sorting, iii) summarising / aggregating, iv) calculating, and v) selection. We also retain reference [10] concentrated on the aggregation stage, broken up into three phases: partition, valuation and reduction. On these bases, five -decision or execution- generic activities in a DAP are identified : partition, valuation, collection, integration and data reduction. However, each centre has its own decision problematic, which leads to adapt the generic model both to decision problems and data types.

✓ Partition

The partition consists in the division of the set of available data into subsets of specific natures: external data, internal data follow-up, technical data. Consequently, each category –family or class- of data has exclusive characteristics.

✓ Valuation

The valuation associates a specific variable (performance, cost, quality, time...) to each data within its family.

✓ Collection

Static data (data describing resources, products and transformation processes) are at each decision maker's disposal. The collection of dynamic data is carried out in various (formal or informal) ways. Internal data acquisition (for example related to production follow-up) is mainly automated, based on the use of bar codes, optical characters, etc. whereas external data collection frequently use Internet, electronics mail, postal or phone.

✓ Integration

Integration consists in associating all collected data to establish the links between elements of the various families. For example, production time t_1 can be associated with load x and cost y.

✓ Reduction

Reduction is the simplification of information to make it interpretable and exploitable by a decision centre. It can be a selection, a filtering operator elements (Max, Min...) or any arithmetic operation (sum, average, etc...). In production management, information reduction leads to aggregate nomenclatures, resources, manufacture tasks, production capacities, etc... For example, it can be judicious to define 'macro tasks' as the reduced views of transformation processes, as well as 'macro resources' as reduced views of production lines [10].

4 Application

In this section, the integration of MIS in the GRAI model is applied to a study case resulting from the simplification of an industrial example. Let (R) be a make-to-order manufacturer of wood made windows. The company is composed (see Figure 3) of an annex unit (A) which produces panes, a principal factory (B) producing left/right-hand windows and assembling final products, and a subcontractor (C) manufacturing window steel handles. Unit (A) supplies the principal factory (B). This factory is composed of a wood machining workshop (B1) and of an assembly workshop (B2). B1 is composed of three cells : sawing (B11), drilling (B12) and painting (B13). B2 is composed of an assembly cell (B21), a seal mounting cell (B22) and a packaging cell (B23). Finally, there are four levels of decision identified in the GRAI grid model of the management system as shown on Figure 4.

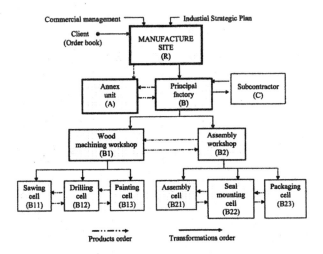

Figure 3. Flow chart of the company R

Let for instance the focus be put on the decision centre "load planning", codified PL20 in the GRAI grid (see Figure 4). Centre PL20 receives a decision frame from centre PL10, and three information flows from MR10, II20 and IE20. All data contained in these decision and information flows must be treated and constitute, in addition to the requirements of PL20, the inputs of the DAP providing PL20 with relevant information (cf. Figure 5, Part A).

The decision frame coming from PL 10 contains the structural frame (objectives, decision variables, constraints and performance indicators) within which PL 20 can make its decisions : competence and capacity subcontracting, critical procurements and workload envisaged over the month-horizon. An operational decision frame, contained in the structural frame, quantifies these data each time PL 10 makes a decision (every month, the duration of its period). Here, the decision frame contains already classified and quantified data and consequently directly reaches the collection activity step. The information flows stemming from EI20 and MR10 (urgent orders, technical or human resources avalaible) is already adapted to PL20 requirements and consequently directly reaches the collection activity step too. Data within the information flow coming from II20, i.e. the follow-up of (B) production and stocks levels, not relevant for PL20, must be adapted and consequently reaches the partition activity step.

Controls are the objectives, horizon and period of PL 20. Indeed, the DAP must respect a total treatment time remaining minimal relatively to the period of PL 20 (i.e. time of treatment << 1 week). DAP outputs will feed the decision-making process of PL 20 (see GRAI Networks figure 5, Part B).

5 Conclusion

GRAI grid and networks allow to capture all interactive decision centres of a production system, as well as the principal information required for decision. However, the information is in general not suitable as such and must be adjusted to the type of decision expected. A Management Information System (MIS) must be defined which provides the relevant information to the appropriate decision makers in the enterprise at the appropriate time, requiring a Data Adjustment Process (DAP), which transforms "rough " data into relevant information, necessary and sufficient for decision-making. The DAP is made up of five activities: partition, information, collection, integration and reduction. These activities are not systematic, some of them being skipped depending on the studied case. The MIS often remains implicit in the modelling of control systems. It is the purpose of the results presented here to clarify its role and to integrate it in the representation of decisional flow as an extension of GRAI formalisms to model production management systems.

6 Bibliography

[1] I. Comyn-Wattiau and J. Akoka. Logistics Information System Auditing Using Expert System Technology. *Expert Systems With Applications*, Vol. 11, No. 4, pp. 463-473, 1996.

[2] J. Rowley. Towards a Framework for Information Management. *International Journal of Information Management*, Vol. 18, No. 5, pp. 359-369, 1998.

[3] G. Doumeingts, B. Vallespir and D. Chen. GRAI Grid: decisional modelling. *Handbook on architectures of information systems*. Edition Springer, 1998.

[4] B.J. Hicks, S.J. Culley, R.D. Allen, G. Mullineux. A framework for the requirements of capturing, storing and reusing information and knowledge in engineering design. *International journal of Information Management*, vol. 22, pp. 263-280, 2002.

[5] A.W. Court. Modelling and classification of information for engineering design. *Ph.D. Thesis, University of Bath*, UK, 1995.

[6] C.A. McMahon, D.J. Pitt, Y. Yang, J.H. Sims Williams. An information management system for informal design data. *Engineering with Computers*, Vol. 11, pp. 123-135.

[7] T. Davenport and D. Marchand. Is KN just good information management? *Information Management*, 6 March, 2-3, 1999.

[8] AMICE. CIMOSA: Open Systems Architecture for CIM, *Springer-Verlag, Berlin*, 1993.

[9] G. Curtis. Business Information Systems : Analysis, Design and Practice. *Addison-Wesley, Wokingham*. 1989.

[10] M. Zolghadri et J.P Bourrières. Data aggregation in Production management systems : from mono-dimensional to bi-dimensional aggregators. *7th World Multiconference on Systemics, Cybernetics and Informatics*. Orlando, July 27-30 2003.

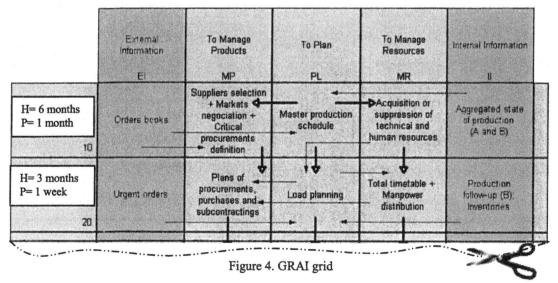

Figure 4. GRAI grid

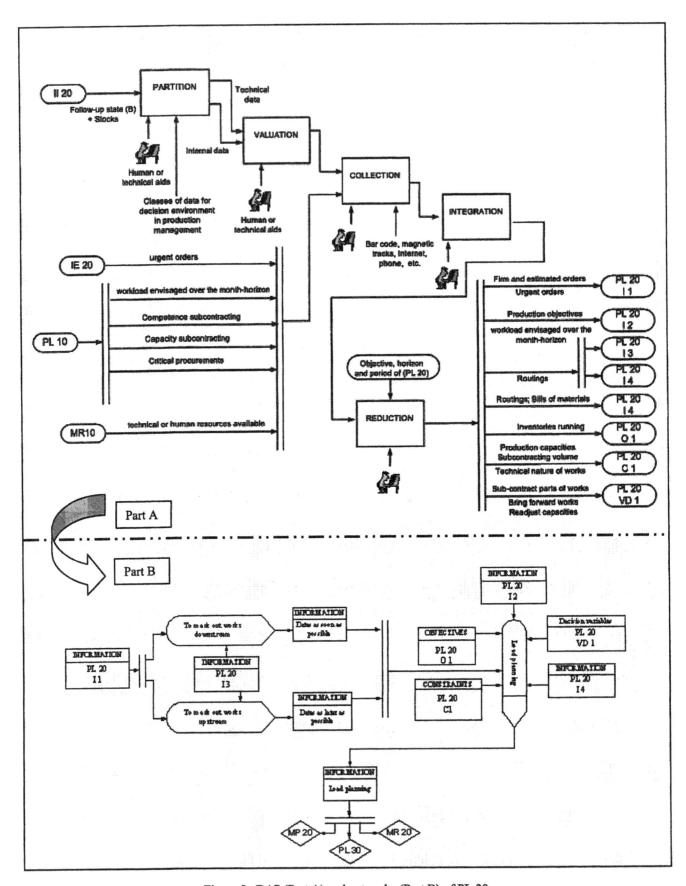

Figure 5. DAP (Part A) and networks (Part B) of PL 20

DIFFERENT WAYS OF SUBCONTRACTING FOR COMPANIES USING "ENGINEER TO ORDER" MANAGEMENT

Thierry LATTANZIO, Aline CAUVIN, Jean-Paul KIEFFER
Laboratoire des Sciences de l'Information et des Systèmes
2, cours des Arts et Métiers - 13617 Aix-en-Provence
France
e-mail: t.lattanzio@vigie.adepa.asso.fr
aline.cauvin@lsis.org
jean-paul.kieffer@aix.ensam.fr

Corresponding author:

    Aline CAUVIN
    LSIS
    Domaine Universitaire de Saint-Jérôme
    Avenue Escadrille Normandie-Niemen
    13397 MARSEILLE Cédex
    FRANCE
    tel.: (33) (0)4 91 05 60 21
    fax: (33) (0)4 91 05 60 33
    e-mail: aline.cauvin@lsis.org; aline.cauvin@voila.fr

Abstract: Nowadays, companies managed in an *Engineer To Order* way use subcontracting to reduce cost prices and to be more reactive to answer customers' requirements. We deal with the issue of subcontracting in order to propose a classification of the different subcontracting forms. We first present the interactions between functions and activities involved in the project life cycle. We then define the role of internal actors like Purchasing, Procurement and Design managers and external actors like suppliers and subcontractors. Finally, we propose a classification of the different types of subcontracting.

Key-words: Subcontracting – Procurement – Purchasing – Design – Engineer To Order

1. INTRODUCTION

In the present economic context, market is variable and the number of competitors is increasing. Consequently, more and more manufacturers deal with subcontractors to reduce cost prices.

Nowadays, companies which are managed in an Engineer to Order way (*ETO companies* in the paper) cannot do without subcontracting. In fact, these companies' stakes and motives are multiple. They must:
- be competitive in terms of cost of sales which are fixed by market demand;
- control and reduce cost prices;
- meet the cost of cyclic overloads and under loads resulting from market fluctuations;
- be more reactive and flexible in relation to customers' demands.

Moreover, the choice of a subcontracting policy also depends on the company's strategic objectives.

Our paper presents the different types of subcontracting used by ETO companies and proposes a classification of subcontracting forms. We first consider the different functions and activities interacting during the project life cycle.

Then we describe the role of internal and external actors in relation to subcontracted products typologies.

2. FUNCTIONS AND ACTIVITIES INTERACTING WITH THE PROJECT LIFE CYCLE

Figure 1 shows how activities are organized in the project life cycle. The project starts with both the Design function and the Purchasing, Procurement and Subcontracting (PPS) functions. Our paper focuses on materials and information flows between the three functions (on the left of figure 1) and underlines that:
- when the manufacturing activity is external, exchanges between Design and Manufacturing functions are rerouted to PPS function. Consequently, the exchanges between Manufacturing function and PPS function disappear;
- essential exchanges occur between Design and PPS functions, in a unidirectional way; between PPS and Assembly functions in a unidirectional way; between Design and Assembly functions in a bidirectional way.

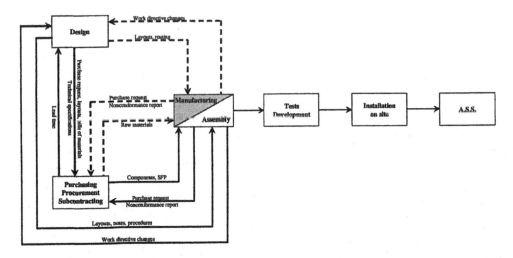

Figure N°1: Project life cycle (Lattanzio, 2004)

Each of these functions develops into a customer-supplier relationship internal to the company. The success of the project will mainly depend on this relationship, in terms of meeting deadlines, accordance to customer needs and margin.

3. TERMINOLOGY AND DEFINITIONS

Figure 2 shows how the Design, Purchasing and Procurement functions interact with suppliers and subcontractors management. Even if purchasing and procurement are complementary, they show differences. In fact, purchasing is outward oriented (Bruel, 1999). Its activities are developed in the medium and long term. Procurement is inward oriented and its activities are developed in the short term. Design together with Purchasing manage subcontracting.

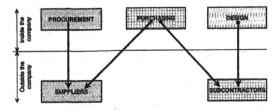

Figure N°2: Relationships between inside functions and suppliers and subcontractors

3.1. DEFINITION OF THE DESIGN FUNCTION

Design consists in developing a project, or parts of a project, from expressed needs, using available means and particularly technological means. Specifications provide functional description for products or services allowing Design to meet the expressed needs.

Design main missions consist in:
- assessing estimate related to each project;
- formalizing customer needs into functional specifications;
- carrying out technical study and feasibility for draft;
- selecting technological solution to implement;
- working out bills of material;
- drawing general layouts;
- developing prototypes;
- performing tests and development, ...

3.2. DEFINITION OF THE PROCUREMENT FUNCTION

Procurement provides raw materials and components required by manufacturing. Moreover, this function develops the planning of needs and it manages daily-stored items. Procurement has a technical dimension positioned towards logistics. It aims at minimizing storage costs by reducing inventory level. Its main missions are:
- processing results from material-requirement planning;
- performing purchase requisitions and submitting them to Purchasing;
- entering storage inputs and outputs;
- defining procurement parameters (inventory level, safety stock);
- managing parts-storage according to procurement policy;
- preparing raw materials and components outputs according to allocations;
- putting raw materials and components at workshop disposal in time, ...

3.3. DEFINITION OF PURCHASING FUNCTION

Purchasing more particularly concerns relationships with suppliers from a financial and commercial point of view. It acts as an interface between the outside and an inside function like Design or

Manufacturing. Purchasing main missions consist in (Baglin et al., 2001):
- developing purchasing strategy according to suppliers ' market;
- providing the required products at a lower cost;
- Working out purchasing policy;
- defining objectives and budgets;
- participating in product and service definition;
- knowing about needs and about suppliers' markets;
- consulting, negotiating and ordering;
- assessing and validating suppliers;
- contacting and following up suppliers;
- settling disputes;
- carrying out technological watch, ...

3.4. DEFINITION OF SUPPLIERS

The suppliers' mission consists in putting products from the catalog at the customer companies' disposal. It corresponds to standardized provision. It is about the selling of equipment that is not directly sold but is integrated in assemblies which will be sold to end customers. (Badot, 1997)

3.5. DEFINITION OF SUBCONTRACTING

Subcontracting consists in making companies partly or entirely process products or services that the prime manufacturer does not want or cannot process itself (Bertrand & Sridharan, 2001). The latter entrusts a subcontractor with a task according to specifications. There are three types of subcontracting (Javel, 2000):
- *Specialty subcontracting*, when a company has got neither know-how nor the means necessary to perform some parts of their products or services. It essentially concerns companies which decide to cut out some activities to develop know-how.
- *Capacity subcontracting*, when a company makes some subcontractors manufacture products because of overload though it has obviously means at its disposal.
- Finally, *"delocalizing" subcontracting* when a company aims at reducing cost prices. Hence, a company relocates some activities, for instance in countries where wages are lower.

4. ROLE OF ACTORS ACCORDING TO THE TYPOLOGY OF PRODUCTS IN AN ETO CONTEXT

Figure 3 presents the three cases involving Purchasing that we can meet in ETO context. Purchasing is involved in the subcontracting process. It is more particularly involved in supplier consulting, negotiation and contract drafting.

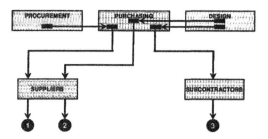

Figure N°3*: Actors involvement according to the products typology in ETO context*

4.1 TYPOLOGY OF THE DIFFERENT CASES

Case n°1: this case deals with the identification of raw materials and components codified by the prime manufacturer. There are two sub-cases:
- purchasing of raw materials and components for internal use. Purchase is connected to restocking or to a particular project;
- purchasing of raw materials for supplying subcontractor who will perform one or several manufacturing operations. Prime manufacturer will receive raw materials and will send them directly to the subcontractor.

Case n°2: this case deals with the purchase of raw materials and specific components for a particular project or for the development of new products. In fact, Design determines the product requirement. Raw materials and components are systematically issued from supplier consulting. They will be codified as soon as possible.

Case n°3: the Engineering and Design department takes part in the process of subcontracting management in the following cases: complex pieces specified by plans, particular machining operations, semi-finished products assembled by the subcontractor. This one also takes part in the co-design with the Engineering and Design department in order to design end products. In this case, the relationship -prime manufacturer / subcontractor- tends to partnership.

4.2. PROJECT DEVELOPMENT ACCORDING TO ITS LIFE CYCLE

Figure 4 shows the different phases of project development according to its life cycle:

1) Generally, Design function (i.e. Engineering and Design department and Planning department) starts up each project.
2) Purchasing and Procurement jointly deal with raw materials and components.
3) Procurement time limit changes according to the types of materials or components and according to procurement sources.

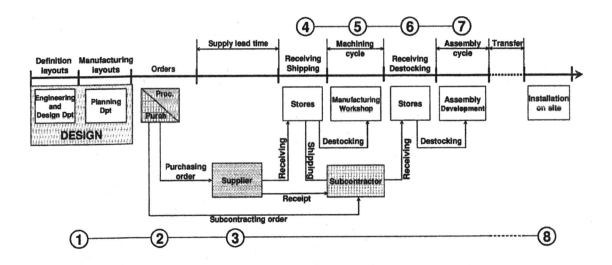

Figure N°4: Project development according to its life cycle

A particular issue concerns components which need long procurement time: procurement time is generally crucial.

4) Raw materials and components are received in store rooms or directly in workshops for the project. Raw materials and components are:
   - sent from stocks to workshops when pieces are manufactured by prime manufacturer
   - sent directly to the subcontractor when pieces are manufactured by the latter.

5) Pieces and subassemblies are manufactured:
   - inside prime manufacturer: this case is more and more frequently given up by ETO companies;
   - by the subcontractor from layouts and manufacturing drawings defined by prime manufacturer and / or subcontractor.

6) Manufactured pieces come from workshop or subcontracting.

7) Pieces (from item (6)) are destocked and assembled. Tests and development are performed to validate equipment. As soon as validation report is ratified, equipment is dismantled to be sent to the customer.

8) From disassembled parts, equipment is assembled and developed at customer's place. Consequently, customer validates equipment and another validation report is ratified.

4.3. CASES N°1 & 2: PURCHASE OF RAW MATERIALS AND COMPONENTS

Figure 5 shows that purchase order is issued by purchasing from two different sources: procurement (case n°1) or design (case n°2).

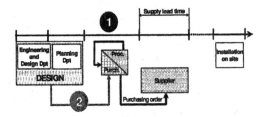

Figure N°5: Purchase of raw materials and components for internal use

In this context, the company processes itself its products. This case is more and more often given up by manufacturers in order to favor know-how during the phases of design, assembly and development.

4.4. CASE N°3: ONE OR SEVERAL SUBCONTRACTING OPERATION(S)

Figure 6 shows the needs from design. There are two possibilities:

- **Possibility 1**: Purchasing places two orders dealing with raw materials and components (a), and with subcontracting provision (b). There are two ways of receiving raw materials and components:
   - by prime manufacturer, subsequently sent to subcontractor;
   - directly sent by supplier to subcontractor; the delivery address refers to subcontractor's.
- **Possibility 2**: purchasing places only one order (c) corresponding to the purchase of a more or less complete provision. Subcontractor will provide a subassembly. Hence, subcontractor provides raw materials and components required by subassemblies (plans are sometimes put

together with order) according to customer's specifications.

In this context, the collaboration between Design from prime manufacturer and Subcontractor is fundamental since the beginning of the study. Prime manufacturer will receive:
- Either an assembled subassembly which can be part of another subassembly;
- Or a non-assembled subassembly which will be later assembled by prime manufacturer.

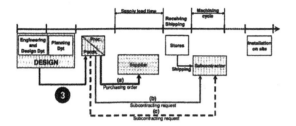

Figure N°6: *One or several subcontracting operations*

5. CLASSIFICATION OF TYPES OF SUBCONTRACTING ACCORDING TO PRODUCT TYPOLOGY

Figure 7 shows the integration of different functions into subcontracting companies according to the typology of products (Hill, 2000).
We can notice that:
- the number of involved functions proper to the subcontractor increases according to the complexity of products;
- Design and Purchasing from prime manufacturer are more and more involved according to the complexity of products (figure 3);
- only case 3 needs trials and tests.

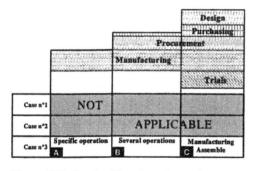

Figure N°7: *Involved functions from subcontractor in relation to the typology of products*

Scenario A:
The aim is to process a **specific operation**: for example, heat treatment, surface treatment, resurfacing ... Prime manufacturer does not have technological means to perform this operation.

Such an operation is called Specialty Subcontracting.

Planning subcontracting operation is difficult, particularly when other operations precede or follow it. Generally, planning is focused on subcontracting operation (figure 8), using backward scheduling for precedent operations and forward scheduling for following operations (Marris, 1994). The aim is to define accurately the date of shipping pieces to subcontractor and the date of receiving pieces after subcontracting operation.

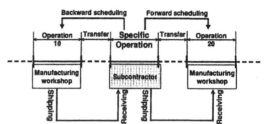

Figure N°8: *Planning centered around subcontracting operation*

Few production and inventory management-packages efficiently deal with this issue as far as planning and scheduling are concerned. We thus appeal to finite loading scheduling and planning packages.

Scenario B:
The aim is to process pieces using several subcontracting operations (for example, turning, milling, reaming ...). Each subcontracting operation can be processed by one or several subcontractors. When several subcontractors are involved in the process, we note that (figure 9):
- during the production cycle, pieces go from one subcontractor to another up to prime manufacturer who receives the finished piece.
- the finished piece ends job-order for subcontracted pieces.

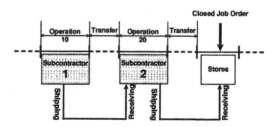

Figure N°9: *A piece processed by several subcontractors*

In this context, there are two cases:
1) capacity subcontracting;
2) relocated subcontracting.

In the case of more complex pieces, Design is highly involved in subcontracting-management in order to give instructions for performing production plans and to give the required technical specifications. We note that few production and inventory management packages provide efficient subcontracting-order management.

Scenario C:

The development of subassemblies is more and more often managed by partnership between prime manufacturer and subcontractor or even only by the latter. There are three cases:

1) Subassemblies are kits that will be later assembled by prime manufacturer (for example, conveyors, mechanized welded base for machine frame, manipulator arm, etc.). Production and inventory management packages manage non-assembled subassemblies as "phantom bill of material". They are included in end product bill of material.

2) Subassemblies are assembled and they are a part of production equipment, like electrical equipment boxes, sections from production equipment (high frequency welded part, fumigation post...). Each subassembly is defined by its own bill of material included in end product bill of material.

3) Complete systems like production equipment (autonomous machines) or parts of production system. This situation corresponds to co-design involving both prime manufacturer and subcontractor.

This scenario requires:

- capacity subcontracting, particularly for processing subassemblies;
- relocated subcontracting, particularly for processing complete systems.

6. CONCLUSION: CLASSIFICATION OF THE DIFFERENT WAYS OF SUBCONTRACTING

From previous scenarios, we classify each way of subcontracting (defined in section 3.5) according to outsourced products in table 10.

	Specific operation A	Several operations B	Manufacturing and assemble C
Specialty Subc.	Heat treatment Machining operations		
Capacity Subc.		Simple or complex pieces (overload)	Electrical equipment boxes Machine frame
Relocated Subc.		Simple or complex pieces (strategy)	Parts from flow line Complete systems

Tableau N°10: Classification of different ways of subcontracting according to product typology

Presently, too many companies tend to select subcontractors according to the cost criterion to the detriment of time and quality. We can notice that:

- time spent in design and purchasing is now transferred to subcontracting management ;
- the number of subcontracted activities is more and more important and consequently synchronizing subcontracting operations is very difficult;
- generally, production costs decrease when subcontracting is involved in manufacturing process;
- co-design between prime manufacturer and subcontractor is more and more predominant.

In the case of a subcontracting operation including the management of an inventory inside a subcontracting company, prime manufacturer has difficulties to know precisely the quantity of non-processed pieces from subcontractor's stock and quantity of processed pieces going from subcontractors to prime manufacturers.

Relocated subcontracting develops because ETO companies tend to subcontract more and more tasks when they have not got the required know-how. In fact, they particularly focus on "noble" tasks such as design, assembly and development.

7. REFERENCES

Badot O., *Théorie de l'Entreprise Agile*, Editions l'Harmattan, 1997.

Baglin G., Bruel, O., Garreau, A., Greif, M., Van Delft, C., *Management industriel et logistique*, Economica, Paris, 2001.

Bertrand J.W.M., Sridharan V., "A study of simple rules for subcontracting in make-to-order manufacturing", *European Journal of Operational Research*, 2001, vol. 128, n° 3, pp 509-531.

Bruel O., *Politique d'achat et gestion des approvisionnement 2ème édition*, Dunod, Paris, 1999.

Hill T., *Manufacturing strategy third edition*, McGraw-Hill, 2000.

Javel G., *Organisation et gestion de la production*, Dunod, Paris, 2000.

Lattanzio T., Cauvin A., Kieffer J.P., "Caractéristiques et spécificités des entreprises organisées en gestion par affaires", *Revue Française de Gestion Industrielle*, 2004, vol 23, n°2, pp 113-124

Marris P., *Management par les contraintes en gestion industrielle : trouver le bon déséquilibre*, Les Editions d'Organisation, Paris, 1994.

ELSEVIER

IFAC

PUBLICATIONS
www.elsevier.com/locate/ifac

A framework based on knowledge management as a support of cooperative work in the SME networks.

Gerardo Gutiérrez Segura[1], Véronique Deslandres[1] and Alain Dussauchoy[1]

[1] UCB-Lyon I, UFR d'informatique, 43 Bd du 11 novembre 1918,
69622 Villeurbanne cedex France
Tel : (33)472431251, Fax : (33)472431537

{ggutierr, deslandres, dussauchoy}@bat710.univ-lyon1.fr
http://prisma.insa-lyon.fr

Abstract. This paper initially introduces specificities of small and medium enterprises networks, then deals with the recent evolution of the Knowledge Management (KM). In convergence with this evolution, we describe a KM process based on a documentary knowledge base, a community of practices and Internet communication which can be applied to these types of groups. The main conclusion is that KM projects provide an interesting way for SME networks to facilitate and increase the collaboration rate as well as to share controlled knowledge, allowing them to make collaborative work more efficient.

Keyword: SME networks, Knowledge management, Ontology, Communities of Practice CoPs, NLP tools (Natural language processing)

1. Introduction

A considerable number of computing and social sciences researches testify the application of knowledge management (KM) on large-enterprises networks. However, the case of SME networks is particular in the sense that generally little time is planned for coordination and collaborative work. Hence, decisions and schedules are made without sufficient consultation and sometimes in a hurry. We have been working for six years on the contribution of information technologies (IT) for the co-operation and knowledge sharing within SME networks. This research fits into the two GRECOPME projects [5], partially financed by the French Rhone-Alpes region.

For more than a decade, many companies have been aiming alliances in order to develop theirs activities within a network. Several kinds of co-operation and integration can be observed, creating a virtual company. The mainly motivations which lead to companies' co-operation could be: offer expansion, introduction on new markets, strategic alliance against competition, etc. For small and medium enterprises, the motivation is still different and can be regarded as defensive or pro-active. In defensive networks, the objective can be: size effect compensation, gap filling, leader's retirement anticipation, etc. Pro-active networks, rarely seen, they are created with the strategic aim of expand the offer, to compete, to co-operate on innovation possibilities within a specific sector [5].

This paper initially introduces features of SME networks, then deals with the recent evolution of the Knowledge Management. We suggest an approach which can be applied to this kind of networks in order to enhance partner integration and therefore ensure a better co-operation. Our approach is based on a KM process, Internet, communities of practice, NLP tools (Natural Language Processing) and an ontology.

2. The uniqueness of SME networks

In SME networks, co-operation intensity strongly depends on the degree of confidence acquired between the companies. For example an information system project development can be an interesting integration vector [5]. Different cases of IT project for the co-operative work have been proposed in [2] and [7]. However, the installation of an information system is based mainly on a strong confidence. Actually, in the new SME networks (often non well-stabilized alliances) there is a low level of confidence between members.
We established that these networks had their own life cycle, with three different phases:
• *Confidence Construction:* The SME networks during this first phase of its life cycle we can find organizations dealing with the same activities (competitors). Then, each partner keeps in secret the information that represent the main activities of their work.
The coordination and running modes of SME networks are made complex due to the fact that

they are not controlled by a single manager but by a group of manager leaders (group created with the leader from each SME of group) and they consider the possibility to leave the network if theirs goals are not reached. In addition, one have noted that even if the SME's wish to belong to a network, they rather remain independent in order to continue with their own projects, customers and providers that they had before the creation of the network (usually, until they get the first results of the network). Often group members create a basic contract stating the conditions to get in or leave the network. However, this document does not contain enough co-operation rules to ensure the partners really trust each other (legal protection). That is the reason why they have not entire confidence between them. It is interesting to realize that during this period the economical (financial) investments into the group are almost impossible (i.e. to develop a computer information system for the network enterprise).

• *Co-operation Test(s):* In this period, the SMEs start some activities together (new activities). Indeed, some common activities of the group at this stage of the co-operation i.e. purchases made together to diminish the costs or professional training of workers on a new technology. We have noted that enhanced confidence starts with any kind of co-operation like common projects that involves constant document exchanges, phone calls, e-mails, etc. This allows to the members to know each other and work together improving collaboration habits and boost confidence between partners. When the co-operation increase between group members, it will take an irreversible character (e.g. company specialization) and more the company will depends of the network.

• *Group stabilization (fusion, fragmentation or new network):* When the group gets to this level of life cycle (objective not always reached), we consider that the confidence level among the members is very strong. We can feel that in this moment the network works and takes decisions as a large enterprise. Indeed, in this moment the network has a very strong confidence mainly based on signed documents, projects documents, experience, etc. In addition, in this kind of network, it is possible to make a fusion to create a large enterprise or to make a new alliance whit other large company. SME networks are organizations where the KM can constitute an important integration vector towards co-operating work. Therefore, the setting up of several SME in a network will produce a vast stock of human and documentary knowledge that will have to be managed in a suitable way. However, at the beginning of such organization, SME do not especially agree to share information, documents or knowledge with the new partners who may become competitors.

We consider that for a SME network, it can be convenient to set up an adapted co-operation system at the very beginning of the network. Our approach is based on the aim to provide the confidence by using a structure based on the KM adapted rather to the managerial limits (few time for coordination and co-operation works, a low confidence level, a few economical investment, etc.) of this kind of networks. I.e., we plan to set up a work co-operation framework mainly focus on the two first phases of the above-mentioned life cycle.

3. The knowledge management (KM)

The methods of acquisition and capitalization were very criticized when they consider knowledge as an isolated object, not located, described apart from any context and giving their interpretation by a random user [14]. Other methods were developed for the knowledge located, but the complexity of the systems of representation used made the update and the evolution adaptation a very complex and difficult activity [11].

Thus, although our perception of the KM models had evolved since its beginnings, no model economically valid shows how knowledge is connected to the tasks and the performance [12]. For this reason the role of the leadership is perceived like a weak point in the economy of the knowledge, as well as the role carried to the attention [18]: "The managers [...] must learn cognition. On one hand, there is the problem of confidence (based on the exchanges) but also the problem of the attention paid to the individuals". The concept of attention (regard) incorporates and extends the confidence. The exchange of tacit knowledge is thus almost impossible without attention. Researchers study for example how to conceive training schemes which would support a behavior of attention. According to our experience, the problem relating to the appropriation by the users of the KM system set up exists indeed, and can be more or less important. Even if the experts having taken part in the project were qualified, the users do not grant confidence in the results obtained only if they estimate that the system does not block their autonomy of decision [4].

4. Our approach

In the current marketing strategy, the software editors do not hesitate to qualify any new functionality which allows the management of documents in KM or like KM systems for the SME. When we talking about of Knowledge Management for SME's, one can notice that it is not widely used in this kind of companies, mostly because small structures can't follow up with big organizations on the technological level. Nevertheless, a small percentage of SME have some KM activities, and a traditionally accepted limit is that KM concerns companies of more than 40 employees [10]. Indeed, current literature states that KM is more needed by large firms rather than by the small ones, due to the fact that in SME, the problem of knowledge exchange and sharing is lessened by the size of the companies (you know immediately who to ask to), the versatility of employees (making them be concerned by more several subjects) and physical proximity (easier meeting possibilities).This is true for a SME, but the SME-networks are out of this context : in these organizations, the partners almost do not know each other, they are not close to each other and they are not always disposed to individually expose their knowledge.

The knowledge sources of an organization are various and may be formalized differently [15]:

- The human capital: this capital represents the knowledge and know-how hold by the company's experts. Mostly, this knowledge is of the tacit nature, and it stays possessed by the experts, in their memory. When some know-how is laid out in a precise form (modeling procedure, quality procedure, etc.) in the externalization process [13], the knowledge becomes the information and a part of human capital is transformed into the documentary or digital capital.

- The documentary capital: it is composed of the all the existent documents of a company: reports, notes and resumes. The majority of this knowledge is textual and non structured format. A great part is digitalized, and those documents are shared within the positions of all company's departments, what makes their exploitation difficult.

- The digital capital: it is composed of the structured documents having a semantic signification shared within a group of persons of one field (i.e. Purchase dpt, Receivables dpt, R&D)

- The external capital: this capital gathers the knowledge of external actors with whom the company works, for instance: the technical contracts, the partners, the clients, the suppliers, but also the strategic and technological observatories.

Our work concerns the knowledge capital that might be shared in SME's, especially the documentary and human knowledge capital of any SME of the group which is quite big but rests unrecognized by the rest of the partners (fig. 1).

This methodology has two phases: The first phase consists in the development of a community of practice (CoPs) like a base of human-knowledge sharing. The second one deals with the creation of a documentary-knowledge base from documents selected in each SME. We chose Internet as

communication support because we consider that this is the best way in order to respect the existing investment restrictions in a SME network above metioned.

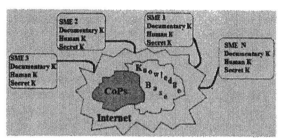

Fig. 1. The cooperative approach based on KM for SME network

Our method will allow the actors to share information, ideas, documents, etc. with the aim of increasing their own knowledge level, and ensure the correct broadcast of technical information to the persons who need it. In other words, rather than study the wealth of information in the channel of communication, it is interesting to make an analysis of the dynamics of the organization. This is particularly true within SME networks which support the co-construction of strategic directions especially based on exchanged information.

4.1. Phase of communities of practice creation

The community of practice (CoPs) is a group of individuals who have specified subjects in common interest, which need to interact around problems, which develops an expertise on a specific field, and which is implied in the objective of collective training [19]. It is the understanding and the management of the tacit knowledge who brought certain authors like Wenger [20] to base CoPs on the relational and social nature of knowledge. This approach has been successfully used in the private sector over the past decade and now being applied in the public sector [16].

At present, this part of methodology was finished and presented previously in the article [8], we will present a summary of this work:

For the creation of our CoPs, we based ourselves on the four stages of development of the communities of practice proposed by Wegner [19]: 1) The field stage: For this point we chose the computing field, in particular on the development of web sites and office applications, field which all partners of group are concerned. 2) The operation stage: A minimum of organization is necessary in the CoPs development. Indeed, we asked at people should be engaged to the community to give him a minimum of stability, this people will also make it possible to have points to locate as well as reference persons. 3)The actions stage: We identified two main objectives for the starting the community, our purpose is to share a knowledge which concerns the actors of the fieldwork, for instance the way of how protecting a computer against a virus propagated by the Net, but also about information about how to manage the project of setting-up a ERP in a society etc. 4)The tools: The objective of our approach is to make the most of the adapted technologies to be easy to use for the participants of the community, it permitted us the specification of the computing tools, support tools: Internet a Wiki and a forum.

In this phase, we consider that knowledge sharing is the result of the interaction between information (information of a customer, an opinion of a colleague) and a person. Indeed by the interpretation process according to the actors that the information is transformed in knowledge.

4.2 Phase of documentary-knowledge-base creation

We have previously pointed out that this phase is based on the use of the existing documents in each SME. Indeed, we consider that the creation of a group will generate a great quantity of documents source of knowledge within the group. This part of our method suggests the creation of a text corpus from which we will extract knowledge (by using tools NPL). We would create afterwards an ontology which would help us to improve the management the corpus of texts.

The process of knowledge management in a company requires different steps, which are referred to in most of the bibliography, i.e. [3] [Grunstein[1]]: detection, preservation, diffusion, updating.

In this part of our work, we will use the steps mentioned above and combining them with Aussenac's suggests [1]: She uses a methodological frame to set up an ontology with NLP tools (Natural language processing). This part of our approach is shown in fig.2.

Phase-1) Detection: This phase consists in identifying the knowledge to be capitalized. The SMEs networks have specific limits linked to the share of information, knowledge, etc. for the reasons mentioned above. In our method, we decided to use a documentary base created with documents from each SME. This phase consists in identifying, locating, characterizing and grading the documents that give information (the company charter, reports, accounts, project memories, etc.) but that the SME's consider (after analyses) they can share. i.e. without any risk of leak or use by a competitor. Once the documentary knowledge

capital is selected, we will set up a corpus from the documents we gathered.

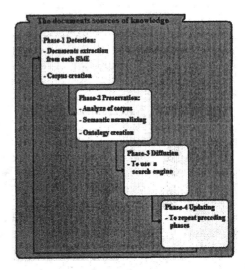

Fig. 2. Creation of a knowledge base in the SME networks

Phase-2) Preservation: Here, we suggest three activities. The first one is the realization of a linguistic analysis of the Corpus using NLP (Natural Language Processing) tools. For that, we chose Nomino[2]. Nomino allows extract NOUNS (we call them MOTs in this paper) and conceptual links from a specific document [17]. Nomino is based on the extensive use of the attraction power of the Nominal Complex Unities UCNs ('Unités Complexes Nominales' in French). These unities are composed of expressions that allow clarifying the sense of some words and structure their sense, particularly in scientific and technical texts.

We analyzed with Nomino some French documents provided by participating companies (SME working in computeing area). As an indication we show results obtained with these first analyses: système d'information (Information system), mécanisme de filtrage (filtering mechanism), moteur de recherche (engine search), apprentissage à distance (remote training), représentation d'ontologie (ontology representation), etc.

The second activity consists in normalizing semantics of terms (MOTs and UCNs), by assigning them a meaning. First, we propose that a (or more, depending on the resources) person belonging to the domain, analyzes and select NOMs by eliminating those that are present only once. A second analysis should be performed to eliminate UCNs that are not relevant with respect to the criteria used by the person performing the analyzing. Finally, we claim that the term (MOTs and UCNs) meanings should be general enough and understandable by most of the actors that will use it, and not only by a group of experts. Indeed, we first propose to categorize the MOTs (i.e. services,

---

[1]

http://perso.wanadoo.fr/michel.grundstein/Homeactivit esF.htm

[2] http://www.ling.uqam.ca/nomino

software platforms, R&D dept, etc.) and let experts of the SME network assign relevant meanings, which has the advantage of situating concepts, and validating them by the community which will manipulate them.

The last activity of Preservation is the ontology creation. We suggest Protégé-2000[3] tool. Protégé-2000 is an extensible editor of ontology's. In addition, it allows to declare and include new ontologies, as well as to build new interfaces for the acquisition of knowledge. We suggest to create the ontology based on categories as a first level (i.e service whom in English means service too), as the second level we suggest to use the MOT's (i.e. gestion that in English it means management) and we suggest put in the last level the UCNs (i.e. "apprentissage à distance" that in English means "Distance training") In addition we are going to include the UCN sources (i.e. CharteAcrobas.txt). We present an excerpt of our ontology created on the first tests fig. 3.

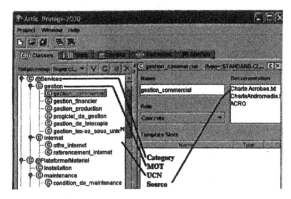

Fig. 3. Ontology excerpt

Phase-3) Diffusion: In this phase, it is necessary to set up some applications which allow to take advantage of knowledge. That is the main objective in the step of diffusion. Indeed, it is necessary to provide the access tools to information, knowledge-objects, projects memory, etc. contained in the knowledge bases as well as search engine or other systems rather advanced which allow to give the relevant knowledge.

We use a search-engine prototype which receives an expression request from the user (Number 1 in fig. 4.). The search engine will compare each UCN in the ontology with the expression request. In this level of work we have tested the first version of the algorithm that we will show: Initially, this application will check out and matches the expression request and it will put equivalence label levels: strong, medium or weak. If it finds out the exact expression request, the label will be set to strong. If the expression does not match, then it will extract each word and it will search any possible combination until at least two words in an UCN

match to turn label to medium equivalence note. Then the algorithm will continue searching in the UCN for every word of the main expression. These last results will turn to weak level label. After searching it will separate and present the results from strong levels until weak ones. Finally it will show at first screen the first six UCN (Number 2 in fig. 4) and the user will be able to see the next ones (Number 3 in fig. 4). The user is able to see the UCN to confirm his request, so if one or more of them satisfy it, he can select them (Number 4 in fig. 4) to know the source(s) document(s) (Number 5 in fig. 4). Finally, the user must search (write identically) these UCNs in the main source(s) documents supported with a text editor. Indeed, our search-engine shows the UCNs (with the source(s) document(s) inside the ontology) more closed to the main request. However, we must say that the UCN founded are current expressions in the working area, thus they make sense to the user compared with his main request and he decides if the UCNs are suitable or not. In fact those expressions are exact in the source documents, hence easy to find.

Fig. 4. Search-engine prototype

Phase 4) Updating: This last stage consists in updating, improving and increasing the documentary knowledge capital, by repeating the previous stages.

5. CONCLUSIONS

We have introduced a methodology for a KM system within SME networks which can run at the beginning of co-operating activities. The objective of this research is to propose a solution closely suitable for the problematic of SME networks towards knowledge sharing. We estimate that the KM field for the SME networks is not explored enough by the scientific communities (principally in the two first phases of the SME life cycle). The

---

[3] http://protege.stanford.edu/index.html

objective of this research is to stimulate the integration and co-operation to increase the confidence within such organization, especially using the existing and scattered knowledge capital of the network.

This research was at first based on the analysis of the problematic of integration and sharing existing in these organizations. Then we applied the feedbacks of related works from the KM community to develop a methodology for the SME networks. We are aware that the mere installation of certain technologies will not guarantee the success of projects of knowledge management in SME networks, because the management and strategic approach remain fundamental. Nevertheless a global co-operative technological culture is appearing due to everyday Internet use by professionals. Our purpose is that the KM within SME network can be applied gradually, aiming to develop a learning organization [9]. Due to their structure -a SME network-, to co-operation imperatives not always precisely formalized, and to the necessary up-date regarding computer technologies, SME networks provide an auspicious natural setting for the use of the new knowledge economics.

6. REFERENCES

1. Aussenac-Gilles, N., Biébow, B., Szulman, S. "Modélisation du domaine par une méthode fondée sur l'analyse de corpus", in IC'2000 (Journées francophones de l'ingénierie des connaissances), Toulouse 10-11 May 2000.

2. Bienner, F. and FAVREL J. (1999) "Organization and management of a distributed information system shared by a pool of enterprises". Acts of the conference IEPM'99. Glasgow, Juillet 1999.

3. Ermine, J.L., " La gestion des connaissances, un levier stratégique pour les entreprises ", Acts in IC2000 (Journées francophones de l'ingénierie des connaissances), Toulouse 10-11 May 2000.

4. Exworth, M., Wilkinson, E. K. McColl, A., Moore, M., Roderick, P., Smith, H. and Gabbay, H. (2003) "The role of performance indicators in changing the autonomy of the general practice profession in the UK", Social Science & Medicine, Vol. 56, Issue 7, 1493-1504.

5. GRECOPME (2000) Vincent, L. et al, "Groupement d'Entreprises Coopérantes : Potentialités, Moyens, Evolutions", Rapport du projet de la Région Rhône-Alpes 1997-2000

6. Guffond, J.L. ; Leconte, G. (1998) "Logistique de chantier, modes d'organisations et outils de pilotage - le cas de l'activité de construction", Acts of the second international meetings of research en logistique, "Logistique et interfaces organisationnelles", ed. N. FABBE-COSTES et C. ROUSSAT, Marseille, January 27-28.

7. Gutierrez-Segura, G. (2001) "ERP pour les groupements de PME/PMI", Rapport du DEA ISCE (Informatique et Systèmes Coopératifs pour l'Entreprises) Univ. Claude Bernard LYON1.

8. Gutierrez-Segura, G., Deslandres V. and Dussauchoy A. (2004) "A KM Based Framework as a way for SME network integration", will appear in 6th IFIP International Conference on Information Technology for BALANCED AUTOMATION SYSTEMS in Manufacturing and Services "BASYS04" 27-29 September 2004, Vienna, Austria.

9. Jacob R. et S. Turcot (2000) "La PME apprenante: Information, connaissance, interaction, intelligence", Rapport de veille, projet Globalisation et PME innovante, Université du Québec à Trois-Rivières, Juillet, 113p.

10. Lim, D. and Klobas J. "Knowledge management in small enterprises" The Electronic Library, volume 18, nombre 6, 2000. http://www.emerald-library.com

11. Lucier, C.E. and Torsilieri, J.D. (1998) "Why Knowledge Programs Fail: A C.E.O.'s Guide to Managing Learning", article web visible sur http://www.it-consultancy.com /extern/extern.html (visible le 26 Juin 2003).

12. Malhorta, Y., (2002) "Why Knowledge Management Systems Fail? Enablers and Constraints of Knowledge Management in Human Enterprises". In Holsapple, C.W. (Ed.) Handbook on Knowledge Management 1: Knowledge Matters, Springer-Verlag, Heidelberg, Germany, 577-599.

13. Nonaka, I. and Takeuchi, H. (1995) "The Knowledge-Creating Company", Oxford, Oxford University Press.

14. Querel, L. (1999) "Action et cognition situées", public Conference 17 juin in Montpellier, France. 1999.

15. Stock, R. " Méthodologie et architecture informatique pour l'acquisition et la gestion des connaissances. Application à une entreprise de fabrication mécanique ", finally report of PHD in computer science, université Reims Champagne-Ardenne. 2002

16. Snyder W. M. et al (2003) "Communities of practice: A new tool for Government Managers", http://www.businessofgovernment.org/pdfs/Snyder_report.pdf

17. Van Campenhoudt M., " Les voies de recherche actuelle en terminologie et en terminotique ", in 7e Université d'Automne en Terminologie, En bons termes, Paris, La Maison du dictionnaire, 1998, pp. 109-119.

18. Von Krogh, G., Ichijo, K. and Nonaka, I., (2002) "Enabling Knowledge Creation: How to Unlock the mystery of tacit Knowledge and Release the Power of Innovation" Oxford university press, N.Y.

19. Wenger, Etienne. (2001). "Supporting Communities of Practice: A Survey of Community-oriented Technologies". Published as "shareware" and available at www.km.gov under "Group Documents", then "Documents and Resources."

20. Wenger, E. McDermott, R. and W. Snyder (2002). "A guide to managing knowledge : Cultivating Communities of Practice", Harvard Business School Press.

PUBLICATIONS
www.elsevier.com/locate/ifac

A STUDY ON MEASURING AND EVALUATING THE PERFORMANCE IN SUPPLY CHAINS

Sílvio R. I. Pires, Dr.
Carlos H. M. Aravechia, MSc.
Production Engineering Post Graduate Program
The Methodist University of Piracicaba (UNIMEP)
Correspondence Address: Prof. Sílvio Pires, sripires@unimep.br, Rod. Santa Bárbara-Iracemápolis, Km 1, Santa Bárbara do Oeste, SP, 13.450-000, Phone: 55-19-31241822/23, Fax: 55-19- 34551361, Brazil

Abstract
During the last decade Supply Chain Management (SCM) has emerged and attracted the attention of the academic and business environment and performance measurement and evaluation has become a constant necessity for managers. However, most traditional performance measurement and evaluation systems were designed in a competitive scenario focused on isolated companies' performance. In this context this paper provides a brief discussion about this current subject as well as presents the main results from a case study conducted within a representative and innovative supply chain in the Brazilian tractor industry.
Key words: Supply Chain Management, Performance Measurement and Evaluation, Case Study.

1. Introduction

During the last decade Supply Chain Management (SCM) has emerged as a new frontier to increase the company's competitiveness in the marketplace. As a consequence of this process, it has created a series of new business opportunities and, on the other hand, a set of new challenges to the production and logistics management. In this context, the supply chain performance measurement and evaluation (PME) has become a critical subject, mainly because the traditional PME models were designed for a competitive scenario focused only on isolated companies' performance. In this direction, the paper's main purpose is to verify the hypothesis that the SCM competitive model implies the re-evaluation and/or the adaptation of the usual PME systems. The hypothesis verification was done through a brief literature review as well as through a case study conducted at a representative and innovative supply chain in the Brazilian tractor industry.

2. Measuring and Evaluating Performance in Supply Chains

2.1. The need for a new approach

During decades companies used to measure and evaluate its performance through traditional PME systems based mainly on financial indicators. Many studies show that the exclusive use of financial indicators does not provide a reliable analysis of the company's health (Kaplan & Norton, 1992; Neely et al., 1995).

In turn, during the last years it was possible to notice a series of significant changes regarding the way that business are accomplished. In connection with the market globalization process, managerial models have been constantly reviewed and improved, as well as new concepts and paradigms, such as the contemporary e-business and mass customization, have quickly emerged. It is in this competitive context that the Supply Chain Management (SCM) should be understood.

With the introduction of the Supply Chain Management competitive model, where is considered that the competition happens not only between isolated companies but between supply chains, including several companies, the PME should take place not only in terms of isolated companies but in terms of the whole supply chain performance. This means that today there is an increasing demand for expanding the concepts of PME to the supply chains. In turn, this new PME approach increases the complexity of the performance issues once supply chains tend to present more complex relationships among independent business units, that is, among its several companies. In following this issue will be more explored.

A supply chain can be represented by a group of integrated logistical processes that originates at the raw material source and comprise several companies, until delivering the products to final customer in the form of goods and aggregated services. In general, the complexity in managing the relationships of these chains increases with the number of companies participating in the supply chain. In Figure 1 some supply chains are illustrated through horizontal arrows running along a supplier, an assembler, a distribution center and a final consumer.

Nowadays it is important to recognize that in most competitive industries (such as in the automotive, electronics and computer industries) competition happens in fact among the supply chains, involving all (or the key) companies that participate in them, different from the established by Porter (1980), who affirmed that the competition takes place among the business units, in an isolated way. Considering the fact that competition is happening more and more among the "virtual business units", that is, among chains constructed by independent companies that are working in a integrated and collaborative way. Then, in order to make possible the management of the companies under this new perspective, recently several practices and new tools have being developed, most of them strongly based on information technology infrastructure. Also it is important that companies of the chain (business units) establish, in a consistent and integrated way, a strategic alignment including the business goals along the entire supply chain.

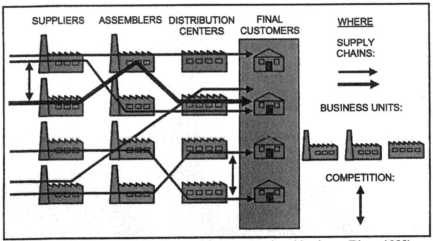

Figure 1: Supply chains and competition among virtual business (Pires, 1998)

Under the managerial perspective, and facing the implications brought by this new competitive paradigm, it is necessary to focus on the supply chain core business through the correct determination of the business unit's core competencies. In this direction, initiatives and practices such as partnership, outsourcing, Early Supplier Involvement (ESI), Efficient Consumer Response (ECR), Vendor Managed Inventory (VMI), In-plant representatives and Postponed Manufacturing have emerged to help on this task. Moreover, it is import to note that in the SCM is desired the development of virtual business units with many of the benefits from the traditional vertical integration but without the common disadvantages regarding the relationship costs and their inherent flexibility loss. In this way, it is also necessary to manage not only the business unit's performance but also the entire supply chain performance, turning the business unit into an effective SCM, that is, more effective and efficient.

2.2. The performance measurement and evaluation issue

From the managerial focus, the performance measurement can be defined as the information regarding the processes and products results that allows the evaluation and the comparison in relation to goals, patterns, past results and with other processes and products. Also, it is important to highlight that a PME system needs to be focused on results, which should be guided by the stakeholder's interests. Based on Beamon & Ware (1998), it is possible to affirm that the adoption of performance indicators should deal with the following questions:

- Which aspects should be measured?
- How to measure these aspects?
- How to use the measures to analyze, improve and control the supply chain as a whole?

It is noticed this is not an easy task, once there are several indicators available and it is necessary to align the used measures with the involved companies' goals. Also, it is notable in the literature that extensive use of costs as a performance indicator is very common among the companies.

This happens because the PME through a single indicator is relatively simple. It should be attempted, even so, to the fact that this practice can provide very superficial information about the reality. In general terms a measurement system should contain indicators to present the following features: inclusiveness (to include the measure of all the pertinent aspects), universality (to allow the comparison under several operational conditions), measurability (to guarantee that the necessary data are measurable) and consistency (to guarantee consistent measures with the objectives of the organization).

On the other hand, it is not recommended to simply discard costs as a performance indicator due to its importance. The alternative would be the adoption of multiple indicators, involving a cost combination with time, flexibility and quality, according to the company competitive priorities.

As pointed earlier, traditionally, the PME is limited to an isolated company or productive process. For an effective SCM it is necessary to expand these concepts beyond the company limits, involving all the supply chain players. It is necessary then, the development of a PME system embracing all the business units. A usual way of doing this is through the adaptation of the traditional PME systems.

2.3. Adapting existing PME systems

In the specific case of a PME system for the SCM, it is necessary to guarantee the compatibility of the used measures along the whole supply chain. This means that the individual measures, for a certain business unit, should be interpreted and compared with all the remaining ones. Figure 2 presents the adapted structure of a PME system to the SCM.

The model illustrated by the Figure 3 presupposes the existence of an individual group of indicators (represented by Ind. 1, Ind. 2 and Ind. 3), used in each one of the business units. Besides, there are some common indicators to the whole supply chain and these common indicators will determine the supply chain performance.

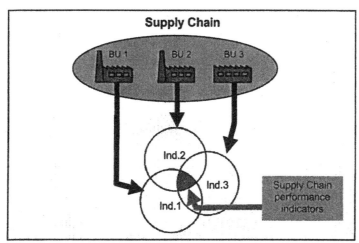

Figure 2: The basic structure of a SCM performance measurement and evaluation (PME) system.

3. A case study in the tractor industry

In order to verify the hypothesis that the SCM competitive model implies the re-evaluation and/or the adaptation of the usual PME systems a case study in a representative supply chain in the tractor industry was conducted.

The research was conducted focusing on the perspective of a large multinational company (company focal) and two of its suppliers, that is, the "supplier 1" (a first tier supplier) and the "supplier 2" (a second tier supplier). It is illustrated in the Figure 3.

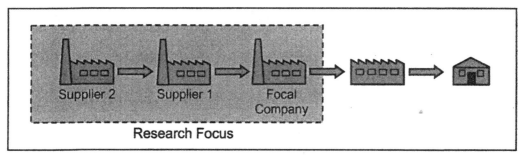

Figure 3: An illustration of the studied supply chain

The focal company of the studied supply chain has been operating in Brazil since the 1950's and is a traditional market leader in tractors to haul materials in the construction, mining, agriculture, forestry, and manufacturing industries. The company recognized as a worldwide leader in innovative initiatives and practices on SCM. The main company's operation in Brazil consists of a single plant located in the state of São Paulo which produces five product lines, with 21 basic models and more than 60 different configurations. The company is ranked among the country's 20 largest exporters and exports approximately 50% of its production in manufactured volume. Also, among more than 50 company's plants operating around the world, the Brazilian factory produces the largest variety of products.

The company plant in Brazil depends more heavily on importation than all the other units of the company in the world. Approximately 40 to 45% of the components used by the Brazilian plant are imported, while others plants of the company in the USA, Europe and Asia import an average of 5% of components. The plant under study has a few more than 100 suppliers located in Brazil, while American suppliers are responsible for 85% of its import volume. In this direction, the company's priority in the last five years has been to purchase from local suppliers and much effort has been dedicated to improving its inbound processes. However, there are still many drawbacks involving the lack of technology and/or the suppliers' low production capacity that need to be solved.

The two suppliers were selected based on prior interviews conducted with the responsible for the focal company logistics and SCM operations. Table 1 presents a brief description of the three companies involved in the research.

Table 1: General data from the studied companies

Company	Industrial Sector	Numbers of Supplier	Main Product	Main Productive Process
Focal	Mechanical	104	Tractors	Assembly
Supplier 1	Mechanical	50	Grinded parts	Grinding
Supplier 2	Metallurgical	4	Thermal Treated parts	Thermal Treatment

The case study was conducted in the main supply chain of the focal company. The method adopted for data collection was an interview based on a structured questionnaire. In a first stage, the study was developed by multiple interviews (together the SCM managers of three companies) and *in loco* observations within the focal company and one of its main first tier suppliers. In a second stage the study was developed within the main supplier of the studied first tier supplier. It is important to note that the study focused only on the inbound supply chain of the company, that is, on its upstream processes. Also, the presence of the interviewer was important to eliminate possible wrong interpretation during the interview phase of the research.

The questionnaire was divided into three parts. The first part was used for the company general characterization including the respondent occupation, the number of employees and the annual income. The second part included questions regarding the SCM initiatives and practices adopted by the company and the relevance of those practices for the overall company performance.

After the company characterization, the last part of the questionnaire presented a list of performance indicators which were categorized into four traditional competitive priorities: cost, quality, flexibility, and delivery performance.

The respondents were asked to fill in the questionnaire regarding each one of the listed performance indicators, determining what were the impacts felt by the company after the adoption of SCM practices. To do so, the respondent was supposed to choose between six possible answers: (1) is not used, (2) stopped to be used, (3) lost emphasis, (4) started to be used, (5) gained emphasis or (6) suffered no change.

The Table 2 brings the list of SCM initiatives and practices adopted by each one of the companies involved in the research as well as the importance attributed to them.

Table 2: Initiatives and practices adopted by the companies.

Practices	Focal Company	Supplier 1	Supplier
Supply base restructuring	4	2	1
Outsourcing	4	5	5
Electronic Data Interchange	4	5	X
Early Supplier Involvement	4	4	X
Efficient Consumer Response	4	x	X
Postponed manufacturing	4	3	X
6 Sigma	4	5	5
Vendor Managed Inventory	3	2	X
In-plant representatives	3	4	X
Just in time	3	5	4
Key:	x - not adopted; 1 - irrelevant; 2 - less important; 3 - important; 4 - very important; 5 - essential		

Considering the Table 2 it is possible to notice that the companies, specially the focal company, ingoing the new competitive model once they present the restructuring (reduction) of their supplier base and the adoption of several SCM initiatives and practices, some of them considered essential for current business.

Further with the case study, the following tables present the results of the impact brought by the SCM paradigm for each one of the three companies as explained before.

Table 3: The impact on cost performance indicators.

Cost performance indicators	Impact		
	Focal Company	Supplier 1	Supplier 2
Cost relative to competitors	5	5	6
Perceived relative cost performance	5	5	6
Manufacturing cost	5	5	6
Capital productivity	5	5	4
Labor productivity	5	5	5
Machine productivity	6	5	5
Total factor productivity	5	5	5
Total product cost as a function of lead time	5	5	1
Direct labor	5	5	1
Repair or rework	5	5	5
Breakeven time	5	5	1
Value added as per cent of total elapsed time	5	5	1
Key:	1 - is not used; 2 - stopped to be used; 3 - lost emphasis; 4 - started to be used; 5 - gain emphasis; 6 - suffered no change		

In Table 3 it is possible to notice that a significant number of cost related performance indicators suffered a positive impact after the adoption of SCM initiatives and practices.

Table 4 – The impact on quality performance indicators.

Quality performance indicators	Impact		
	Focal Company	Supplier 1	Supplier 2
Perceived relative quality performance	5	5	5
Quality relative to competitors	5	5	5
Product reliability relative to competitors	5	5	5
Product durability relative to competitors	5	5	5
Percentage of surveyed customers satisfied	5	5	5
Customer satisfaction	5	5	5
Percentage conform to targets	6	5	6
Assembly line defects per 100 units	5	5	6
Percentage with no repair work	5	5	6
Percentage defect reduction	5	5	6
Percentage reduction in time between defect detection and correction	5	5	6
Percentage scrap	5	5	5
Percentage scrap value reduction	5	5	6
Repairman per assembly line direct labour	5	5	6
Percentage of inspection operations eliminated	6	5	6
Cost of quality	5	5	5
Vendor quality	6	5	6
Percentage supplier reduction	5	5	5
Key : 1 - is not used; 2 - stopped to be used; 3 - lost emphasis; 4 - started to be used; 5 - gain emphasis; 6 - suffered no change			

In Table 4 it is also possible to notice that some quality related performance indicators suffered a positive impact after the adoption of SCM practices.

Table 5: The impact on flexibility performance indicators.

Flexibility performance indicators	Impact		
	Focal Company	Supplier 1	Supplier 2
Perceived flexibility	5	5	5
Flexibility relative to competitors	5	5	5
Process flexibility relative to competitors	5	5	5
Extent to which quality is unaffected by mix/volume changes	6	5	6
Extent to which cost is unaffected by mix/volume changes	6	5	6
Extent to which delivery performance is unaffected by mix/volume changes	6	5	6
Perceived relative product flexibility	5	5	6
How quickly plant responds to product mix changes	5	5	5
Number of part types process simultaneously	5	5	5
Production cycle time	5	5	5
Cycle time (make time/total time)	5	5	5
Set-up time	5	5	5
Time to replace tools, change tools, assemble or move fixtures	5	5	5
Percentage increase in average number of set-ups per day	5	5	6
Perceived relative volume flexibility	6	5	5
Vendor lead time	5	5	1
New product introduction versus competition	5	5	1
Time for idea to market	5	5	1
Average time between innovations	5	5	1
Key: 1 - is not used; 2 - stopped to be used; 3 - lost emphasis; 4 - started to be used; 5 - gain emphasis; 6 - suffered no change			

Regarding Table 5 it is possible to notice that the major impacts brought by the SCM practices occurred at the flexibility performance indicators. Finally, Table 6 presents the results of the impacts brought by the SCM practices at the delivery performance indicators.

Table 6: The impact on delivery performance indicators

Delivery performance indicators	Impact		
	Focal Company	Supplier 1	Supplier 2
Perceived relative reliability	6	5	5
Reliability relative to competitors	5	5	5
Percentage on-time delivery	5	5	5
Due date adherence	5	5	5
Percentage increase in portion of delivery promises met	5	5	5
Percentage of orders with incorrect amount	3	5	6
Schedule attainment	5	5	6
Average delay	3	5	6
Percentage reduction in lead time per product line	5	5	5
Percentage improvement in output/desired output	5	5	6
Percentage reduction in purchasing lead time	5	5	5
Percentage reduction in average service turnaround per warranty claim	6	5	6

Key: 1 - is not used; 2 - stopped to be used; 3 - lost emphasis; 4 - started to be used; 5 - gain emphasis; 6 - suffered no change

4. Final Remarks

The study also showed that during the last years the focal company have got significant progress in managing its internal supply chains. These efforts have been conducted with base mainly on an international logistical certification provided by a consulting company. In addition it is notable the progress achieved together its key first tier suppliers, especially with the "supplier 1" here studied. On the other hand, the relationship of the focal company with its second tier suppliers is still very incipient, even with its recognized key second tier suppliers (as the supplier 2 here studied).

Based on the conducted research also it is possible to conclude that the paper's basic hypothesis was proved for the studied case. In this situation the SCM competitive model implied in the re-evaluation and/or the adaptation of the usual PME systems.

In addition, it was possible to note that there was no alignment regarding the impacts on the performance indicators along the analyzed supply chain. This lack of alignment is evident when analyzing the answers originated at Supplier 1, were all indicators gained emphasis. It was not possible to identify the causes of this lack of alignment but it is possible to speculate that the performance indicators used in the research were not totally adequate for the supply chain performance evaluation. Other possibility is the lack of formal procedure to guarantee a strategy alignment and consequent performance evaluation system alignment among the supply chain.

Finally, the study highlighted the need for the development of contemporary performance evaluation systems capable of measuring and evaluating the real performance of a supply chain.

References

Beamon, B. M., Ware, T. M., A process quality model for the analysis, improvement and control of supply chain systems. *International Journal of Physical Distribution & Logistics Management*, v.28, n.9/10, p. 704-715, 1998.

Kaplan, R. S., Norton, D. P., Balanced Scorecard: measures that drive performance. *Harvard Business Review*, p.71-79, jan/feb, 1992.

Neely, A., Gregory M., Plattes K. Performance measurement systems design: a literature review and research agenda. *International Journal of Operations & Production Management*, v.15, n.4, p.80-116, 1995.

Pires, S. R. I., Managerial implications of the modular consortium model in a Brazilian automotive plant. *International Journal of Operations & Production Management*, p.221-232, 18: 3 1998.

Porter, M. E. *Competitive Strategy: Techniques for Analyzing Industries and Competitors*, Free Press, New York, NY, 1980.

ELSEVIER
IFAC
PUBLICATIONS
www.elsevier.com/locate/ifac

A MODEL OF CONSIGNMENT DECISION FOR FOREIGN MATERIAL IN SUPPLY CHAIN MANAGEMENT

Cristiano Morini, Dr.
College of Management and Business
Methodist University of Piracicaba (UNIMEP), Brazil. Rodovia do Açúcar, Km. 156, Piracicaba, SP, Brazil, 13.400-911 - 0055-19-3124 1507 (phone/fax) - cmorini@unimep.br

Sílvio R. I. Pires, Dr.
Production Engineering Post Graduate Program
Methodist University of Piracicaba (UNIMEP), Brazil. Rodovia SP 306, Km. 1, Santa Bárbara D'Oeste, SP, Brazil, 13.450-000 - 0055-19-3124 1822 (phone/fax) - sripires@unimep.br

Abstract:

Globalization of international business, reduced number of international transportation routes through South America and customs bureaucracy make pressures on industrial enterprises concerning the reduction of time of response to customers. This paper proposes a model of decision for consignment management of foreign materials in supply chains. A key element of the proposed model is the definition of a set of decision variables. In order to achieve this, a set of variables initially proposed was tested and adjusted to the reality of importer companies those consigning foreign goods to reduce time of response.
Key words: consignment, supply chain management, foreign goods.

1. Introduction

Since the turn of Twenty-first century and even before, Supply Chain Management (SCM) has been considered the new forefront for competitive advantages in a world that globalizes faster. In South American countries, the impact of globalization can also be noticed in far distances of the biggest demanding and consuming centers for the majority of goods and services, including in the United States of America, in Europe, and in East Asia.

The adhesion of new countries in a competitive scenario of international business was reached by strategies of manufacturing costs reduction whereas new logistics challenges were added, as delays caused by customs clearance bureaucracy, bigger transit times, bigger supplying cycles, and reduction of flexibility (Larrañaga, 2003).

The poor availability of maritime and air routes through South America make both supplying and distribution time longer than the worldwide average. That is why this paper searched to contribute to the proposition of a decision model at the SCM, looking for the decrease of the customer's time of response. With more agility, the availability of items is increased to fulfil customer's demand. The definition of time of response assumed by this work is a synonym of supplying time, which is understood as the process of material ordering, period from the order emission until the effective moment of material receiving at the industrial customer.

2. Supply Chain Management in a Global Perspective

The SCM concept encompasses the management of the total chain of materials from the primary supplier to the final customer. According to the SCM perspective the chain is focused as a whole and not as a fragmented structure or fragmented groups (Gentry, 1996).

The bottom line, in terms of objective, is to maximize the integration and become reality potential cooperation among chain parts in order to fulfil more efficiently demands of the final customer, reducing costs and adding value to final products (Vollmann and Cordon, 1996).

As an integrated part of the SCM, a business unit can be understood as a distinctive group of an enterprise that makes part of a specific chain. Pires (1998, 2004) stated that each unit of this virtual business unit must focus on the product competitiveness to the final customer and must be concerned with the performance of the supply chain as a whole. Consequently, this demands an integrated management of the chain.

Authors such as Larrañaga (2003) stated that organizations are becoming more concerned on competition focused on time, and time reduction to provide goods and/or services to the final customers is one of the major forces that pressure for the revision of order cycle, mainly for imported goods.

Following, inventory management is presented as a potential opportunity to process revision in companies with global operations.

2.1. Inventory Management

Low levels of inventory can cause risks by lack of items and high costs in obtaining them. Huge stocks can cause additional invests in warehousing and its financial cost of maintenance, and loss for obsolescence or deterioration. (Carretoni, 2000).

From a global perspective, demand uncertainty and customs barriers have justified security stocks in many managerial situations. On the other hand, the re-supplying time depends directly on the availability of air and maritime routes.

Dornier et al. (2000) stated that the delivery of goods from foreign countries can suffer unpredictable delay and other problems due to bureaucratic custom procedures. Christopher and Towill (2002) pointed out that the reduction of time of response is related to the ability of supply chain to react faster due to the market demand variations related to volume or variety.

3. Inventories and its Place in a Supply Chain

The place of foreign goods at the international supply chain can provide gains in terms of increasing agility and decreasing time of response.

SCM with opportunities of customs legislation can add more value than cost due to the stock and its place and the increase of response agility in terms of demand variation.

In some cases, foreign goods can entry into a country been admitted and not imported by the strict sense of the word. In admission, the applicant of goods do not assume the role of buyer or importer, because material is admitted "no dollar remittance", that is, in consignment only. This type of operation is called "no money remittance", with values "for customs only". In this case, the material is admitted no money remittance and it is stocked in a warehouse allowed by surveillance agency. The warehouse is called credited dry port.

In a dry port, the material which has been admitted in a country is not property of applicant enterprise, but its ownership is of the supplier which maintains goods in foreign country, that is, in a country of the buyer (the consignee), near than the consumption market.

Therefore, there is not stock cost for the applicant, but only warehousing costs. Supplier is the owner of the material held in stock, warehousing costs are a concern of applicant, and dry port is responsible of material security and stock accuracy.

As the material requester, the applicant becomes the buyer with the nationalization procedure by the fulfillment of the importation and customs legislation procedures. In that moment, supplier receives the material payment related to that materials were at applicant's country, in consignment, like the Vendor Managed Inventory (VMI). At this time the buyer also pays taxes to the government in regime of suspension until the nationalization being completed.

Supplier also have benefits from the consignment practice, because part of its stock is maintained at the buyer's country with warehousing expenses being paid by the applicant, that is the future buyer, opening space at the vendor's stock and having the guarantee of sale. If demand of item is not confirmed, there are some procedures to follow, according to the Brazilian law. For example: export back to origin, with expenses paid by the applicant; re-export to other countries, in order to accomplish foreign demand; and scrap followed by surveyors.

4. Research Methodology

According to Martins (1999), models search specification of nature of events and the importance of variable relations. Model is an abstraction that characterizes ideas supported by concepts which are familiar.

Zendejas (2000) pointed out that there are different instruments to qualitative measures: Likert scale, semantic difference, and Guttman scale. This paper uses Likert scale. According to Pereira (2001), the success of the Likert scale consists in that it is sensible to rescue Aristotelian concepts relating quality attributes: it recognizes the opposition between opposites, recognizes a range, and recognizes an intermediary situation, in a qualitative approach. He added also that the arithmetic representation of a qualitative event is a strategy of processing and analysis, and the interpretation of results makes that researcher to come back to original meaning of his/her measures.

It was utilized a questionnaire to fulfil the objectives, that is to verify the feasibility of variables which are part of the suggested model. The questionnaire contains closed questions and it was addressed to enterprises listed in a file of the importer companies of the Association of Industry (FIESP) in Campinas, important city of the State of São Paulo, in Brazil. The original sample contains 142 companies and the research got 35 answers, which represents 24.6% of return tax. Results are commented further.

5. Development of the Proposed Model

The model presents variables that once utilized under specific circumstances, suggests consignment of foreign materials, with the objective of making more available foreign materials in shorter cycle time. The conditions are detailed ahead. The logic of consignment makes reference to VMI in the point that the consignment makes possible the strength of the partnership between the supplier and the buyer, in this case, the industrial company, having in mind the increase of competitiveness in a supply chain. The model does not make reference to the VMI as the supplier managing the customer's demand, but only related to the consignment. The particular contribution related to the application of part of the VMI concept with foreign goods, which is uncommon in SCM literature.

5.1. Guidelines of the Model's Variables The model considers the following variables for the utilization of consignment of foreign material, in accordance with current literature of performance indicators: value added, demand, volume (in terms of physical dimensions), life cycle, weight, expiration date, "criticity" (condition of item of being critical to the, in terms of manufacturing process continuity), fixed assets, air freight value (in case of non-availability), turns, availability of final product, supplying complexity (in terms of logistics procedures), supplying lead time, and supplier-industrial buyer relationship. Selected variables are identified and presented together in radar format (Figure 1)

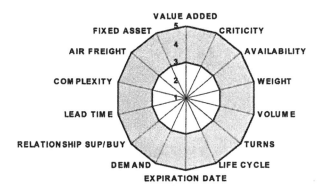

Figure 1 – Variables of the Model (as Radar Presentation)

A radar graphic can show several dimensions presented simultaneously, easy comparative visualization, and standardization of measures of different indicators. This structure allows compare different types of indicators. This standardization of measure units can be considered as the strongest point of this graphic.

The model proposes that variables be scored near than 5 (five) or between 3 (three) and 5 (five) to have acceptable results for admission in consignment at credited dry port. The model construction allows that scores be in white area, considering the final average of each results be inside the indicated interval, that is, between 3 and 5, because 3 represents "reasonable importance" and 5 represents "huge importance" related of each indicator, as is clarified ahead.

For items and chains with scores out of related range, the consignment at dry port is not desirable, because of the increase of costs in process due to (1) high warehousing expenses, (2) the non-acceptance of the supplier in providing items in consignment, or (3) the choose of air freight for availability of goods, considering viable this cost.

For each one of the analyzed items there is a range from 1 (one) to 5 (five) to check its feasibility on industrial enterprises (Figure 1). The number 1 means that the analyzed indicator was considered very low or very small, depending on each indicator, that is, with no importance. Number 2 means that the indicator was analyzed and it was considered small or low, with minimal importance. Number 3 means that the indicator has reasonable importance or medium importance, and it means also a neutral point of range. In number 4, the indicator is considered important, of the high level of big one. Finally, number 5 is the place that indicator assumed its maximum importance. This model does not have mathematical aspects, so the choice for the best position of indicator, from very low to very high is the attribution of the people who answered the questionnaire.

Related to the 1-5 range, there is a variety of other Likert ranges (1-7, 1-9, 0-4, for example). The criteria to determine the ideal range can be the capacity of the responsible people has of discriminating the object. As the questionnaire was sent to companies and these ones determine the specific people to answer it, the study considered that the questionnaire was answered by professional people of different status in hierarchical structure, that is, with different

approaches of evaluation. Therefore, the study has opted to 1-5 Likert scale, because the interpretation margin in this scale is restricted.

When variables show indicators as "big" or "high", the model proposes that is feasible the utilization of consignment of imported material. It is also important to point out that there is a big variety of relationship possibilities among variables, when analyzed two-two, three-three, four-four, and so on. In the web of variable relationship, it is possible that some situations be feasible the consignment, even though not being considered "big" or "high" for all of them. In these cases, final results of the average of indicators must be figure out and have to be inside the range 3-5.

5.2. Reduction of Customer's Time of Response

The supplier time of the imported material can be understood as the sum of each time indicated as T1, T2, T3, T4, T5, T6, and T7:

• T1 is the processing order time from the customer to the supplier at the foreign country, which also can be understood as purchasing lead time;

• T2 is the processing time of material, invoicing and packaging;

• T3 is the time of picking up material at supplier or in other place indicated by him;

• T4 refers to time of container or box stuffing, export customs procedures, ship or airplane waiting, cargo loading, and ship or airplane departure;

• T5 is the international transit time;

• T6 is the time of cargo unloading and import customs procedures;

• T7 refers to inbound transit time, from customs to plant, at the country of destination.

It is possible to identify three situations in terms of availability and costs. First, at the process of direct importation, under common customs regime, T1 to T7 must be added. Second, in case of items be in stock available at plant, supplier has already been paid and also government has already been paid by taxes. This situation is ideal in terms of availability, but it is not economically feasible in many cases. Third, is the situation of consignment at the country of destination. It looks for avoiding disadvantages:

on the one hand, costs of stock maintenance inside plant, related to payment to supplier and taxes to government at the admission of material, on the other, huge supplying lead time due to stock being maintained at supplier located at foreign country. In situation of consignment, T2, T4 and T5 are disregarded to the fulfillment of industrial customer, because goods have already been stocked at the warehouse of importer's country, near than the importer.

In situation of consignment, dry port can play an important role, because it means the possibility of stock be maintained strategically near than the industrial customer, only with warehouse expenses. Demand uncertainties can be focused on different manner, having in mind the reduction of time of response to customers.

5.3. Checking and Analyzing Model's Variables

This section analyses the feasibility of proposed variables and the collected data from a questionnaire sent to the universe of 142 industrial companies that also are importers. These companies are part of a file of the Association of Industry, in Campinas city, Brazil. The file is dated of 2003, July. The main objective of the questionnaire was to verify how many companies make consignment and for these ones, how the behavior of variables is.

6. Research Results

Accordingly to industrial activity of sample, vehicle and spare part manufacturers are companies that more practice consignment of foreign goods. From 35 companies which answered the questionnaire, 12 of them consignee foreign material at dry port. Those 12 companies have attributed the following scores to the 14 variables:

• Related to the variable "fixed assets", those 12 companies do not consider it as an important aspect. The final score obtained was 2.5. The score is exactly between 2 and 3, that is, between "low importance" and "of reasonable importance". So, this variable was fire of the proposed model, because its result was not conclusive. It was the first adjust of the initial model;

• Seven indicators (turns, availability, value added, volume, weight, life cycle and "criticity") have their scores in a range of 2.6 and 3.5. In this interval, was considered that 2.6 is near than 3. So, it was considered important or of "reasonable importance" scores between 2.6 and 3.5, considering 3 at the middle of the interval. Variables of this interval have received weight 1 (one) for model application. The determination of different weights to the variables of the model constitutes second adjust to be promoted at the model;

• Six indicators (complexity, lead time, demand, expiration time, air freight, and relationship with supplier) have gotten their averages between 3.6 and 4.5, having the 4 (four) as a middle of the interval, that is, they were considered "very important" to the consignment of foreign material at dry port. So, these indicators have received weight 2 (two) to application of the model.

The equation below presents the way of calculus of the average of each of the 13 remaining variables. These variables have to be tested in a case study for verification of its feasibility.

$$MSV = \frac{(Weight_1 \bullet V1) + (Weight_2 \bullet V2) + ... + (Weight_n \bullet Vn)}{(Weight_1 + Weight_2 + ... + Weight_n)}$$

where:
MSV is the Medium Score of Variables (all of them), and V1, V2, Vn are obtained scores at the respective variables.

The final average of each obtained variable of the research is showed at the Figure 2.

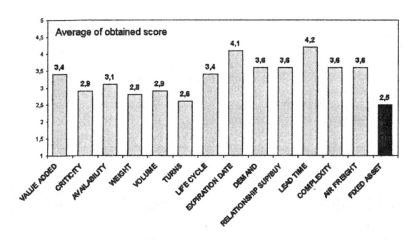

Figure 2 – Average of Results by Variable

It is important to mention that the interval initially idealized by the model, for each variable, is the range from 3 to 5, and the interval showed at the Figure 2 is considering the average of results that have gotten at the research with 12 companies of the sample that consignee foreign material. The study have considered important to adjust the interval from 3-5 to 2,6-4,5 for each variable, becoming closer than the reality of the analyzed companies, but maintaining the interval of 3-5 to the final score (MSV).

7. Final Remarks

The model proposes consignment of foreign material due to reduction of time of response to industrial customer. For this, the model idealized initially fourteen variables. One of them, "fixed assets" was discharged due to the uncertainty of obtained score in research, because in the sample it was exactly between the interval of 2 and 3, that is, it not possible to consider it of "few importance" (related to score 2) or of "reasonable importance" (related to score 3).

The model considers that if those thirteen variables be used and the final average of obtained scores from variables be inside the interval 3-5, it means that material and the supply chain will be benefit adopting the decision model. In case of the final result be out of the interval 3-5, consignment is

unfeasible for some material or some supply chain, because the item and the chain present characteristics that allow to import the material directly from the supplier, when demand indicates it. In this case, there will not be pressure on costs of warehousing at the country of destination, being possible to import by air when necessary.

The benefits associated with this model are that it brings thirteen variables of decision to industrial companies, aims the reduction of time of response in the international supply chain, can be feasible if the supplier accepts to consignee material, and can use the available legislation of customs regimes in Brazil and some other countries.

The proposed model was tested in manufacturing companies in Brazil. After that, the model was adjusted to the reality of importer companies that consignee foreign materials. The feasibility of the model must be checked in case studies in Brazil and other South America countries those allow consignment of foreign goods in regime of importation.

The theoretical model is focused on manufacturing companies, independently of the industrial sector. Further researches can show its usefulness in specific sectors of economy and the set of variables can be adjusted to specific supply chains worldwide.

REFERENCES

Carretoni, E. (2000), Administração de Materiais: uma abordagem estrutural, Alínea, Campinas, SP.

Christopher, M. and Towill, D. R. (2002), "Developing market specific supply chain strategies", The International Journal of Logistics Management, Badford-Ohio, Vol. 13 No. 1, pp. 1-12.

Dornier, P. P. et al. (2000), Logística e Operações Globais: texto e casos, Atlas, São Paulo, SP.

Gentry, J. J. (1996), "The role of carriers in buyer-supplier strategic partnerships: a supply chain management approach", Journal of Business Logistics, Oak Brooks, Vol. 2 No. 17, pp. 35-53.

Larrañaga, F. A. (2003), A Gestão Logística Global, Aduaneiras, São Paulo, SP.

Martins, G. A. (1999), "Teorias e Modelos nas Ciências Administrativas", Business Seminars FEA-USP, Vol. 6, Universidade de São Paulo, São Paulo, SP.

Pereira, J. C. R. (2001), Análise de Dados Qualitativos: estratégias metodológicas para as ciências da saúde, humanas e sociais, 3rd. ed., Edusp, São Paulo, SP.

Pires, S. R. I. (1998), "Managerial Implications of the Modular Consortium Model in a Brazilian Automotive Plant", International Journal of Operations & Production Management, Manchester, Vol. 18 No. 3, pp. 32-41.

Pires, S. R. I. (2004), Gestão da Cadeia de Suprimentos: conceitos, estratégias, práticas e casos, 312 pp., Atlas, São Paulo, SP.

Vollmann, T. E. and Cordon, C. (1996), Making Supply Chain Relationships Work, IMD, Lausanne, No. 8.

Zendejas, V. S. (2000), "Escala Likert: Preguntas cerradas", available at: www.orion2000.org/documentos (accessed 22 January 2004).

Distributed Information System for Cooperative Supply-Chain Management

Dahane M. and Monteiro T.

MACSI PROJECT – INRIA LORRAINE

LGIPM/AGIp, Université de Metz, Ile du Saulcy, F-57045 METZ Cedex 1

E-mail: {mohammed.dahane, thibaud.monteiro}@loria.fr

Abstract: The purpose of this article is to propose a multi agent approach dedicated to distributed information system for cooperative supply-chain management.
This approach is based on an elementary component called Virtual Enterprise Node (VEN). This VEN models each enterprise of the network. This VEN is composed by four agents: the Planner Agent, the Negotiator Agent, the Communicator Agent and the Configurator Agent. This article details Communicator Agent and Configurator Agent behaviors which are dedicated to a distributed information system.

Keywords: Multi-Agent System, Distributed control, Cooperative work, Supply-chain management

1. INTRODUCTION

The Industrial Architecture (IA) notion was introduced for controlling and analyzing relationships among companies involving in a supply-chain (Haurat, & Monateri, 1999). To generate a better productivity, these companies need to coordinate actions which are distributed among autonomous partners (Altersohn, 1992; Rota, 1998; Kjenstad, 1998; Huhns, 1987). One approach to realize this coordination is the cooperation one. The important number of works dealing with this approach shows a growing interest in cooperation relationships among the multiple actors of an IA (Axelrod, 1992; Rapoport, 1987; Ferrarini, 2001; Monteiro, & Ladet, 2001). In our work, cooperation is defined as collaboration between partners each having equivalent decisional capacity and acting together towards a common objective. This cooperation is used to coordination and synchronization of operations carried out by independent actors (Malone, & Crowston, 1994; Monteiro, & Roy, 2003;). The intrinsic distribution of the AI results of this independence. So, each partner has a limited decision power that corresponds to its action field

(Camalot, Esquirol, Huguet, & Erschler, 1997; Camalot, 2000; Huguet, 1994).

To support cooperative supply-chain management, information system has to be able to take into account specific configurations as: informational and decisional distribution, dynamic architecture and flexible integration (Ouzounis, 2001).

As the IA is distributed, it cannot be managed by a single and centralized data-processing application. Indeed, the exchanges of information and the behaviors specific to operations of the network members are so complex that they ask for CPU time which can only be shared. Accordingly, it seems to us that Multi-Agent approach can support this need (Ferber, 1995; Patriti, Schäfer, Ramos, Charpentier, Martin, & Veron, 1997).

In this presentation, we are interested in a distributed information system for cooperative supply-chain management. First, the distributed cooperative decision is presented. After, we present the information system organization, and we illustrate our approach by two examples. And finally we present some conclusions and future researches.

2. DISTRIBUTED COOPERATIVE DECISION

In a distributed decision-making structure, each member-company of the IA is considered an independent decision center and is therefore capable of modifying the details of the industrial process, which it has elected to include in its activity. This type of architecture is also called Virtual Enterprise (VE) (Hardwick, & Bolton, 1997).In a cooperative context, a member-company's internal decisions have to be considered. Cooperation can be described as a will to act collectively towards a common goal. In the case of inter-firm industrial flow control, cooperation can be considered to be the coordination of the means of each company composing the IA, to produce goods that in the end, generally optimize an equilibrium among cost, quality and delay. This coordination allows the making of consistent individual decisions and aims at synchronizing them.

Decision-making becomes necessary when the environment changes. A change can be external to the member-company (direct decision maker's environment), as is the case of receiving an additional order from a client or when a problem occurs with a provider; or it can be the result of an internal evolution, when an unexpected event appears in the company itself. Here, cooperation promotes both collective and distributed decision-making, aiming to synchronize actions shared among the different partners. We are in a co-decision context which intends to coordinate individual actions.

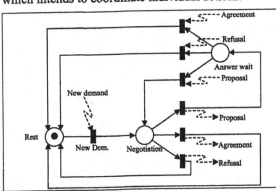

Figure 1. Evolution of negotiation

The Petri net (Figure 1) illustrates the mechanism of the decision making. When a demand is initially detected, the Rest token, which indicates there is no negotiation, is consumed. The negotiation phase creates three possibilities of firing. Depending on a feasibility study, decision-makers can either further negotiation by replying with a proposal, or stop it by replying with an agreement or a refusal. If he receives a proposal, the second partner will determine his possibility of firing. Decision-making is interactive. When a company has to deal with situations beyond its control, conducting a feasibility study causes the company to make new demands on providers who will in their turn have to start negotiations (Monteiro, Ladet, & Bouchriha, 2004). Decision-makers have to take partners' prerogatives into account during the decision making. By this

process each partners' internal constraints have been propagated among the IA.

So, by this decisional mechanism which respects the propagation of constraints, we guarantee at the same time, center's autonomy of a decision and the coherence of decision making between centers (Monteiro, 2001). A group of bilateral decisions acting in concert with two protagonists can guarantee general coherence in an IA having a treelike structure, as illustrated in Figure 2.

3. INFORMATION SYSTEM ORGANIZATION

In our study, the supply chain is seen like a network of enterprises (Figure 2) in a VE form. This network consists of row made up of several entities which acts in the same logistic tier. Each one of these companies is represented by Virtual Enterprise Node (VEN) which has relations (information and matter flows) with the tier adjacent VEN.

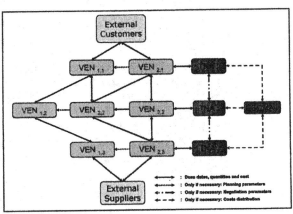

Figure 2. Information flow and architecture

The VEN forms the basic element in our architecture for that, it is necessary to have a good control of its internal architecture. Thus, modeling in Multi-Agent system defines the VEN like an entity made of following agents:

• Planner Agent (PA): is the agent in charge of planning at the enterprise level. Its objective is to decide if an order can be supported by the production process (Ouzizi, Anciaux, & Portmann, 2003).

• Negotiator Agent (NA): it receives orders, production modifications... It transmits the requests to the PA and receives the answers (decisions) of its share. Then it conveys the results with the various partners.

In the case of problems which haven't a local solution, the NA contacts the TNA to obtain its assistance.

• Communicator Agent (COMM): This agent is in charge of the various communication tasks between several agents, both the internal VEN communications, and VEN communications with others agents of the adjacent VEN. Generally; its role is to ensure information routing between two communicating agents under the best conditions.

- Configurator Agent (PARAM): Its mission is necessary parameter setting and reconfiguration after each change in the structure of the VEN, like the management of IP addresses of the machines which support the various agents.

3.1 Functionalities of the VEN

In this communication, we are interested more precisely in specify the distributed informational system. The informational system is the core of the supply chain performance and coordination (Gavirneni, Kapuscinski, & Tayur, 1999; Cachon, & Lariviere, 2001; Camarinha-Matos, Afsarmanersh, & Rabello, 2003). So the following paragraph detailed PARAM and COMM agents.

3.1.1 Functionalities of the COMM

COMM agent constitutes the core of all communications and the interactions between agents. COMM agent is used also at the internal level (at the interior of the VEN), and at the external level with agents belonging to the external environment (other VEN, or the TNA). That implies that COMM agent constitutes a layer adding to the total architecture. The advantage of this addition is justified by the fact that this agent avoids having an advanced part of communication for each agent (PA, NA, PARAM...); however, its competences are summarized in the following points:

1. Information routing: is the most elementary role, which consists in receiving exchanged data between the agents and sending them.

2. To mask the heterogeneity of the various communicating entities, because of the multitude of the companies and the partners makes difficult to have only one configuration for all agents, since the virtual company includes partners of various natures using of the tools and basing on distinct planning strategies.

3. To concentrate information necessary to the communication (environmental and other agents information). Thus, all updated will be carried out quickly and effectively.

4. To absorb the impact of the changes in VEN internal architecture. That means that any modification at the interior of VEN (i.e. change of the agent situation) is not reflected on global architecture. In that case, it is enough to carry out reconfigurations necessary on corresponding COMM agent.

In the communication point of view, the COMM offers the following types of communications:

3.1.1.1 Internal communications:

In fact communications are carried out at the interior of the VEN. We distinguish there:

- Communications PA/NA: All Information concerning orders, production planning changes, production cost....

- Communications PARAM/PA or NA: Each modification concerning the internal architecture of the VEN must be announced to PARAM via COMM, to carry out necessary updates.

3.1.1.2 External communications:

- Communications inter VEN (between PA of two different VEN): Information flows among various partners of the supply chain (i.e. orders, there modifications...).

- Communications NA/TNA: is the communication of a NA with the TNA of its row, which can take place for example when the NA encounters a problem, the TNA is called to try to find a solution.

- Communications between TNA: The TNA has a sight that on its own tier and in certain cases it needs to cooperate with the TNA of the adjacent tier in order to obtain useful information in local decision-making.

- Communications TNA/SCMA: The SCMA has the capacity to have a global sight of the supply chain while communicating with various TNA. The communications between the TNA and the SCMA is used, for example, to find a global resolution of a problem that could not be solving locally.

3.1.1.3 Communications related to VE configuration:

A modification of a VEN involves an information flow among various partners (other VEN, TNA, …). Those modifications can be at the interior of the VEN (ex: PA modification) or at the VEN globally (VEN withdrawal from the VE). This flow generates communications between the PARAM agents which are given the responsibility to carry out necessary reconfigurations.

3.1.2 Functionalities of the PARAM

Considering the dynamic aspect of the supply chain, our approach is based on a flexible and dynamic distribution of all the principal agents, in particular the PA and NA, thus and for a company, its agents PA and NA can be geographically distributed. For that, agent PARAM proceeds to the updates and reconfigurations each time it there is a change.

Some changes likely to launch the PARAM are:

The changes which can launch the PARAM are classified in two kinds:

- The modifications in VEN internal architecture, for example machine change of one agent (PA or NA).

- Changes in the global architecture of the supply chain, for example adhesion of new partner, or the rejection of another.

Thus, the PARAM functionalities are summarized in the following points:

1. To locate modifications in VEN internal architecture.

2. To communicate the changes with the PARAM of the adjacent VEN and TNA agent of corresponding tier in order to guarantee the information coherence for each part.

3. To listen to and receive changes from other VEN.

4. To modify local information of COMM following the locally changes located or on the total level of VE.

5. To follow the reconfiguration prescriptions of the TNA and the SCMA in the case of an important transformation in the supply-chain architecture. In this case cooperation between PARAM is not enough to detect changes. Then to update proves very delicate, for that an intervention of TNA and SCMA is necessary (i.e. the exclusion and the adhesion at the same time of several partners).

4. ILLUSTRATIONS

Now; let us give two examples to illustrate the functionalities of each one of these two agents. The first example exposes the interactions produced at the time of an order of a customer. These interactions are based on agents COMM of the various implied VEN. The second example exposes a case among several cases where agent PARAM is called. We will follow the stages necessary to reconfigure the supply chain after a departure of an enterprise.

4.1 Order life cycling

Basing on the functionalities of the COMM; the Figure 3 presents the customer order life cycling, while following his various stages:

1: the order is initiated by agent PA of the customer, which transmits the order towards the agent NA by the means of the COMM, in order to find the supplier who can satisfy it.

2: the COMM transmits the request towards its agent NA.

3: NA identifies the request and starts the phase of negotiation with available suppliers, through the COMM.

4: The two agents COMM of the customer and the supplier constitute the support of this negotiation with the exchange of various corresponding information.

5: The COMM of the supplier transmits the request to NA to establish the process of negotiation.

6: NA calls the PA through the COMM, to evaluate the load induced by this order.

7: The COMM transmits information to PA.

8: The PA evaluates the order while being based on its strategies of planning, and make the adequate decision.

9/10: Generally, this supplier cannot satisfy an order alone, and consequently it calls upon other suppliers by the means of the COMM and NA

11: the various COMM convey information exchanged between the agents.

At the interior of the VEN of the other suppliers the process of communication and interaction between the COMM, PA and NA is identical.

12: After the evaluation of the order, the second supplier sends his response to the agent NA of the first supplier via agents COMM.

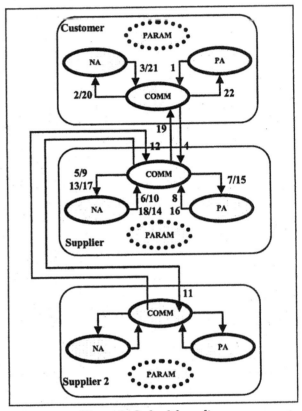

Figure 3. Order life cycling

13/.../18: In the same way, the internal process of the VEN consists of several interactions between PA and NA by the means of the COMM. This process produces the final decision of the supplier.

19/.../22: The decision will be forwarded to the customer who produced the order through agents COMM, and arrives at the end of the cycle to the agent PA.

4.2 Dynamic supply chain modification

In this part, we will illustrate an example of the exclusion of a partner of the chain. In more of the modifications on the logistic and economic level, this change in total architecture involves modifications on the functional level of the adjacent VEN. Thus, Figure 4 shows the cycle of process to eject a partner:

1: The VEN which leaves the virtual company informs the TNA of the corresponding row, of its departure.

2: the TNA propagates the information of modification towards the adjacent NEV, in more of the TNA of the close rows, which constitute the entities directly concerned by the updates to be carried out.

3: Agents PARAM receive the orders of modifications.

4: The PARAM proceed to the various updates on respective agents COMM.

5: Confirmations of the modifications, and consequently the final adoption of the new configuration.

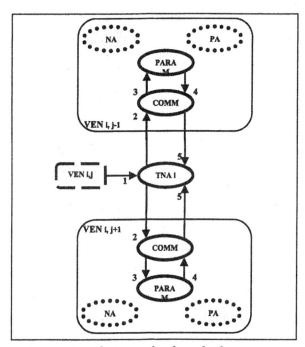

Figure 4. exclusion cycle of supply chain partner

5. CONCLUSIONS AND FUTURE RESEARCHES

The dynamics of the supply chain involves a flexibility problem in suggested solutions. For that, it is necessary to take into account the fundamental characteristics of the supply chain, and on the other hand, the advantages offered by the used tools.

In our work, we were based our architecture on the multi-agents system technology which provided aspects of cooperativeness, coherence and flexibility, necessary for the distributed information system proposed for the supply chain management.

This approach is based on a generic entity called Virtual Enterprise Node. This VEN is completed with two others entities: the TNA and the SCMA. Those entities are used to limit information flows and to aid to find solutions involving several partners.

Network communication and network configuration are managed by agents (COMM and PARAM) which are distributed among autonomous partners (VEN).

Some evolutions of this approach which is concentrated essentially on the VEN, supply chain elementary entity, will concern behavior development of TNA and SCMA. Those

developments are essential to have a complete and perform supply chain management applications.

Moreover, PARAM behaviors will be completed to take into account TNA and SCMA improvements.

6. REFERENCES

Altersohn, C. (1992). De la sous-traitance au partenariat industriel, dynamiques d'entreprises. Édition l'Harmattan.

Axelrod, R. (1992). Donnant, donnant : une théorie du comportement coopératif. Édition Odile Jacob.

Cachon, G.P., & Lariviere, M.A. (2001). Contracting to assure supply: how to share demand forecasts in a supply chain. *Management Science*, 47(5), 629-647.

Camalot, J.P. (2000). *Aide à la décision et à la coopération en gestion du temps et des ressources.* Thèse de l'Institut National des Sciences Appliquées de Toulouse.

Camalot, J.P., Esquirol, P., Huguet, M.J., & Erschler, J. (1997). Aide à la décision et à la négociation dans un problème de gestion de production distribuée. *Journées du Groupement de Recherche en Productique.* Cachan, France.

Camarinha-Matos, L.M., Afsarmanersh, H., & Rabello, R. (2003). Infrastructure developments for agile virtual enterprises. *IJCIM*, 16(4/5), 235-254.

Cauvin, A., Véron, P., & Martin, P. (2003). Analyse de projets pilotes de conception collaborative multisites – Mise en oeuvre entre différents sites de formation. *Proceedings of the 5th International Industrial Engineering Conference*, Quebec, Canada.

Ferber, J. (1995). Les Systèmes Multi-Agents – Vers une Intelligence Collective. Paris: InterEditions.

Ferrarini A., Labarthe, O., & Espinasse, B. (2001). Modélisation multi-agents de chaînes logistiques. *Proceedings of the 4th International Industrial Engineering Conference* (pp. 1165-1174), Aix-Marseille, France.

Gavirneni, S., Kapuscinski, S.R., & Tayur, S. (1999). Value of information in capacitated supply chains. *Management Science*, 45(11), 16-24.

Haurat A., & Monateri J.C. (1999), Dynamique des relations durables entre entreprises – Architectures industrielles – coordination, pilotage, performance, *Proceedings of the 3th International Industrial Engineering Conference* (pp. 505-518), Montréal, Canada.

Hardwick, M., & Bolton, R. (1997). The industrial virtual enterprise. *Communications of the ACM*, 40(9), 59-60.

Huhns, M., editor (1987) "Distributed Artificial Intelligence". Pilman Publishing/Morgan Kaufman Publishers, San Marco, CA.

Huguet, M.J. (1994). Approche par contraintes pour l'aide à la décision et à la coopération en gestion de production. Thèse de l'Institut National des Sciences Appliquées de Toulouse.

Kjenstad, D. (1998). *Coordinated supply chain scheduling*. Thèse soutenue au department of Production and quality engineering, Norwegian university of science and technology.

Malone, T. W., & Crowston, K. (1994). The interdisciplinary study of coordination. *ACM Computing Survey*, 26(1).

Monteiro, T. (2001), Conduite distribuée d'une coopération entre entreprises, le cas de la relation donneurs d'ordres – fournisseurs, thèse soutenue à l'Institut National Polytechnique de Grenoble.

Monteiro, T., & Ladet, P. (2001). Formalisation de la coopération dans le pilotage distribué des flux interentreprises, Application à une entreprise de production de biens. *Numéro spécial - pilotage distribué, JESA*, 35(7-8), 963-989.

Monteiro, T., & Roy, D. (2003). Architecture de pilotage semi-distribuée de l'entreprise virtuelle. *Proceedings of the 5th International Industrial Engineering Conference*, Quebec, Canada.

Monteiro, T., Ladet, P., & Bouchriha, H. (2004). Multi-criteria negotiation for a distributed control of a client/provider relationship. *Journal of Decision System*, 13(1), 63-89. Paris: Hermès.

Ouzizi, L., Anciaux, D., Portmann, M.C., & Vernadat, F. (2003). A model for co-operative planning using a virtual enterprise. *Proceedings of the 6th international conference on Industrial Engineering and Production Management*, Porto, Portugal.

Ouzounis, E.K. (2001). An agent-based platform for the management of dynamic virtual enterprises. Thesis in elektrotechnik und informatik des technische Universität Berlin.

Patriti, V., Schäfer, K., Ramos, M., Charpentier, P., Martin, P., & Veron, M. (1997). Multi-agent and manufacturing: a multilevel point of view. *Proceedings of the Computer Application in Production and Engineering*, Detroit, Michigan.

Rapoport, A. (1987). Game theory as a theory of conflict resolution. Kluwer Academic Press.

Rota, K. (1998). Coordination temporelle de centres gérant de façon autonome des ressources – Application aux chaînes logistiques intégrées en aéronautique. Thèse de l'Université Paul Sabatier de Toulouse et de l'ONERA.

ELSEVIER
IFAC
PUBLICATIONS
www.elsevier.com/locate/ifac

Supply-Chain Management Based on Cost

Anciaux D.[1], Roy D.[2] and Monteiro T.[1]

MACSI PROJECT – INRIA LORRAINE

[1]LGIPM/AGIp, Université de Metz, Ile du Saulcy, F-57045 METZ Cedex 1
E-mail: {anciaux, monteiro}@agip.sciences.univ-metz.fr

[2]LGIPM/AGIp, ENIM, Ile du Saulcy, F-57045 METZ Cedex
E-mail: roy@enim.fr

Abstract: The purpose of this article is to propose a cost based approach dedicated to distributed supply-chain management. This approach is based on multi agent concept and on distributed enterprise networks. Each enterprise is autonomous and must perform, in the same time, local and global goals. Those goals are cost leaded. The base component of our approach is a Virtual Enterprise Node (VEN). This VEN has to plan its production respecting internal and external constraints. This approach allows enterprise network management which is completely transparent seen from simple enterprise of the net. The used of MAS allows physical distribution of the decisional system. A mathematical model is presented. This model is based on cost sharing among supply-chain. Its goal is double, first each enterprise has to respect partners' needs and second total benefit is maximized.

Keywords: Multi-Agent System, Distributed control, Decision support systems, Supply-chain management

1. INTRODUCTION

Product design, their manufacture and conditioning, or their marketing do not result of isolated and autarkical companies but of increasingly complex corporate networks. These networks can take various forms which are as much of "Industrial Architectures".

Architecture development and effectiveness, answering in their constitution to criteria as well economic as geographical or technical, need a rigorous approach of their definition and their capacity to answer performance criterion which becomes much linked with relation between companies than each company separately considered. This performance takes into account two aspects: the information and material flows.

So, to improve its reactivity and to better manage its costs, a company which involves in such architecture might consider subcontracting and partners (Cachon, & Lariviere, 2001; Gavirneni, Kapuscinski, & Tayur,

1999). It involves a "make or buy" decision (Monteiro, Ladet, & Bouchriha, 2004). This may imply various time frames of decision: a long run decision specifies for the company the whole of its external and internal nodes of production, distribution and supply; a medium run specifies contracts (average quantities, delays, prices and penalties, etc.) that the company is likely to have with its internal and/or external providers in order to carry out a production program which reaches the best balance between cost and delay; or a short run decision considers, for example, subcontracting simply for overload capacity in order to absorb the temporary fluctuation of demand.

In those three time frames of decision, strategic, tactic and operational, we are interested, in this paper, in operational one.

We are interested in multi site resources planning management based on cost. This management has to be controlled by performance criterion which is declined as much local as global criteria. We made the choice to manage network following a "win-win" policy (Ouzizi, Anciaux, Portmann, & Vernadat, 2003).

First, the cost criteria approach and the management architecture are presented. After, we present the mathematical model. And finally we present some conclusions and future researches.

2. COST CRITERIA APPROACH

Enterprise network objective is to minimize purchasing and production costs and also to ensure a positive global benefit (Anciaux, Ouzizi, & Portmann, 2003).

The global benefit of the network is:

$$\sum_{all\ partners} selling - \sum_{all\ partners} costs \geq 0 \qquad (1.)$$

To ensure these two objectives, each enterprise may be able to establish direct and backward planning for each planning exercise.

For direct (respectively, backward) planning:

- The release time for the purchased items (respectively, deadline for the selling products) are fixed.

- The ideal due date for selling products (respectively, the ideal release time for purchased items) are computed.

- Penalties, for one time unit and one product unit, depending on the importance of some delays (respectively, advances), can be used to optimize the planning.

Two criteria are important for each enterprise planning:

- minimizing the cost of production,

- minimizing the sum of penalties that are caused by the difference between the ideal due date of the selling product (respectively, purchased item) and the due date computed.

Penalty parameters and importance for the two criteria can be set for each enterprise, so that each of them can propose several production plans. Other penalties can be asked linked to the difference between the demands and the suppliers' contracts.

3. MANAGEMENT ARCHITECTURE

The supply chain considered in our work can be summarized as follows (Monteiro, & Roy, 2003):

The supply chain is viewed as a set of tiers (Figure 1), in which each partner, called Virtual Enterprise Node (VEN), is in relation with customers and suppliers on the adjacent tiers. We assume that each VEN is only in relationship with its adjacent VENs (no loop between the VENs allowed). So, each VEN belongs to one tier. The VEN is the base component of this architecture.

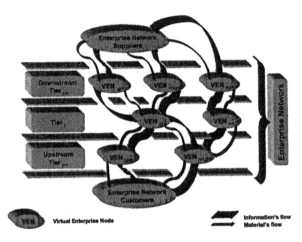

Figure 1. Supply chain architecture

The concept of virtual enterprise (VE) was introduced with an aim of widening the concept of the extended enterprise towards a concept of less centralized organization (Camarinha-Matos, Afsarmanersh, & Rabello, 2003). Contrary to the extended enterprise which is arranged around a decision center, the VE characterizes an independent consortium which links their resources for grow their reactivity regarding the unpredictable environment (Hardwick, & Bolton, 1997).

It is assumed that any component can be provided to each VEN by only one supplier VEN without the possibility of changing it by another supplier, except in the case of a long-term disagreement.

To generate a better productivity, these companies need the coordination of the actions which are distributed among autonomous partners (Altersohn, 1992; Rota, 1998; Kjenstad, 1998; Huhns, 1987). Recent research shows a growing interest in studying cooperation relationships among the multiple actors of an industrial architecture (Axelrod, 1992; Rapoport, 1987; Ferrarini, 2001; Monteiro, & Ladet, 2001b). Cooperation can take various forms. It can be defined as collaboration between partners, each having equivalent decisional capacity and acting together towards a common objective. In our approach, cooperation is used to coordinate and synchronize operations carried out by independent actors (Malone, & Crowston, 1994; Monteiro, & Ladet, 2001a). Then, each VEN's local decision might be validated with adjacent partners (client and suppliers). A tier j decision may be validated with tier j+1 and tier j-1.

3.1 Multi-agent system organization

In such architecture and cooperation organization, each partner has a limited decision power that corresponds to its action field (Camalot, Esquirol, Huguet, & Erschler, 1997; Camalot, 2000; Cauvin, Véron, & Martin, 2003; Frayret, 2004; Huguet, 1994). In our approach, the enterprise network is modeled as a multi-agent system (Ferber, 1995; Patriti, Schäfer, Ramos, Charpentier, Martin, & Veron, 1997). Agents use cooperative negotiation to establish a global consistent planning.

If planning problem is detected, an agent called Tier Negotiator Agent (TNA) is activated. The purpose of this agent is to limit the negotiation process in terms of iterations and to facilitate cooperation between VEN agents.

If the TNA cannot solve the problem at the tier level, a Supply Chain Mediator Agent (SCMA) involving whole enterprise network will be used.

Furthermore, each VEN is made of:

• A Planner Agent (PA): It is used to establish the VEN planning with different parameters (costs, penalties, etc.). In function of the enterprise planning strategy, different tools for planning could be used.

• A Negotiator Agent (NA): It receives orders, modifications of product demands and component deliveries from partner agents. It transmits requests to the planner agent and receives responses from it. Then, it transmits results to different partners. In case of local problem, this agent contacts its TNA to obtain some help.

Figure 2 illustrates the multi-agent system used in our architecture. VEN, TNA and SCMA are described in following parts.

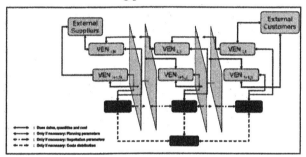

Figure 2. Agent architecture representation

3.2 VEN

In principle, each VEN is faced to internal constraints, related to its capacity limits, and to external constraints, related to:

• on the one hand, its customer VENs which require products with, for example, due date or maximum cost,

• on the other hand, its supplier VENs which also have constraints of lead times, costs, etc.

The VEN could be in two different situations depending on its capability to make or not the request. First, none consistency problem occurs. In this case, VEN is used only to propagate client needs to supplier requests. Second, local problem occurs, so a negotiation process has to be initiated.

Agents use internal constraints to make decision. They depend, in the same time, on production process and management, and on strategic assessment.

3.3 Tier Negotiator Agent (TNA)

When a VEN_{ij} of a considered tier is unable to find alone a valid planning, this VEN_{ij} forwards problem

to its TNA_j. The TNA has a total sight on its level. Thus, the TNA can be informed, by questioning its VEN, of the system state and the local blocking causes. The TNA first objective is to solve on its own tier the considered problem. For example, for first goal, TNA_j can propose a load distribution between several VENs able to carry out the blocking production source. If any solution could not be find, the second objective is to initiate and manage negotiations with its direct TNA partners. For this, TNA_j is linked with TNA_{j+1} and TNA_{j-1}. This set of negotiations allows, by recursion, problem propagation until problem resolution.

If the TNA_j cannot solve the problem, the Supply-Chain Mediator Agent (SCMA) involving whole enterprise network will be used.

3.4 Supply-Chain Mediator Agent (SCMA)

SCMA goal is to solve conflicts by relaxing constraints. This relaxation is based on global cost defined in part #2 "cost criteria approach".

While preserving benefit, SCMA authorizes local deficit on one or more VEN responsible of the blocking point. SCMA distributed this deficit on the whole of the partners, and manages penalties induced by no respecting contracts. Inference supports on the "win-win" game concept in the long run. This makes it impossible to penalize always the same company and to preserve advantages in a durable matter relation.

4. MATHEMATICAL MODEL

The planning model consists in maximizing the profit of a VEN by maximizing the sales and minimizing the various production costs (i.e. in terms of normal working hours, or over-time hours), the unemployed hours, the costs of subcontracting, provisioning, transport and the penalties of delay. The penalties of delay are true penalties only for the first period and for the customers of supply chain, and of the fictitious penalties for the future. This mathematical model is used in each VEN of the supply chain.

In addition, it is assumed that:

• The provisioning carried out for period t-1 will be useful for the production of the period t.

• Products take exactly a unit of time for moving between two nodes (i.e. what is produced, stored or moved by a node for period t can be used by the following node for the period t+1, without worrying about the exact moment when transformations are carried out for period t).

• Times of change of manufacture are negligible.

• The production is carried out in a complex workshop in which the production lines are in series (in the event of parallel production lines, the capacity of the resources are multiplied)

• The process plans are known as well as the occupancy rate of each product on each machine. In our case, our interest is on the components at the entry, the times spent on each machine and finally the products at the exit, without worrying about the intermediate products.

First of all, let us denote the notations used. Then, the planning model is presented.

4.1 Notations used:

Definition of sets:

- P: set of products identified by p ($1 \leq p \leq Np$)

- Np_i: Number of supplied products (VEN raw material) identified by i ($1 \leq i \leq Np_i$),

- Np_o: Number of delivered products (VEN finished product) identified by j, ($Np_{i+1} \leq j \leq Np$),

- C: set of VEN customers identified by c;

- S: set of VEN suppliers identified by s;

- R: set of resources identified by r;

- T: planning horizon;

- T': planning horizon plus one period.

Various costs taken into account:

- $cs_{j,c}(q)$: selling price of q units of j product for c customer (decreasing function according to the quantity q).

- $ctr_{j,c}(q)$: transport cost of q units of j finished product for c customer (decreasing function according to the quantity q).

- $cpn_{j,r}$: production cost in normal hour of one unit of j product on r resource.

- $cot_{r,t}$: overtime hour cost on r resource during t period.

- $csc_{j,t}$: one unit subcontract cost of j product during t period.

- $chu_{r,t}$: unemployed hour cost on r resource during t period.

- $csp_i(q)$: purchasing cost of q units of i product (decreasing function according to the quantity q)

- ch_j: holding cost of one unit of j finished product

Information concerning the system of production

- $g_{i,j}$: required i product quantity to manufacture one unit of j product

- $C_{r,t}$: usual production capacity, expressed in hours, of r resource for t period

- $b_{j,r}$: one unit production time of j product on r resource

- $MaxST_{p,t}$: p product maximum stock during t period

- $MaxSC_{j,t}$: j product maximum sub-contracted quantity during t period

- $MaxOH_{r,t}$: overtime hours maximum on r resource during t period

- t_0: the first period of the plan

Information concerning the customers and the suppliers

- $Pp_{i,t}$: i product quantity purchased at the beginning of t period [1]

- $D_{j,c,t}$: j product demand to be delivered to c customer at the end of t period [2]

- $ct_{j,c,t}$: unit penalty (in €/day) on the not delivered j product (differed delivery) to c customer at t period

- $MaxC_{j,c,t}$: maximum quantity of j product that a VEN is assumed to be able to deliver to c customer at t period (contract related information)

- $MinC_{j,c,t}$: minimal quantity of j product that a VEN is assumed to be able to deliver to c customer at t period (contract related information)

- $MaxS_{i,t}$: maximum quantity of i product that the VEN is assumed to receive at the beginning of t period (contract related information)

- $MinS_{i,t}$: minimal quantity of i product that the VEN is assumed to receive at the beginning of t period (contract related information)

Decision variables

- $QPp_{i,t}$: i product purchased quantity at the beginning of t period (equal to $Pp_{i,t}$ for direct planning)

- $QD_{j,c,t}$: j product delivered quantity to c customer at the end of t period (equal to $D_{j,c,t}$ for retrograde planning).

- $X_{j,t}$: j product produced quantity during t period

- $I_{p,t}$: stock level of p product at the end of t period

- $SC_{j,t}$: j product quantity subcontracted for t period

- $O_{r,t}$: overtime hours on r resource at t period

- $U_{r,t}$: unemployed hours on r resource at t period

4.2 Objective function

To take into account the time effect, i.e. the reliability of the data which the VEN possesses, an up-dating rate $\dfrac{1}{(1+\alpha)^{t-t0}}$ is introduced, where α is an actualization factor

---

[1] This quantity is a known parameter which corresponds to a strict constraint when computing a direct planning or a soft constraint when computing a retrograde planning.

[2] This quantity is a known parameter which corresponds to a strict constraint when computing a retrograde planning or a soft constraint when computing a direct planning.

$$Max \left[\sum_{t=t_0}^{T} \left(\frac{1}{(1+\alpha)^{t-t_0}} \times \left[\sum_{j \in Np_o} \sum_{c \in C} \left[\begin{array}{l} (cs_{j,c}(q) - ctr_{j,c}(q)) \times QD_{j,c,t} \\ -(ch_j \times I_{j,t}) - (csc_{j,t} \times SC_{j,t}) \\ -\sum_{r \in R}(X_{j,t} \times cpn_{j,r}) \\ -ct_{j,c,t} \times [D_{j,c,t} - QD_{j,c,t}]^+ \end{array} \right] \right] - \sum_{i \in Np_i} (csp_i(q) \times QPp_{i,t} + ch_i \times I_{i,t}) \\ - \sum_{r \in R} (cot_{r,t} \times O_{r,t} + chu_{r,t} \times U_{r,t}) \right) \right] \quad (2.)$$

where

$$[D_{j,c,t} - QD_{j,c,t}]^+ = \begin{cases} 0 & \text{if } D_{j,c,t} \leq QD_{j,c,t} \\ D_{j,c,t} - QD_{j,c,t} & \text{if not} \end{cases} \quad (3.)$$

4.3 Constraints

The constraints of the planning model are for each $i \in Np_i, j \in Np_o, t \in T, c \in C, s \in S$ and $r \in R$.

$$I_{i,t} = I_{i,t-1} + QPp_{i,t} - \sum_{j \in Np_o} X_{j,t} \times g_{i,j} \quad (4.)$$

The level of stock of products purchased (supplied components) at the end of period t is equal to the sum of the level of the stock of the preceding period and the supplied quantities at the beginning of period t less the quantities used for the production (which are due to the nomenclature on several levels).

$$I_{j,t} = I_{j,t-1} + X_{j,t} + S_{j,t} - \sum_{c \in C} QD_{j,c,t} \quad (5.)$$

The level of stock of finished products at the end of period t is equal to the sum of the stock at the preceding period, the quantities produced during the period, the quantities sub-contracted during the period less the quantities delivered to the customers.

$$\sum_{t \in T'} QD_{j,c,t} = \sum_{t \in T} D_{j,c,t} \quad (6.)$$

The VEN must satisfy the whole of the orders on the horizon with a possible delay of one period, $T' = T+1$ period.

$$\sum_{t \in T} QPp_{i,t} = \sum_{t \in T} Pp_{i,t} \quad (7.)$$

The VEN must purchase quantities of products (components or raw material) necessary for the production of finished products.

$$I_{p,t} \leq MaxST_{p,t} \quad (8.)$$

The level of stock should not exceed a maximum limit.

$$QPp_{i,t} \geq MinS_{i,t} \quad (9.)$$

This constraint ensures a minimal quantity purchased from each supplier, according to the contracts signed between the various VEN of the virtual enterprise, one recalls that the contracts obliged each VEN to give its forecasts for each period to its various suppliers.

$$QPp_{i,t} \leq MaxS_{i,t} \quad (10.)$$

This constraint ensures a maximum quantity of products purchased from each supplier, according to the contracts.

$$QD_{j,c,t} \geq MinC_{j,c,t} \quad (11.)$$

This constraint ensures a minimal quantity to deliver to each customer, according to the contracts established between each VEN and its customers.

$$QD_{j,c,t} \leq MaxC_{j,c,t} \quad (12.)$$

This constraint ensures a maximum quantity to deliver to each customer, according to the contracts.

$$O_{r,t} \leq MaxO_{r,t} \quad (13.)$$

This constraint makes it possible to limit overtime available to a maximum value.

$$SC_{j,t} \leq MaxSC_{j,t} \quad (14.)$$

This constraint makes it possible to limit the subcontracted quantities with a maximum value.

$$\sum_{j \in Np_o} b_{j,r} \times X_{j,t} + U_{r,t} - O_{r,t} = C_{r,t} \quad (15.)$$

This constraint ensures that the capacity of the resources available is equal to the sum of the operational durations and the unemployed hours minus overtime.

$$QD_{j,c,t}, QPp_{i,t}, I_{p,t}, U_{r,t}, O_{r,t}, SC_{j,t} \geq 0 \quad (16.)$$

This constraint indicates that all variables of decision are positive or null.

5. CONCLUSIONS AND FUTURE RESEARCHES

Cost based management used in our approach ensures in the same time:

- to respect at best client needs in terms of cost, quantity and delay

- to minimize global production cost among whole supply chain

Then this approach, allowing the maximization of the global benefits, shares lost and profit between partners.

This cost management algorithm uses a multi-agent architecture modeling a distributed decision system. With this architecture, enterprise decision autonomy is respected.

The mathematical model presented here, could be drifted in more specifically models to fit particular supply chain needs or management strategy.

Moreover, for now, only one supplier by product is accepted in the model. One of our future works will be to manage a multi-supplier system. In the same way, real transportation time (different of one planning time unit) could be taken into account.

Some evolutions of this system witch only concerns a supply chain, could be extended to the distribution net. Some particular architecture, as skill center, has to be studied.

6. REFERENCES

Altersohn, C. (1992). De la sous-traitance au partenariat industriel, dynamiques d'entreprises. Édition l'Harmattan.

Anciaux, D., Ouzizi, L., & Portmann, M.C. (2003). Planification d'une chaine logistique par négociation. *Proceedings of the 5th International Industrial Engineering Conference*, Quebec, Canada.

Axelrod, R. (1992). Donnant, donnant : une théorie du comportement coopératif. Édition Odile Jacob.

Cachon, G.P., & Lariviere, M.A. (2001). Contracting to assure supply: how to share demand forecasts in a supply chain. *Management Science*, 47(5), 629-647.

Camalot, J.P. (2000). *Aide à la décision et à la coopération en gestion du temps et des ressources*. Thèse de l'Institut National des Sciences Appliquées de Toulouse.

Camalot, J.P., Esquirol, P., Huguet, M.J., & Erschler, J. (1997). Aide à la décision et à la négociation dans un problème de gestion de production distribuée. *Journées du Groupement de Recherche en Productique*. Cachan, France.

Camarinha-Matos, L.M., Afsarmanersh, H., & Rabello, R. (2003). Infrastructure developments for agile virtual enterprises. *IJCIM*, 16(4/5), 235-254.

Cauvin, A., Véron, P., & Martin, P. (2003). Analyse de projets pilotes de conception collaborative multisites – Mise en oeuvre entre différents sites de formation. *Proceedings of the 5th International Industrial Engineering Conference*, Quebec, Canada.

Ferber, J. (1995). Les Systèmes Multi-Agents – Vers une Intelligence Collective. Paris: InterEditions.

Ferrarini A., Labarthe, O., & Espinasse, B. (2001). Modélisation multi-agents de chaînes logistiques. *Proceedings of the 4th International Industrial Engineering Conference* (pp. 1165-1174), Aix-Marseille, France.

Frayret, J.-M., D'Amours, S., & Montreuil, B., (2004). Co-ordination and control in distributed and agent-based manufacturing systems. *Production Planning and Control*, 15(1), 1-13.

Gavirneni, S., Kapuscinski, S.R., & Tayur, S. (1999). Value of information in capacitated supply chains. *Management Science*, 45(11), 16-24.

Hardwick, M., & Bolton, R. (1997). The industrial virtual enterprise. *Communications of the ACM*, 40(9), 59-60.

Huhns, M., editor (1987) "Distributed Artificial Intelligence". Pilman Publishing/Morgan Kaufman Publishers, San Marco, CA.

Huguet, M.J. (1994). Approche par contraintes pour l'aide à la décision et à la coopération en gestion de production. Thèse de l'Institut National des Sciences Appliquées de Toulouse.

Kjenstad, D. (1998). *Coordinated supply chain scheduling*. Thèse soutenue au department of Production and quality engineering, Norwegian university of science and technology.

Malone, T. W., & Crowston, K. (1994). The interdisciplinary study of coordination. *ACM Computing Survey*, 26(1).

Monteiro, T., & Ladet, P. (2001a). Formalisation de la coopération dans le pilotage distribué des flux interentreprises, Application à une entreprise de production de biens. *Numéro spécial - pilotage distribué, JESA*, 35(7-8), 963-989. Paris: Hermès.

Monteiro, T., & Ladet, P. (2001b). Modélisation d'une architecture industrielle pour le pilotage distribué des flux interentreprises. *Proceedings of the 4th International Industrial Engineering Conference*, Aix-Marseille, France.

Monteiro, T., & Roy, D. (2003). Architecture de pilotage semi-distribuée de l'entreprise virtuelle. *Proceedings of the 5th International Industrial Engineering Conference*, Quebec, Canada.

Monteiro, T., Ladet, P., & Bouchriha, H. (2004). Multi-criteria negotiation for a distributed control of a client/provider relationship. *Journal of Decision System*, 13(1), 63-89. Paris: Hermès.

Ouzizi, L., Anciaux, D., Portmann, M.C., & Vernadat, F. (2003). A model for co-operative planning using a virtual enterprise. *Proceedings of the 6th international conference on Industrial Engineering and Production Management*, Porto, Portugal.

Patriti, V., Schäfer, K., Ramos, M., Charpentier, P., Martin, P., & Veron, M. (1997). Multi-agent and manufacturing: a multilevel point of view. *Proceedings of the Computer Application in Production and Engineering*, Detroit, Michigan.

Rapoport, A. (1987). Game theory as a theory of conflict resolution. Kluwer Academic Press.

Rota, K. (1998). Coordination temporelle de centres gérant de façon autonome des ressources – Application aux chaînes logistiques intégrées en aéronautique. Thèse de l'Université Paul Sabatier de Toulouse et de l'ONERA.

PUBLICATIONS
www.elsevier.com/locate/ifac

PLANNING STABILITY AND DECISIONS ROBUSTNESS FOR TACTICAL PLANNING OF INTERNAL SUPPLY CHAIN

P. GENIN[1] et S. LAMOURI[1] et A. THOMAS[2]

[1] LISMMA : Laboratoire d'Ingénierie des Systèmes Mécaniques et des Matériaux
Equipe Optimisation des Systèmes de Production et Logistique (OSIL)
3, rue Fernand Hainault
93407 Saint-Ouen CEDEX France
Samir.Lamouri@ismcm-cesti.fr

[2] CRAN : Centre de Recherche en Automatique de Nancy
Faculté des sciences – BP239 –
54506 – VANDOEUVRE les NANCY.
Andre.Thomas@cran.u-nancy.fr

SUMMARY: *Supply Chain Management, considered as a tool for the synchronization of material flow, aims to control the bullwhip effect. This phenomena is mainly caused by demand uncertainty and is propagated by decision rules at the different stages of the supply chain.*
The tactical planning level is the most appropriate level to limit this demand amplification. In particular, Sales and Operations Planning (S&OP) is the first time that demand is taken into account in the planning process. This paper compares different decision rules at the S&OP level and proposes a new policy which we have called "the reference plan". According to different indicators we show that "stability" and "robustness" can be obtained at this level and will lead to a dampening of the bullwhip effect stimulus.

KEYWORDS: *Tactical planning, Performance Indicators, Stability, Robustness, Optimization, Decision rules, Linear Programming.*

In today's industrial and economic context, companies are affected by globalization. They are attempting to acquire a global view of their production-distribution system that will allow them to control the flows of material and the reduction of their global costs, and also lead to the creation of new revenues and higher profit.

Supply Chain Management (SCM) mostly focuses on using information flow relative to material, sites and market to optimize material flow through the successive steps across the SC [3]. Since the beginning, MRP II, JIT, TOC have also aimed at the same goal which was already proposed by Closed-loop MRP II to coordinate the material and information flows in the production system [25]. The centralization of data though the use of ERP (Enterprise Resources Planning) revealed the need for optimized tools to manage the logistic chain (APS – Advanced Planning Systems) [23]. But this data coordination through the supply chain strengthens the need to deal with uncertainty properly [10].

The importance and impact of uncertainty in the supply chain has been frequently discussed in published journals. Uncertainty is identified as a major influence in the behavior of the supply chain (SC). The volatility of customer demand is often identified as a root cause of uncertainty [4]. Demand is considered (in these systems) as an exogenous variable. When used in an aggregated manner, its reliability should increase [16]. Reactivity and flexibility imposed by customers increase the market-based information flow, both in frequency and in level of detail. Consequently this information is hard to take into account in classic

MRP II systems with their sequential and hierarchical structures [21].

The problem is to know how best to obtain reactivity and flexibility in mid or short term (Master Production Schedule (MPS) and Purchase and Distribution Plans) whilst ensuring commitment to the global goals defined in a tactical plan [25].

Moreover, one of the most disturbing characteristics discovered in supply chains is the bullwhip effect (BWE). The concept describes the demand amplification which arises when even small disturbances in customer demand increase as they are transferred along the SC. Forrester first describes this phenomena [8] and later Metters [17], Wilding [26], and others. One of the root causes of the BWE is the introduction of the demand, especially the use of inadequate forecasting methods [2]. Another crucial cause of the BWE is the way the demand information is processed and the planning is elaborated [14], [22]. Consequently, SCM looks for a better synchronization amongst the entities, supported by cooperation and communication reinforced between the entities of the logistic network to decrease the impact of BWE [22].

Van Landeghem and Vanmaele suggested that tactical planning is the most appropriate level to provide buffers against uncertainty based on the time period over which they fluctuate [13]. On the one hand the SC infrastructure is fixed by strategic level. On the other hand, because of the planning constraints, at the operational level there is often insufficient time to react to variations. They conclude that demand uncertainty can be handled best at the tactical level.

Their robust planning approach applies risk assessment to the tactical planning level. The authors employ Monte-Carlo simulation to evaluate different methods (periodic order review, economic order quantity, and modified period order review models) [13]. Their approach is based on the risk a particular demand occurs, changing it from exogenous to endogenous data. They show that it is possible to achieve more effective SCs with less re-planning and less safety stock. They apply it at MPS level where material flows are adjusted based on demand mix [25].

At tactical level, Sales and Operations plan (S&OP) directly adjusts material flow characteristics based on the demand volume of families for the mid to long term. As we have previously mentioned, it is the first planning level where demand forecasts are taken into account to establish supply chain plans and transmit supply forecast to suppliers [25]. This level plays a key role in bullwhip effect, propagating demand information. Consequently, the goal at this level is to determine stable and robust plans.

In this paper, we compare, at S&OP level, different decision rules to achieve robust and stable supply chain

plans considering that demand is exogenous to our model.

Alternatively, we propose another way to achieve this objective. We propose a special S&OP policy named "the reference plan". The main idea in our approach is to introduce a plan at the S&OP level which constrains plan changes so as to make it stable and robust.

A definition for "stability" and "robustness" and a performance indicator for each property will be proposed first. The description of a simplified model typically used for S&OP planning follows. It will be used to illustrate "the reference plan" approach. The different decision rules under evaluation will be explained in paragraph 4. Paragraph 5 compares them.

1. DEFINITION - *STABILITY*

The term stability is generally associated, by opposition, with that of nervousness [1], [11], [18]: De Kok and Inderfurth have defined "nervousness" as "a lack of stability in the material requirements planning" [5]. Donselaar and al compare the nervousness of the plan generated by MRP with that of their heuristics [6]. The indicator considered is the number of "re-plannings" encountered. The instability studied is the change within the periods. If a quantity appears or disappears during a period, the indicator will be incremented. However, if quantities are only modified, the indicator remains at the same level. The instability of a plan is defined by the number of modifications made on the levels of decision parameters between two successive versions of the plan.

The term stability is thus related to the number of changes in a plan from one generation to the next. Depending on planning typology, the stability indicator can be linked to one or more different variables. In our example, it is the number of iterations in calculating the plan when the "supplied quantity" has been changed.

2. DEFINITION - *ROBUSTNESS*

The term robustness is generally associated with that of risk and decision-making [12]. The underlying idea of system robustness is that the measured functions do not diverge significantly from a given value [19], [27].
According to Lee and al, robustness is relative to one or several functions results and to their dispersion caused by uncertain parameters and costs [15].
Robustness will be calculated by the standard deviation of each of the followed indicators.
In our application, the decision robustness will be calculated by standard deviation of costs of implemented decisions. Other performance indicators could have been chosen according to different planning typology:
- Net margin,
- Capital in inventories or safety stock level,
- Service level.

Taguchi reminds us that minimizing a function by setting control parameters can cause a lack of robustness [24]. He proposes a compromise evaluated by the signal/noise ratio. It can be calculated in different ways depending of the situation: function to minimize, to maximize, to get to a target.

In our problem, the objective is to minimize the cost occurred by the decisions put into practice. The signal/noise ratio, S/N, will be measured by formula 1 where y is the indicator to be robust [24].

$$S/N = -10\log(\sum \frac{1}{n} y_i{}^2) \qquad (1)$$

3. BASE MODEL DESCRIPTION

The decisional situation is the following. On a product group or family, the manager seeks to answer the following questions:

· How much to produce to smooth seasonality?
· How many to keep in inventory in anticipation of consumption?
· How many operators for each month?
· How much overtime to authorize?

Doing so, he is looking for minimizing the plan cost. Another objective is to stabilize the supply plan transmitted to its supplier. Indeed a change in it can have disturbing effects in the transport plan and its organization, because of the transfer of production from one site to another.

3.1. Notation

n: number of periods incremented during simulation,
h: horizon of the tactical plan,
p: the current period index of simulation p = 1, 2... n,
t: index of the period of the plan t = p, p+1 ..., p+h-1,
$F_p(t)$: demand forecast for period t calculated in p,
D_p: real demand for period p,
CR(t): maximum resource capacity usable in period t in production unit,
CS(t): maximum storage capacity usable in period t in production unit,
u(t): quantity of production units in period t per operator,
uo(t): quantity of production units per overtime hour in period t,

CO(t): maximum of overtime hours per operator in period t,
I_0: Beginning inventory (positive or null),
B_0: Beginning backorders (positive or null),
IM_0: Raw material beginning inventory (positive or null),
O_0: beginning number of operators (positive or null),
I_h: target level of inventory (positive or null),
R_h: target level of backorders (positive or null),
O_h: target numbers of operators (positive or null),
$C_S(t)$: storage cost per production unit in period t,
$C_B(t)$: cost of backorders per production unit for period t,
$C_H(t)$: cost of hiring one operator in period t,
$C_L(t)$: cost of one layoff in period t,
W(t): wages per operator in period t,
$C_O(t)$: cost of one overtime hour in period t,
$C_{SU}(t)$: supply cost of one unit in period t,
$C_{MS}(t)$: storage cost per unit of raw material in period t,
$OH_p(t)$: numbers of overtime hours in period t planned in period p,
$O_p(t)$: numbers of operators in period t planned in p,
$H_p(t)$: number of hired operators in period t planned in p,
$L_p(t)$: number of layoffs in period t planned in p,
$I_p(t)$: level of inventory at the end of period t planned in p,
$B_p(t)$: level of backorders at the end of period t planned in p,
$IM_p(t)$: level of raw material inventory at the end of period t planned in p,
$P_p(t)$: numbers of production units manufactured in period t planned in p,
$S_p(t)$: numbers raw material units to be supplied in period t planned in p.

3.2. Linear model

The manager seeks to optimize resource use, minimizing the costs over the whole horizon of his plan CT(p) at each period p (1) under constraints of the equations (2)-(10).

These costs are the storage cost, the cost of backorders, the costs of hiring and layoffs, wages, overtime, storage cost and supply costs of raw materials.

$$CT(p) = \sum_{t=p}^{p+h} \begin{array}{l} C_S(t) \times I_P(t) + C_B(t) \times B_P(t) + C_H(t) \times H_P(t) + C_L(t) \times L_P(t) + W(t) \times O_P(t) \\ + C_O(t) \times OH_P(t) + C_{MS}(t) \times IM_P(t) + C_{SU}(t) \times S_P(t) \end{array} \quad -\forall p \qquad (1)$$

Inventories balance:
$$I_P(t-1) - B_P(t-1) + P_P(t) = I_P(t) - B_P(t) + F_P(t) \qquad \forall p,t$$
$$IM_P(t-1) + S_P(t) = IM_P(t) + P_P(t) \qquad \forall p,t \qquad (2)$$

Material availability:
$$P_P(t) \le IM_P(t-1) \qquad \forall p, \forall t \qquad (3)$$

Number of operators balance:
$$O_P(t-1) + H_P(t) = L_P(t) + O_P(t) \qquad \forall p,t \qquad (4)$$

Resource constraint:
$$0 \le P_P(t) \le CR(t) \qquad \forall p,t \qquad (5)$$

Targets for inventory, backorders, and number of operators:
$$I_P(p+h-1) = I_h \qquad \forall p$$
$$B_P(p+h-1) = B_h \qquad \forall p$$
$$O_P(p+h-1) = O_h \qquad \forall p \qquad (6)$$

Labor constraint:
$$0 \leq P_P(t) \leq u(t) \times O_P(t) + uo(t) \times OH_P(t) \qquad \forall p,t \quad (7)$$

Storage constraint:
$$0 \leq I_P(t) + IM_P(t) \leq CS(t) \qquad \forall p,t \quad (8)$$

Overtime constraint:
$$0 \leq OH_P(t) \leq CO(t) \times O_P(t) \qquad \forall p,t \quad (9)$$

Positive variables:
$$0 \leq O_P(t), H_P(t), L_P(t), B_P(t), P_P(t), S_P(t), OH_P(t) \quad \forall p,t \quad (10)$$

The costs induced by cancellations or additional purchases are not taken into account when elaborating the plan but are calculated as a result of the decision in current period p.

3.3. Simulation process

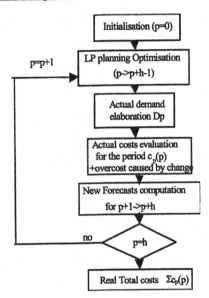

Figure 1. simulation Process

At each period p, the manager establishes his plan according to the simulated decision rule. When this plan is calculated, the first period, p, of this plan is implemented (level of production, of hiring or overtime). The actual demand for period p is established from the estimated demand from period F_{pp}, according to a normal distribution and to which we add a bias. This bias is a random number that follows a normal distribution with a mean equals to 0 and standard deviation 6 with an upward trend of 6 to model a growth in demand. Knowing the actual demand, the costs of having implemented period p are calculated.

To do the next iteration, forecasts have to be calculated. A single exponential smoothing has been used. Another forecasting method can lead to better forecasts. But since we are seeking to study the robustness of the various practices, we must create a sufficient degree of uncertainty in order to generate some variability. The smoothing constant α was fixed at 0.3. By fixing the constant at this level, the forecasts are sensitive to changes in the demand [15].

This loop is made n times doing one simulation. 10'000 simulations have been run for each decision rules.

4. DECISIONS RULES EVALUATION

In order to stabilize planning, management rules have been established. They are traditionally called "Frozen zone", in which the decision variables do not change in first time periods when re-planning [25]. This decision rule is modeled writing constraints on the variable in the first periods.

Management by exception is a new decision rule which comes with new decision support tools. Some indicators are continuously evaluated and alerts are generated when a bound is reached. In this case, an optimization of the whole planning is performed if a threshold is reached. In our model, the only existing variation is demand. The indicator is the difference between the forecast demand and the real demand. This indicator is the Forecast bias [25].

At each period and for each optimization of the basic model, all the decisions variables are adjusted according to the modifications in demand. It does not take into account the decisions taken the previous period in supplied quantities, levels of production, the possible recruitment. The new plan can therefore cancel decisions from one period to the other without taking into account the costs associated with these changes.

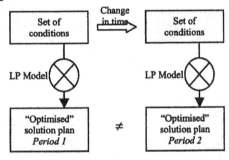

Figure 2. Common process optimization

In this way, we can reach the desired reactivity while maximizing profitability, but without being concerned with modifications made in the production-distribution and supply plans. However these practices induce strong disturbances on the productive system and on the partners in the supply chain by generating the bullwhip effect. The over-costs created by these variations are not taken into account in the system global optimization.

The reference plan (Figure 3) is the tactical plan validated by the concerned departments, production, commercial, supply chain during the S&OP meeting and implemented at the previous period. It is used as a framework to set up the new tactical plan of current period p. This plan remains a reference as long as the control factors influencing it have not evolved significantly.

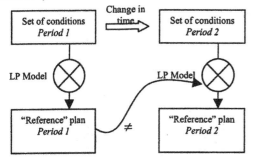

Figure 3. process optimization with reference plan

Because the manager seeks to stabilize supplies, the reference plan is the supply plan in our case. The manager prefers to keep it unchanged except for a large change in the set of conditions, i.e. the demand varies to much from the forecast.

Costs of changing the supply plan have been added in the model to take into account the reference plan in the evaluation of the new plan.

5. REFERENCE PLAN DECISION RULE ADVANTAGE

For each policy, the different indicators have been calculated. Figure 4 shows the position of each policy according of the indicators

Besides robustness and stability indicators, service performance was computed. For each period p, the service level is the quantity delivered divided by the quantity to be delivered, I_{pSP}. The quantity to be delivered is the backorders of the previous period plus the demand of the period. The service performance indicator, I_{SP}, is the mean over n simulated periods of I_{pSP}.

$$if\ I_P(t) \geq 0\ then\ I_{pSP} = 1\ \forall p$$

$$if\ I_P(t) < 0\ then\ I_{pSP} = \frac{B_{p-1}(t) + D_p - B_p(t)}{B_{p-1}(t) + D_p}\ \forall p$$

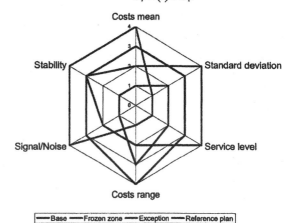

Figure 4: Simulation result synthesis

The range of variation is the difference between the lowest total costs ant the highest costs.

: best approach : second approach	Base	Frozen zone	Exception	Reference Plan
Costs mean	250 562	241 715	249 027 / 247 338	245 677
Standard deviation	32 391	36 673	28 706 / 28 540	31 820
Costs range	229 150	243 804	231 511 / 230 467	219 236
Signal/Noise	-108,050	-107,765	-107,982 / -107,923	-107,880
Service level	62%	51%	78% / 84%	71%
Stability	11	0	1 / 2*	2

Figure 5: Simulation result synthesis table

Using a frozen time fence reduces reactivity, which leads to higher costs but increases stability because of no change in decision variables. However this rule gives a low service level, a lack of robustness. Stability can lead to bad performances.

Decision rule by exception leads to the best service level and increases robustness but also costs and instability. Costs are roughly equal to base costs for the two different alert levels.

The reference plan approach is a compromise between both previous rules: low costs, good stability, robustness and signal/noise ratio, intermediate service level. Reference plan offers a short range in costs decreasing the risk undertaken by the manager.

6. CONCLUSION

In this paper we stated that S&OP is the first planning level which can impact the bullwhip effect. S&OP robustness and stability are consequently key attributes for the performance of an internal supply chain.

In our problem of S&OP performance, robustness is not an intrinsic property of a plan but described more the property of the decision rule.

We propose to modify traditional decision rules at S&OP level to take into account a reference plan, gathering variables the manager wish to stabilize. This policy delivers a better compromise between exception rules and frozen zones.

It must be noted that traditional S&OP models remain purely deterministic. Tactical plans are based on almost certain data on a short term horizon. They can correctly be calculated by deterministic approaches. However, the further one looks into the future, the more imprecise the data. As the future is uncertain, it is worthless to base decisions on forecasts used as deterministic data. At that horizon, it is often sufficient to give an approximate decision. Fuzzy logic is an approach which models uncertainty in a different manner that allows qualitative descriptions to be used such as "forecasts will be around 20 but certainly not upper 25 or below 17". Different authors have already studied fuzzy schedule problems but at the operational level [7], [9], [20]. At tactical level where uncertainty has a major impact, fuzzy rules can also be used to describe such qualitative variability. Future work will apply this method to the tactical planning model used in this article in order to obtain a "fuzzy" plan.

REFERENCES

[1] J.D. Blackburn, H.D. Kropp, R.A. Millen, "Comparison of strategies to dampen nervousness in MRP systems", *Management Science*, 32(4) :413-429, 1986.

[2] F. Chen, Z. Drezner, J. Ryan, D. Simchi-Levi, The bullwhip effect: managerial insights on the impact of forecasting and information on variability in a supply chain. In: Quantitative Models for Supply Chain Management, *Kluwer Academic Publishers*, Dordrecht :418–437, 1998.

[3] M. Christopher, "Logistics and Supply Chain Management", *Financial Times*, Pitman, London, 1992.

[4] T. Davis, "Effective supply chain management", *Sloan Management Review*, 35-46, 1993.

[5] A.G. De Kok, K. Inderfurth, "Nervousness in inventory management: comparison of basic control rules", *European Journal of Operational Research*, 103 :55-82, 1997.

[6] K. Van Donselaar, J. Van Den Nieuvenhof, J. Visschers, "The impact of material coordination concepts on planning stability in supply chains", *International Journal of Production Economics*, 68 :16-176, 2000.

[7] D. Dubois, H. Fargier, P. Fortemps, Fuzzy scheduling: Modelling flexible constraints vs. coping with incomplete knowledge, *European Journal of Operational Research*, 147 :231-252, 2003.

[8] J.W. Forrester, "Industrial Dynamics", Wiley, New York, 1961.

[9] L. Geneste, B. Grabot and A. Letouzey, Scheduling uncertain orders in the customer–subcontractor context, *European Journal of Operational Research*, 147 :297-311, 2003.

[10] P. Genin, « Planification tactique robuste avec usage d'un APS », *Thèse de Doctorat de l'Ecole des Mines de Paris*, 2003

[11] C. Ho, "Evaluating the impact of operating environments on MRP system nervousness", *International Journal of Production Research*, 27 :1115-1135, 1989.

[12] J. P.C. Kleijnen, E. Gaury, "Short-term robustness of production management systems: A case study", *European Journal of Operational Research*, 148 :452-465, 2003.

[13] H. Landeghem, H. Vanmaele, "Robust planning: a new paradigm for demand chain planning", *Journal of Operations Management*, 319 :1-15, 2002.

[14] H.L. Lee, V. Padmanabhan, S. Whang, "Information distortion in a supply chain: The bullwhip effect", *Management Science*, 43 (4) :546-558, 1997

[15] J. H. Lee, Z. H. Yu, "Worst-case formulations of model predictive control for systems with bounded parameters", *Automatica*, 33 :765-781, 1997.

[16] H. Lee, Ultimate enterprise value creation using demand-based management, *Stanford Global Supply Chain Management*, Forum Report No. SGSCMF-W1-2001 :1–12, 2001.

[17] R. Metters, Quantifying the bullwhip effect in supply chains, *Journal of Operations Management*, 15 :89–100, 1997.

[18] J.R. Minifie, R.A. Davis, "Interaction effects on MRP nervousness", *International Journal of Production Research*, 28 :173-183, 1990.

[19] J.M. Mulvey, R.J. Vanderbei, S.A. Zenios, 'Robust optimization of large-scale systems", *Operations Research*, 43(2) :264-281, 1995.

[20] E. C. Özelkan, L. Duckstein, Optimal fuzzy counterparts of scheduling rules, *European Journal of Operational Research*, 113 (3) :593-609, 1999.

[21] J. Shapiro, "Bottom-up versus top-down approaches to supply chain modelling", In: Quantitative Models for Supply Chain Management, *Kluwer Academic Publishers*, Dordrecht :739-759, 1998.

[22] D. Simchi-Levi, P. Kaminsky, E. Simchi-Levi, "Designing and Managing the Supply Chain", *Irwin, Homewood/McGraw Hill*, Boston, 2000.

[23] H. Stadtler, C. Kilger, "Supply Chain Management and Advanced Planning: Concepts Models, Software and Case Studies", *Stadtler H., Kilger C. (ed)*, Springer-Verlag, Berlin, 2000.

[24] G. Taguchi, "Orthogonal arrays and linear graph", *American Supplier Institute press*, 1987.

[25] T.E. Vollman, W.L. Berry, D.C. Whybark, "Manufacturing planning and control systems", *4th ed., New York et al.*, 1997.

[26] R. Wilding, The supply chain complexity triangle, *International Journal of Physical Distribution and Logistics*, 28 (8) :599–616, 1998.

[27] C.-S. Yu, H.-L. Li, "A robust optimization model for stochastic logistic problems", *Int. J. Production Economics*, 64 :385-397, 2000. 205, 1993.

PUBLICATIONS
www.elsevier.com/locate/ifac

ABOUT THE EFFECT OF COORDINATION AND INFORMATION SHARING ON THE PERFORMANCE IN A TYPICAL SUPPLY CHAIN.

J.P. SEPULVEDA and Y. FREIN

Laboratoire GILCO – Gestion Industrielle Logistique et COnception
ENSGI-INPG. 46 avenue Félix Viallet
38031 Grenoble CEDEX – France
Phone : (33) +4 76 57 43 20. Fax : (33) +4 76 57 46 95
{JuanPedro.Sepulveda,Yannick.Frein} @gilco.inpg.fr

ABSTRACT: This paper considers the effect of coordination and information sharing on the performance in a typical supply chain. For this, we have studied a typical generic situation. We have modeled the supply chain used in the MIT "Beer Game". We compare the performance of the supply chain through different scenarios. These scenarios are characterized by the knowledge of the customer demand, the type of coordination and the information sharing among the members. We have used linear programming methods to measure the performance of the supply chain. Then, we have quantified the gains of a global optimization in relation to the local one. We showed that a global coordination is necessary to have interesting performances, and that the coordination is much more important when there is little available information. Another result is that the advanced knowledge of the final customer demand is not so important if there is no order information sharing and coordination between the members of the supply chain.

KEYWORDS: *Information Sharing, Coordination, Supply Chain, Beer Game, Linear Programming Models.*

1. INTRODUCTION

It is not a secret that there is a direct link between the performance of supply chains and the availability and quality of timely information, Li et al (2001). Several examples from industrial practices show the positive impact of information sharing and coordination on supply chain performance. Dell utilizes online information sharing to leverage the logistics capability that can create excellent customer service, Simatupang (2001). Benetton electronically receives orders and sales information from hundreds of company agents located around the world, Simatupang (2001). Wal-Mart and Procter & Gamble share information regarding the retail sales of P&G products at Wal-Mart stores.

Within any supply chain there are many systems, including various manufacturing, storage, transportation, and retail systems. And all of these systems are connected. Specifically, the outputs from one system are the inputs to the next. So we need to consider the entire system and coordinate decisions, Simchi Levi et al (2000). If there is no coordination among the members, we have local optimizations. Each member of the supply chain optimizes its own operation without due respect to the impact of its policy on other members in the supply chain. The alternative approach is global optimization, which implies that each member identifies what is best for the entire system. One important mechanism for coordination in a supply chain is the information flow among the members, Lee et al. (1997).

The literature about information sharing and coordination in the business press is proliferating. Nevertheless, although the benefits are intuitively clear, the literature is scant on the quantifications of the benefits as well as the drivers of the magnitudes of these benefits, Lee et al. (2000).

Which is the impact of coordination and information sharing? The answer is not simple. It should be apparent that having accurate information throughout the supply chain should not make the managers of a supply chain less effective than if this information was not available, Simchi Levi et al (2000). Unfortunately, using this information effectively does make the management of the supply chain more complex because many more issues must be considered, Simchi Levi et al (2000). For instance, let us consider Electronic Data Interchange (EDI): while some firms were very happy with improved information, others were disappointed at the benefits, see Gavirneni et al. (1996). We would like to know when it is more beneficial and when it is only marginally useful.

This work aims to provide quantitative analysis of the influence of information sharing and coordination in supply chains. For this, we have studied a typical generic situation. We have modeled the supply chain used in the MIT "Beer Game". The "Beer Game", an exercise developed at MIT in 1960's, offers an easy-to-use tool for creating a common knowledge of the fundamental issues in a supply chain. In the original "Beer Game", there is no available information about the future, there is no coordination and there is no information sharing. Some consequences are: information on the demand is distorted as one moves up in the supply chain, the variability of the orders is amplified upstream in the chain (inventories and backlogs increases), the information that an upstream member "sees" is not the same that the one seen by the retailer. Lee et al (1997) called this phenomenon, the "Bullwhip" effect. In the MIT Beer Game only the Bullwhip effect can be demonstrated, while the effects of supply chain strategies cannot be shown, Hieber and Hartel (2003).

Our motivation is to demonstrate the effects on supply chain performance of the types of coordination, the advance knowledge of the customer demand and order information sharing among the members. Hieber and Hartel (2003) dealt with a similar objective by simulation analysis. In our paper we consider analytical methods of optimization, specifically linear programming.

We define different scenarios and we consider different types of advance knowledge of the customer demand. For example, in one of the scenarios only the first member in the chain knows the final costumer's demand during the studied horizon and upstream members have to use forecast. And for each of these situations we consider global or local coordination strategies. In our work, we analyze quantitatively the influence of these scenarios on the total cost of the supply chain.

First (section 2), we present in detail the studied model. We present then (section 3) the different scenarios. For each scenario we describe in section 4 the algorithms used to solve the different optimizations problems. We provide the used data in section 5. We give the different results in section 6. Finally the conclusions are presented in section 7.

2. THE MODELING FRAMEWORK

The Beer Game is a role-playing simulation developed at MIT in the 1960's to clarify the advantages of taking an integrated approach to managing the supply chain.

It's a supply chain consisting of a single retailer, a single wholesaler that supplies the retailer, a single distributor that supplies the wholesaler, and a single factory with unlimited raw materials that makes the beer and supplies the distributor. Each member in the supply chain has unlimited storage capacity, and there is a fixed supply lead time and order delay time between each member. Also, players cannot share any information beyond what's conveyed by orders and shipments. All four participants know only what they have on hand, and the orders they have to fulfill. Each week, each member in the supply chain tries to meet the demand of the downstream member. Any orders that cannot be met are recorded as backorders and met as soon as possible. At each period, each member of the supply chain is charged

a shortage cost of 1.0 per backordered item. Also, at each period, each member is charged an inventory holding cost of 0.5 per inventory item that it owns. The goal of the retailer, wholesaler, distributor and factory is to minimize the total cost, either individually or for the system. Some typical behaviors of the supply chain can be seen in Lee et al (1997).

Our model does not differ greatly from the original Beer Game in structure and operation. We have changed the unit holding and shortage costs and we allow information sharing and coordination among the members. Also, and maybe the most important difference, is that our approach is analytic.

In order to explore the behavior of a supply chain more deeply, a linear programming model of the Beer Game has been developed. This model allows the development of different information sharing and coordination scenarios for optimizing the system. We have used CPLEX 9.0 for the optimization.

3. DEFINITION OF THE SCENARIOS

There are several alternatives of information sharing and coordination for our supply chain. We have proposed different scenarios. Each scenario is defined by three characteristics. (See figure 1):

- First, we distinguished different levels of advanced knowledge of the customer demand for the retailer at the instant t=0. Demand information sharing exists when the retailer shares this information with the others members.
- Then, we divided the scenarios according to the type of coordination among the members: a global optimization of the supply chain or a local optimization of each one. (Type of coordination)
- Finally, in the case of local optimizations, we divided the scenarios according to the advanced knowledge about the orders from the immediate downstream member. If a member shares his information in advance we have order information sharing. This means, order information sharing is when each member of the supply chain knows the orders from its immediate downstream member in advance.

All the members of the chain have some knowledge about the behavior of the demand: that it is uniformly distributed between 4 and 8. This information is known and shared by all the members in the 8 different scenarios.

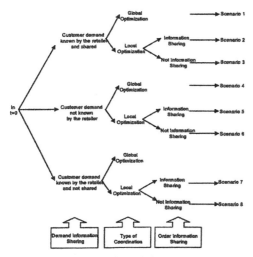

Figure 1. Definition of the scenarios

These 3 levels lead therefore to the 8 scenarios presented below:

a) Scenarios where the customer demand during the studied horizon is known in advance by the retailer and shared with all.

Scenario 1:
- Complete coordination among the members of the supply chain.

Scenario 2:
- Each member of the supply chain optimizes its own operations with transmission of information about the orders to the upstream member at t=0 (order information sharing).

Scenario 3:
- Each member of the supply chain optimizes its own operations, but without order information sharing among the members. Each upstream member uses the customer demand during the entire time horizon to make their forecasts.

b) Scenarios where the customer demand is unknown for all. In scenarios 4, 5 and 6 each member uses the demand available information to make their forecasts (only that the demand is uniformly distributed between 4 and 8). In this paper, we are not interested in the forecast techniques. It is for this reason that in our work we use simple forecast techniques. We only use a constant number during the entire time horizon. This number may be 4 or 6 or 8.

Scenario 4:
- Complete coordination among the members of the supply chain. We use forecast to estimate the customer demand.

Scenario 5:
- Each member of the supply chain optimizes its own operations and there is order information sharing. Only the retailer uses forecasts during the entire time horizon. The others use the partial information transmitted from downstream members about the orders and the forecast for the remaining periods.

Scenario 6:
- Each member of the supply chain optimizes its own operations and there is no order information sharing among the members.

c) Scenarios where the customer demand is known by the retailer and it is not shared. For each scenario each upstream member uses the demand available information to make their forecasts. For this level, it is not necessary to study the global optimization scenario, because if the retailer doesn't share the information of the customer demand, the global optimization doesn't have much sense. Also, if we have order information sharing among the members, the scenario 7 will be the same one that the scenario 2. The retailer at time t=0 send the information about the orders for the entire time horizon to the wholesaler, after the wholesaler at time t=0 makes the same thing with the distributor and so on.

Scenario 7:
- Each member of the supply chain optimizes its own operations and there is order information sharing.

Scenario 8:
- Each member of the supply chain optimizes its own operations and there is no order information sharing among the members.

4. MODEL FORMULATION AND ALGORITHMS

4.1. Linear programming model.

For our work, we have used a linear programming proto-type to solve the different scenarios.

We present here the model for the scenario 1, global optimization with total knowledge of the customer demand. The models of the other scenarios use the same principles of this one.

The objective function of our problem is the minimization of the costs. We consider the inventory and backlog costs for each member of the supply chain in each period of time. The objective function is:

$$MinCt = \sum_{t=1}^{N}\sum_{i=1}^{4} CS_{i,t} * S_{i,t} + \sum_{t=1}^{N}\sum_{i=1}^{4} CB_{i,t} * b_{i,t} \quad (1)$$

Where,
$CS_{i,t}$: Unit inventory holding cost for the member i of the chain in the period t.
$CB_{i,t}$: Unit backlog cost for the member i of the chain in the period t.
$S_{i,t}$: Units in stock of the member i at the end of the period t.
$b_{i,t}$: Units in backlog for the member i at the end of the period t.
i=1 (Retailer), 2 (Wholesaler), 3 (Distributor), 4 (Factory).
N= number of periods.

We have divided the constraints in the following way. For the Wholesaler and the Distributor:

$$S_{i,t} = S_{i,t-1} + Sh_{i+1,t-Lp} - Sh_{i,t} \qquad (2)$$

$$Sh_{i,t} \le S_{i,t-1} + Sh_{i+1,t-Lp} \qquad (3)$$

$$Sh_{i,t} \le X_{i-1,t-Li} + b_{i,t-1} \qquad (4)$$

$$b_{i,t} = X_{i-1,t-Li} + b_{i,t-1} - Sh_{i,t} \qquad (5)$$

Where,

L_i: Information lead time.

L_p: Production and transport lead time.

$Sh_{i+1,t-Lp}$: Received units at the instant t from the member i+1 of the supply chain that was send in the period t-L_p.

$Sh_{i,t}$: Units sent by the member i of the chain in the period t.

$X_{i-1,t-Li}$: Demand of the member i-1 received in the instant t that was send in the period t-L_i.

For the retailer, we maintain the equations 2 and 3, and we change in the equations 4 and 5 the term $X_{i-1,t-Li}$ by D_t, where D_t is the customer demand in the period t.

For the Factory, we maintain the equations 4 and 5, and we change in the equations 1 and 2 the term $Sh_{i+1,t-Lp}$ by $X_{i,t-Lp}$.

4.2. Resolution methods

We describe here the resolution methods used to solve the different scenarios. For the scenario 1 we used the linear programming model of the section 4.1. For the scenarios with local optimization we solve a linear programming model for each member.

In local optimization scenarios, the model outputs (quantity orders) of a member are the inputs for its upstream member.

Each member solves an optimization problem with the available information. When the member knows only that the demand is uniformly distributed between 4 and 8, he must use forecasts. In all the scenarios with forecast each member uses the following algorithm:

1. *In $t=t_o$ at the beginning of the period, we solve the optimization problem for the entire or remaining time horizon with the forecasts.*
2. *We obtain the order quantity ($X_{i,to}$) for the upstream member in $t=t_o$.*
3. *In $t=t_o$ at the end of the period, we know the realization of the customer demand (or the orders from the downstream member) for t_o. We actualize the stock ($S_{i,to}$), the shipments ($Sh_{i,to}$) and the units in backlog ($b_{i,to}$).*
4. *$t=t_o+1$*
5. *Repeat until $t=N$ periods.*

In scenario 5 each member solves a local optimization problem with order information sharing. Each member has more available information to make his optimizations. Because, the order quantity of the member "i" ($X_{i,to}$) is transmitted instantly at time $t=t_o$ to the upstream member although information lead time exists.

5. USED DATA AND EXPERIMENT DESIGN

5.1. Used Data

We consider unit inventory holding and backlog cost equal to 1(€/week) and 2(€/week) respectively for all the members. The information and production/transport lead time is equal to 2 weeks. We use the same data for all the members of the supply chain. The time horizon is 36 weeks (N=36 weeks).

The costumer demand in the first four periods is equal to 4 (units/week). In the next periods, we work with a customer demand during the entire time horizon obtained from an uniform distribution with values between 4 (units/week) and 8 (units/week). This information is known by all the members in each scenario.

We work with and without initial conditions. In the situation with initial conditions we suppose that the inventory units are equal to 12 (units/week) during the first four periods. The shipments and quantity orders in the same periods are equal to 4 (units/week). Finally the backlogs orders are 0 (units/week). These data are the same for all the members of the supply chain. In the situation without initial conditions, the values of the variables in the first periods are determined by the results of the optimization.

We want to know if there are qualitative differences among the results with and without initial conditions.

5.2. Experiment Design

To validate our results, we have tested our model using 100 different sets of data for each one of our 8 scenarios with and without initial conditions.

We have realized different applications. First, a base application that considers the data mentioned in the last section, called "application 1".

The "application 2" is an application that consider the same data as in the application 1 except for the unit backlog cost of the retailer which is of 0.1 (€/week) instead of 2(€/week). In "application 3" we changed the unit inventory and backlog cost of the retailer making it of 10 (€/week) instead of 1(€/week) for the inventory cost and of 1 (€/week) instead of 2(€/week) for the backlog cost. We want to know if there are qualitative changes in the results when we use different unitary costs in the supply chain. In previous tests we have found that the unitary costs having a bigger impact on the performance of the supply chain are those of the retailer. For that reason, the applications 2 and 3 consider only changes in the unit costs of the retailer.

In "application 4" we used limited production capacity for each member of the supply chain. The capacities are: 7 (units/week) for the retailer, 6 (units/week) for the wholesaler, 5 (units/week) for the distributor and 5 (units/week) for the factory. We do it this way because we want to work with a more real scenario where each member has particular constraints in his own production system.

6. RESULTS

In the following tables there are a summary of the obtained results for the 8 scenarios in the 4 applications.

Base application: The results correspond to the average of the 100 different sets of data.

Scenarios	Applic.1
Scenario 1	0
Scenario 2	0
Scenario 3	1184
Scenario 4	146
Scenario 5	441
Scenario 6	512
Scenario 7	0
Scenario 8	498

Table 1. Summary of results without initial conditions for application 1.

The total cost of the supply chain is zero when we have global coordination and total knowledge about the customer demand. (Sc1=0)

When there is total knowledge about the customer demand, global coordination is equal to local coordination with order information sharing. This means that, if the customer demand during the entire time horizon is known by all the members and there is order information sharing among the members, there is no difference between a global and local optimization. (Sc1=Sc2)

In the other cases, global coordination is always better than a local one. We obtain savings between 20% and 70% on the costs. (Sc1 better than Sc3, Sc4 better than Sc5, Sc4 better than Sc6).

If we have total knowledge about the customer demand, there is no difference between global and local coordination, but if we don't know this information, a global coordination implies at least savings of 67% in relation to the local coordination. It is better to work together when we don't have information than when we have total information. Said with other words, coordination is much more important when there is little or no information. This is maybe, the most important result we obtained. (Sc1=Sc2 vs. Sc4 better than Sc5)

The performance of the supply chain is marginally better if the retailer knows the demand but he doesn't share it with the other members. The savings are only of 3%. (Sc8 better than Sc6)

Order information sharing is a good strategy when no one knows in advance the customer demand. The savings are of 13%. (Sc 5 better than Sc6)

Modification of the evaluation criteria: We consider the applications 2 and 3. In the application 2, Cbacklog_retailer=0.1 (€/week) and in the application 3, Cbacklog_retailer=1(€/week) and Cstock_retailer= 10 (€/week).The results correspond to the average of the 100 different sets of data.

Scenarios	Applic.2	Applic.3
Scenario 1	0	0
Scenario 2	0	0
Scenario 3	878	1023
Scenario 4	55	366
Scenario 5	311	589
Scenario 6	366	648
Scenario 7	0	0
Scenario 8	381	436

Table 2. Summary of results without initial conditions for applications 2 and 3.

If we change the evaluation criteria, in our case, the retailer's unitary costs, there are not qualitative changes in the results. Even the global coordination is equal to local coordination with order information sharing when we have total knowledge about the customer demand. (Sc1=Sc2)

Limited production capacity: The capacities are 7 (units/week), 6 (units/week), 5 (units/week) and 5 (units/week) for the retailer, wholesaler, distributor and factory respectively. The results correspond to the average of the 100 different sets of data.

Scenarios	Applic.4
Scenario 1	254
Scenario 2	482
Scenario 3	1307
Scenario 4	393
Scenario 5	1135
Scenario 6	1197
Scenario 7	482
Scenario 8	1192

Table 3. Summary of results without initial conditions for application 4.

The total cost of the supply chain is bigger than zero when we have global coordination and total knowledge about the customer demand. (Sc1>0)

Global coordination is better than local coordination with order information sharing and total knowledge about the customer demand. This is an interesting result because in all the other applications we had found that there was not difference among these scenarios. Therefore, in a limited production capacity environment, global coordination is always better than local coordination (Sc1 better than Sc2)

In the other scenarios, there are not qualitative changes in the results.

When we work with initial conditions, we find that there are not qualitative changes in the results in all the applications. Except that the total cost of the supply chain is bigger than zero in scenarios 1 and 2. Anyway the supply chain reaches a permanent régime in a few periods with zero stock and backlog.

7. CONCLUSIONS AND PERSPECTIVES

In this work we were interested in the influence of the coordination and information sharing between supply chain members.

We were interested in 2 types of information sharing: demand information sharing and order information sharing. Demand information sharing is when each member has, in advance, full information about the customer demand for the entire time horizon, and order information sharing is when each member of the supply chain knows in advance the orders from its immediate downstream member. The main results are:

A global coordination among the members is better than a local coordination, especially if we have limited production capacity.

The coordination is much more important when there is little or no information. If we work together, the savings are bigger when we have no information than when we have total information.

Information sharing improves the supply chain performance, but we must make a distinction in which type of information. Sometimes having more information is not reflected as an improvement in the performance of the supply chain. To have more information about costumer demand is not important or significant if there are no other types of collaboration among the members (a global coordination or order information sharing).

Our work will continue with a generalization of these results for different structures and configurations of supply chains.

REFERENCES

Gavirneni et al. Value of Information in Capacited Supply Chains. Management Science, 45(1):16-24, 1999.

Grean M. and M. Shaw. Supply-Chain Integration through Information Sharing: Channel Partnership between Wal-Mart and Procter & Gamble. Working paper, Center for IT and e-Business Management at University of Illinois at Urbana-Champaign.

Hieber R. and I. Hartel. Impacts of SCM order strategies evaluated by simulation-based "Beer Game" approach: the model, concept, and initial experiences. Production planning and control. 14(2):122-134,2003.

Lee et al. Information Distortion in a Supply Chain. The Bullwhip Effect. Management Science, Vol.43, N°4, April 1997, pp.546-558.

Lee et al . The Value of Information Sharing in a Two-Level Supply Chain. Management Science, 46(5):626–643, 2000.

Li J.et al. The Effects of Information Sharing Strategies on Supply Chain Performance. Working paper, Center for IT and e-Business Management at University of Illinois at Urbana-Champaign, October 2001.

Simatupang T. and R. Sridharan. A Characterisation of Information Sharing in Supply Chains. ORSNZ Conference Twenty Naught One, University of Canterbury, Christchurch, NZ, 2001.

Simchi-Levi, Kaminsky et Simchi-Levi. Designing and Managing the Supply Chain. McGraw-Hill. 2000.

Sterman, J.D. Teaching Takes Off - Flight Simulators for Management Education. OR/MS Today 19(5):40-44, 1992.

ELSEVIER
IFAC
PUBLICATIONS
www.elsevier.com/locate/ifac

Quality Management and Metrology in Modern Production Plants Supported by Artificial Intelligence and Information Technology

A. Afjehi-Sadat [1], P.H. Osanna [1], N.M. Durakbasa [1], J.M. Bauer [2]

[1] Vienna University of Technology, Karlsplatz 13/3313, A-1040 Vienna, Austria
[2] National University of Lomas Zamora, Buenos Aires, Argentina

Abstract: To meet market demands in present and future global industrial world, manufacturing enterprises must be flexible and agile enough to quickly respond to product demand changes. With support of artificial intelligence and modern information technology it is possible to realise modern cost-effective customer-driven design and manufacturing taking into account the importance and basic role of quality management and metrolog.y. This will be especially possible on the basis of an innovative concept and model for modern enterprises the so-called Multi-Functions Integrated Factory (MFIF) that makes possible an agile and optimal industrial production.

Keywords: Quality engineering, quality management, metrology, information technology, artificial intelligence, MFIF, MFP, IPS

1 PREAMBLE

To meet high-level demands both from industrial and from private customers in the future, manufacturing enterprises must be flexible and agile enough to respond quickly to product demand changes. New models and alternative configurations for future enterprises and all kind of industrial organisations in general which are usually applied need to be investigated.

Those can be developed on the basis of intelligent production technology, information highway, distributed computing environment (DCE) technology, parallel-processing computing and advanced engineering data exchange techniques (Osanna & Si, 2000). By this means global competitive factories with intelligent, associative, concurrent, interactive, collaborative, modular, integrative, learning, autonomous, self optimising and self organising functions are already under development and will be used world wide in the future.

2 DIVISION OF LABOUR AND COLLABORATION IN MODERN PRODUCTION PLANTS

An innovative concept and model for future enterprises is the Multi-Functions Integrated Factory MFIF which is initiated with the aim to provide cost-effective, agile and optimum ways to produce customer-driven Multi-Functional Products (MFPs) in the near future (see Figure 1). By means of information technology and artificial intelligence, factories which for instance produce cars, aircrafts and ships respectively, for instance, could be linked to each other to form a new kind of factory with all three functions according to needs. The product - MFP - will be produced in such a way that the different function tasks of the product should be manufactured in adequate function factory, and then assembled and integrated to realize the combination of the functions. The factory works by using its advantages of multi-functions, and produces high efficiently and agilely low cost customer-driven multi-functional products (Si & Osanna, 1995).

Such MFIF has the potential to improve industrial competitiveness, fully manufacturing automation and optimally to manufacture the customer-driven MFP worldwide. Intelligent manufacturing systems (IMS) are the basis for realization of MFIF. In MFIF individual functional enterprises are functionally and configurationally integrated with other functional enterprises located in different parts of the world to produce MFPs respectively. This concept of MFIF will come into existence in the near future and will be realised step by step. One feature of MFIF is the use of cross-functional design and manufacturing teams, in which engineering staffs with different skills and expertise work together on a MFP project concurrently, collaboratively and interactively.

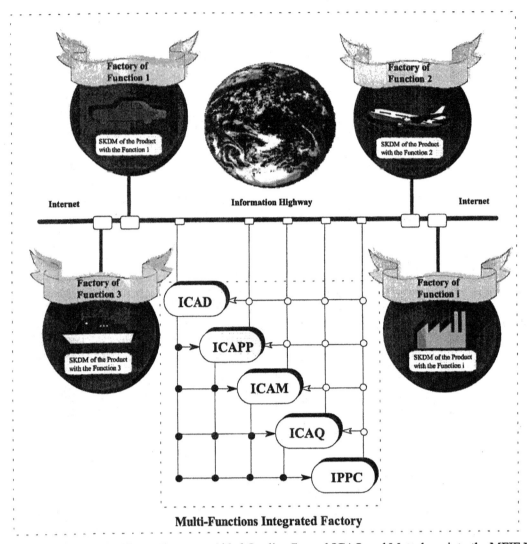

Figure 1. Integration of Intelligent Computer Aided Quality Control ICAQ and Metrology into the MFIF Model

MFIF is based on the assumption that it works only under the condition that the each single-functional factory has a possible full-scale IMS working environment and is an integration of intelligent manufacturing machines, cells and systems. Concurrent, interactive, collaborative, modular, integrative, learning, autonomous, self optimising and self organising functions are the main features of MFIF.

The factories are reconfigurable to take advantages of agile manufacturing production for the MFPs. The MFIF provides a function-business-shared feature to create new customer-driven markets. MFIF is controlled and arranged by collaborative activities between the individual factories. Cooperative activities between all units of the factories can also be concurrently on-line carried out. Learning is carried out step to step by using the methods of evolution, and used to optimise process control. It is possible that all the systematisation knowledges for design and manufacturing (SKDM) of each factory in MFIF and all production

information between ICAx systems of MFIF can be exchanged simultaneously on-line.

Failurefree concurrent exchange of production information and data, concurrent processing and executing of production processes through distributed computing environment DCE, STEP (ISO, 1994) or new standards, and learning of the all collaborative production processes are also features of MFIF.

3 ARTIFICIAL INTELLIGENCE BASED METROLOGY AND QUALITY MANAGEMENT

The quality assurance process will be carried out simultaneously in all product realisation steps in future factories - from the design to the assembly. In order to realise automatic quality assurance and to deal with complex, variable and dynamic quality control problems of the production processes in this environment, the quality assurance system will be enhanced comprehensively through self-optimising processes thrust in the design system and all manufacturing production processes in MFIF. Quality management (QM) and quality assurance (QA) with intelligent, associative, concurrent, interactive, collaborative, modular, integrative, learning, autonomous, self optimising and self organising functions will be realized in MFIF.

The assembly process is the final manufacturing step for MFPs in MFIF. This process will be carried out in one of the individual factories which is near to the customer. The coupling areas of the different function parts of MFP are produced for ease-assembling, and they can be very easily combined and put together. By means of quality management and quality assurance used in all manufacturing steps - from design to assembly - it will not be necessary to arrange a final quality assurance for completed MFPs.

Learning with self improving ability makes possible the way to "Zero Error" production. A method has been developed enabling the supervision of quality in the process chain as well as its optimization by means of a knowledge and neuronal-network-based learning management system, and a self-learning system with neuronal networks has been realised (Westkaemper, 1994). The method can be used in MFIF and permits to learn stepwise from deviations and to improve the processes continuously.

In the ICAQ system Fuzzy Logic will be applied for Quality Function Deployment (QFD) and for monitoring and forecasting of maintenance of measuring instruments, furtheron for CAD and for a high-level knowledge-based expert system for tolerancing and quality planning.

In order to get concurrent, interactive and associative quality management and quality assurance and to optimise the quality assurance process through itself, learning method for quality assurance processes must be used, and the ICAQ components of the factories in MFIF link all process steps to each other from design model to all other systems, so the quality management and assurance informations for MFP's production are automatically regenerated when the design is changed or quality assurance activities are modified in one of the processes for MFP production.

4 QUALITY MANAGEMENT LINKED TO INTELLIGENT DESIGN AND PRODUCTION

In MFIF, the design tasks of MFP can be carried out intelligently, interactively, concurrently and collaboratively. The quality and quality assurance data exchange of the design work of MFP can be guaranteed by using the following technologies

Distributed design system (Krause, 1994) can be used for multi-functional product design and quality assurance for the design. Using DCE and parallel-processing technologies, the Engineers can work on parts of a design task, but the content of the design as a whole is a corporate resource to be managed and secured. This system makes it possible that the product designers of different function factories can work parallelly on the all subtasks of the product.

Collaborative working method in MFIF has the goal to realise not only electronic data exchange function but also an interactive working function on-line, and to work at the different places and at heterogeneous systems out the same product model. Transmission of words, figures and sound by means of multimedia will also be integrated in such interactive CAD systems which could recognise manual drawings, learn the design process of the product, even understand the natural language instruction for the design, and optimise the design process and design quality. This kind of process would be guaranteed by modern data communication technologies and intelligent quality management and quality assurance as described in above. The design quality and quality assurance data exchange of the design work of MFP can be guaranteed by using the already mentioned techniques.

An effective use of analysis, simulation and visualisation tools gives several advantages for MFP design and quality assurance of the design. In this system the designers from the different single-functional enterprises use typically parallel-processing, virtual reality and virtual prototyping technologies, to design and simulate the customer-driven MPFs and their systematical function activities as well as to create a fully digital MFP production and programs for theentire manufacturing process.

In the ICAM system, the quality management and quality assurance information and programs will also be on-line exchanged and modified concurrently, interactively and collaboratively. It runs autonomously according to the adequate functional qua7lity assurance tasks and organises all manufacturing quality activities and units optimally in adequate factories. Learning the processes from the processes, the quality assurance parts of the ICAM units improve the all manufacturing process quality assurance parameters continuously.

The implementation of all these properties in an intelligent quality management and assurance architecture is a great challenge, and distributed, decentralised, self-organised and self-optimised concepts will be the main approach for this goal.

5 MEASUREMENT AND QUALITY MANAGEMENT IN GLOBAL FACTORIES

5.1 Intelligent Metrology and Intelligent Quality Assurance

Every enterprise in MFIF has a computer-integrated and intelligent manufacturing environment, and utilises the integrated ICAQ system with intelligent metrology with e.g. ICMM (intelligent co-ordinate measuring machine) to test the product or to scan and digitise complex product models with freeform surfaces, in order to obtain the digital model of the product and to modify it in ICAD system and then to create a new modified freeform surface model and NC programs for manufacturing the end product by machine center in the workshop.

ICM (Intelligent Co-ordinate Metrology) in MFIF is a very important tool to solve various problems of quality management and quality assurance in MFP production especially when high flexibility and high accuracy are demanded. This way of metrology is the uptodate measuring method for complex dimensional and geometrical measuring problems.

Figure 2 shows as nonconventional example the evaluation of measuring data of an artificial tooth for human teeth prosthesis delivering as result a complex dimensional and geometrical measurement model.

In MFIF, CNC-controlled ICMMs are connected by using networks with design and manufacturing. The goal is to mutually use the data stored in ICAD, ICAM and ICAQ systems, and to realise data parallel-processing. For the concurrent production

and the quality management system, it is suitable to use off-line programming technique, through which CNC inspection programs can be worked out without using the CMMs and the products. By means of this technique, the quality assurance data and inspection CNC program for ICMMs can be simultaneously generated during the product design.

Because of world wide needs for customer-driven MFPs, a global concurrent quality assurance system must be used with the support of internet and parallel processing computer technology. Internet makes it possible to establish a global quality assurance information highway for simultaneous on-line MFP quality assurance data exchange in MFIF environment, and to interact with suppliers and customers world wide.

Off-line programming for ICMM in ICAD system and in special programming software is the basis for simultaneous quality assurance in the individual enterprises but also in global MFIF. Many off-line programming packages as well as ICAD/ICAM/ICAQ system architecture are typical combinations in the integrated factory. ICAD/ICAQ data communication technique will be widely used in MFIF. On the basis of computer aided measurement technique and especially co-ordinate metrology quality management is integrated in the production information network.

An off-line programming package based on 3D-CAD model that represents nominal data of products can be used for the application. The probe configurations can be selected through the created probe database. The operator can call all regular element measuring functions and the actual data

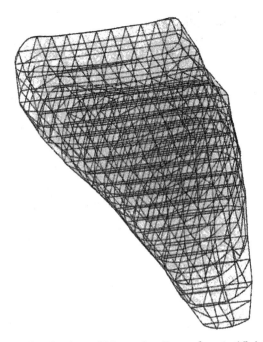

Figure 2. Evaluation of Measuring Data of an Artificial Tooth

evaluation functions, using main dialog menu of the package. On this basis measuring programs and the probe paths can be simulated, edit and optimised. During the simulation a CNC measuring program is generated in a specific format. Additionally a collision control function is realised through simulating the measuring processes on the computer monitor.

5.2 Application of Optoelectronic and Nanotopographic Methods

Besides co-ordinate metrology modern optoelectronic methods are important measurement tools in computer integrated production plants and also as basic tools for quality management and assurance activities in MFIF. Their efficient use and correct calibration are crucial requirements for quality management in this environment. Figure 3 gives an overview on optoelectronic methods for dimensional and geometrical measurements (Pfeifer & Moehrke, 1991).

Presently exists the general development from micro technology to "nano technology". Nano technology describes new innovative manufacturing technologies, finishes, tolerances and especially measurement technique in the nanometer range (Taniguchi, 1974; Whitehouse, 1991).

In persecution of this aim since about 1982 new high resolution and high precision measuring devices have been developed, especially Scanning Tunnelling Microscopy (STM) (Binnig & Rohrer, 1982) and Atomic Force or Scanning Probe Microscopy (AFM, SPM). For highest demands these methods make it possible to explore atomic structures and in general very accurate and small industrially produced parts and structures (Ichida & Kishi, 1993). With scanning tunnelling and scanning probe microscopes lateral resolutions up to 10 nm and in vertical direction up to atomic resolution are achieved (Figure 4).

As additional example the following Figure 5 shows the measurement data of a small part of an integrated electronic circuit.

It is emphasised, that in this respect applications in micro electronics do not stand in the focal point. Rather instruments of mechanical engineering and particularly precision engineering are addressed in the first hand. Extremely high accuracy demands deposit presently already at highly developed instruments for everyday use as there are VCRs or CD-players and in the sensor technique in automotive engineering and even in the home appliance if we think on one-hand mixing taps which demand ultra precision form tolerances.

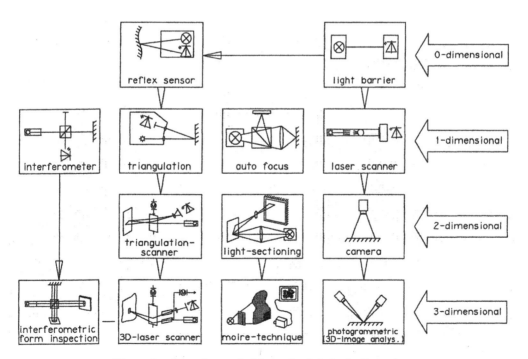

Figure 3.   Optoelectronic Measuring Methods, Overview

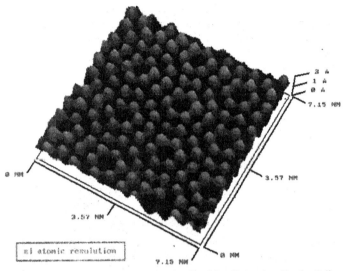

Figure 4. Atomic Structure Topography Obtained by Scanning Probe Microscopy

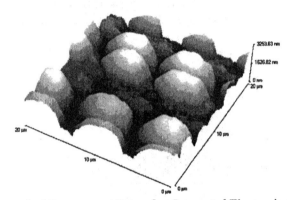

Figure 5. Measurement Data of an Integrated Electronic Circuit

6 PROPOSED CONCEPT OF AN INTELLIGENT QUALITY ASSURANCE SYSTEM

For the intelligent flexible automation of quality management and assurance, data collection and evaluation in single functional enterprises a proposed system in the form of an intelligent measuring cell can solve the following tasks:

- automatic intelligent measurement by using CNC measurement programs,
- off-line CNC programming of measurement devices,
- automatic changing of workpieces,
- automatic probe changing,
- automated evaluation of measuring results.

Figure 6 shows the principal structure of such an Intelligent Quality Assurance Cell according to the above given definition. It consists of a series of devices and components:

- a local area network of various PCs especially for ICAD, ICAE and ICAQ evaluation,
- a precision intelligent CNC dimensional measuring instrument with control computer,
- a probe changer with interface and control computer,
- a robot for workpiece manipulation,
- various measuring instruments, for instance a small CMM and other devices,
- a scanning probe microscope to evaluate surfaces in the submicrometer and atomic range,
- printers for data and graphic output,
- database systems for construction data, measuring results and quality data etc.

The proposed solution can be seen as a further step with the goal to achieve intelligent and economical MFP manufacturing, inspection and quality management in MFIF, especially in small and medium sized multi-function integrated entprises, and to find flexible solutions for all kinds of measurement problems in an automated intelligent manufacturing environment in MFIF.

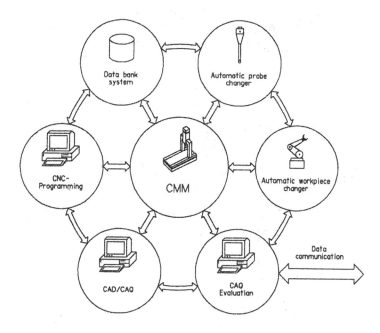

Figure 6. Configuration of an Intelligent Quality Assurance Cell.

7 SYSTEM FOR GLOBAL CONCURRENT QUALITY MANAGEMENT

A data communication model as described in paragraph 2 is the ideal solution for future solutions in global and intelligent manufacturing environment. By means of a common data model and powerful communication network established among ICAD/ICAQ and all other manufacturing processes, it is possible to realise the concurrent quality management and quality assurance activities and other production activities in MFIF, e.g. all production processes, for example design and development, process planning, manufacturing, quality assurance and management etc., which are traditionally carried out sequentially, can be parallelly carried out in MFIF. The quality management and assurance production knowledge can be stepwise parallelly established and refined. If a modified quality activity is made in a process, a correspondent quality assurance activity change will be simultaneously carried out in all other intelligent CAx systems.

Internet makes it possible to establish a global information highway for simultaneous on-line exchange of production data for collaboration on the design, quality assurance and all the production processes, for concurrent communication of all the systematization knowledges for design, manufacturing and quality assurance of global and intelligent production, and to interact with the factory's suppliers and customers worldwide. In (Fu & Raja, 2000) an example is described for an appropriate internet based application of production metrology and ICAQ. Basic studies for that system have been carried out in the course of a University collaboration (Steindl, 1999).

A global information system has to be investigated to fulfil the global information connection in MFIF environment. The development of information highway, DCE technology and advanced engineering data exchange technique make global information systems possible. Such systems can be realised by utilising the mentioned technologies, by means of which whole collaborative, interactive and concurrent design and manufacturing processes of products in global and intelligent production environment can be achieved. STEP provides an unambiguous representation and an exchange mechanism for computer-interpretable product data throughout the whole life cycle of a product, independent from any particular system.

Through global information connection a quality assurance process could be so carried out, that during the CAD modelling the quality assurance planning, modelling, programming and simulating processes which cooperate with the customers and suppliers could also be simultaneously carried out. The design, quality assurance planning and the quality assurance programming can be carried out at one place and the quality assurance simulating, measuring and evaluating processes can concurrently be carried out at another place in the world.

8 SUMMARIZING AND CONCLUDING REMARKS

Multi-functions integrated factory is a innovative concept and a new model for future enterprises developed to meet demand for cost-effective customer-driven design and manufacturing, to realize agile and optimal manufacturing production. The quality assurance process will be used in all product production processes in MFIF - from the design to the assembly. QM and QA in individual activities of different function enterprises in MFIF play a basic role to ensure the realisation of MFIF, e.g. through the intelligent production systems based on quality management and quality assurance in MFIF to create, to realize and to present the features, such as, concurrent, interactive, collaborative, modular, integrative, learning, autonomous, self optimising and self organising functions.

In this presentation, the intelligent quality assurance system in MFIF, an off-line programming technique for ICMM as basis for simultaneous quality assurance and an intelligent measuring cell for the flexible automation of quality assurance and management, data collection and data evaluation in multi-function integrated factory was proposed and discussed. Optoelectronic and Nanotopographic quality assurance methods and a global data communication model for the future that is investigated to realise the concurrent quality management and quality assurance activities and other production activities in MFIF, e.g. all production processes, for example design and development, process planning, manufacturing and quality assurance and management etc., which are traditionally carried out sequentially, can be parallelly carried out in MFIF in global and intelligent manufacturing environment, are introduced.

Quality management systems with intelligent, associative, concurrent, interactive, collaborative, modular, integrative, learning, autonomous, self optimising and self organising functions will be realized in MFIF in the near future.

9 REFERENCES

[1] Binnig, H.; Rohrer, H. (1982), Scanning Tunnelling Microscopy. Helv. Phys. Acta 55, 726-731

[2] Fu, S.; J. Raja (2000). Internet Based Roundness and Cylindricity Analysis. Proceedings of IMEKO 2000 World Congress Vol. VIII, Editors: M.N. Durakbasa; P.H. Osanna; A. Afjehi-Sadat, Wien: Austauschbau und Messtechnik, ISBN 3-901888-10-1, 83-88

[3] Ichida, Y; Kishi, K. (1993). Nanotopography of Ultraprecise Ground Surface of Fine Ceramics Using Atomic Force Microscope. Annals of the CIRP, 42/1, 647-650

[4} ISO (1994). ISO 10303-1: Industrial automation systems and integration - Product data representation and exchange - Part 1: Overview and fundamental principles. ISO Standard

[5] Krause, F.-L.; T. Kiesewetter; S. Kramer (1994). Distributed Product Design, Annals of the CIRP, 43(1), 149-152

[6] Osanna, P.H.; Si, L. (2000). Multi-Functions Integrated Factory MFIF - a Model of the Future Enterprise. Conference Proceedings: Internet Device Builder Conference, Sta. Clara, May 2000, 6pp

[7] Pfeifer, T.; Moehrke, G. (1991). Optoelectronic Measuring Techniques for Application in Automated Production. Proceedings of IMEKO XII, Vol. IV, 637-642

{8] Si, L.; P.H. Osanna (1995). Multi-Functions Integrated Factory. Proceedings of 11th ISPE/IEEE/IFAC International Conference on CARS&FOF'95, Colombia, 578-586

[9] Steindl, K.G. (1999).Development of a Software Package for the Internet Based Analysis of Roundness Data. Master Thesis, TU-Wien, A, and UNC Charlotte, USA.

[10] Taniguchi, N. (1974). On the Basic Concept of Nanotechnology. Proc. Int. Conf. Prod. Eng., Tokyo: JSPE, part 2, 18-23

[11] Westkaemper, E. (1994). "Zero-defect" manufacturing by means of a learning supervision of process chains, Annals of the CIRP, Vol. 43., 1, 405-408

[12] Whitehouse, D.J. (1991). Nanotechnology Instrumentation. Measurement + Control, 24 (2), 37-46

AUTHORS:

Ass. Professor Dr.techn. Ali **AFJEHI-SADAT** *)
Professor Dr. DDr. h.c. Prof. h.c. P. Herbert **OSANNA** *)
Professor Dr.techn. Prof. h.c. Numan M. **DURAKBASA** *)
Prof. Dr.techn. Eng. Jorge M. **BAUER** **)
*) Department for Interchangeable Manufacturing and Industrial Metrology (Austauschbau und Messtechnik) at the Institute of Production Engineering, Vienna University of Technology, Karlsplatz13/3113, A-1040 Vienna, Austria
**) Faculty of Engineering, National University of Lomas Zamora, Bunenos Aires, Argentina
Phone: +43 1 58801 31105
Fax: +43 1 58801 31196
E-mail: afjehi@mail.ift.tuwien.ac.at &
osanna@mail.ift.tuwien.ac.at

AUTHOR INDEX

Title/Year of publication	Editor(s)	ISBN
2002 continued		
Periodic Control Systems (W)	Bittanti & Colaneri	0 08 043682 X
Modeling and Control in Environmental Issues (W)	Sano, Nishioka & Tamura	0 08 043909 8
Computer Applications in Biotechnology (C)	Dochain & Perrier	0 08 043681 1
Time Delay Systems (W)	Gu, Abdallah & Niculescu	0 08 044004 5
Control Applications in Post-Harvest and Processing Technology (W)	Seo & Oshita	0 08 043557 2
Intelligent Assembly and Disassembly (W)	Kopacek, Pereira & Noe	0 08 043908 X
Adaptation and Learning in Control and Signal Processing (W)	Bittanti	0 08 043683 8
New Technologies for Computer Control (C)	Verbruggen, Chan & Vingerhoeds	0 08 043700 1
Internet Based Control Education (W)	Dormido & Morilla	0 08 043984 5
Intelligent Autonomous Vehicles (S)	Asama & Inoue	0 08 043899 7
2003		
Proceedings of the 15th IFAC World Congress 2002 (CD + 21 vols)	Camacho, Basanez & de la Puente	008 044184 X
Modeling and Control of Economic Systems (S)	Neck	0 08 043858 X
Mechatronic Systems (C)	Tomizuka	0 08 044197 1
Programmable Devices and Systems (W)	Srovnal & Vlcek	0 08 044130 0
Real Time Programming (W)	Colnaric, Adamski & Wegrzyn	0 08 044203 X
Lagrangian and Hamiltonian Methods in Nonlinear Control (W)	Astolfi, Gordillo & van der Schaft	0 08 044278 1
Intelligent Control Systems and Signal Processing (C)	Ruano, Ruano & Fleming	0 08 044088 6
Guidance and Control of Underwater Vehicles (W)	Roberts, Sutton & Allen	0 08 044202 1
Analysis and Design of Hybrid Systems (C)	Engell, Gueguen & Zaytoon	0 08 044094 0
Intelligent Manufacturing Systems (W)	Kadar, Monostori & Morel	0 08 044289 7
Control Applications of Optimization (W)	Gyurkovics & Bars	0 08 044074 6
Fieldbus Systems and Their Applications (C)	Dietrich, Neumann & Thomesse	0 08 044247 1
Intelligent Components and Instruments for Control Applications (S)	Almeida	0 08 044010 X
Modelling and Control in Biomedical Systems (S)	Feng & Carson	0 08 044159 9
2004		
Advances in Control Education (S)	Lindfors	0 08 043559 9
Robust Control Design (S)	Bittanti & Colaneri	0 08 044012 6
Fault Detection, Supervision and Safety of Technical Processes (S)	Staroswiecki & Wu	0 08 044011 8
Technology and International Stability (W)	Kopacek & Stapleton	0 08 044290 0
System Identification (SYSID 2003) (S)	Van den Hof, Wahlberg & Weiland	0 08 043709 5
Control Systems Design (C)	Kozak & Huba	0 08 044175 0
Robot Control (S)	Duleba & Sasiadek	0 08 044009 6
Time Delay Systems (W)	Garcia	0 08 044238 2
Control in Transportation Systems (S)	Tsugawa & Aoki	0 08 0440592
Manoeuvring and Control of Marine Craft (C)	Batlle & Blanke	0 08 044033 9
Power Plants and Power Systems Control (S)	Lee & Shin	0 08 044210 2
Automated Systems Based on Human Skill and Knowledge (S)	Stahre & Martensson	0 08 044291 9
Automatic Systems for Building the Infrastructure in Developing Countries (Knowledge and Technology Transfer) (W)	Dimirovski & Istefanopulos	0 08 044204 8
Intelligent Assembly and Disassembly (W)	Borangiu & Kopacek	0 08 044065 7
New Technologies for Automation of the Metallurgical Industry (W)	Wei Wang	0 08 044170 X
Advanced Control of Chemical Processes (S)	Allgöwer & Gao	008 044144 0

Customers wishing to obtain details of all available IFAC volumes, should contact their nearest Elsevier office or check the IFAC Publications website (www.elsevier.com/locate/ifac).

Printed and bound by CPI Group (UK) Ltd, Croydon, CR0 4YY

03/10/2024

01040319-0020